STATISTICAL THEORY
FOURTH EDITION

STATISTICAL THEORY
FOURTH EDITION
Bernard W. Lindgren

Stephanie L. Pham
(612) 321-9739

CHAPMAN & HALL

First published in 1993 by
Chapman & Hall
29 West 35th Street
New York, NY 10001-2299

Published in Great Britain by
Chapman & Hall
2-6 Boundary Row
London SE1 8HN

© 1993 Chapman & Hall, Inc.

Printed in the United States of America

All rights reserved. No part of this book may be reprinted or reproduced or utilized in any form or by any electronic, mechanical or other means, now known or hereafter invented, including photocopying and recording, or by an information storage or retrieval system, without permission in writing from the publishers.

Library of Congress Cataloging-in-Publication Data

Lindgren, B. W. (Bernard William), 1924-
    Statistical theory / Bernard Lindgren. — 4th ed.
       p.   cm.
    ISBN 0-412-04181-2
    1. Mathematical statistics.   I. Title.
QA276.L546 1993
519.5—dc20   93-1042
                                                                             CIP

British Library Cataloguing in Publication Data also available.

Contents

Preface xi

1 Preliminaries   1
1.1 Counting Outcomes   2
1.2 Set Algebra   8
1.3 Events   14
1.4 Functions on a Sample Space   15

2 Probability   19
2.1 Experiments with Symmetries   19
2.2 The Law of Large Numbers   21
2.3 Subjective Probability   24
2.4 Probability Axioms   27
2.5 Discrete Probability Spaces   30
2.6 Densities   31
2.7 Conditional Probability   33
2.8 Bayes' Theorem   37

3 Random Variables   43
3.1 Discrete Random Variables   43
3.2 Distribution Functions   47

3.3 Continuous Random Variables 54
3.4 Continuous Random Vectors 58
3.5 Functions of Random Variables 63
3.6 Conditional Distributions 69
3.7 Independence 77
3.8 Exchangeable Variables 85

4 Expectations 89

4.1 Discrete Random Variables 90
4.2 Continuous Random Variables 93
4.3 Functions of Random Variables 99
4.4 Moments 104
4.5 Variance 105
4.6 Chebyshev and Related Inequalities 110
4.7 Covariance 113
4.8 Correlation 116
4.9 Generating Functions 120
4.10 Characteristic Functions 128
4.11 Multivariate Generating Functions 131

5 Limit Theorems 135

5.1 Convergence 135
5.2 Law of Large Numbers 137
5.3 The Central Limit Theorem 139

6 Some Parametric Families 143

6.1 Bernoulli Trials 143
6.2 The Binomial Distribution 145
6.3 Hypergeometric Distributions 150

6.4 Negative Binomial Distributions   154
6.5 The Poisson Process   160
6.6 Exponential and Related Distributions   166
6.7 Gamma and Beta Distributions   172
6.8 Normal Distributions   178
6.9 Chi-square Distributions   182
6.10 Multinomial Distributions   186
6.11 Exponential Families   188
6.12 The Cauchy Family   190

7 Sampling and Reduction of Data   195
7.1 Random Sampling   195
7.2 Describing Samples   198
7.3 Sampling Distributions   206
7.4 Sampling Distributions of Sample Moments   208
7.5 Sampling Normal Populations   211
7.6 Sampling Distributions of Order Statistics   215
7.7 Likelihood   220
7.8 Sufficiency   228
7.9 Minimal Sufficiency   235
7.10 Information in a Sample   240
7.11 Prior to Posterior   246
7.12 Principles of Inference   250

8 Estimation   253
8.1 Mean Squared Error   254
8.2 Efficiency   260
8.3 Using Sufficiency for Variance Reduction   265
8.4 Consistency   270

8.5 Method of Moments   272
8.6 Maximum Likelihood Estimators   274
8.7 Bayes Estimators   279
8.8 Interval Estimates   281
8.9 Estimating Normal Population Parameters   285

9 Testing Hypotheses   291

9.1 Some Generalities   292
9.2 The Likelihood Ratio   296
9.3 Tests: Assessing Evidence   300
9.4 Tests: Decision Rules   307
9.5 Bayesian Testing   313
9.6 Normal Populations—the Variance   317
9.7 Normal Populations—the Mean   322
9.8 Simple H_0 vs. Simple H_A 329
9.9 The Neyman-Pearson Lemma   334
9.10 The Power Function   340
9.11 Power Functions of t-tests   346
9.12 Uniformly Most Powerful Tests   350

10 Analysis of Categorical Data   357

10.1 The Bernoulli Model   357
10.2 Pearson's Chi-square for Goodness of Fit   361
10.3 Likelihood Ratio for Goodness of Fit   366
10.4 Two-way Contingency Tables   369
10.5 Independence and Homogeneity   371
10.6 Large- and Small-sample Tests   376
10.7 Measures of Association   380
10.8 McNemar's Test   381

10.9 Log-linear Models for 2×2 Tables 385
10.10 Three-way Classifications 386

11 Sequential Analysis   395
11.1 The Sequential Likelihood Ratio Test 395
11.2 Expected Sample Size 402
11.3 Power of the SLRT 404

12 Multivariate Distributions   409
12.1 Transformations 409
12.2 The Bivariate Normal Distribution 417
12.3 Correlation and Prediction 424
12.4 Multivariate Transformations 429
12.5 The General Linear Transformation 432
12.6 Multivariate Normal Distributions 435
12.7 A Decomposition Theorem 440

13 Nonparametric Tests   445
13.1 Distribution of the Order Statistic 446
13.2 The Transformation $F(X)$ 449
13.3 Testing Randomness 452
13.4 The Sign Test 459
13.5 The Signed-rank Test 461
13.6 Asymptotic Relative Efficiency 465
13.7 Confidence Intervals 467
13.8 Two-sample Tests 470
13.9 Goodness of Fit 479

14 Linear Models and Analysis of Variance — 489

14.1 The Likelihood Function 491

14.2 Simple Linear Regression—Estimation 493

14.3 Simple Linear Regression—Testing and Prediction 501

14.4 Testing Linearity 505

14.5 Multiple Regression 508

14.6 Nested Models 512

14.7 A Single Classification Model 517

14.8 A Random Effects Model 524

14.9 Two-way Classifications 527

14.10 Other Designs 535

15 Decision Theory — 539

15.1 Convex Sets 539

15.2 Utility and Subjective Probability 542

15.3 Loss, Regret, and Mixed Actions 547

15.4 Minimax Actions 549

15.5 Bayes Actions 554

15.6 Admissibility 558

15.7 The Risk Function 560

15.8 Using the Posterior 565

15.9 Randomized Decision Rules 568

15.10 Monotone Problems and Procedures 570

Tables 575

References and Further Reading 601

Answers to Problems 603

Index 627

Preface

Like the earlier editions, this textbook is intended for a year's course in the theory of statistics. No previous knowledge of statistics is assumed. However, a study of the theory is perhaps better motivated and more appreciated by students who have had some exposure to statistical methodology. I assume that the student has a good command of first-year calculus as well as some knowledge of multivariable calculus and at least an introduction to linear algebra.

In the seventeen years since the third edition, the body of knowledge we think of as statistical theory has of course expanded considerably, but the material covered in earlier editions remains important as a foundation, both for a practitioner's better understanding of basic statistical methods and for a theoretician's further excursions into the theory. Thus, I have not changed the coverage in any significant way.

What I have done is to eliminate certain material that made the text seem more formidable than it was—material that was usually passed over in our own teaching from the text. There is still ample material, however, for a course of three lectures per week. The major change in organization is the deferring of statistical decision theory (per se) to a final chapter, although the Bayesian approach is now introduced earlier—in the chapters on estimation and testing, and before that, as one motivation for studying likelihood.

Without a background of measure theory, a mathematically rigorous treatment (definition-theorem-proof) is not feasible. For this reason I avoided stating formal theorems in the earlier editions. This had the effect of tending to hide important and useful information, so in this revision I have stated results as "theorems," with the understanding that perhaps both statement and proof might be wanting in mathematical rigor.

In the process of rewriting I have naturally tried to clarify the exposition, and generally to make the text more accessible. And I have increased the number of problems for the student by about forty percent. Answers, for problems that have answers, are given at the back of the book. An instructor's manual, with problem solutions and additional problems, will be available.

I am indebted to students and colleagues in the School of Statistics at the University of Minnesota whose reactions, comments, corrections, and criticisms—based on their experience with earlier editions and their classroom use of a preliminary version of this fourth edition—have been most helpful. I am especially indebted to Ms. Wei Shen, whose careful reading of the text and solving of all of the problems have eliminated numerous errors.

Bernard W. Lindgren

1
Preliminaries

An **experiment** is doing something or observing something happen under certain conditions, resulting in a final state of affairs or **outcome**. The experiment may be physical, chemical, social, medical, industrial, educational, etc. It may be conducted in the laboratory or in real-life surroundings—in the factory, the economy, the hospital, the classroom.

An experiment may be one whose outcome is completely determined by what one knows of the conditions under which it is carried out. In practice, however, one often finds that the outcome varies from one trial to another. This will happen when there are conditions or factors affecting the outcome that cannot be measured or taken into account or (perhaps) even identified. We say that such experiments involve "chance," and refer to them as **experiments of chance**. In such cases the usual type of deterministic model, involving equations of motion and/or equations of state that embody physical or other "laws," is inadequate for describing or predicting what will happen at a given trial. A different type of mathematical structure is called for—a **probability model**.

For understanding how things work, for aid in making decisions or plans or predictions, we need to know the appropriate model. Statistical practice involves conducting one or more trials of an experiment and analyzing the results for evidence about the model. Drawing conclusions about the model from what we see in experimental data is **statistical inference**. Before taking up the topic of inference, however, we need to study probability models.

The basic ingredients of a probability model are a **sample space**, a collection of **events**, and **probabilities** of events. The sample space of an experiment is simply the set of possible outcomes, and events are subsets of the sample space. We begin with a review of some useful devices for counting outcomes, and of some elementary algebra of sets.

1.1 Counting outcomes

Counting the outcomes of an experiment is easy when we can make a list of them—one that is not too long. But in many instances this is simply not feasible. If the experiment can be thought of as a composition of simpler experiments, we can count outcomes by appealing to the following simple principle:

COUNTING PRINCIPLE   *When an experiment $\mathcal{E}$ consists of an experiment $\mathcal{F}$ with m outcomes followed by an experiment $\mathcal{G}$ with n outcomes, the number of outcomes in $\mathcal{E}$ is the product mn.*

In applying this principle it is important that $\mathcal{G}$ have the same number of outcomes no matter which of the m outcomes of $\mathcal{F}$ has occurred. When it is physically possible, the second experiment can be carried out at the same time as the first; but the outcomes of $\mathcal{E}$ are *ordered* pairs of outcomes of $\mathcal{F}$ and $\mathcal{G}$.

The principle is really a theorem, whose proof consists of counting the number of ordered pairs in which the first element is one of the outcomes f_i of $\mathcal{F}$ and the second element, one of the outcomes g_j of $\mathcal{G}$. Laying these out in a two-way array makes the count obvious:

$$(f_1, g_1) \quad (f_1, g_2) \quad \cdots \quad (f_1, g_n)$$
$$(f_2, g_1) \quad (f_2, g_2) \quad \cdots \quad (f_2, g_n)$$
$$\vdots \qquad \vdots \qquad \qquad \vdots$$
$$(f_m, g_1) \quad (f_m, g_2) \quad \cdots \quad (f_m, g_n).$$

By induction, the principle extends to an experiment composed of k subexperiments. The outcomes of the composite experiment are ordered k-tuples of outcomes of the subexperiments, and the count is obtained by multiplication.

EXAMPLE 1.1a   Three ordinary dice are rolled; how many outcomes are possible? The multiplication rule will give a count of ordered sequences, so the three dice are effectively distinguished. Perhaps they are of different colors, or numbered die 1, die 2, die 3. Indeed, we may have only one die and roll it three times to obtain a sequence of three outcomes. Since each die can turn up in one of six ways, the number of outcomes in the composite experiment is $6 \cdot 6 \cdot 6 = 216$.

Similarly, in a roll of n dice, the number of ordered sequences of outcomes of the individual dice is 6^n. ∎

report that there are about seven million combinations.

1-10. A delegation of three of the ten members of a city council is to be chosen to represent the council at a convention. In how many ways
(a) can the delegation be chosen?
(b) can it be chosen, if two particular members refuse to go together?
(c) can it be chosen if two certain members will either both go or neither?

1-11. In how many ways can a group of eight individuals be divided into
(a) subcommittees of sizes three, three, and two (with distinct tasks)?
(b) four pairs, the individuals in a pair to play each other in tennis, on four indistinguishable tennis courts?

1-12. Show: For any positive integer n and any positive integer $k \leq n$,
$$\binom{n}{k-1} + \binom{n}{k} = \binom{n+1}{k}.$$

1-13. Consider the following identity:
$$\binom{n}{m}\binom{n-m}{r}\binom{n-m-r}{k-m} = \binom{n}{k+r}\binom{k+r}{k}\binom{k}{m}.$$
(a) Establish this by expressing each combination count in terms of factorials.
(b) Interpret each side as a count of the outcomes in a sequence of three selections and deduce the equality from this reasoning.
(c) Express the common value as a quantity of the form $\binom{n}{a,\ b,\ c,\ d}$.

1-14. The *binomial theorem* is the following identity:
$$(x+y)^n = \sum_{0}^{n} \binom{n}{k} y^k x^{n-k},$$
(a) Show this using induction on n: First verify it for $n = 1$. Then, assuming it to be true for some n, multiply both sides by $(x+y)$ and exploit the result in Problem 12.
(b) Show it by arguing that each term in the expansion must involve n factors—either an x or a y from each of the factors $(x+y)$. The number of terms with a given number of x's as factors is the number of ways of selecting that many factors $(x+y)$ from which to use the x.
[The symbol $\binom{n}{k}$ is often referred to as a *binomial coefficient*.]

1-15. Use the binomial theorem to show that $\sum_{0}^{n} \binom{n}{k} = 2^n$. This is the number of subsets of a set of n elements (including the set with no elements). Also, obtain the same result by counting the ways of deciding, for each element, whether or not to use it in the subset.

1-16. The *trinomial theorem* says that

$$(x+y+z)^n = \sum_{i+j+k=n} \binom{n}{i,\,j,\,k} x^i y^j z^k,$$

where the indices i, j, and k range over the integers 0 to n. Obtain this result paralleling the reasoning in Problem 14(b).

1.2 Set algebra

Because we'll be dealing with *sets* of outcomes, combining and relating them in various ways, we need to be familiar with some simple algebra of sets.

Consider a sample space Ω—a set of elements (outcomes) ω. If Ω is finite, with n elements, there are finitely many subsets, namely, 2^n (from Problem 1-15). When a set A of these elements is countable, we may define the set by listing its elements: $A = \{\omega_1, \omega_2, ...\}$. Sometimes it's convenient to define a set by giving a condition satisfied by each of its elements. For example, if $\Omega = \Re^1$, the set $E = [x > 2]$ is the set of points x with the property $x > 2$. We may also use the more complete and standard notation $\{x: x > 2\}$.

Sets can be related by **inclusion** and by **equality**. A set A is included in a set B when each element of A is also an element of B, and we write $A \subset B$. If both $A \subset B$ and $B \subset A$, then we say A equals B. To prove $A = B$ we need to show that each element of A is in B, and that each element of B is in A.

DEFINITION The *complement* of a set A is the set of all elements in Ω that are *not* in A.

The complement of A is denoted variously by A', A^c, $\bar{A}$. In this text we use A^c. Since there are no elements that are not in the whole sample space Ω, its "complement" is empty. The symbol $\emptyset$ is used to denote the empty set—a "set" with no elements. Thus, we write $\Omega^c = \emptyset$, and $\emptyset^c = \Omega$.

DEFINITION The set of elements that are common to all the sets in a collection of sets (finite or infinite) is their *intersection*.

For a finite number of sets, we use a juxtaposition of their names as the symbol for their intersection; thus, ABC denotes the intersection of sets A, B, and C. (Another common notation is $A \cap B \cap C$.) Although we necessarily write the names of the sets in some sequence, the definition of intersection does not involve order, so the operation of intersection is *commutative*. It is also *associative*: $(AB)C = A(CB) = ABC$. For an infinite collection of sets $\{B_\alpha\}$, the intersection is denoted by $\bigcap B_\alpha$, in a usage like that of $\sum$.

EXAMPLE 1.2a Let Ω denote the points in a plane, each identified by its rectangular coordinates (x, y). The condition $x^2 + y^2 = 4$ defines the set E of points on a circle of radius 2 centered at the origin. The condition $x > 4$ defines the set F of points to the right of the line $x = 1$. The intersection EF is the set of points to the right of $x = 1$ that are within the circle.

The intersection AB of the sets $A = [|x| < 2]$ and $B = [|y| < 1]$ is the set of points (x, y) within the rectangle with corners at the points $(\pm 2, \pm 1)$. ∎

DEFINITION The set of elements consisting of all elements contained in any one or more of a collection of sets is their *union*.

For a finite collection, we may use the symbol $\cup$. Thus, the union of A, B, and C is written $A \cup B \cup C$. Like "intersection," the operation "union" is commutative and associative. We write the union of sets in an infinite collection of sets $\{B_\alpha\}$—countable or uncountable—as $\bigcup B_\alpha$.

Some special intersections and unions should be noted:

$$A\emptyset = \emptyset, \quad A\Omega = A, \quad A \cup \emptyset = A, \quad \text{and} \quad A \cup \Omega = \Omega.$$

Moreover, since A and its complement have no elements in common,

$$AA^c = A^cA = \emptyset, \quad A \cup A^c = \Omega, \quad \text{and} \quad (A^c)^c = A.$$

You should verify each of these (see Problem 1-17).

In many ways, the intersection and union of sets behave, respectively, like multiplication and addition of numbers. So when both operations appear in an expression, we use parentheses and the corresponding conventions for $+$ and $-$ from algebra to determine which operations are performed first. Thus, the expression $A \cup BC$ means the union of A and BC. The intersection of $A \cup B$ and C is written $(A \cup B)C$.

EXAMPLE 1.2b Consider the sample space $\Omega = [0, 1]$, the closed unit interval on the real line, and the subintervals $I_n = [0, 1/n]$ for $n = 1, 2, \ldots$. The intersection of these subintervals is the only point they have in common: $\bigcap I_n = \{0\}$. The union of their complements is $\bigcup I_n^c = \bigcup [1/n, 1] = (0, 1]$. ∎

The distributive law for numbers has a counterpart in set algebra:

$$A(B \cup C) = AB \cup AC. \tag{1}$$

To see this, observe that any element both in A and in B or C is in either AB or AC, and conversely. But there is another distributive law, with no counterpart in the realm of numbers, obtained by interchanging intersection and union in (1) (see Problem 1-17):

$$A \cup BC = (A \cup B)(A \cup C). \tag{2}$$

A kind of set "subtraction" can be defined that behaves in some ways like number subtraction: $A - B = AB^c$. This is the set of points in A that are not in B. It is then true that $C(A - B) = CA - CB$ (Problem 1-17). However, although $(A - B) + B = A$ when A and B are numbers, it is *not* always true that $(A - B) \cup B = B$ when A and B are sets. But the latter formula is correct for sets A and B such that $B \subset A$ (see Problem 1-20).

It can be helpful to have schematic pictures in understanding set operations. A **Venn diagram** uses a plane region, such as the points within a rectangle, as the sample space; subsets are then sets of points within that rectangle. Figure 1-1 is a Venn diagram showing sets E (interior of a circle), F (interior of a rectangle), and G (interior of a triangle). The set G is *disjoint* from E and from F—the intersections EG and FG are empty.

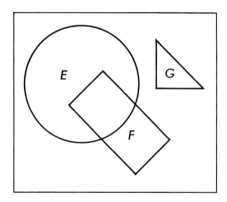

Figure 1-1 A Venn diagram

Although a Venn diagram is useful for illustration, relying on them for proofs can be risky if it is not perfectly clear that the figures chosen to represent events are chosen in a way that is quite general. That is, a relationship may hold for certain configurations of events that do not hold in general.

Figure 1-2 shows the intersection EF, the union $E \cup F$, the complement of E, and the difference $E - F = EF^c$ as shaded regions in a Venn diagram.

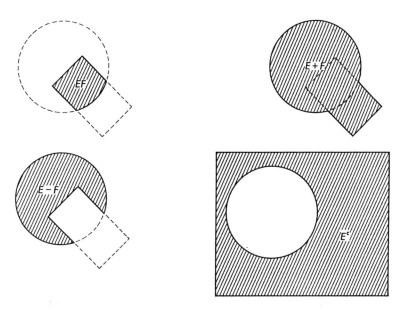

Figure 1-2

We say that the pair (A, A^c) constitutes a **partition** of Ω, a division into disjoint pieces. Moreover, it effects a partition of any set B. For, the sets BA and BA^c are disjoint, and (according to the distributive law)

$$BA \cup BA^c = B(A \cup A^c) = B\Omega = B. \qquad (3)$$

More generally, we may define a partition of a sample space by any countable collection of nonoverlapping sets that fill out the whole space:

DEFINITION A *partition* of Ω is any collection of disjoint sets $\{A_\alpha\}$ that fill out the whole space:

$$\bigcup A_\alpha = \Omega, \text{ and } A_\alpha A_\beta = \emptyset \text{ for } \alpha \neq \beta. \qquad (4)$$

A partition of Ω serves to partition an event in Ω. Thus, it $\{A_\alpha\}$ is a partition of Ω, the sets BA_α constitute a partition of B:

$$\bigcup BA_\alpha = B, \text{ and } (BA_\alpha)(BA_\beta) = \emptyset \text{ when } \alpha \neq \beta.$$

Figure 1-3 illustrates the partitioning of an event F by an event E and its complement E^c.

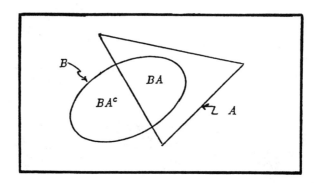

Figure 1-3

The notion of a partition is useful in establishing an important result concerning sequences of events:

THEOREM 1 *If $\{A_n\}$ is an ascending sequence: $A_1 \subset A_2 \subset \cdots$, the sequence $\{A_{n+1} - A_n\}$ constitutes a partition of their union, $A = \bigcup A_n$.*

To see this, suppose $\omega \in A$, and let n be the largest integer such that $\omega \notin A_n$. Then $\omega \in A_{n+1}$, so $\omega \in A_{n+1} - A_n$, and $\omega \in \bigcup (A_{n+1} - A_n)$. Conversely, if $\omega \in A_{n+1} - A_n$, then $\omega \in A_{n+1}$ and hence $\omega \in A$. Therefore the differences $A_{n+1} - A_n$ fill out A: $\bigcup (A_{n+1} - A_n) = A$. Since these differences are disjoint (verify this!), they constitute a partition of A, as asserted in the theorem.

EXAMPLE 1.2c Consider the disks $\{B_n\}$ defined by the inequalities $x^2 + y^2 \leq 1 - 1/n$, for $n = 1, 2, \ldots$. The sequence $\{B_n\}$ is clearly ascending. With $B_0 = \emptyset$, the differences $B_n - B_{n-1}$ $(n = 1, 2, \ldots)$ are the annuli or rings between concentric circles, and these annuli constitute a partition of their union—the closed unit disk: $x^2 + y^2 \leq 1$. ∎

In defining a probability model we'll need to work with a family of sets, combining them and operating on them using intersections, unions, and complements. The results of such operations must then also be in the family. A *field* is a family of sets closed under complementation, finite unions, and finite intersections. In the case of a finite sample space, say one with k elements, the collection of *all* subsets is finite in number (2^k), and this is a field. More generally, to allow countable operations, we need a "σ-field" of sets—a collection of sets which is a field but which is also closed under countable unions and countable intersections.

An important sample space is the real line—the set of all real numbers. Sets of interest in this sample space are individual values (points) and intervals of values, as well as sets obtained by countable operations on these. The *Borel sets* on the line are the sets constituting the smallest σ-field containing the half-lines $[x \leq \lambda]$. These include half-open intervals $(a, b]$ as the difference $(-\infty, b] - (-\infty, a]$, single points x as the intersection of the intervals $(x - 1/n, x + 1/n]$ for $n = 1, 2, 3, ...$, and therefore closed and open intervals as well. However, not all subsets of the real line are Borel sets; but it turns out that one cannot define a model in which *all* subsets are assigned probabilities in a useful way. Non-Borel sets are difficult to construct, and (fortunately) these pathological sets are not needed as events in models for the kinds of real experiments with which we are concerned.

PROBLEMS

1-17. Show the following, for any events E, F, and G in a sample space Ω:
(a) $\emptyset E = \emptyset$, and $\emptyset \cup E = E$.
(b) $E\Omega = E$, and $E \cup \Omega = \Omega$.
(c) $(EF)^c = E^c \cup F^c$.
(d) $(E \cup F)^c = E^c F^c$.
(e) $E \subset F$ implies $F^c \subset E^c$
(f) $G(F - E) = GF - GE$.
(g) If $E \subset F$, $EF = E$ and $E \cup F = F$.
(h) $(E - F) \cup F = E \cup F$.
(i) $E \cup FG = (E \cup F)(E \cup G)$.
(j) $E \cup FGH = (E \cup F)(E \cup G)(E \cup H)$.

1-18. Simplify:
(a) $E^c FG \cup EFG$.
(b) $(E^c FG)(EFG)$.
✓ (c) $(E \cup F)(E \cup F^c)$.

1-19. Use Venn diagrams to illustrate the distributive laws (1) and (2).

1-20. Show that $(A - B) \cup B = A$ when $B \subset A$, and (with the aid of a Venn diagram) show that the equality does not always hold.

1-21. Show: $A \cup B = AB^c \cup A^c B \cup AB$.

1-22. Generalizing (c) and (d) of the Problem 17, obtain the dual relations known as *DeMorgan's laws*:

$$\left(\bigcup E_\alpha\right)^c = \bigcap E_\alpha^c, \quad \text{and} \quad \left(\bigcap E_\alpha\right)^c = \bigcup E_\alpha^c.$$

1-23. List all subsets of a sample space with four elements: $\{a, b, c, d\}$.

1-24. Consider the sample space $\Re^2$, with elements (x, y). Identify and describe the sets defined by each of the following conditions:
(a) $A = [x^2 \leq 4y]$.
(b) $B = [x + y \leq 1]$.
(c) $C = [x + 2y > 3]$.
(d) $D = [x \leq 2y]$.
(e) $E = [x^2 + y^2 = 1]$.
(f) $F = BD$.
(g) $G = (A \cup C)^c$.

1.3 Events

The sample space Ω of an experiment of chance is the set of all possible outcomes of the experiment. Except for some hedging to be discussed below, a subset of the sample space is called an **event**. An event E can be defined by making a list of its outcomes, if they can be listed. We can also define E by imposing some condition—a condition satisfied by the outcomes in E but not by any outcomes not in E.

Some conditions are expressed verbally, in language that implies certain set operations. Thus, for given conditions E and F, the condition "E or F" defines the set which is the **union** of the sets E and F. The condition "E and F" defines the set which is the **intersection** of the sets E and F. And the condition "*not* E" defines the **complement**, E^c:

"or" $\leftrightarrow$ union
"and" $\leftrightarrow$ intersection
"not" $\leftrightarrow$ complement.

EXAMPLE 1.3a Consider the selection of a card from a standard deck of playing cards. The 52 cards in the deck constitute the sample space. The condition that the card selected is an ace (the event "ace") defines the set of four aces. The condition that the card is red (the event "red") defines the set of (26) red cards. The condition that the card is not an ace or a face card [(ace or face)^c] defines the set of (36) numbered cards (2's,..., 10's). The event "red ace," which means "red and ace," is the intersection of "ace" and "red." ■

It would make things simpler if every subset of a sample space could be called an event. However, we want to assign a probability to each event, and sometimes (as suggested at the end of §1.2) it is not possible to do this in a coherent fashion when *all* subsets are events. In the case of a finite sample space, one with n outcomes, there is no problem when we define each of the 2^n subsets to be an event.

In the case of the real line, we usually take the Borel sets described in §1.2 as events. These include single values, as intersections of descending sequences of intervals surrounding and closing in on a single value.

When the sample space is $\Re^2$, we take events to be subsets of the smallest σ-field containing all half-planes $[x \leq a]$ and $[y \leq b]$. These are the two-dimensional Borel sets and include single points as well as straight lines and other "nice" curves and regions.

1.4 Functions on a sample space

The elements of a sample space may have various characteristics that are of interest. Thus, if a certain population of people is the sample space for an experiment, one might be interested in the age, weight, hair color, etc., of the person selected from the population. The *age* of the person selected varies with the person selected—it is a function of ω: age = $X(\omega)$. Similarly, weight = $Y(\omega)$, hair color = $Z(\omega)$, and so on. Age and weight together define the vector function $[X(\omega), Y(\omega)]$.

Consider a sample space Ω with a family of its subsets considered to be events. Any particular function X is a mapping from Ω (the "domain" of the function) to the set of possible values of X (the "range" of the function). The value of $X(\omega)$ is the *image point* of ω, and for a given value of X, say x, the set of ω's such that $X(\omega) = x$ is the *pre-image* of x. And a set B of values of X defines a pre-image set in Ω—the set of ω's such that $X(\omega) \in B$. These pre-image sets may or may not be events. When the range space is the real line, the function X is called *measurable* if the pre-image of the set $[X(\omega) \leq x]$ is one of the sets called an event, for *every* x.

A function of a function of ω, $Y(X(\omega))$, is a function of ω, say $Z(\omega)$. We usually shorten this and write $Z = Y(X)$. But Z may or may not be measurable, even when X and Y are. Having mentioned this pathological possibility, we'll not concern ourselves further with measurability; we'll only be working with measurable functions.

A function defined on a sample space induces a *partition* of the sample space. Each value of the function defines a partition set: For each particular value a, the condition $X(\omega) = a$ defines a set of ω's which are assigned the value a by $X(\cdot)$; these pre-image sets, defined by the various possible value of X, are disjoint and fill out Ω.

EXAMPLE 1.4a Given any event A, we define its **indicator function** to have the value 1 on A and the value 0 on A^c:

$$I_A(\omega) = \begin{cases} 1, & \omega \in A, \\ 0, & \omega \notin A. \end{cases} \tag{1}$$

The partition defined by this function is simply the pair (A, A^c). (The pre-image of $[x \leq 0]$ is the set A, and the pre-image of the set $[x \leq 1]$ is Ω, so the function I_A is measurable.) ∎

When the sample space is $\Re_2$, a function $z = g(x, y)$ defines a surface. The points on this surface at a given constant height define a *level curve* or *contour curve* as their projection on the xy-plane. A "contour map" is one that shows a selected set of these plane curves. (Perhaps you have seen contour maps of lake bottoms or mountainous regions.) The level curves are the partition sets of the plane defined by the function g.

EXAMPLE 1.4b Suppose Ω is the set of points $\omega = (x, y)$ on the unit disk—that is, points in the plane that lie on and within the unit circle $x^2 + y^2 = 1$. The function $Z(\omega) = x^2 + y^2$ partitions Ω into circles $x^2 + y^2 = a^2$, for each $|a| \leq 1$. These circles constitute a partition of Ω. ∎

EXAMPLE 1.4c Let ω denote a face of an ordinary six-sided die. The number of dots on a face is a function, $X(\omega)$. But we could equally well take the numbers 1, 2, 3, 4, 5, 6 that identify the faces as the ω's. Then $X(\omega) = \omega$. Another function defined on the sample space is the *indicator function* of an event, say of $\{3, 6\}$:

$$Y(\omega) = \begin{cases} 1, & \omega = 3 \text{ or } 6, \\ 0, & \text{otherwise.} \end{cases}$$

The function Y partitions the sample space into the two sets $\{3, 6\}$ and $\{1, 2, 4, 5\}$. ∎

EXAMPLE 1.4d A certain material is thought of as being composed of tubes or fibers, each assumed to have a certain orientation with respect to a given reference frame. Let Ω denote the set of possible undirected orientations. A particular orientation can be identified with the point where a line through the origin with that orientation intersects a unit hemisphere. A convenient way of specifying the orientation is to give the angular coordinates of that point—its colatitude and longitude in the given reference frame: $\omega = (\theta, \phi)$.

1.4 Functions on a sample space

If a force of magnitude K is applied to the material, the direction of the force is a directed orientation, identified with a point on the unit sphere. The components of the force, in terms of rectangular coordinates (X, Y, Z), would be $(K\sin\theta\cos\phi, K\sin\theta\sin\phi, K\cos\theta)$—a vector function on the sample space of directed orientations. ■

PROBLEMS

1-25. Four coins are tossed, with 16 possible outcomes. Of these, how many are in the event
(a) A: exactly two heads turn up?
(b) B: at least two heads turn up?
(c) C: at most three heads turn up?
(d) $D = B^c$?

1-26. A bowl contains five chips numbered 1, 2, 3, 4, 5. We pick one chip and then another from those that remain. How many outcomes in
(a) the sample space?
(b) the event that the first chip is a 4?
(c) the event that the second chip is a 4?
(d) the event that the first chip or the second chip is a 4?
(e) the event that the 4 is not included?

1-27. For the random selection of a card from a deck of playing cards (as in Problem 1-8), define the following events:

F: the card is a face card $\quad$ S: the card is a spade
A: the card is an ace $\quad$ C: the card is a club
R: the card is red $\quad$ D: the card is a diamond
B: the card is black $\quad$ H: the card is a heart.

Count the outcomes in each of the following events:
(a) F and H and R.
(b) A and not F.
(c) B or D.
(d) S or F or A.
(e) Not RB.
(f) A or $(R$ and $F)$.
(g) F and $(C$ or $D)$.
(h) R, but not F or A.

1-28. Consider the sample space of Problem 25 (sequence of four coin tosses). Define the function $X(\omega) =$ number of heads that turn up.
(a) Give the partition sets defined by $X(\,\cdot\,)$.
(b) List the outcomes in the event $X(\omega) \geq 3$.
(c) Give the partition defined by $(X - 3/2)^2$.

1-29. Consider an unordered sample ω of two chips from the bowl of five numbered chips in Problem 26. Define $Y(\omega)$ to be the sum of the digits on the chips selected.
(a) Identify the partition defined by $Y(\,\cdot\,)$.

(b) Give the pre-image of the event $Y \leq 5$.

1-30. Consider the sample space $\Omega = \Re^1: -\infty < x < \infty$, and $Y(x) = x^2$, a function on Ω. Let $Z(y) = \dfrac{y}{1+y}$, a function on the value space of Y.
 (a) What is the range space of the function $Z(Y(x))$?
 (b) What is the pre-image in the x-space of the event $[Z < 1/2]$?

1-31. Describe the level curves (that is, the partition sets) of these functions defined on $\Re^2$, with outcomes identified by rectangular coordinates (x, y):
 (a) $x^2 + y^2$
 (b) $2x - y$
 (c) xy
 (d) $|x - y|$.

1-32. Consider the sample space $\Re^3$, with points identified by their rectangular coordinates (x, y, z). Define U, V, and W, respectively, as the smallest, second smallest, and largest of the coordinates.
 (a) What is the range space of the vector function (U, V, W)?
 (b) What is the pre-image of the event $\{U = 1, V = 2, W = 3\}$?
 (c) What are the partition sets determined by (U, V, W)?

1-33. Consider a sample space $\Omega = \{a, b, c, d, e\}$.
 (a) Given that $\{a, b\}$ and $\{b, c, d\}$ are events, list all sets obtainable from these two events by at most countably many unions, intersections, and complementations. Are any of the 32 subsets of Ω *not* in your list?
 (b) Define a real-valued function which defines the partition sets $\{a\}$, $\{b\}$, $\{c, d\}$, $\{e\}$.

1-34. Let $X = \sin\theta \cos\phi$, $Y = \sin\theta \sin\phi$, and $Z = \cos\theta$, where (θ, ϕ) are the colatitude and longitude of a point ω on the unit sphere (Example 1.4d). Describe the pre-image of the sets defined by $X^2 + Y^2 = r^2$, for $0 \leq r \leq 1$.

2
Probability

After specification of a sample space Ω and a collection $\mathcal{F}$ of its subsets called *events*, the final step in constructing a probability model is the assignment of a number P called "probability" to each event. A **probability space** is defined by these elements: $(\Omega, \mathcal{F}, P)$. The probabilities P are numbers intended to embody our intuitive notions of probability in real-life situations. There have been various approaches to probability; perhaps the earliest is the following.

2.1 Experiments with symmetries

In a **lottery**, tickets called *chances* are sold, and stubs of these tickets are then mixed thoroughly for the drawing of a winning ticket. The drawing is carried out in such a manner that all tickets are viewed as having the same chance of being drawn. The tickets are the possible outcomes of the experiment and the collection of all the tickets constitutes a sample space. We say that the outcomes are *equally likely*. The selection of a lottery ticket is said to be *random* if the outcomes are considered equally likely.

If there are N tickets in a lottery, a person holding a single ticket has "one chance in N" of winning, and we say that the "probability" that that ticket wins is the ratio $1/N$. If a person holds k tickets, we say that he or she has "k chances in N" of winning, and assign the probability k/N to that event. In this way, each subset of tickets or sample space outcomes—that is, each event—is assigned a probability.

When there are k chances in N of winning, there are of course $N - k$ chances of losing. The **odds** on winning are said to be k to $N - k$. And if odds on winning are given, say $k:m$, the corresponding probability of winning is the ratio $k/(k + m)$.

We denote the probability assigned to the event E by $P(E)$. [Some use the symbol $Pr(E)$.] Let $\#(E)$ denote the number of outcomes in E.

When $\#(E) = k$ and $\#(\Omega) = N$,

$$P(E) = \frac{k}{N} = \frac{\#(E)}{\#(\Omega)}. \qquad (1)$$

This definition implies certain basic, noteworthy properties of $P(\cdot)$:

THEOREM 1 *The set function P as defined by (1) has these properties:*
1: *For every event E, $0 \leq P(E) \leq 1$.*
2: *$P(\Omega) = 1$.*
3: *If $EF = \emptyset$, then $P(E \cup F) = P(E) + P(F)$.*

Property 1, which says that the ratio $P(E)$ is a number on the unit interval, follows from the fact that $0 \leq \#(E) \leq \#(\Omega)$. Moreover, when E is the whole sample space, $\#(E) = \#(\Omega)$. Finally, if $EF = \emptyset$, we observe that $\#(E \cup F) = \#(E) + \#(F)$. Thus, as defined by (1), probability is an additive set function defined on the events of a sample space, scaled in such a way that the whole space has probability 1.

Besides lotteries, there are other physical experiments in which the finitely many outcomes are usually viewed as "equally likely." Tossing or rolling symmetrical objects such as coins and dice, and spinning wheels of fortune and roulette wheels are common examples. It is the symmetry of the objects, and the equal divisions together with many rotations of wheels that lead people to feel that, ideally, there is no reason to expect one outcome more than any other. Any experiment in which symmetries imply equally likely outcomes can be replaced by a lottery with the appropriate number of tickets.

It is important to realize that in these various experiments the equal likelihood of outcomes is *assumed*, not derived. Thus, a model with equal probabilities for heads and tails describes, if anything, an *ideal* or mathematical experiment. We can only *assume* that it describes an actual coin-toss.

EXAMPLE 2.1a A roulette wheel has 38 equal compartments numbered 0, 00, 1, 2,..., 36. The 0 and 00 are green, and of the other numbers, half are red and half are black. In one play, the wheel is spun and a ball tossed onto the wheel; as the wheel slows, the ball ultimately ends up in one of the 38 compartments. These 38 outcomes are usually considered equally likely. (This assumption defines a "fair" wheel.)

One possible bet is a bet on the "first dozen"—that is, on the numbers 1 through 12. The probability of this (which follows from the *assumption* of equal likelihood of the 38 compartments) is 12/38—you have 12 chances in 38 of winning this bet. Similarly, $P(\text{5-split}) = 5/38$, $P(\text{red}) = P(\text{black}) = 18/38$, and $P(\text{green}) = 2/38$. ■

EXAMPLE 2.1b When a hand of five cards is dealt from a thoroughly shuffled deck of playing cards, people generally view the $\binom{52}{5} = 2{,}598{,}960$ combinations as equally likely. With this assumption, to find the probability of any event we simply count the number of hands (outcomes) in that event and divide by 2,598,960.

For instance, to find the probability of a full house (three cards of one denomination and two of a different denomination), we count the number of such hands: First pick a denomination for the three of a kind, then pick a different denomination for the pair, and then pick the three and the two from the four cards in these respective denominations:

$$13 \cdot 12 \cdot \binom{4}{3}\binom{4}{2} = 3744.$$

Dividing by 2,598,960 we obtain $P(\text{full house}) \doteq .00144.$ ■

In the above example, we assumed that the possible combinations resulting from a deal of five cards are equally likely. We think of the hand as a "random selection" of five cards. In general, when selecting a sample of n from N objects, we'll use the phrase **random selection** to mean that the $\binom{N}{n}$ possible selections are assumed to be equally likely. Whether or not this is a reasonable assumption depends on how the experiment being modeled is carried out—on what steps are taken so that there will be no favored combinations. In the case of hands of cards, people shuffle the deck several times before a deal. (It has been found, incidentally, that the two or three shuffles one usually sees are not really adequate for the assumption that the next hand dealt is a "random selection" from the deck.) In drawings from the collection of stubs of tickets sold in a lottery, the stubs are usually mixed rather thoroughly, sometimes in a rotating drum.

An important property of random selections from a finite population is that every population member has the same chance of being included. To see this, let N and n denote the population and sample sizes, respectively, and count the number of selections in which a particular member *is* included: $\binom{N-1}{n-1}$. Then, dividing by the number of unrestricted selections: $\binom{N}{n}$, we find that the probability of including that member is n/N—*the same for each population member.*

2.2 The law of large numbers

Another approach to defining probabilities of events stems from a phenomenon known as the "law of large numbers." When an experiment is repeated in a long sequence of trials, where it is clear that the results of any set of trials cannot depend on the results of other trials, it has been observed that the proportion of occurrences of an event tends to stabilize "in the long run," tending to some limiting value as the number of trials becomes infinite.

EXAMPLE 2.2a It is perhaps a bit simplistic to think of a baseball player's successive times "at bat" as successive trials of the same experiment. Yet, those who follow the game have probably observed that a player's batting average—the ratio of number of hits to the number of times at bat from the beginning of the season—fluctuates quite a bit at the beginning and gradually begins to settle down, tending toward some limiting value. Of course, the season (fortunately) is not infinitely long. But it is long enough for the tendency to be seen. ∎

Gamblers will tell you that a similar phenomenon is observed with sequences of dice rolls or roulette wheel spins. Indeed, with coins, dice, and spinning wheels, it turns out that the long run limit of the success ratio for any event appears to be very close to the probability one would calculate under the assumption of equally likely outcomes.

EXAMPLE 2.2b A computer was programmed to simulate successive tosses of a coin and to keep track of the ratio of the number of heads to the number of tosses. A plot of this ratio vs. the number of trials is shown in Figure 2-1 for the first 1,000 tosses. The tendency to stabilize is apparent, but there is still some fluctuation. Figure 2-2 shows the segment of the record for the last 1000 of 10,000 trials. There is less local fluctuation (the vertical scale is expanded), but it is still not obvious that the long-run limit is 1/2. If it is not 1/2, we'd say the coin is biased. ∎

There are of course many experiments whose outcomes are *not* viewed as equally likely. If an experiment is repeatable, one can approach the assignment of probabilities to events in terms of long-run limits. At any rate, it is not unreasonable to imagine that there *is* an appropriate probability for a given event—a number between 0 and 1. (And then one can define a lottery which is approximately equivalent.)

EXAMPLE 2.2c A thumbtack thrown on a table will land either with point up or point down. But there is no reason to feel that one outcome is as likely as the other. Yet the experiment can be repeated, and in a long sequence of trials one would see a tendency of the ratio of the number of points up to the number of trials towards a limiting value. We assume that there is such a limiting value and might take this as the probability of the outcome "point up." Moreover, we could take a rational number approximating that probability as defining an equivalent lottery in which there are lottery tickets marked "point up" in a proportion defined by the rational number. ∎

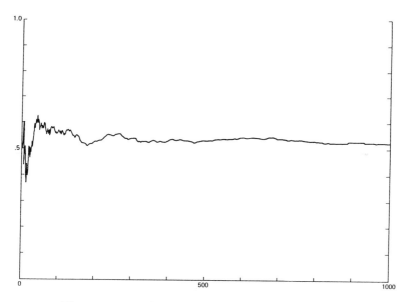

Figure 2-1 Relative frequency—1000 coin tosses

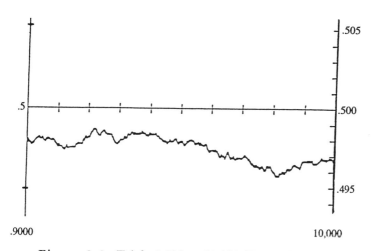

Figure 2-2. Trials 9,000 to 10,000 (Example 2.2b)

If one conceives of the probability of an event as the long-run limit of proportions of successes, certain properties are evident. Being a ratio of a number of successes y to a number of trials n, where $0 \leq y \leq n$, a probability in this new sense is a number on the unit interval, as in the preceding section. And the number of times Ω occurs in n trials is n, so the probability of Ω is 1. Moreover, if $EF = \emptyset$, the number of trials in which the event $E \cup F$ occurs is the sum of the number in which E occurs and the number in which F occurs, so this new probability is at additive. Thus, all three of the basic properties given in §2.1 hold when probabilities are conceived of as long-run limits.

2.3 Subjective probability

People bet on experiments with symmetries, as well as on other repeatable experiments. They also bet on experiments that one can hardly imagine being repeated and where there is nothing to suggest probabilities *a priori*: Will your house burn down during the coming year? Will the flight you are about to take get to its destination without accident? People bet against their insurance companies about the answers to these questions. Both parties assume that there is a certain chance of winning their bets—that there *are* probabilities for outcomes even though there are no symmetries and no past experience with repeated trials to be exploited in calculating them.

Will there be a woman president in the next twenty years? Will it rain tomorrow? Will a prescribed operation be successful for a particular patient? Will the horse you favor in the next race win it? Is there life on Mars? People have feelings and beliefs that can be summarized in probabilities, even when (as in the case of life on Mars) the matter may never be settled. But again, two persons encountering the same experiment will generally have different notions as to the probability of a particular event. When there is nothing even seemingly objective to latch on to (as opposed to the case of symmetrical objects), probabilities must be personal or subjective.

Some would argue that, really, *all* probabilities are personal. They'd say that what seems to be objective simply reflects a tendency for people's personal probabilities to agree in those cases.

EXAMPLE 2.3a Suppose you stand a quarter on end with the heads side showing and the face upright, and then flick it to make is spin. (See Problem 2-11.) It will slow down and fall flat with one side up. Are heads and tails equally likely? Maybe you will say they are, as you probably would if the same quarter were tossed in the usual way. Others say heads is the more likely. That is, even for the spin of a nearly symmetrical coin not every one agrees on the probability of heads. (We once did the experiment 20 times and got 19 heads; in another sequence with a different coin we got 60 heads in 100 trials. Do you still say the probability is 1/2?) ∎

To concede that one can have a personal probability of an event E does not say what that probability is. How should one deal with such subjective elements? How can we elicit a person's probability of E? One approach is to offer a choice between (a) a reward (say $10) if E occurs, and (b) the same reward if a coin falls heads. A preference for option (a) would say that the personal probability is greater than 1/2. By determining such choices, using coins with varying biases, one can bracket and close in on the personal probability of E.

It can be shown that if a person's assessment of the relative likelihoods of events are consistent in certain naturally desirable ways, then that person's personal probabilities are uniquely determined—and determined so that the properties of $P(\cdot)$ in Theorem 1 hold for them as well.

PROBLEMS

2-1. A committee of three is chosen at random from a group of five men and five women. Find the probability that
(a) the committee will consist of all men or all women.
(b) at least one man and at least one woman are on the committee.
(c) at least one man is on the committee.
(d) two particular people are not both on the committee.

2-2. Suppose we take the 36 outcomes in the sample space for the toss of two dice as equally likely. Find the probability that
(a) the sum is divisible by 4.
(b) both numbers are even.
(c) the numbers match (a "double").
(d) the numbers differ by at least 4
(e) the sum is at most 10.

2-3. The ballot order of the names of six candidates for three judgeships is determined by lot. Assuming all possible orders are equally likely, find the probability that the three incumbent judges (running for re-election) are at the head of the list.

2-4. The hats of four sailors (A, B, C, and D) get mixed up (they all look alike) and are passed out at random. Find the probability that

(a) C gets his own hat.
(b) A and D get their own hats.
(c) B, C, and D get their own hats.

2-5. Find the probability that when three couples (A, B, C) are seated at random at a round table, Mr. & Mrs. A sit next to each other.

2-6. In a contest involving an element of skill, the top twelve contestants turn out to be tied, and the winners of the four top prizes are chosen from these twelve by lot. If you are one of the twelve, what are your chances of winning one of the prizes?

2-7. On the basis of a random selection of four articles from a lot of 20 articles, the lot will be accepted if at most one of the four in the sample is found to be defective, and otherwise rejected. Find the probability
(a) that a lot with two defectives is rejected by this procedure.
(b) that a lot with one defective article is accepted.

2-8. Find the probability that in a random selection of four cards from a deck of playing cards there is one card of each suit. [See Problem 1-8 on page 6 for a list of cards in the deck.]

2-9. Five cards are dealt (selected at random) from a deck of playing cards. Find the probability of each of these events:
(a) The hand contains exactly one pair (two cards of one denomination and one card of each of three other denominations).
(b) The hand contains exactly two pairs (and a card of yet a different denomination).
(c) The hand has two cards of one denomination and three of another.
(d) The hand consists of five cards all of the same suit.
(e) The five cards can form a sequence (e.g., 8, 9, 10, jack, queen), with no restriction on the suit except that they are not of the same suit.
(f) The five cards can form a sequence in the same suit.

2-10. When the Minnesota state lottery was instituted in 1990, a newspaper item stated that matching six numbers chosen at random from the integers 1 to 54 was "a 1-in-12.9 million shot." Do you agree? There were to be smaller payoffs for matching four of the numbers or five of the numbers. Find the odds on these events.

2-11. Try this experiment: Use a finger of one hand to stand a quarter on end on a table with the heads side showing and the face upright; then flick it with a finger of the other hand to make is spin. When the coin slows, it will fall with one side showing. Do this 50 times, record the results, and plot the relative frequency of heads as a function of the number of trials.

2.4 Probability axioms

Whatever view one might adopt in assigning probabilities to events in real experiments, they must conform to certain rules, for consistency. In view of the properties we have seen to hold when experiments have equally likely outcomes and when probability is thought of as a long-run limit, we require that, for any sample space Ω and family of events E, an assignment of numbers to be called probabilities should satisfy these properties. We take them as axioms for a probability space.

PROBABILITY AXIOMS *A set function P defines a probability space provided it has these properties:*

Axiom 1. For every event E, $P(E) \geq 0$.
Axiom 2. $P(\Omega) = 1$.
Axiom 3. If $EF = \emptyset$, then $P(E \cup F) = P(E) + P(F)$.

These imply important and useful properties of probability:

THEOREM 2 *For any events E, F, G in a sample space Ω, the following properties hold:*

(1) $P(E^c) = 1 - P(E)$.
(2) $P(\emptyset) = 0$.
(3) $P(E \cup F) = P(E) + P(F) - P(EF)$.
(4) $P(E \cup F \cup G) = P(E) + P(F) + P(G) - P(EF) - P(EG)$
$ - P(FG) + P(EFG)$
(5) If $E \subset F$, then $P(E) \leq P(F)$.
(6) If $P(E) = 0$, then $P(EF) = 0$.
(7) $P(E) = P(EF) + P(EF^c)$.
(8) $P(EF) \leq P(E \cup F) \leq P(E) + P(F)$.
(9) If $E \subset F$, then $P(F - E) = P(F) - P(E)$.
(10) If $E_i E_j = \emptyset$ for $i \neq j$, then $P\left(\bigcup_1^n E_i\right) = \sum_1^n P(E_i)$.

Property (1) follows from finite additivity, together with the fact that $E \cup E^c = \Omega$, and Axiom 2. Property (2) follows from Axiom 2 and (1) with $E = \Omega$. Property (3) follows from additivity and the result of Problem 1-21, upon adding the probabilities of $E = EF \cup EF^c$ and $F = EF \cup E^c F$. The problems to follow ask you to obtain properties (4)-(10) from Axioms 1-3. Observe that the property $P(E) \leq 1$, given in Theorem 1 (§2.1) as part of "property 1" is not included here as part of Axiom 1. It follows from (5) of Theorem 2.

Realize that you cannot establish facts about probabilities solely with Venn diagrams, because a Venn diagram doesn't represent probabilities. However, a Venn diagram may help in understanding some of the results of a theorem. For, when we represent events as regions in the plane (as we do in a Venn diagram), we see that probability is an area-like measure, having some of the properties of areas.

EXAMPLE 2.4a Property (1) is useful in finding the probability of an event when it is easier to find the probability of its complement. Thus, for a random selection of five cards from a deck of playing cards, we can find P(at least one ace) by first counting hands with *no* aces:

$$\#(\text{no aces}) = \binom{48}{5} = 1{,}712{,}304.$$

Since "none" is the complement of "at least one,"

$$P(\text{at least one ace}) = 1 - P(\text{no aces}) = 1 - \frac{1{,}712{,}304}{2{,}598{,}960} \doteq .341. \blacksquare$$

In dealing with Euclidean spaces as sample spaces we'll find it convenient to have an extension of Axiom 3 (finite additivity) for the case of *countable* unions:

Axiom 3a: For any sequence $\{E_i\}$ such that $E_i E_j = \emptyset$ for $i \neq j$,

$$P\left(\bigcup_1^\infty E_i\right) = \sum_1^\infty P(E_i). \tag{1}$$

We'll refer to this as the axiom of *countable additivity*. The following two theorems require the assumption of countable additivity:

THEOREM 3 (Law of total probability) *If $\{E_n\}$ is a countable partition of Ω, then*

$$P(F) = \sum_1^\infty P(F E_n). \tag{2}$$

THEOREM 4

(a) *If $\{E_n\}$ is an ascending sequence:* $E_1 \subset E_2 \subset E_3 \subset \cdots$,

$$P\left(\bigcup_1^\infty E_n\right) = \lim_{n \to \infty} P(E_n). \tag{3}$$

(b) *If $\{E_n\}$ is a descending sequence:* $E_1 \supset E_2 \supset E_3 \supset \cdots$,

$$P\left(\bigcap_1^\infty E_n\right) = \lim_{n \to \infty} P(E_n). \tag{4}$$

Theorem 3 follows upon application of Axiom 3a to the partition $\{FE_n\}$. We'll often have occasion to appeal to this law of total probability either in the form given by Theorem 3, or to its expression for finite partitions as given by Property (7) of Theorem 2.

To establish Theorem 4(a), we use the result of Example 1.2a:

$$\bigcup_1^\infty E_n = \bigcup_1^\infty (E_n - E_{n-1}),$$

where the events $E_n - E_{n-1}$ are disjoint, and $E_0 = \emptyset$. From Axiom 3a it follows that

$$P\left(\bigcup_1^\infty E_n\right) = \sum_1^\infty P(E_n - E_{n-1}) = \lim_{n \to \infty} \sum_1^n P(E_i - E_{i-1}).$$

But the events $E_1, E_2 - E_1, \ldots, E_n - E_{n-1}$ constitute a partition of E_n, so

$$P(E_n) = \sum_1^n P(E_n - E_{n-1}).$$

Passing to the limit we get Theorem 4(a). Problem 2-24 asks you to prove Theorem 4(b).

PROBLEMS

2-12. Given $P(A) = .59$, $P(B) = .30$, and $P(AB) = .21$, find $P(A \cup B^c)$.

2-13. In a *fair* bet on event with probability 2/7 you would put up $2 to your opponent's $5. Suppose you are offered a bet on a certain boxer in an upcoming fight at betting odds of 2 to 5 (you put up $2 and your opponent, $5). Explain how it is possible that both you and your opponent might consider this a favorable gamble.

2-14. Obtain the formula for $P(E \cup F \cup G)$ in Property (4) of Theorem 2 by a two-fold application of the general addition law [(3) in that theorem]. Also, conjecture a formula for the case of n events.

2-15. Use the result of the preceding problem (for 4 events) to find the probability that none of the four sailors in Problem 2-4 gets the right hat. [Hint: Find the probability of the complement, namely, of the event that A gets the right hat or B does or C does or D does.]

2-16. Repeat the preceding problem but with five hats instead of four. Can you deduce the limiting probability of no matches (as the number of hats becomes infinite)?

2-17. Show: If $E \subset F$, then $P(E) \leq P(F)$ [Property (5) of Theorem 2].

2-18. Show: If $P(E) = 0$, then $P(EF) = 0$ [Property (6)].

2-19. Show: $P(E) = P(EF) + P(EF^c)$ [Property (7)].

2-20. Show: $P(EF) \leq P(E \cup F) \leq P(E) + P(F)$ [Property (8)].

2-21. Show: If $E \subset F$, then $P(F - E) = P(F) - P(E)$ [Property (9)].

2-22. Show using mathematical induction:
(a) The finite additivity property, (10) of Theorem 2.

(b) $P(E_1 \cup E_2 \cup \cdots \cup E_n) \leq \sum_1^n P(E_i)$.

2-23. Use Problem 2-22(b) and DeMorgan's laws [Problem 1-22] to show the *Bonferroni inequality*:

$$P(E_1 E_2 \cdots E_n) \geq 1 - \sum_1^n P(E_i^c).$$

2-24. Show how Theorem 4(b) follows from Theorem 4(a).

2.5 Discrete probability spaces

When a sample space is finite, $\Omega = \{\omega_1, \omega_2, ..., \omega_n\}$, we can define a probability model satisfying Axioms 1-3 of §2.4 by assigning probabilities $p(\omega_i) \geq 0$ such that $\sum p(\omega_i) = 1$. The collection of all subsets of the sample space is finite, 2^n in number. Taking each subset to be an event E, we define its probability to be the sum of the probabilities assigned to the elements of E:

$$P(E) = \sum_{\omega \text{ in } E} p(\omega).$$

It is clear that $P(\Omega) = 1$ and that $0 \leq P(E) \leq 1$. It is almost as clear that P is finitely additive. (Countable additivity is not an issue here.)

An important model for a finite sample space is that in which the outcomes are equally likely. When Ω has n outcomes, the **uniform** distribution is defined by $p(\omega) = 1/n$. In §2.1 we used such models as *a priori* models for certain experiments with symmetries.

If Ω is countably infinite, $\Omega = \{\omega_1, \omega_2, ...\}$, we can use the same approach—assigning nonnegative probabilities p to the elements ω such that $\sum p(\omega) = 1$. With $P(E)$ again defined as the sum of the probabilities of its elements, probability axioms 1-3a are satisfied. (Infinite sums are needed for some events.) A "uniform" model is not possible when Ω is not finite.

EXAMPLE 2.5a Let Ω be the set of positive integers, $\{1, 2, ...\}$, and assign probabilities $p(n) = 1/2^n$. These are nonnegative and sum to 1. The probability of the event "odd" $= \{1, 3, 5, ...\}$ is the sum

$$P(\text{odd}) = \tfrac{1}{2} + \tfrac{1}{8} + \tfrac{1}{32} + \cdots = \tfrac{2}{3}.$$

The probability of the event $\{4, 5, 6, ...\}$ is

$$P(\text{at least 4}) = \tfrac{1}{16} + \tfrac{1}{32} + \tfrac{1}{64} + \cdots = \tfrac{1}{8}$$

$$= 1 - P(1, 2, \text{ or } 3) = 1 - (\tfrac{1}{2} + \tfrac{1}{4} + \tfrac{1}{8}). \blacksquare$$

2.6 Densities

We'll encounter uncountably infinite sample spaces mainly as Euclidean spaces ($\Re^1$, $\Re^2$, ...). Other possibilities include the set of points on a circle and the set of points on a sphere. Suppose we have a measure of content defined for such spaces (length, area, volume), defined by an appropriate integral: an ordinary integral on $\Re^1$, a double integral on $\Re^2$, a line integral on a circle, or a surface integral on a sphere. A probability model can be defined the way we define a mass distribution on such spaces, in terms of a **density function**.

When mass is distributed over such spaces, one defines average mass density in a region as the amount of mass divided by the content of the region. The density at a point is then defined as the limiting average density over a sequence of elemental regions converging to the point. In terms of this density we find the mass in a region E as the (appropriate) integral of the density function over E. The total mass is the integral of the density over the whole space. If this is finite, we can define a *relative* density by dividing by the total mass. Defining the probability of a region E as the relative mass in E, we find that this satisfies the axioms of probability. Thus, a nonnegative function $f(\omega)$ with the property that its integral over the whole sample space is 1 will serve as a density function for defining a probability measure.

EXAMPLE 2.6a Consider how one might model the spin of a pointer, or a wheel of fortune with no dividing pegs. Putting a scale from 0 to 1 around the rim, we find that the pointer (or wheel) stops at a number between 0 and 1. If there are many revolutions before it stops, we might view all directions—numbers between 0 and 1—as "equally likely." But because there are infinitely many of these numbers, we can't assign equal positive probabilities to all directions. Instead, we assign probabilities using a uniform density. This means that the probability of an arc on the circumference of the circle traced by the pointer (or at the rim of the wheel) is proportional to the length of the arc. Thus, the probability that the pointer stops, say, between .1 and .7 is .6, because the arc between .1 and .7 is sixth-tenths of the total circumference.

What is the probability that the pointer stops at a particular value x? To find this we take x to be the intersection of the descending sequence of intervals $(x - 1/n, x + 1/n)$ for integers $n \geq N$, where N is large enough that the intervals are included in the sample space. Then, according to Theorem 4(b) of the preceding section,

$$P(\{x\}) = \lim_{n \to \infty} P(\{x - 1/n, x + 1/n\}) = \lim_{n \to \infty} (2/n) = 0.$$

(If x is 0 or 1, use a sequence of intervals with x at one end.) ■

EXAMPLE 2.6b In Example 1.4d we discussed a model for a certain material, thought of as being composed of tubes or fibers, each assumed to be oriented in a certain direction with respect to some given reference frame. Each orientation can be identified with a point on the unit sphere. The sample space is a full sphere if the orientation is directed, otherwise a hemisphere.

A fiber selected "at random" has an orientation that can be any of infinitely many possible orientations. If the fiber is indeed chosen at random, the symmetry implied in this phrase would suggest a uniform density on the sphere. The probability of an event—a set of points on the sphere—is then proportional to the area of that set. If Ω is the hemisphere, the probability of E is the area of E divided by 2π, the area of the hemisphere. For instance, if we use angular coordinates (ϕ, θ), the probability of the event $0 < \phi < \pi/4$, which is 1/8 of the sphere, is just 1/8. The probability of the interval $0 < \theta < \pi/4$ is the relative area of a polar cap, in this case $1 - 1/\sqrt{2}$ or about .293. ■

PROBLEMS

2-25. Let $\Omega = \{0, 1, 2, ...\}$, and define $p(\omega) = (1 + \omega)/2^{\omega+2}$. Find the probabilities of the following events:
(a) $\{0, 1, 2, 3, 4, 5\}$ (b) $\{2, 3, 4\}^c$. (c) $\{3, 4, 5, ...\}$.

2-26. A geneticist (G. Mendel) crossed round yellow and wrinkled green peas. According to his theory, the peas of the progeny plants should have the characteristics round-yellow (RY), round-green (RG), wrinkled yellow (WY), and wrinkled green (WG) in the proportions 9:3:3:1. Give a probability distribution in the sample space {RY, RG, WY, WG} that reflects those proportions, and find the probability that a plant has yellow peas.

2-27. Let Ω be the set of points with positive integer coordinates: $\omega = (i, j)$, and define a probability distribution by $p(\omega) = 1/2^{i+j}$.
(a) Show that $\sum p(\omega) = 1$.
(b) Find the probability of the event $\{(i, j): i + j \leq 4\}$.

2-28. Suppose probability is distributed uniformly in the unit disk in the plane: $\|\omega\| \leq 1$, where $\omega = (i, j)$ and $\|\omega\|$ is the distance from the origin to (i, j). Find the following:
(a) $P(\|\omega\| \leq 1/2)$.   (b) $P(i > 0)$.   (c) $P(i + j > 0)$.

2-29. Let Ω be the set of nonnegative real numbers, and define a density at ω by $f(\omega) = e^{-\omega}$. Find the probability of each of these events:
(a) $\omega < 4$.   (b) $\omega > 2$.   (c) $1 < \omega < 2$.

2-30. Given the uniform distribution on the unit hemisphere in Example 2.6b, find the probability of the polar cap $0 \leq \theta \leq \theta_0$ as a function of θ_0.

2-31. Consider a uniform distribution of probability in the (plane) square with corners $(0, 0)$, $(0, 2)$, $(2, 0)$, and $(2, 2)$. Determine the probabilities of the event determined by each of the following conditions on $\omega = (x, y)$:
(a) $x + y < 2$.   (d) $x^2 + y^2 > 1$.   (g) $|x - y| > 1$.
(b) $x + y < 1$.   (e) $x < 1$ and $y > 3/2$.   (h) $x - y < 1$.
(c) $x > 1/2$.   (f) $x > 1$ or $y > 1$.   (i) $x - y > 2$.

2.7 Conditional probability

In a sense, all probabilities are conditional—dependent on the information one has about the experiment at the moment. Given a probability model that describes an experiment as viewed at one point in time, suppose one acquires further information about the experiment. This will generally require changing the model. We call the new probabilities **conditional**.

Consider first a finite sample space, with probabilities $P(\omega)$ defined for its outcomes ω. We then learn that an event F has occurred. To take this into account we need a new model, one with a reduced sample space, namely, F. Within F, we have no reason to alter the relative likelihoods of outcomes, but probabilities must sum to 1. Dividing each $p(\omega)$ by $P(F)$, we have new probabilities:

$$p^*(\omega) = \frac{p(\omega)}{P(F)},$$

for a model in which the axioms of probability hold. The probabilities of events E within F are probabilities, to be sure, but to distinguish them from those previously assigned we call them **conditional probabilities** and use the notation $P(E \mid F)$. (This is read "probability of E, given F.")

One may want to consider outcomes and events of the original sample space that do not happen to lie within F, and we can extend the new probability model to the original sample space by defining $p^*(\omega) = 0$ for any ω not in F. If an event G is not entirely within F, only its outcomes inside F contribute probability when we sum, so

$$P(G \mid F) = \sum_{\omega \in G} p^*(\omega) = \sum_{\omega \in GF} \frac{p(\omega)}{P(F)} = \frac{P(GF)}{P(F)}. \tag{1}$$

These conditional probabilities satisfy the probability axioms of §2.4.

EXAMPLE 2.7a Suppose two dice are rolled so that the 36 pairs (i, j) are equally likely. The probability that the sum is even is 18/36, and the probability that the smaller of i and j is 3 is 7/36. The pairs $(3,3)$, $(5,3)$, $(3,5)$ have even sums, so the probability of the intersection [smaller = 3 and even sum] is 3/36. Thus,

$$P(\text{smaller} = 3 \mid \text{even}) = \frac{P(\text{smaller} = 3 \text{ and even sum})}{P(\text{even})} = \frac{3/36}{18/36} = \frac{3}{18}. \blacksquare$$

Suppose we are thinking of probability as the long-run limit of relative frequency of occurrence. The proportion of trials in which F occurs is $P(F)$. The proportion in which both G and F occur, among all trials, is $P(FG)$. So among the trials in which F occurs, the proportion in which G occurs is $P(GF)/P(F)$. Thus, we arrive at the same ratio for a conditional probability as before. [Of course, the ratio $P(GF)/P(F)$ is not defined for events F with probability 0.]

DEFINITION When $P(F) \neq 0$, the *conditional probability of E given F* is the ratio

$$P(E \mid F) = \frac{P(EF)}{P(F)}. \tag{2}$$

Probabilities in Ω defined by (2) satisfy the axioms of probability: $P(E \mid F) \geq 0$, $P(\Omega \mid F) = 1$, and additivity is shown in Problem 2-35(b). Multiplying through by $P(F)$ in (2), we obtain

$$P(EF) = P(E \mid F) \, P(F). \tag{3}$$

We refer to (3) as a **multiplication rule**. Its usefulness is illustrated in

2.7 Conditional probability

EXAMPLE 2.7b From a bowl with 3 white and 5 blue chips, draw one at random and then another at random from those that remain. What is the probability that you get one of each color?

First we find the probability that the first chip is white and the second blue. The phrase "at random," for the first selection, means that the eight chips are equally likely; so $P(\text{1st white}) = 3/8$. Since the second selection is also random, the seven remaining chips are equally likely. Given that the first was white, there are 2 white and 5 blue ones left, so $P(\text{2nd blue} \mid \text{1st white}) = 5/7$. Then

$$P(\text{1st white, 2nd blue}) = P(\text{2nd blue} \mid \text{1st white}) \cdot P(\text{1st white}) = \frac{5}{7} \cdot \frac{3}{8}.$$

Similarly,

$$P(\text{1st blue and 2nd white}) = \frac{3}{7} \cdot \frac{5}{8} = \frac{15}{56}.$$

Adding, we find the probability of drawing one of each color:

$$P(\text{one blue, one white}) = \frac{15}{56} + \frac{15}{56} = \frac{15}{28}.$$

Suppose, looking at the problem another way, we *define* the selection of the two chips to be random if all 56 possible sequences are equally likely. Then all of the 28 combinations of 2 chips are equally likely, and

$$\#(\text{one white, one blue}) = \binom{3}{1}\binom{5}{1} = 15.$$

So the probability of one white and one blue is $15/28$—as we found above using the multiplication rule. ■

The multiplication rule (3) extends to the intersection of several events. Thus, for three events E, F, and G,

$$P(E) \cdot P(F \mid E) \cdot P(G \mid EF) = P(E) \cdot \frac{P(EF)}{P(E)} \cdot \frac{P(EFG)}{P(EF)} = P(EFG). \quad (4)$$

We can use the extension of this idea to k events to obtain a general property, of which the calculation in the preceding example is a special case: In making n random selections in succession from a population of N individuals or objects (without replacement) the number of possible sample sequences is $(N)_n$. The probability of any particular sequence, by the extended multiplication rule, is

$$\frac{1}{N} \cdot \frac{1}{N-1} \cdots \frac{1}{N-n+1} = \frac{1}{(N)_n}.$$

That is, the $(N)_n$ possible sample sequences are equally likely, and from this it follows that the $\binom{N}{n}$ combinations of n are equally likely. So for events that do not involve order, assuming successive random selections one at a time (without replacement) is equivalent to assuming that the possible combinations are equally likely.

EXAMPLE 2.7c As Example 2.7b, we draw chips from a bowl with 3 white and 5 blue chips, one at a time, each time selecting at random from those that remain. If we are to draw three chips, what is the probability that the third chip drawn is blue?

One's first reaction may be to say that this depends on what we got on the first two draws—as indeed it would, if we knew what they were. Thus,

$$P(\text{3rd B} | \text{2 W's}) = \tfrac{5}{6}, \quad P(\text{3rd B} | \text{W \& B}) = \tfrac{4}{6}, \quad P(\text{3rd W} | \text{2 B's}) = \tfrac{3}{6}.$$

From the law of total probability,

$$P(\text{3rd B}) = P(\text{W, W, B}) + P(\text{B, B, B}) + P(\text{W, B, B}) + P(\text{B, W, B}).$$

And applying (4) in each term on the right, we find

$$P(\text{3rd B}) = \left(\tfrac{5}{6} \cdot \tfrac{3}{28}\right) + \left(\tfrac{3}{6} \cdot \tfrac{10}{28}\right) + \tfrac{4}{6}\left(\tfrac{15}{56} + \tfrac{15}{56}\right) = \tfrac{5}{8}.$$

So the probability that the third chip drawn is blue is simply the ratio of the number of blue chips to the number of chips—the same as on the first, or indeed, on any draw (up to eight). ∎

The calculation in Example 2.7c can be generalized. In random selections, one at a time, from a population of size N, the probability of picking one from a particular set E of M population members at the kth selection is the same as at the first selection, when nothing is known about the results of the earlier selections: Since the $(N)_k$ sequences of length n are equally likely, we count those in which the kth element is one of the M elements of E:

$$P(E) = \frac{(N-1)_{k-1} \cdot M}{(N)_k} = \frac{M}{N}.$$

And this is the proportion of M-elements in the population.

It can happen that the probability of an event E is unchanged with the information that F has occurred: $P(E|F) = P(E)$. The events are then said to be **independent**, and when this is the case, the probability of their intersection factors: $P(EF) = P(E)P(F)$.

EXAMPLE 2.7d Consider a sample space $\{a, b, c, d, e\}$ with probabilities assigned to the individual values as follows: $(.2, .2, .3, .2, .1)$, respectively. The intersection of the events $E = \{a, b, e\}$ and $F = \{a, d\}$ is $EF = \{a\}$, with probability .2. This is the product of $P(E) = .5$ and $P(F) = .4$. So E and F are independent. ∎

Suppose an experiment is a composite of experiment $\mathcal{E}_1$ and $\mathcal{E}_2$. If probabilities are assigned in the composite experiment in such a way that every event in $\mathcal{E}_1$ is independent of every event in $\mathcal{E}_2$, then $\mathcal{E}_1$ and $\mathcal{E}_2$ are said to be independent experiments. We'll have much more to say about independence in the next chapter.

2.8 Bayes' theorem

The multiplication rule [(3) of §2.7] gives us two ways to calculate the probability of an intersection, according to which event is the conditioning event:

$$P(EF) = P(E\,|\,F)P(F) = P(F\,|\,E)P(E).$$

Of course, this assumes that $P(E)$ and $P(F)$ are not zero, so that the conditional probabilities involved are defined. Dividing by $P(E)$ we obtain the essence of **Bayes' theorem:**

$$P(F\,|\,E) = \frac{P(E\,|\,F)P(F)}{P(E)}. \tag{1}$$

With the law of total probability [(7) of Theorem 1, §2.4] we can replace $P(E)$ in the denominator by a useful formula for calculating it:

$$P(F\,|\,E) = \frac{P(E\,|\,F)P(F)}{P(E\,|\,F)P(F) + P(E\,|\,F^c)P(F^c)}. \tag{2}$$

The indicator functions I_E and I_F of E and F, respectively, constitute a random pair with values $(0, 0)$, $(0, 1)$, $(1, 0)$, $(1, 1)$. The probability distribution is defined in the following table:

	F	F^c	
E	$P(EF)$	$P(EF^c)$	$P(E)$
E^c	$P(E^cF)$	$P(E^cF^c)$	$P(E^c)$
	$P(F)$	$P(F^c)$	1

The probabilities in a row (or in a column) give the conditional odds, given that row (or column). Thus, for example, the odds on F given E

are the entries in the first row, $P(EF):P(EF^c)$.

EXAMPLE 2.8a Suppose that 10 percent of the undergraduates at a certain university are foreign students, and that 25 percent of the graduate students are foreign. If there are four times as many undergraduates as graduate students, what fraction of the foreign students are undergraduates? This question can be phrased in terms of probabilities: If a student selected at random is found to be a foreign student, what is the probability that the student is an undergraduate?

With the equal likelihoods implied in the phrase "at random," the given information implies these probabilities: $P(U) = .8$, $P(G) = .2$, $P(F \mid U) = .10$, and $P(F \mid G) = .25$. What's wanted is $P(U \mid F)$. From Bayes' theorem, we have

$$P(U \mid F) = \frac{P(F \mid U)P(U)}{P(F \mid U)P(U) + P(F \mid G)P(G)} = \frac{.10 \times .8}{.10 \times .8 + .25 \times .2} = \frac{8}{13}.$$

These and similar computations are summarized in the following table of joint probabilities of the indicator functions of U and F:

indicator function →

	F	F^c	
U	.08	.72	.8
U^c	.05	.15	.2
	.13	.87	1

The conditional odds on U given F are .08 to .05, or 8 to 5. ■

For a more general formulation, suppose the event "F" in (2) is an event in a partition more general than $\{F, F^c\}$: Consider a countable partition $\{F_k\}$ of Ω. This induces a partition $\{EF_k\}$ of E, for which (by the law of total probability)

law of Total pro.   *proportion Rate*

$$P(E) = \sum_k P(EF_k) = \sum_k P(E \mid F_k)P(F_k). \tag{3}$$

This is the denominator of the fraction in (1). Bayes' theorem states that, given E, the probabilities of the events F_k are proportional to the products $P(E \mid F_k)P(F_k)$, the terms in (3) which comprise the denominator of $P(F_k \mid E)$.

EXAMPLE 2.8b A research study focused on the possible relationship between birth order and delinquency. It classified a group of high school girls as delinquent (D) or nondelinquent (N), and as either eldest child, (E), middle child (M), youngest child (Y), or only child (O). Suppose, in one population, the proportions of E, M, Y, O were .3, .4, .2, and .1, respectively; and suppose further that the corresponding delinquency rates are 5%, 10%, 15%, and 20%. What proportion of delinquents are "only" children (O)? That is, what is $P(O \mid D)$?

The proportion of delinquents in the population (or the probability that a girl selected at random is delinquent) is given by (3):

$$P(D) = (.3)(.05) + (.4)(.10) + (.2)(.15) + (.1)(.20)$$
$$= .015 + .040 + .030 + .020 = .105.$$

The terms in this sum give us the conditional odds on (E, M, Y, O), 3:8:6:4. These are converted to probabilities by (2): $P(O \mid D) = 4/21$. ∎

PROBLEMS

2-32. For a fair die, find the probability
(a) that it shows 4, given that the outcome is even.
(b) that it shows 6, given that the outcome is divisible by 3.

2-33. Given $P(E) = .3$, $P(E \mid F) = .5$, and $P(F^c) = .4$, find $P(E \cup F)$.

2-34. A sample space has just five outcomes: $\Omega = \{a, b, c, d, e\}$. Given $P(\{a\}) = 1/4$ and $P(\{a, b, c\}) = 1/2$,
(a) find the probabilities of all events whose probabilities are determined by those given.
(b) find $P(\{a\} \mid \{a, b, c\})$.
(c) find $P(\{a, d, e\} \mid \{b, c, d, e\})$.

2-35. Given that the events E and F are mutually exclusive,
(a) determine $P(E \mid E \cup F)$ in terms of $P(E)$ and $P(F)$.
(b) show that $P(E \cup F \mid G) = P(E \mid G) + P(F \mid G)$.

2-36. Two cards are drawn at random, one at a time without replacement, from a deck of playing cards. Determine the probability that
(a) both are face cards, given that they are of the same suit.
(b) both are face cards.
(c) both are hearts, given that both are red.
(d) the second card drawn is an ace.

2-37. Six husband-wife couples gather for three tables of bridge (four at a table). Twelve score cards are passed out at random, the twelve cards consisting of six pairs marked, respectively, "Table i, Couple j," where $i = 1, 2, 3$, and $j = 1, 2$. Find the probability of each of the following:
(a) Table 1 has all men.
(b) The four players at Table 1 happen to consist of two married couples.

(c) A particular couple, Mr. and Mrs. X, draw each other as partners.
(d) Mr. and Mrs. X are partners, given that they are at the same table.
(e) All six couples are paired as game partners.

2-38. Consider $\Omega = \Re^1$, with probability defined by the following density: $f(\omega) = [\pi(1+\omega^2)]^{-1}$. Find the probability that the outcome ω
(a) exceeds 1, given that it is positive.
(b) is positive, given that it is between -1 and $+1$.

2-39. I have four University keys that are hard to tell apart (without my glasses). One of them opens my office door. Suppose I try them in my door, one at a time (without "replacement"). What is the probability that it takes three tries to open the door?

2-40. Let Ω denote the unit disk in the plane, and let probability be uniformly distributed over the disk. In terms of polar coordinates, $\omega = (\rho, \theta)$, where $0 < \rho < 1$ and $0 \leq \theta < 2\pi$. Find the probability of the event $[\theta < \pi/4]$ given the event $[\rho > 1/2]$.

2-41. As in Problem 2-28, consider the uniform distribution of probability in the (plane) square with corners (0, 0), (0, 2), (2, 0), and (2, 2). Find the probability
(a) of the set $[x < 1]$, given that $y < 1$.
(b) of the set $[x^2 + y^2 < 1/2]$, given that $x^2 + y^2 < 1$.
(c) of the set $[x < 1]$, given that $x + y > 1$.

2-42. Suppose we select one chip at a time, at random, from a bowl of M blue and $N - M$ white chips.
(a) Find the probability of the sequence (BBWBWB), in six such random selections.
(b) Show that in a selection of n chips the probability of a particular sequence of colors in which there are k blue and $n - k$ white chips is

$$\frac{\binom{M}{k}\binom{N-M}{n-k}}{\binom{n}{k}\binom{N}{n}}.$$

[Hint: The $(N)_n$ possible permutations of n chips are equally likely; count those that result in the particular color sequence.]

2-43. For the distribution in Problem 2-41, let E denote the lower half of the square, and F the left half of the square. Show that E and F are independent events.

2-44. Show the following (assuming the quantities involved are defined):
(a) $P(AC \mid B) = P(A \mid BC)P(C \mid B)$.
(b) $P(EFGH) = P(E \mid FGH)P(F \mid GH)P(G \mid H)P(H)$.

2-45. A certain disease is present in about 1 out of 1000 persons in a population, and a testing program uses a detection device which gives a positive reading for a healthy person with probability .05, and a negative reading for a diseased person with probability .01. Find the probability that a person whose reading turns out positive actually has the disease.

2-46. Machines A, B, and C turn out, respectively, 10, 30, and 60 percent of the total production of a certain item. Suppose items made on machine A are one percent defective, and those made on machines B and C are 5 percent defective. Among defective items, what proportions are made on machines A, B, and C, respectively?

2-47. Suppose that the proportions of "lemons" among automobiles made on a Monday is .03, of those made on a Friday is .04, and of those made on Tuesday through Thursday is .01. If equal numbers of cars are made on the five weekdays, what is the probability that a car found to be a lemon was made on Monday?

2-48. Political prisoners Matthew, Mark, and Luke are told that two of the three are to be executed. Matthew takes the jailer aside and says, "We both know that at least one of the others will be executed; if you know, tell me the name of *one* who will be executed." Upon reflection, the jailer sees no harm in revealing one name and tells Matthew that Mark will be executed. Matthew takes heart because, since (as he sees it) it is now between him and Luke, so his probability of being executed has decreased from 2/3 to 1/2. Is his reasoning correct? Assume that if the jailer has a choice of two answers, he chooses according to the toss of a coin.

3
Random Variables

In selecting a population member at random, or performing any experiment of chance for which a probability structure has been postulated, it is common for one to focus on and observe some quantity or quality associated with the outcome. Since what is observed depends on the outcome of an experiment of chance, we call it a **random variable.** Thus, the random variable X is simply a function X defined on a probability space Ω.

What one observes about an outcome ω may be a *number*, or possibly a vector of numbers. Some reserve the terms "random variable" and "random vector" for these cases, respectively. However, qualitative aspects of ω are also variable, and there is no reason not to think of them as random variables. Thus, when ω denotes an individual drawn at random from some population, the age of that individual is a numerical random variable, and eye color is a qualitative or categorical random variable.

3.1 Discrete random variables

Consider first the case of discrete Ω, in which all subsets are "events" and have assigned probabilities. The value-space of a function X can only be discrete, and we can define probabilities in this space by assigning probabilities to the individual possible "values." (These can be either numbers or category labels.) Thus, the probability of $[X = a]$ is the probability of the set of ω's such that $X(\omega) = a$:

$$P(X = a) = P(\{\omega \colon X(\omega) = a\}) = \sum_{X(\omega)=a} P(\omega).$$

With probabilities of individual values so defined, the value space of X is a discrete probability space.

When Ω is not discrete, the value space of a function X may or may not be discrete. When the value-space of X is discrete (whether Ω itself is discrete or not), we say that X is a **discrete random variable**. The possible values are countable: $\{x_1, x_2, ...\}$. This value-space is a discrete probability space when we define the probability of a set of values as the sum of the probabilities of its elements. The distribution of probability in this space, also referred to as the distribution of X, is defined by the **probability function (p.f.)**:

$$f(x_i) = P(X = x_i) = P(\{\omega \colon X(\omega) = x_i\}).$$

THEOREM 1 When $X(\omega)$ is discrete, its p.f. has these properties:

1. $f(x_i) \geq 0,$

2. $\sum_{1}^{\infty} f(x_i) = 1.$

Properties 1 and 2 follow easily from the properties of probabilities in the probability space Ω. Moreover, in the value space $\{x_1, x_2, ...\}$, the assignment of probabilities to the individual values by the p.f. f makes the value space a probability space. The p.f. may be given as a formula, or as a table listing all the possible values x and the corresponding probabilities $f(x)$.

It is often convenient to *define* the distribution of a random variable X by assigning probabilities to its possible values using any function f that satisfies Properties 1 and 2 above.

EXAMPLE 3.1a Consider a student drawn at random from the students at an undergraduate college. The student is classified as freshman (F), sophomore (So), junior (J), or senior (Sr). The category X of the student selected is a random variable with value space {F, So, J, Sr}. The probability of a category is the proportion of students in that category. Thus, for example,

$$P(X = \text{J}) = P(\{\omega \colon X(\omega) = \text{J}\}) = \frac{\#(\text{juniors})}{\#(\text{students})}. \blacksquare$$

EXAMPLE 3.1b A useful two-valued random variable is the indicator function of an event E [see (1) in Example 1.4a], in an arbitrary sample space Ω: Let $I(\omega) = 1$ for $\omega \in E$, and $I(\omega) = 0$ for $\omega \notin E$. The value space of the discrete variable I is $\{0, 1\}$, with p.f.: $f(0) = P(E^c)$ and $f(1) = P(E)$. $\blacksquare$

For a *vector* of two discrete random variables (X, Y), the distribution is defined by its **joint p.f.**: $f(x, y) = P(X = x, Y = y)$. When the value space is finite, a sometimes convenient way to exhibit a p.f. is to give the joint probabilities in a two-way table.

EXAMPLE 3.1c Let ω denote the card drawn in a random selection from a deck of playing cards, and define random variables $X(\omega) = $ suit, $Y(\omega) = I(\omega)$, the indicator function of the event $F = $ "face card." The table of joint probabilities is as follows:

	S	H	D	C	
1	3/52	3/52	3/52	3/52	12/52
0	10/52	10/52	10/52	10/52	40/52
	13/52	13/52	13/52	13/52	

Observe that (according to the law of total probability—Theorem 3 of §2.4) the table implies distributions for X and Y separately, in the column and row totals, respectively; these are called "marginal" distributions. ∎

The joint p.f. of more than two variables is defined similarly. For the random k-vector $(X_1, X_2, ..., X_k)$, the joint p.f. is

$$f(x_1, ..., x_k) = P(X_1 = x_1, ..., X_k = x_k). \qquad (1)$$

The term **marginal distribution** will refer to the distribution of any proper subset of the variables in a random vector. Thus, a random vector (X, Y, Z) with joint p.f. $f(x, y, z)$ has univariate *and* bivariate marginal p.f.'s, such as

$$f_{X,Y}(x, y) = P(X = x, Y = y) = \sum_z f(x, y, z), \qquad (2)$$

and

$$f_X(x) = P(X = x) = \sum_{y, z} f(x, y, z). \qquad (3)$$

EXAMPLE 3.1d Consider a random vector with six possible "values," points (x, y, z), as follows:

$$(1, 0, 0), \ (0, 0, 1), \ (1, 1, 1), \ (0, 1, 0), \ (1, 2, 0), \ (0, 2, 1).$$

Suppose probability 1/6 is assigned to each. The marginal distribution of the pair (X, Z) is easily determined by counting the points that have a given pair of x- and z-values:

$$f_{X,Z}(x, z) = \begin{cases} 1/6 & \text{at } (0, 0) \text{ and } (1, 1) \\ 2/6 & \text{at } (1, 0) \text{ and } (0, 1). \end{cases}$$

[marginal dist of fail]

The marginal of Y is given by the p.f. $f_Y(y) = 1/3$ for $y = 0, 1, 2$. ∎

PROBLEMS

3-1. When a coin is tossed three times, the possible outcomes ω are the eight sequences HHH, HHT, HTH, THH, TTH, THT, HTT, TTT. Suppose these are equally likely. Let $X(\omega)$ be defined as the number of heads (H) in the outcome ω. Give the probability table for the p.f. of X.

3-2. Consider the function $f(x) = 1/2^x$, defined for $x = 1, 2, 3,...,$
(a) Show that this is a probability function for a random variable X.
(b) Find $P(X \geq 3)$.
(c) Find $P(X \text{ is odd})$.

3-3. In a selection of two from the integers 1 to 4, one at a time at random and without replacement, let X denote the first and Y the second number drawn. Let S denote the smaller and L the larger of the two numbers.
(a) Obtain the joint distribution of X and Y in table form.
(b) Find the marginal distributions of X and Y.
(c) Find the distribution of Z, the sum of the two numbers.
(d) Obtain the joint distribution of S and L in table form.
(e) Find the marginal distributions of S and L.
(f) Find $P(L - S > 1)$.

3-4. A tester is to identify three colas after tasting one can of each of three different brands, A, B, and C, assigning brand name A to one, B to another, and C to the third. Suppose the tester really cannot tell them apart and assigns names at random. Let X denote the number of colas correctly identified and give its probability distribution as a table of values and corresponding probabilities.

3-5. Consider the roll of two dice and, as in Problem 2-2, assume that the 36 outcomes (ω_1, ω_2) are equally likely. Give the probability function in table form for the random variables
(a) $X = \omega_1 + \omega_2$. (b) $Y = \omega_1 - \omega_2$.

3-6. A carton of a dozen eggs includes two that are rotten. In a random selection of three, let Y denote the number of rotten eggs among the three. Determine the p.f. of Y.

3-7. Consider the random vector (X, Y) whose distribution is defined by the following joint probability function:

$$f(x, y) = \begin{cases} 1/12 \text{ at } (1,1), (3,1), \\ 1/6 \text{ at } (2,1), (1,2), (3,2), \\ 1/3 \text{ at } (2,3). \end{cases}$$

$\frac{1}{12}$
$\frac{2}{12}$
$\frac{4}{12}$

(a) Determine the marginal p.f.'s.
(b) Find $P(X = 2 \text{ or } Y = 2)$. $5/6$
(c) Find $P(X = Y)$. $\frac{1}{12}$
(d) Find $P(X + Y \leq 4)$.
(e) Find $P(X = 1 \mid X + Y \leq 4)$.

3-8. Referring to the model in Problem 3-5 for one roll of a pair of dice, let U denote the number of dice showing a 5 and V, the number of dice showing a 6. Construct the table of joint probabilities for (U, V) and calculate:
(a) $P(U + V \geq 1)$.
(b) The marginal p.f.'s of U and V.

3.2 Distribution functions

$F_\lambda = (X \leq \lambda)$ also called c.d.f

We'll often be dealing with *numerical* random variables. The value-space of a single numerical variable X is a subset of the real line, $\Re^1$. It is often convenient to take it as the whole of $\Re^1$, by assigning probability 0 to the set of values not assumed by X. The important sets in this space are intervals and countable combinations of intervals; these are the Borel sets—sets "generated" by the half-infinite intervals $X \leq \lambda$. (In particular, a set with only one point is such a set.) When the pre-images of all such sets are events in Ω, each set can be assigned the probability of the corresponding event in Ω. In such case we say that $X(\omega)$ is a *measurable* function, as discussed in §1.4. Technically, we require that a random variable be measurable, so that probabilities are defined in its value-space. We shall assume the variables we deal with are measurable, without verifying this. Similar considerations apply to random vectors, for which the value-space is $\Re^k$, Euclidean space of finite dimension k.

The distribution of probability in the value space $(\Re^1)$ of a *numerical* random variable X is characterized by a **distribution function**:

$$F(\lambda) = P(X \leq \lambda). \tag{1}$$

When X is measurable, this is defined for all real λ. Since it gives the accumulated probability from $-\infty$ up to λ, it is also called the "cumulative" distribution function and referred to familiarly as the **c.d.f.** A c.d.f. F satisfies certain important properties, as a consequence of its definition:

THEOREM 2 *The c.d.f. of a continuous random variable $X(\omega)$ has these properties:*
1. $0 \le F(x) \le 1$ for all x.
2. F is nondecreasing: $F(x_1) \le F(x_2)$ whenever $x_1 < x_2$.
3. F is right-continuous: $\lim_{x \to a+} F(x) = F(a)$.

The first property is clear, since $F(x)$ is a probability. The second follows from the fact that the interval $[X \le x_1]$ is a subset of the interval $[X \le x_2]$. The third, we reason as follows: Because F is monotonic and bounded, it must have a right-hand limit at $x = a$; and this limit can be calculated by finding the limit over a descending *sequence* with intersection $(-\infty, a]$. Applying Theorem 4(b) of §2.4 (page 28) yields the desired result.

A c.d.f. implies probabilities of intervals and single points in $\Re^1$ as follows: The finite half-open interval $(a, b]$ can be expressed in terms of semi-infinite intervals: $(a, b] = (-\infty, b] - (-\infty, a]$. So from (9) in Theorem 2 of §2.4 (page 27), we have

$$P(a < X \le b) = P(X \le b) - P(X \le a) = F(b) - F(a). \quad (2)$$

Consider next a single value, a. As in Example 2.6a, we can express this as the intersection of the intervals $(a - 1/n, a + 1/n]$, for $n = 1, 2, \ldots$, a *descending* sequence. Again applying Theorem 4(b) of §2.4, we obtain

$$P(X = a) = \lim_{n \to \infty} P(a - 1/n < X \le a + 1/n)$$

$$= \lim_{n \to \infty} [F(a + 1/n) - F(a - 1/n)] = J(a). \quad (3)$$

We refer to $J(a)$ as the **jump** in the c.d.f. at the point a. For the closed interval $[a, b]$ we have

$$P(a \le X \le b) = F(b) - F(a) + J(a),$$

and for the open interval (a, b),

$$P(a < X < b) = F(b) - F(a) - J(b).$$

EXAMPLE 3.2a Suppose the c.d.f. of a random variable X is given by

$$F(x) = \begin{cases} 1 - .8e^{-x} & \text{if } x > 0 \\ 0, & \text{otherwise,} \end{cases}$$

shown in Figure 3-1. For this variable,
$$P(1 < X < 4) = F(4) - F(1) = .8(e^{-1} - e^{-4}) \doteq .28.$$
A single value has probability 0, except that $P(X = 0) = .2$, the jump at 0. Also, $P(X \leq 0) = F(0) = .2$, and $P(X > 0) = 1 - P(X \leq 0) = .8$. A distribution such as this might be used to describe the lifetime of a piece of equipment with a positive probability of not functioning when it is plugged in. ∎

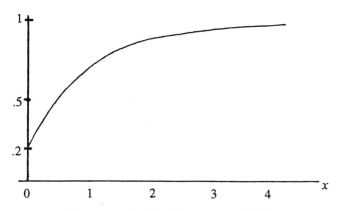

Figure 3-1. C.d.f. for Example 3.2a

As in the discrete case, so in more general cases, we'll often define a probability model for the random variable X by specifying the distribution of probability in its value-space. It can be shown that one can use a function F that satisfies properties 1-3 of Theorem 2 to define $\Re^1$ as a probability space. This is done by defining probabilities of semi-infinite intervals as $F(x) = P(X \leq x)$, and using these to define probabilities for Borel sets in $\Re^1$—probabilities that satisfy Axioms 1-3 at the beginning of §2.4. The function $X(x) = x$, for $x \; \varepsilon \; \Re^1$, is then a random variable with value-space $\Re^1$.

When X is discrete, a distribution in the value-space is defined by the p.f. (probability function), $f(x)$. The c.d.f. is then obtained from f as a sum:
$$F(x) = \sum_{u \leq x} f(u).$$
This is a *step-function*, horizontal between the possible values of X and jumping an amount $f(x)$ at each possible value x.

EXAMPLE 3.2b Let X be a random variable with the discrete distribution defined by the following table of probabilities:

x	0	1	2	3
$f(x)$	.4	.3	.2	.1

The c.d.f. is 0 for $x < 0$, jumps to height .4 at $x = 0$, remains at that height until $x = 1$, where it jumps to height .7, and so on until it reaches height 1 at $x = 3$. (See Figure 3-2.) ■

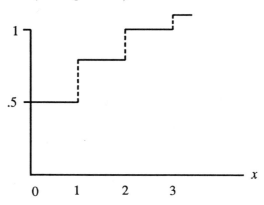

Figure 3-2. C.d.f. for Example 3.2b

Some important and often used c.d.f.'s are not readily definable with formulas involving familiar functions. It is common practice to give these in a table, showing the value of $F(x)$ for each value of x in some suitably fine grid. Table I in the Appendix is of this type. When F is continuous and strictly monotonically increasing, it is sometimes more convenient to use tables that give the values of x corresponding to selected values of F. When these probabilities are given as so many hundredths, the corresponding x-values are called **percentiles**. Thus, the 35th percentile is a value of X, called $x_{.35}$, such that $P(X \leq x_{.35}) = .35$. In particular, the 25th, 50th, and 75th percentiles are called **quartiles**. The second quartile or 50th percentile has its own special name, the **median** of the distribution, so when we speak of "the quartiles," we usually mean the first and third, denoted by $Q_1 = x_{.25}$ and $Q_3 = x_{.75}$. A general term for values of a random variable corresponding to a specified value of its c.d.f. is *quantile*.

When F is not continuous and strictly monotonic, quantiles may not be uniquely defined—there may be many values of x, or none, with the property that $F(x) = p$. Any such value would be a p-quantile. Thus, with $p = 1/2$:

3.2 Distribution functions

DEFINITION A *median* of X is any number m such that
$$P(X \le m) \ge 1/2 \text{ and } P(X \ge m) \ge 1/2.$$

Three possibilities are illustrated in Figure 3-3. When the increase of a c.d.f. is *strictly* monotonic, the median is uniquely defined. Other quantiles are defined (not necessarily uniquely) in similar fashion.

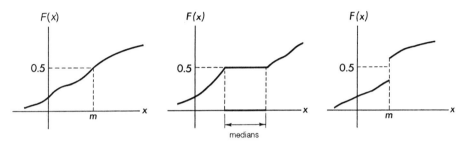

Figure 3-3. Medians: three cases

EXAMPLE 3.2c Consider the random variable X in Example 3.2a, whose c.d.f. is $F(x) = 1 - .8e^{-x}$, $x > 0$. An exponential function table (or hand calculator key) gives the values of F. The median of X is the number m such that $1 - .8e^{-m} = .5$, or $m = -\log(5/8) \doteq .47$. Similarly, to find the 95th percentile $x_{.95}$, set $F(x) = .95$, or $.8e^{-x} = .05$, and solve: $x = -\log(5/80) \doteq 2.77$. ■

A c.d.f. is also defined for *vectors* of numerical random variables. For the pair (X, Y), we define the **joint c.d.f.**, a *bivariate* distribution function:

$$F(x, y) = P[X \le x, Y \le y] = P[\omega: X(\omega) \le x, Y(\omega) \le y]. \quad (4)$$

THEOREM 3 *A bivariate distribution function F satisfies the following conditions:*
1. $F(x, \infty)$ and $F(\infty, y)$ are univariate c.d.f.'s,
2. $F(\infty, \infty) = 1$.
3. $F(c, d) - F(a, d) - F(c, b) + F(a, b) \ge 0$ for $a < c, b < d$.

The "second difference" in property 3 is simply the probability of the half-open rectangle defined by the vertices $(a, b), (c, d)$, which, as a probability, must be nonnegative [see Figure 3-4].

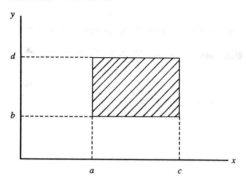

Figure 3-4.

If a function F has properties 1-3 in Theorem 3, it can serve as the c.d.f. of a random vector (X, Y), for it can be shown that any such function F defines $\Re^2$ as a probability space, one in which the two-dimensional Borel sets are "events."[1]

Given a bivariate c.d.f. f for (X, Y), marginal distributions are distributions for X and Y separately, defined as univariate c.d.f.'s which are obtained from the joint c.d.f. as follows:

$$F_X(x) = P(X \leq x) = F(x, \infty), \quad F_Y(y) = P(Y \leq y) = F(\infty, y). \quad (5)$$

(These are univariate c.d.f.'s, according to property 1 of Theorem 3.)

EXAMPLE 3.2d The function $F(x, y) = (1 - e^{-x})(1 - e^{-y})$ for $x > 0$ and $y > 0$ (and equal to 0 elsewhere) clearly satisfies properties 1 and 2 of Theorem 3. The second difference in property 3 is

$$(1 - e^{-c})(1 - e^{-d}) - (1 - e^{-a})(1 - e^{-d})$$
$$- (1 - e^{-c})(1 - e^{-b}) + (1 - e^{-a})(1 - e^{-d})$$
$$= (e^{-a} - e^{-c})(e^{-b} - e^{-d}).$$

This is nonnegative when $a < c$ and $b < d$. So F is a bivariate c.d.f. The marginal c.d.f. of X is $F(x, \infty) = 1 - e^{-x}$ for $x > 0$. ∎

Joint distributions of several numerical random variables are defined by multivariate c.d.f.'s. Given a probability space Ω, the joint c.d.f. of $[X_1(\omega), ..., X_k(\omega)]$ is a probability in Ω:

$$F(x_1, x_2, ..., x_k) = P(\{\omega: X_1 \leq x_1, ..., X_k \leq x_k\}). \quad (6)$$

It is convenient and conventional to use boldface letters for vectors, and

---

[1] See H. Cramér, *Mathematical Methods of Statistics*, Princeton U. Press (1946).

write $\mathbf{X} = (X_1, ..., X_k)$. Thus, we denote the joint c.d.f. in (6) by $F_\mathbf{X}(\mathbf{x})$.

THEOREM 4 *The joint c.d.f. of $X_1(\omega), ..., X_k(\omega)$ defined on Ω has the the following properties:*
1. *$F(\infty, ..., \infty) = 1$.*
2. *The function of j variables obtained by setting all but j arguments equal to ∞ $(j < k)$ is a j-dimensional c.d.f.*
3. *The nth difference $\Delta^{(n)} F$, the probability of a "rectangle" defined at an arbitrary point $\mathbf{x}$ by a vector increment $\Delta\mathbf{x}$, is nonnegative.*

(The "nth difference" is the numerator of an increment quotient defining the nth-order mixed partial derivative of F as its limit.)

Given a function F satisfying the properties in Theorem 4, it can be shown to define a probability distribution in $\Re^n$ (as the sample space). This is often a convenient way of defining a multivariate distribution.

Marginal distributions are the distributions of the various possible proper subsets of the variables, and their c.d.f.'s are obtainable from the joint c.d.f. of all of the variables. For instance, the marginal distribution of the first two components is defined by their joint c.d.f., obtained from $F_\mathbf{X}$ as follows:

$$F_{X_1, X_2}(x_1, x_2) = P(X_1 \le x_1, X_1 \le x_2) = F_\mathbf{X}(x_1, x_2, \infty, ..., \infty).$$

PROBLEMS

3-9. Show that if $A > 0$, the function $F(x) = \frac{1}{2A}(x + A)$ for $-A \le x \le A$ is a c.d.f., when suitably extended outside that interval.

3-10. Show that the following are c.d.f.'s:
(a) $(1 + e^{-x})^{-1}$, $-\infty < x < \infty$.
(b) $1 - 1/y$, $1 < y < \infty$ [and 0 for $y < 1$].

3-11. Let $F(x) = 1 - \frac{1}{2x}$ for $x \ge 1$, and 0 for $x < 1$. Verify that this is a c.d.f. and find
(a) $P(X < 1)$.
(b) $P(X = 1)$.
(c) $P(2 < X < 3)$.
(d) $P(X = 2)$.

3-12. Define the c.d.f. of X (in part) as follows:

$$F(x) = \begin{cases} x/2, & 0 \le x < 1, \\ 3/4, & 1 \le x < 2, \\ 1 + (x - 3)/4, & 2 \le x \le 3. \end{cases}$$

Complete the definition of F and find the following:

(a) $P(X = 1/2)$.
(b) $P(X = 1)$.
(c) $P(X < 1)$.
(d) The quartiles.

(e) $P(X \leq 1)$.
(f) $P(X > 2)$.
(g) $P(1/2 \leq X \leq 3/2)$.

3-13. Given the c.d.f. $F(x) = x^2$ for $0 \leq x \leq 1$, deduce the value of $F(x)$ for x outside the unit interval, and find
(a) $P(X = .5)$
(b) $P(X \leq .5)$
(c) $P(X > .2)$
(d) $P(.2 < X < .8)$
(e) The median value of X.

3-14. Given that X has the c.d.f. $1 - \cos x$ for $0 < x < \pi/2$, find
(a) the median.
(b) $P(X \leq \pi/4)$.

3-15. Given that F_1 and F_2 are (univariate) c.d.f.'s, show that
(a) $F = \alpha F_1 + (1 - \alpha) F_2$ is also a c.d.f. when $0 \leq \alpha \leq 1$.
(b) $F(x, y) = F_1(x) F_2(y)$ is a bivariate distribution function.

3-16. Referring to Problem 2-40 (uniform distribution on the unit disk), define the random pair (R, Θ), where $R = R(\omega)$ and $\Theta = \Theta(\omega)$ are the polar coordinates of ω.
(a) Obtain the joint c.d.f. of (R, Θ).
(b) Obtain the marginal c.d.f.'s of R and Θ.

3-17. Let $F(x, y) = 1 - e^{-x-y}$, for $x \geq 0$ and $y \geq 0$ (and 0 elsewhere). Show that F satisfies Properties 1 and 2 of a bivariate c.d.f. but is *not* a bivariate c.d.f.

3.3 Continuous random variables

When a c.d.f. $F(x)$ is *continuous* at each x, the probability assigned to an individual value is 0, since the jump at each x is 0. One might think to define a continuous random variable as one that has such a c.d.f., but as it turns out, the class of *all* continuous F's satisfying properties 1–3 of Theorem 3 (page 51) includes more functions than are needed or realistic for modeling continuous phenomena. So we restrict the class of models for continuous random variables to include only functions F that are continuous and differentiable at all but a finite set of points. In fact, we could widen the class a bit, but this is wide enough for our purposes. The essential requirement is that the F be recoverable from its derivative by integration:

3.3 Continuous random variables

$$\int_{-\infty}^{x} F'(u)\, du = F(x). \tag{1}$$

The class of functions satisfying (1) includes those that are differentiable except at finitely many points.

The derivative F' has an intuitive interpretation: At the point x it is the usual limit of an increment quotient:

$$F'(x) = \lim_{\Delta x \to 0} \frac{\Delta F(x)}{\Delta x}.$$

The numerator ΔF is the "mass" (probability) in the interval Δx, so the ratio $\Delta F/\Delta x$ is the *average* density in the interval Δx. The limit of this average density is the *point* density at x, termed the *probability density*. The derivative function F' is called the **probability density function** and abbreviated **p.d.f.** It is (usually) denoted by f, sometimes with a subscript for identification. The p.d.f. may be undefined at finitely many points—or even redefined at finitely many points without changing the value of an integral such as (1).

When a distribution is given by specifying a c.d.f., the properties of a c.d.f. imply corresponding properties of the p.d.f.: Because a c.d.f. is nondecreasing, its derivative can't be negative, and because $F(\infty) = 1$, the area under the p.d.f. must be 1. Thus, we have:

THEOREM 5 *The p.d.f. of a random variable X with c.d.f. F has the following properties*:

1. $f \geq 0$,
2. $\int_{-\infty}^{\infty} f(x)\, dx = 1.$

The graph of $f(x)$ is a curve lying above the x-axis, and property 2 says that the total area under the curve and above the axis is 1.

It is often convenient to define a probability distribution for X by giving its p.d.f. Suppose a function f has the properties of Theorem 5 (and is defined at all but at most a finite number of points). Its integral,

$$F(x) = \int_{-\infty}^{x} f(u)\, du, \tag{2}$$

satisfies the required properties of a c.d.f., and its derivative is $F' = f$. (It is the "fundamental theorem of calculus" that gives us this derivative as the integrand evaluated at the upper limit.) Geometrically, $F(x)$ is the *area* above the x-axis under the p.d.f. to the left of the point x. [See Figure 3-5.]

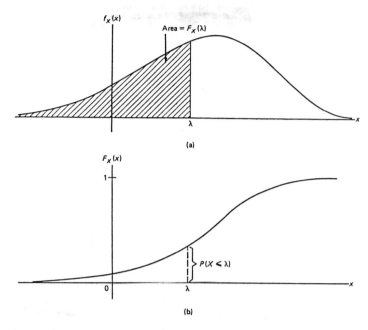

Figure 3-5

The probability of an interval is the increment in F over that interval:

$$P(a < X < b) = F(b) - F(a) = \int_a^b f(x)\,dx. \qquad (3)$$

This is the *area* under the p.d.f. above the interval (a, b). [Figure 3-6.]

When we deal with continuous distributions, single points have probability 0, so intervals have the same probability whether open, closed, or half-open:

$$P(a < X < b) = P(a \leq X < b) = P(a < X \leq b) = P(a \leq X \leq b).$$

Moreover, the integral (3) does not change its value when the p.d.f. is altered or even undefined at any finite number of points.

EXAMPLE 3.3a When X has a density f which is constant on an interval (c, d), we say that its distribution is **uniform** on that interval and write $X \sim \mathcal{U}(c, d)$. The c.d.f. is a *ramp function*:

$$F(x) = \frac{x-c}{d-c}, \quad c < x < d,$$

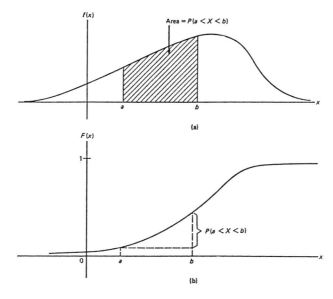

Figure 3-6.

(Of course $F = 0$ when $x \leq c$, and $F = 1$ when $x \geq d$.) The probability of any subinterval (a, b) is proportional to its length:

$$P(a < X < b) = \frac{b-a}{d-c}. \blacksquare$$

If an interval is very short, there is a convenient way of approximating its probability in terms of the p.d.f. Using (3), we find the probability of the infinitesimal interval $(x, x + dx)$ to be

$$P(x < X < x + dx) = F(x + dx) - F(x) \doteq dF(x) = f(x)dx.$$

We'll refer to the differential $f(x)dx$ as the **probability element**. The probability of the interval (a, b) as given by (3) is thought of intuitively as the "sum" of the probability elements in the interval.

EXAMPLE 3.3b The function F defined to be 0 for $x \leq 0$, 1 when $x \geq 1$, and x^2 when $0 < x < 1$ is a c.d.f: It is monotone nondecreasing from 0 to 1 and differentiable except at $x = 0$ and 1. Where the derivative exists, it is the p.d.f.:

$$f(x) = \begin{cases} 2x, & 0 < x < 1, \\ 0, & \text{otherwise}. \end{cases} \qquad (4)$$

The probability of an interval can be found by either summing the probability element $2x\,dx$ or substituting in F to find the increment in the c.d.f. For example,

$$P(.2 < X < .5) = F(.5) - F(.2) = \int_{.2}^{.5} 2x\,dx = .5^2 - .2^2.$$

[See (3) above and (1) of §3.2.] ■

Henceforth, we'll adopt the convention of giving density formulas such as (4) only for those points of the value space where the p.d.f. is positive. It is to be understood that the density is zero outside that set, as of course it *must* be, if the integral of f over $(-\infty, \infty)$ is to be 1. Thus, we'll write (4) in the form $f(x) = 2x$ for $0 < x < 1$, without mentioning specifically that $f(x) = 0$ for other values of x.

DEFINITION The *support* of a distribution is the set of points where the p.f. or p.d.f. is positive.

Because most problems we encounter involve either discrete random variables or continuous random variables, we may give the erroneous impression that there are no other types; but of course there are. A c.d.f. can increase in discrete jumps at some points and continuously at others. In §3.2 we saw distributions which are neither discrete nor continuous but involve both discrete and continuous components. Example 3.2a and Problem 3-12 involve mixture distributions, which we can represent in the form of the c.d.f. in Problem 3-15:

$$F(x) = \alpha F_1(x) + (1-\alpha) F_2(x), \quad 0 \leq \alpha \leq 1,$$

where F_1 (say) is continuous and F_2 is discrete.

3.4 Continuous random vectors

Suppose the joint distribution of (X, Y) has a twice differentiable c.d.f. whose second mixed partial derivative f has the property that the c.d.f. is recoverable from f by integration:

$$F(x, y) = \int_{-\infty}^{x} \int_{-\infty}^{y} f(u, v)\,du\,dv. \tag{1}$$

We then say that (X, Y) has a *continuous* bivariate distribution, and

$$\frac{\partial^2}{\partial x \partial y} F(x, y) = f(x, y). \tag{2}$$

joint prob. dens. func t. of bivariate funct.

3.4 Continuous random vectors

This is the limit of the ratio of the second difference $\Delta^{(2)}F$ corresponding to increments Δx and Δy to the product of those differences, as the larger of them tends to 0. So f is the point *density* of probability at (x, y), which we refer to as the **joint p.d.f.** (joint probability density function) of the bivariate distribution. The geometrical representation of a joint p.d.f. is a surface above the xy-plane, whose height at a given point (x, y) is the value of f at that point.

A bivariate p.d.f. has characteristic properties resulting from corresponding properties of the c.d.f.

THEOREM 6 The p.d.f. of (X, Y) with c.d.f. (1) has these properties:

1. $f(x, y) \geq 0$. $\quad\quad\quad\quad\quad\quad\quad\quad\quad\quad\quad\quad\quad\quad\quad\quad\quad\quad\quad\quad$ (3)

2. $\displaystyle\int_{-\infty}^{\infty}\int_{-\infty}^{\infty} f(x, y)\, dx\, dy = 1.$ $\quad\quad\quad\quad\quad\quad\quad\quad\quad\quad\quad$ (4)

The first property follows from the fact that the second difference of a c.d.f. is nonnegative; the second is equivalent to $F(\infty, \infty) = 1$: The total volume under the surface $f(x, y)$ and above the xy-plane is 1.

We'll call $f(x, y)\, dx\, dy$ the bivariate *probability element*, and think of (4) as saying that we "sum" the probability elements over the whole plane to obtain its probability. To find the probability of any region R in the plane, we sum the probability elements over R:

$$P(R) = \iint_R f(x, y)\, dx\, dy \quad\quad (5)$$

A nonnegative function $f(x, y)$ with the property that the volume under the surface f is finite can be made into a joint p.d.f. by dividing it by that volume (so that the total volume under the surface is 1). In particular, the function $f(x, y) = 1/A$ for (x, y) in a plane region of finite area A (and 0 outside the region) satisfies the properties of a bivariate density and defines a **uniform** distribution over that region. For such distributions, the probability of any subset of the support region is just the area of that subset divided by the area A.

EXAMPLE 3.4a Consider the p.d.f. $f(x, y) = K(x + y)$ for (x, y) in the triangle in the first quadrant bounded by the line $x + y = 1$ (and 0 elsewhere). This is nonnegative, and the volume under $z = x + y$ is

$$\int_0^1 \int_0^{1-x} (x + y)\, dy\, dx = 2 \int_0^1 x \int_0^{1-x} dy\, dx \left(\frac{1}{3}\right).$$

(Formulas from geometry would give this volume without a formal integration.) To satisfy (3) we take $K = 3$ so that the given f is a bivariate p.d.f.

To find the probability of the region where $x > y$, say, we integrate the joint p.d.f. over that region:

$$P(X > Y) = \int_0^{.5} \int_y^{1-y} 3(x+y)\,dx\,dy = \frac{1}{2}.$$

This could have been predicted in view of the symmetry of the distribution about the line $y = x$. ∎

In §3.2 we saw how to find the marginal c.d.f.'s of X and Y. The marginal p.d.f. of Y is the derivative of the marginal c.d.f.:

$$f_Y(y) = \frac{d}{dy}F(\infty, y) = \frac{d}{dy}\int_{-\infty}^{y}\int_{-\infty}^{\infty} f(x, u)\,dx\,du = \int_{-\infty}^{\infty} f(x, y)\,dx. \quad (6)$$

As in the discrete case [cf. (3) of §3.1], we've "summed" over the x corresponding to the unwanted variable X. (Clearly, there is a parallel formula for the p.d.f. of X.) Observe that the integral in (6) is the inner integral of (4), so that the integral of f_Y is 1 (as it must be).

The geometric interpretation of $f_Y(y_0)$ is the area of the region under the cross-section curve $z = f(x, y_0)$ in the plane $y = y_0$—defined by the intersection of that plane with joint density surface.

EXAMPLE 3.4b Consider again the p.d.f. of the preceding example:

$$f(x, y) = 3(x+y) \text{ for } x > 0,\, y > 0,\, x+y < 1.$$

The marginal p.d.f. of X is

$$f_X(x) = 3\int_0^{1-x}(x+y)\,dy = 3[x(1-x) + (1-x)^2/2],$$

$$= 3(1-x^2)/2,\ 0 < x < 1.$$

And because $f(x, y)$ (including its support region!) is a symmetric function of x and y, the p.d.f. of Y is $3(1 - y^2)/2,\ 0 < y < 1$. ∎

A random k-vector has a continuous k-variate distribution when its c.d.f. is obtainable as the integral of its mixed partial derivative. That derivative is the joint density function $f(\mathbf{x})$, with the properties that it is

3.4 Continuous random vectors

nonnegative and its integral over $\Re^k$ is 1. Marginal densities for one or more components of the random vector are obtained from the joint p.d.f. by integrating out the dummy variables corresponding to the other components. And the probability of a region $R \subset \Re^k$ is the integral of the joint p.d.f. over that region:

$$P(R) = \int_R f(\mathbf{x}) \, d\mathbf{x}. \tag{7}$$

In particular, when a distribution is uniform over a region of finite content, the probability of a subregion is R proportional to its content.

PROBLEMS

3-18. Suppose $X \sim \mathcal{U}(-1, 1)$.
(a) Find the c.d.f.
(b) Find $P(-1/4 < X < 1/4)$.

3-19. A random variable X has the p.d.f. $f(x) = 1 - |x|$ for $|x| < 1$.
(a) Graph this density function and find $P(|X| > 1/2)$, shading the corresponding region on your figure.
(b) Determine and graph the c.d.f., indicating on this graph the quantity that represents the probability in (a).

3-20. Given that X has this c.d.f.: $F(x) = \frac{1}{2} + \frac{1}{\pi}\text{Arctan } x$,
(a) find the p.d.f.
(b) determine the median and quartiles.
(c) find $P(X > 2)$.

3-21. Given that Y has the p.d.f. $f(y) = 2e^{-2y}$ for $y > 0$,
(a) find the c.d.f.
(b) find $P(Y > a)$. (This is the "tail probability" beyond a.)

3-22. Suppose Z has the p.d.f. $f(z) = [\exp(-z^2)]/\sqrt{\pi}$. Estimate the probability that $|Z| < .1$ without integrating. [Use the probability element.]

3-23. Show that if a p.d.f. f is *symmetrical* about $x = a$ [that is, if $f(a - x) = f(a + x)$ for all x], then the median of the distribution is a. [Hint: Without loss of generality let $a = 0$. Then break the integral of f over $(-\infty, \infty)$ into integrals over $(-\infty, 0)$ and $(0, \infty)$.]

3-24. In each case determine whether the given function, with suitable choice of a constant k, can serve as a density function with support as shown:
(a) kx, $0 < x < 2$.

(b) $kx(1-x)$, $0 < x < 1$.
(c) $ke^{-|x|}$.

3-25. Find and sketch the c.d.f. for each of the densities in the preceding problem, and calculate $P(0 < X < 1/2)$ in each case.

3-26. Suppose (X, Y) is uniformly distributed on the disk $x^2 + y^2 < 4$. Give the p.d.f. and find the following:
(a) $P(Y > kX)$.
(b) The marginal densities of X and Y.
(c) $P(X^2 + Y^2 > 1)$.
(d) $P(X^2 + Y^2 < \lambda)$. [This is the c.d.f. of $Z = X^2 + Y^2$.]

3-27. Consider the joint p.d.f. $f(u, v) = e^{-(u+v)}$ for $u > 0$, $v > 0$. Find
(a) the marginal p.d.f.'s.
(b) $P(U + V) \leq 4)$.
(c) the joint c.d.f.
(d) $P(U = V)$.

3-28. Suppose (X, Y) has the bivariate c.d.f. $F(x, y) = \frac{1}{2}(x^2 y + xy^2)$ on the unit square: $0 < x < 1$, $0 < y < 1$.
(a) Show that the unit square has probability 1 and deduce the value of the c.d.f. outside the first quadrant.
(b) Obtain the p.d.f.
(c) Find $F(x, y)$ for $0 < x < 1$ and $y > 1$.
(d) Find the marginal distribution of X and of Y. (They're the same; why is this?)
(e) Find $P(X + Y < 1)$.

3-29. Show the following:
(a) If f_1 and f_2 are univariate p.d.f.'s, then $f(x, y) = f_1(x) f_2(y)$ is a bivariate p.d.f.
(b) If f_1, f_2, and f_3 are univariate c.d.f.'s, then $f_1(x_1) f_2(x_2) f_3(x_3)$ is a trivariate p.d.f.

3-30. Given the uniform distribution on a hemisphere described in Example 2.6b, define $\Theta(\omega)$ and $\Phi(\omega)$ to be the colatitude and longitude angles of the point ω. Determine the joint c.d.f. of (Θ, Φ).

3-31. Express the c.d.f. F of Problem 3-12 as a mixture, in the form

$$F(x) = \alpha F_1(x) + (1 - \alpha) F_2(x), \quad 0 \leq \alpha \leq 1,$$

where F_1 is of discrete and F_2, of continuous type.

3.5 Functions of random variables

We considered functions of functions on a sample space in §1.4 and stated that measurability of the composite function cannot be guaranteed, even when the functions involved are individually measurable. Fortunately, the functions we'll encounter offer no trouble in this regard, and we proceed as though the composite function $g(X(\omega))$ does define a random variable Y. Suppressing the dependence on ω in the notation, we write $Y = g(X)$. Just as the function $X(\omega)$ defines a distribution in $\Re^1$ (its value space), so the mapping via g from $\Re^1$ to the value space of Y ($\Re^1$ or a subset thereof) defines a distribution in the Y-space.

The question we take up here is how, given the distribution of X, one can determine the implied distribution of $Y = g(X)$ from the distribution of X. The connection is as follows:

$$F_Y(y) = P[g(X) \leq y] = P[g(X(\omega)) \leq y], \qquad (1)$$

where the first P is a probability in the X-space and the second, a probability in Ω. When probabilities are assigned directly in the X-space, by specification of a p.d.f. or c.d.f., the function $g(X)$ is simply a random variable on this probability space.

In discrete cases we can find the probability of the value $Y = a$, where $Y = g(X)$, by the usual kind of calculation—as the sum of the probabilities of the x's assigned the value $y = a$ by g:

$$P(Y = a) = P(\{x: g(x) = y\}) = \sum_{g(x) = y} f(x_i).$$

Such calculations can be quite straightforward, as in the next example.

EXAMPLE 3.5a Let X denote the number of points showing when an ordinary die is rolled. In the model for a "fair" die, the faces are equally likely:

$$f(x) = 1/6, \quad x = 1, 2, 3, 4, 5, 6.$$

Consider the variable $Y = (X - 3)^2$. The possible values and corresponding probabilities are as follows:

y	0	1	4	9
$f(y)$	1/6	2/6	2/6	1/6

[You should check each of the entries for $f(y)$.] ∎

EXAMPLE 3.5b Consider the random variable X from Example 3.2d, whose c.d.f. is $F(x) = 1 - e^{-x}$ for $x \geq 0$. [Since $F(0) = 0$, the value $F(x) = 0$ is implied for any negative x.] Suppose $Y = \sqrt{X}$. The c.d.f.

of Y, for $y > 0$, is

$$F_Y(y) = P(Y \le y) = P(\sqrt{X} \le y)$$
$$= P(X \le y^2) = F_X(y^2) = 1 - \exp(-y^2).$$

The p.d.f. then follows upon differentiating with respect to y:

$$f_Y(y) = \frac{d}{dy}[1 - \exp(-y^2)] = 2y\exp(-y^2), \quad y > 0. \blacksquare$$

An important and useful type of function of a random variable is a **linear transformation**. Given constants a and b, define $Y = a + bX$. We calculate the c.d.f. of Y, first for the case $b > 0$:

$$F_Y(y) = P(a + bX \le y) = P\left(X \le \frac{y-a}{b}\right) = F_X\left(\frac{y-a}{b}\right).$$

If X is a continuous random variable, so that F_X is continuous and differentiable except at (at most) finitely many points, then F_Y has these properties, and Y is a continuous random variable, with p.d.f.

$$f_Y(y) = F_Y'(y) = f_X\left(\frac{y-a}{b}\right) \cdot \frac{1}{b}.$$

[The "chain rule" for differentiation accounts for the factor $1/b$, the derivative of $(y-a)/b$.] When $b < 0$, the last multiplier is $-1/b = 1/|b|$.

THEOREM 7 *The p.d.f. of a linear function of X: $Y = a + bX$, is*

$$f_Y(y) = f_X\left(\frac{y-a}{b}\right) \cdot \frac{1}{|b|}. \tag{2}$$

Next we consider transformation by a function g which is *strictly monotonic* on the support of f. For an *increasing* function we have

$$P(Y \le y) = P[g(X) \le y] = P[X \le g^{-1}(y)] = F_X[g^{-1}(y)].$$

When g^{-1} is differentiable, we can differentiate with respect to y to obtain the p.d.f. of Y:

$$f_Y(y) = f_X[g^{-1}(y)] \cdot \frac{d}{dy}(g^{-1}(y)).$$

If g is a *decreasing* function, we need absolute value signs around the second factor.

3.5 Functions of random variables

THEOREM 8 When Y is a monotone function of X, its p.d.f. is

$$f_Y(y) = f_X[g^{-1}(y)] \cdot \left| \frac{d}{dy}(g^{-1}(y)) \right|. \tag{3}$$

A more intuitive way of seeing (3) is to realize that for any increment dx and corresponding increment dy (see Figure 3-7), the probability assigned to dy is just the probability of the x-interval dx:

$$f_Y(y)\,dy = P(dy) = P(dx) \doteq f_X(x)\,dx.$$

Since $y = g(x)$, division by dy yields (3). Although the same probability is in the interval dx as in dy, the densities differ because the lengths differ—by the factor dx/dy.

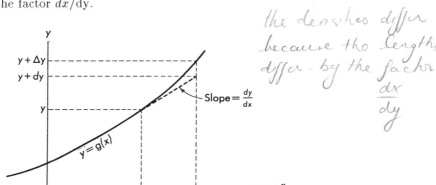

Figure 3-7.

When the function g is not monotonic, one can obtain the correct relationship between the X- and the Y-density by exercising care when tracking down the various possible sources of the probability in an increment dy, as in the following:

EXAMPLE 3.5c The function $g(x) = x^2$ has two inverses, so to each element dy on the y-axis there correspond two elements dx, one on the positive and one on the negative x-axis. The value space of $Y = X^2$ is $[0, \infty)$, and for $y > 0$,

$$F_Y(y) = P(X^2 \leq y) = P(-\sqrt{y} \leq X \leq \sqrt{y}) = F_X(\sqrt{y}) - F_X(-\sqrt{y}).$$

Differentiating, we find

$$f_Y(y) = f_X(\sqrt{y}) \cdot \frac{1}{2\sqrt{y}} - f_X(-\sqrt{y}) \cdot \frac{-1}{2\sqrt{y}}.$$

For example, suppose $X \sim \mathcal{U}(-1, 1)$. Its p.d.f. is $f(x) = \frac{1}{2}$, $-1 < x < 1$, and the p.d.f. of X^2 is then constant

$$f_Y(y) = \frac{1}{2} \cdot \frac{1}{2\sqrt{y}} + \frac{1}{2} \cdot \frac{1}{2\sqrt{y}} = \frac{1}{2\sqrt{y}}, \quad 0 < y < 1. \blacksquare$$

The transformation $Y = F_X(X)$ is particularly useful in connection with "order statistics." When F_X is strictly monotonic, its inverse function is uniquely defined, so

$$F_Y(y) = P(Y \leq y) = P[F_X(X) \leq y]$$
$$= P[X \leq F_X^{-1}(y)] = F_X[F_X^{-1}(y)] = y,$$

for $0 \leq y \leq 1$. Thus, $Y \sim \mathcal{U}[0, 1]$. (The transformation F_X spreads out probability where it was dense and crowds it where it was sparse, just in the right amounts to make it uniform on the y-axis.) If F_X is not strictly monotonic, let x_y denote the largest of the possible inverses $F_X^{-1}(y)$; we then find (for $0 < y < 1$):

$$F_Y(y) = P[F_X(X) \leq y] = P(X \leq x_y) = F_X(x_y) = y,$$

since the difference between the event $[X \leq x_y]$ and the event $[F(X) \leq Y]$ is a set of x-values on which F is not increasing, which has probability 0.

THEOREM 9 *The random variable $Y = F_X(X)$ is uniformly distributed on the interval $(0, 1)$.*

An intuitive notion of what a transformation does to a distribution can sometimes be helpful in finding the distribution of a transformed variable, as in the following:

EXAMPLE 3.5d Let X have the distribution whose p.d.f. is shown in Figure 3-8(a), and consider the transformation $Y = 2(|X| - 1)$. Figure 3-8(b) shows the p.d.f. of $|X|$, obtained by simply reflecting the negative part of the distribution onto the positive part, doubling the density at each positive x. Subtraction of 1 from each possible value of $|X|$ moves the distribution to the left 1 unit, as shown in Figure 3-8(c). Finally, multiplication by 2 stretches the distribution to the range $(-2, 2)$, decreasing the density by a factor of 2, as shown in the final result, Figure 3-8(d). ∎

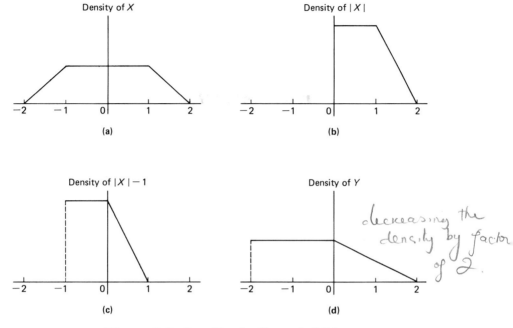

Figure 3-8. Densities for Example 3.5d

A function of two or more random variables also defines a random variable. For instance, $U(\omega) = g[X(\omega), Y(\omega), Z(\omega)]$ is a function on Ω and, with suitable (mild) restrictions on g, defines a random variable, which we write $U = g(X, Y, Z)$. Its distribution is implied by the distribution in Ω, or by the distribution of (X, Y, Z) in $\Re^3$.

EXAMPLE 3.5e Suppose the random vector (X, Y) has the p.d.f.

$$f(x, y) = K \exp\left\{-\tfrac{1}{2}(x^2 + y^2)\right\},$$

and define $Z = \sqrt{X^2 + Y^2}$. The c.d.f. of Z, for $z > 0$, is

$$F_Z(z) = P(Z \leq z) = P(X^2 + Y^2 \leq z^2),$$

which we can evaluate using (5) of §3.4:

$$F_Z(z) = \iint_{x^2+y^2 \le z^2} K \exp\left\{-\tfrac{1}{2}(x^2+y^2)\right\} dx\, dy.$$

Changing to polar coordinates, we obtain

$$F_Z(z) = K \int_0^{2\pi}\!\!\int_0^z \exp(-\tfrac{1}{2}r^2)\, r\, dr\, d\theta = 2\pi K\, [1 - \exp(-\tfrac{1}{2}z^2)].$$

Since this is a c.d.f., $F_Z(\infty) = 2\pi K = 1$; so $K = 1/(2\pi)$. The p.d.f. of Z is

$$f_Z(z) = z \exp(-\tfrac{1}{2}z^2)], \quad z > 0. \blacksquare$$

PROBLEMS

3-32. Find the distribution of $Y = X^2 - 7X + 10$, where X denotes the number of points thrown with a fair die.

3-33. Suppose $X \sim \mathcal{U}(-1, 1)$. Determine the distribution of
(a) $2|X|$.   (b) X^2.   (c) $(X+1)/2$.

3-34. Let $F(x) = x^2$ for $0 < x < 1$, with F suitably defined elsewhere so as to be the c.d.f. of a random variable X. Obtain the c.d.f. of $Y = X^2$.

3-35. Suppose X has a strictly increasing, continuous c.d.f. F, and let G be another strictly increasing, continuous c.d.f. Show that c.d.f. of the random variable $G^{-1}[F(X)]$ is G.

3-36. Let X_1 and X_2 be the numbers on first and second chips selected at random from a bowl of chips numbered from 1 to 4. Define $U = X_1 + X_2$ and $V = X_1 - X_2$. Find the distributions of U and V, and also the joint distribution of (U, V).

3-37. Given the p.d.f. $f(x) = \theta x^{\theta - 1}$, $0 < x < 1$, where θ is a positive constant, find the p.d.f. of $V = -\log X$.

3-38. Let (X, Y) have the distribution of Problem 3-27, with joint p.d.f. $f(x, y) = e^{-x-y}$ for $x > 0$, $y > 0$. Find the distribution of $Z = X + Y$.

3-39. Let (X, Y) be uniformly distributed on the unit square: $0 < x < 1$, $0 < y < 1$. Find the distribution function of $W = \max(X, Y)$, the larger of X and Y. [Hint: $W \le w$ if and only if both X and Y are not larger than w.]

3-40. Find the distribution of $U = X + Y$, where (X, Y) has the distribution of the preceding problem.

3.6 Conditional distributions

As probabilities are apt to change, generally, with the acquisition of further information, so probability distributions for random variables may change when one learns something additional about their values. Thus, given an event E with positive probability, the new distribution for X is given by a *conditional* c.d.f.:

$$F_{X|E}(x) = P(X \leq x \mid E) = \frac{P(X \leq x \text{ and } E)}{P(E)}.$$

The distribution defined by $F_{X|E}$ is called "conditional" because it is derived from an initial distribution by conditioning; but it is a distribution nonetheless. If it is of the discrete type, there is a corresponding conditional p.f.; if it is continuous, there is a conditional p.d.f. These might be denoted by $f_{X|E}(x)$ or by $f_X(x \mid E)$.

EXAMPLE 3.6a Let X be the number of "honor points" assigned to a playing card in a certain system of bidding in the game of bridge: $X(\text{ace}) = 4$, $X(\text{king}) = 3$, $X(\text{queen}) = 2$, $X(\text{jack}) = 1$, and $X(\omega) = 0$ for any other card ω. The p.f. of X is then

$$f(x) = \begin{cases} 9/13, & x = 0, \\ 1/13, & x = 1, 2, 3, 4. \end{cases}$$

Given E: the card is a black face card, probabilities must be revised. Since

$$P(E) = 6/52, \text{ and } P(E \text{ and } X = x) = \begin{cases} 2/52, & x = 1, 2, 3, \\ 0, & x = 0, 4, \end{cases}$$

the new probabilities are

$$f(x \mid E) = \begin{cases} 1/3, & x = 1, 2, 3, \\ 0, & x = 0, 4. \end{cases}$$

These are obtained by dividing $P(X = x \text{ and } E)$ by $P(E)$. ∎

EXAMPLE 3.6b Consider the joint p.d.f. $f(x, y) = 24xy$ for $x > 0$, $y > 0$, $x + y < 1$. If it is given that the event $E = [X + Y < 1/2]$ has occurred, the joint distribution must change. An intuitive derivation of the new distribution is as follows: The new probability element $f(x, y \mid E)\, dx\, dy$ is 0 outside E. Within E, it is obtained from the old simply by dividing by the probability of the condition, $P(E) = 1/16$. So

$$f(x, y \mid E) = \frac{24xy}{1/16} = 384xy \text{ for } x > 0,\ y > 0,\ x + y < 1/2. \ \blacksquare$$

The most commonly encountered kind of condition on a bivariate distribution is that the value of one of the variables is known. Again we consider first the easier, discrete case. Given that $Y = y$, a value of Y with positive probability, the definition of conditional probability yields

$$f_{X\mid Y=y}(x) = P(X = x \mid Y = y)$$
$$= \frac{P(X = x \text{ and } Y = y)}{P(Y = y)} = \frac{f(x, y)}{f_Y(y)}. \tag{1}$$

The notation $X \mid Y = y$ in a subscript is cumbersome, so (1) is usually written as in the following:

DEFINITION The conditional p.f. of X given $Y = y$ is

$$f(x \mid y) = \frac{f(x, y)}{f_Y(y)}. \tag{2}$$

In using the notation $f(x \mid y)$ it must be remembered that this is not usually the same function as $f(y \mid x)$.

Conditional probabilities for X defined by (2) have the important property that weighting them with marginal probabilities and summing produces the marginal or *unconditional* probability of $X = x$:

$$f(x) = \sum_j f(x \mid y_j) f_Y(y_j). \tag{3}$$

This is an instance of the law of total probability (Theorem 3, §2.4).

EXAMPLE 3.6c In Problem 3-7, the discrete random vector (X, Y) was assigned a distribution given by the following table of joint probabilities:

	X:	1	2	3	
Y:	1	1/12	1/6	1/12	1/3
	2	1/6	0	1/6	1/3

3.6 Conditional distributions

3	0	1/3	0	1/3
	1/4	1/2	1/4	1

Given $Y = 2$, the new probabilities for X are proportional to the joint probabilities in the second row, obtained by dividing them by the row sum—the marginal probability of $Y = 2$:

x	1	2	3
$f(x \mid 2)$	1/2	0	1/2

Division by the row sum has simply renormalized the probabilities in the row so that they sum to 1. Similarly, conditional probabilities for the values of Y, given X, are proportional to the columns, obtained by dividing the joint probabilities in a column by the column total. ■

Turning to the case of *continuous* bivariate distributions, we find the situation complicated by the fact that any given value of Y, say, has probability zero. Attempting to define a conditional c.d.f. for X given $Y = y$ results in

$$P(X \leq x \mid Y = y) = \frac{P(X \leq x \text{ and } Y = y)}{P(Y = y)} = \frac{0}{0}.$$

Still, single values of Y do occur, and we need a model for the conditional distribution of X when it becomes known that Y has a particular value.

Suppose we know Y is in the infinitesimal interval $(y_0, y_0 + \Delta y)$. The conditional probability element for X is

$$P(x < X < x + \Delta x \mid y_0 < Y < y_0 + \Delta y)$$
$$\doteq \frac{f(x, y_0) \Delta x \Delta y}{f_Y(y_0) \Delta y} = \frac{f(x, y_0)}{f_Y(y_0)} \Delta x.$$

This suggests the following:

DEFINITION The conditional p.d.f. of X given $Y = y$, at a point y where $f_Y \neq 0$, is

$$f(x \mid y) = \frac{f(x, y)}{f_Y(y)}. \tag{4}$$

As a function of x, $f(x \mid y)$ has the properties of a p.d.f.: It is nonnegative, and

$$\int_{-\infty}^{\infty} f(x \mid y) \, dx = \int_{-\infty}^{\infty} \frac{f(x, y)}{f_Y(y)} \, dx = 1.$$

Moreover, if we multiply (4) by the probability element $f_Y(y)\,dy$ and "sum" (integrate) over y, as in (3) above, we obtain the unconditional p.d.f. of X:

$$\int_{-\infty}^{\infty} f(x\mid y)\,f_Y(y)\,dy = \int_{-\infty}^{\infty} f(x,y)\,dy = f_X(x). \tag{5}$$

Although one might be led to other definitions of a conditional p.d.f., it can be shown that (4) is essentially the only one with intuitively desirable properties such as (5). (See also Problem 3-47.)

EXAMPLE 3.6d Consider again the p.d.f. $f(x,y) = 3(x+y)$, $x > 0$, $y > 0$, $x + y < 1$. In Example 3.4b we found the marginal p.d.f. of X: $f_X(x) = 3(1-x^2)/2$ for $0 < x < 1$. Given $X = 1/2$, the conditional p.d.f. of Y, from (4), is

$$f(y\mid \tfrac{1}{2}) = \frac{f(\tfrac{1}{2},y)}{f_X(\tfrac{1}{2})} = \frac{3(\tfrac{1}{2}+y)}{\tfrac{9}{8}} = \tfrac{4}{3}(1+2y),\quad 0 < y < 1/2.$$

Similarly, and for general x,

$$f(y\mid x) = 2\,\frac{x+y}{1-x^2},\quad 0 < y < 1-x.$$

This conditional p.d.f. varies in proportion to the joint p.d.f. along the curve which is the intersection of the joint p.d.f. with the plane $z = x$. The divisor $3(1-x^2)/2$ has served to make the area under the conditional p.d.f. equal to 1. [The analog in the discrete case is that conditional probabilities are in proportion to the joint probabilities in a row (or column), and division by the row (or column) total yields conditional probabilities that sum to 1.] ∎

When (X,Y) is *uniformly* distributed over a region R in the plane, its p.d.f. $f(x,y)$ is constant over that region, and the graph of f is a flat surface at a constant height above the xy-plane over R (and 0 elsewhere). Its intersection with the plane $y = y_0$ defines a function which is constant above the chord of R determined by $y = y_0$, and zero elsewhere. Thus, when the joint distribution of (X,Y) is uniform, the conditional distribution of X given $Y = y_0$ is uniform, for any y_0. Similarly, the conditional distributions of Y given $x = x_0$ are uniform on the intersections of R with the lines $x = x_0$.

THEOREM 10 *If a p.d.f. $f(x,y)$ is constant in its support R, the conditional variables $X\mid y$ and $Y\mid x$ are also uniformly distributed.*

EXAMPLE 3.6e Suppose (X, Y) is uniformly distributed within the unit circle. Given $X = x$, the conditional p.d.f. is proportional to the cross-section of the joint p.d.f., which is constant on its support, with the plane $X = x$. Thus, the conditional distribution of X given $Y = y$ is *uniform* on the interval between $y = -\sqrt{1-x^2}$ and $y = +\sqrt{1-x^2}$. ∎

The definition of conditional p.d.f. (4) implies a multiplication rule for densities:
$$f(x, y) = f(x \mid y) f_Y(y) = f(y \mid x) f_X(x). \tag{6}$$
This in turn gives a form of Bayes' theorem:
$$f(x \mid y) = \frac{f(y \mid x) f_X(x)}{f_Y(y)}, \tag{7}$$
when $f_Y(y) \neq 0$. The denominator would ordinarily be calculated as
$$f_Y(y) = \int_{-\infty}^{\infty} f(y \mid x) f_X(x)\, dx.$$

EXAMPLE 3.6f Given that X has the p.d.f. $2x$, $0 < x < 1$, and that $Y \mid x$ is uniform on the interval $(0, x)$, the *joint* p.d.f. [from (7)] is the product
$$2x \cdot \tfrac{1}{x} = 2, \ \ 0 < y < x < 1.$$
To obtain the marginal p.d.f. of Y we integrate this over x:
$$f_Y(y) = \int_y^1 2\, dx = 2(1-y), \ \ 0 < y < 1.$$
So, from (7), we find
$$f_X(x \mid y) = \frac{2}{2(1-y)}, \ \ y < x < 1. \ \blacksquare$$

In later chapters we'll have occasion to consider joint distributions in which, marginally, one variable is discrete and the other, continuous. To handle such cases rigorously requires a more general notion of integral. However, we'll usually get correct results when we manipulate formally using equations (4)-(7), interpreting "f" as a p.f. or a p.d.f., as appropriate.

EXAMPLE 3.6g Suppose X has one of two distributions, according to the toss of a coin:

74   Chapter 3   Random Variables

$$f(x \mid y) = \begin{cases} 2x, & 0 < x < 1, \ y = \text{heads}, \\ 3x^2, & 0 < x < 1, \ y = \text{tails}. \end{cases}$$

If we observe $X = .2$, what is the probability that $Y =$ heads?

What is asked for is the conditional probability of heads (H), given $X = .2$. First, we find the marginal p.d.f. of X at $x = .2$:

$$f_X(.2) = f(.2 \mid H) \cdot \tfrac{1}{2} + f(.2 \mid T) \cdot \tfrac{1}{2} = .20 + .06 = .26.$$

Then, using Bayes' theorem (8), we obtain

$$f(H \mid .2) = \frac{f(.2 \mid H) \cdot f_Y(H)}{f_X(.2)} = \frac{.20}{.26} = \frac{10}{13}. \ \blacksquare$$

EXAMPLE 3.6h   Suppose $Y \sim \mathcal{U}(0,1)$ and that, given $Y = y$, X is discrete with p.f.

$$f(x \mid y) = y(1-y)^x, \quad x = 0, 1, 2, \ldots$$

If we observe $X = 4$, the conditional p.d.f. of Y is

$$f(y \mid 4) = \frac{f(4 \mid y) f_Y(y)}{f_X(4)},$$

where

$$f_X(4) = \int_0^1 y(1-y)^4 \cdot 1 \, dy = \frac{1}{30}.$$

So then

$$f(y \mid 4) = \frac{y(1-y)^4}{1/30} = 30y(1-y)^4, \quad 0 < y < 1. \ \blacksquare$$

We define conditional distributions for continuous random vectors of more than two components in much the same way as for two. For instance, in the case of (X, Y, Z, W), the bivariate conditional p.d.f. of (X, Y) given $Z = z$ and $W = w$ is

$$f(x, y \mid z, w) = \frac{f(x, y, z, w)}{f_{Z,W}(z, w)}.$$

Again observe that if the joint distribution of all four variables is *uniform* over some region, the conditional distribution of (X, Y) is also uniform, over the xy-region defined by the original support set and the values of the conditioning variables.

3.6 Conditional distributions

EXAMPLE 3.6i Consider a uniform distribution on the set $\mathcal{B}$, the unit ball in $\Re^3$: $x^2 + y^2 + z^2 \leq 1$. For $(x, y, z) \in \mathcal{B}$, the joint p.d.f. is the reciprocal of the volume of the ball: $f(x, y, z) = 3/(4\pi)$. The conditional distribution of (X, Y), given $Z = 1/2$, is *uniform* on $x^2 + y^2 \leq 3/4$, the disk which is the projection of the intersection of the plane $z = 1/2$ with the ball. The constant value of the conditional density on that disk is $4/(3\pi)$, the reciprocal of the area of the disk. To obtain this formally, we first find the marginal p.d.f. of Z by integrating over x and y:

$$f_Z(z) = \iint_{x^2+y^2 \leq 1-z^2} \frac{3}{4\pi} \, dx \, dy = \frac{3}{4}(1-z^2), \quad 0 < z < 1.$$

Dividing $f(x, y, 1/2)$ by $f_Z(1/2)$ yields the $4/(3\pi)$ we found above. ■

PROBLEMS

3-41. Let (X, Y) have the discrete distribution shown in the following table of joint probabilities:

	0	1	2	3
0	1/27	3/27	3/27	1/27
1	3/27	6/27	3/27	0
2	3/27	3/27	0	0
3	1/27	0	0	0

(By symmetry, marginal labels X and Y are interchangeable.)
 (a) Find the marginal p.f.'s.
 (b) Find the conditional p.f. $f(x \mid 2)$.
 (c) Find the distribution of $U = X + Y$.
 (d) Find the conditional distribution of X given $X + Y = 3$.

3-42. Let (X, Y) have the discrete distribution defined by the following table of joint probabilities shown below.

		Y			
		1	2	3	
	1	1/12	1/6	1/4	1/2
X	2	1/18	1/9	1/6	1/3
	3	1/36	1/18	1/12	1/6

| | 1/6 | 1/3 | 1/2 | 1 |

(a) Find the distributions of $X \mid y$ and $Y \mid x$; how do these conditional distributions compare with the marginal distributions?
(b) Find the p.f. of $V = Y - X$.
(c) Find the p.f. of $X + Y$, given $X - Y = 1$.

3-43. Suppose (X, Y) is uniformly distributed on the triangle bounded by the coordinate axes and the line $x + y = 1$.
(a) Find the marginal p.d.f.'s.
(b) Find the conditional p.d.f.'s, $f(x \mid y)$ and $f(y \mid x)$.

3-44. Let (X, Y) have a uniform distribution on $x^2 + y^2 < 4$.
(a) Find the distribution of $X \mid Y = 1$.
(b) Find $P(|X| < 1 \mid Y = 1)$.

3-45. Given the joint p.d.f. $f(x, y) = 6(1 - x - y)$ for $0 < y < 1 - x$ and $x > 0$. Find the conditional p.d.f.'s, $f(x \mid y)$ and $f(y \mid x)$.

3-46. Given the c.d.f.: $F(x, y) = \frac{1}{2}(x^2 y + xy^2)$ for $0 < x < 1$, $0 < y < 1$, find the conditional p.d.f. of X given $Y = y$. (Cf. 3-28.)

3-47. Let $\{E_i\}$ denote a countable partition of the sample space. Show:
$$f_X(x) = \sum_i f_{X \mid E_i}(x) P(E_i),$$
where the "f" can mean either a probability function or a probability density function. (Consider both cases.)

3-48. Let (X, Y) have a continuous bivariate distribution, and let A and B denote sets of X-values and Y-values, respectively. Show the following properties of a conditional distribution:

(a) $P(A) = \int_{-\infty}^{\infty} P(A \mid Y = y) f_Y(y) \, dy$.

(b) $F_X(x) = \int_{-\infty}^{\infty} F(x \mid y) f_Y(y) \, dy$.

(c) $P(X \in A, Y \in B) = \int_B P(A \mid Y = y) f_Y(y) \, dy$.

3-49. Suppose the number of a baseball player's "at bats" in a game is a random variable X with p.f. $f(3) = .2$, $f(4) = .7$, $f(5) = .1$. Suppose further that, given $X = x$, the number of hits he will get in a game is a random variable with p.f. $f(y \mid x) = \binom{x}{y}(.3)^y(.7)^{x-y}$, $y = 0, 1, ..., x$.
(a) Find $f_Y(2) = P(Y = 2)$.

(b) Given that the player got two hits in a certain game, find the probability that he was at bat four times in the game.

3-50. Suppose Y is uniformly distributed on $(0, 1)$, and $X \mid y$ is uniform on the interval $(0, y)$.
(a) Find the joint p.d.f. of (X, Y).
(b) Find the marginal p.d.f. of X.
(c) Find the conditional p.d.f. of Y given $X = x$.

3-51. Suppose that $f(y \mid x) = \binom{5}{y} x^y (1 - x)^{5-y}$, $y = 0, 1, \ldots, 5$, and that $f(x) = 4x^3$, $0 < x < 1$. Find the conditional p.d.f. of X, given $Y = 4$.

3.7 Independence

Two experiments are independent, in the stochastic sense, when the model for one does not change with any knowledge about the outcome of the other. For random variables, this means that the conditional distribution of one, given a value of the other, is the same as the *un*conditional distribution of the one. In the following definition, "f" can denote either a p.f. or a p.d.f.

DEFINITION *Discrete variables X and Y are **independent** provided*

$$f(x \mid y) = \frac{f(x, y)}{f_Y(y)} = f_X(x) \tag{1}$$

for all x and y such that $f_Y(y) \neq 0$, or alternatively,

$$f(x, y) = f_X(x) \cdot f_Y(y) \quad \text{for all } x \text{ and } y. \tag{2}$$

Observe that (2) implies (1), as well as the corresponding formula with roles of x and y reversed.

The two-way table of joint probabilities for independent discrete variables has a very special structure: Each entry in the body of the table is the product of the probabilities in the corresponding right- and left-hand margins.

EXAMPLE 3.7a Example 3.1c gave the joint distribution of the variable "suit" and the indicator of the event F = face card:

	S	H	D	C	
F	3/52	3/52	3/52	3/52	12/52

$$F^c \quad 10/52 \quad 10/52 \quad 10/52 \quad 10/52 \quad 40/52$$
$$13/52 \quad 13/52 \quad 13/52 \quad 13/52$$

Each entry within the body of the table is the product of corresponding marginal probabilities; so the variables represented are independent. ∎

EXAMPLE 3.7b Consider an urn with four balls numbered from 1 to 4. Let X denote the number on a first ball chosen at random and Y, the number on a second ball chosen at random from those that remain. The probability table is as follows:

	1	2	3	4	
1	0	1/12	1/12	1/12	1/4
2	1/12	0	1/12	1/12	1/4
3	1/12	1/12	0	1/12	1/4
4	1/12	1/12	1/12	0	1/4
	1/4	1/4	1/4	1/4	1

Clearly, the entries 1/12 are not the products of marginal 1/4's, so the variables are not independent. We could construct a table of joint probabilities with the same marginals in which the variables are independent simply by using (2): $f(x, y) = 1/16$, for each pair (x, y). Indeed, this is the appropriate model for two random selections when the first ball is *replaced* and mixed with the rest before the second selection. ∎

Since the factorization (2) must hold for *every* pair (x, y), independence is disproved by any single entry for which the joint probability does not factor. As in the above example, a zero in the table is an immediate tip-off: When the joint p.f. factors as in (2), a zero in the body of the probability table can only be the product of marginal probabilities when at least one of them is 0. If none of the marginal entries is 0, the presence of a 0 in the table indicates dependence!

Two *events* A and B are said to be independent if and only if their indicator variables I_A and I_B are independent, i.e., when [from (2)]

$$P(I_A = i, I_B = j) = P(I_A = i) \cdot P(I_B = j) \text{ for } i = 0 \text{ or } 1, \ j = 0 \text{ or } 1.$$

All of the following then hold:

$$P(AB) = P(A)P(B), \quad P(AB^c) = P(A)P(B^c),$$

$$P(A^c B) = P(A^c)P(B), \quad P(A^c B^c) = P(A^c)P(B^c).$$

Indeed, any one of these four equalities implies the other three and could be taken as the definition of independence of events A and B. For, with the joint p.f. exhibited in a two-way table with given marginal probabilities, specifying just one cell probability determines the others:

	A	A^c	
B	$P(A)P(B)$		$P(B)$
B^c			$P(B^c)$
	$P(A)$	$P(A^c)$	

The (A^c, B)-cell must be $P(B) - P(A)P(B) = P(B)P(A^c)$, and so on. Of course, this definition of independent events agrees with that given in Chapter 2.

THEOREM 11 *In a bivariate distribution for (X, Y),*

(a) X and Y are independent if and only if the events $X \in A$ and $Y \in B$ are independent events for all A and B.

(b) numerical variables X and Y are independent if and only if

$$F_{X,Y}(x, y) = F_X(x) \cdot F_Y(y) \text{ for all } x, y. \tag{3}$$

(c) functions $g(X)$ and $h(Y)$ of independent variables X and Y are also independent random variables.

For (a), if A and B are arbitrary events in the sample spaces of independent discrete variables X and Y, respectively, then

$$P(X \in A, Y \in B) = \sum_{x \in A} \sum_{y \in B} f(x, y)$$

$$= \sum_{x \in A} f_X(x) \sum_{y \in B} f_Y(y) = P(X \in A)P(Y \in B). \tag{4}$$

Moreover, when (4) holds for all A, B, this implies (2), since individual values x_i and y_j are events. In the case of continuous variables, the argument for showing the independence of A and B is similar to the above, with sums replaced by integrals. To show that independence of arbitrary A and B implies independence of X and Y, we need (b).

To show (b) of the theorem, we first apply (a) with the special events $A = [X \leq x]$ and $B = [Y \leq y]$, for arbitrary x and y, to see that their independence implies factorization of the joint c.d.f. To show that the factorization of c.d.f.'s in the discrete case implies (2), we first calculate the joint p.f. Let the values of X be $x_1 < x_2 < \cdots$ and of Y, $y_1 < y_2 < \cdots$. Then

$$f_{X,Y}(x_i, y_j)$$
$$= F(x_i, y_j) - F(x_{i-1}, y_j) - F(x_i, y_{j-1}) + F(x_{i-1}, y_{j-1}).$$

By assumption, each of the terms on the right factors into the product of marginal c.d.f.'s, and the four products together factor into the product of $f_X(x_i) = F_X(x_i) - F_X(x_{i-1})$ and $f_Y(y_j) = F_Y(y_j) - F_Y(y_{j-1})$.

To show that factorization of a continuous bivariate c.d.f. implies independence of arbitrary A and B (in the X- and Y-value spaces, as above), we first show that it implies independence of X and Y. To do this, we differentiate the joint c.d.f. (in its factored form) to obtain the joint p.d.f.:

$$f(x, y) = \frac{\partial^2}{\partial x\, \partial y} F(x, y) = \frac{\partial^2}{\partial x\, \partial y} F_X(x) F_Y(y) = f_X(x) f_Y(y).$$

Then, for arbitrary A and B,

$$P(X \in A, Y \in B) = \int_A \int_B f_X(x) f_Y(y)\, dy\, dx$$
$$= \int_A f_X(x)\, dx \int_B f_Y(y)\, dy = P(X \in A) P(Y \in B),$$

so A and B are independent events.

To show (c) of the theorem define, for arbitrary events A and B in the value spaces of $g(X)$ and $h(Y)$, respectively, the events

$$A' = \{x\colon g(x) \in A\},\ B' = \{y\colon h(y) \in B\}.$$

Then because X and Y are independent, so are the events A' and B':

$$P[g(X) \in A \text{ and } h(Y) \in B)] = P(X \in A' \text{ and } Y \in B')$$
$$= P(X \in A') P(Y \in B') = P[g(X) \in A] P[h(Y) \in B],$$

which [by Theorem 11(a)] shows independence of $g(X)$ and $h(Y)$.

EXAMPLE 3.7c The joint c.d.f. given in Example 3.2d (page 52) is the product of the marginal variables: For x and y positive,

$$F(x, y) = (1 - e^{-x})(1 - e^{-y}) = F_X(x) \cdot F_Y(y).$$

For other (x, y) the factorization holds because both sides are zero, so and Y are independent. The joint p.d.f. also factors: If $x > 0$ and $y > 0$,

$$f(x, y) = e^{-x-y} = e^{-x} e^{-y} = f_X(x) f_Y(y).$$

[For any other pair (x, y), both sides are 0.] ∎

Some care must be taken to be sure that not only the "formula" for a joint p.d.f. factors, but that the function as a whole factors—for all points in its domain. For this to be so, the support of the joint p.d.f. must be a *product set*—the intersection of sets A on the X-axis and B on the Y-axis. If the support is not a product set, there will be regions in the plane where points (x, y) have joint density zero, and the marginal p.d.f.'s $f_X(x)$ and $f_Y(y)$ are not zero, which contradicts (5).

EXAMPLE 3.7d Consider the bivariate distribution defined by the p.d.f. $f(x, y) = 24xy$ on the triangle bounded by the positive axes and the line $x + y = k$. The function f may look like a product, but the support is not a product set (see Figure 3-9). The density is 0 in the product of the sets labeled A and B, the support regions of X and Y, respectively. The product of the marginal p.d.f.'s cannot be zero, so X and Y are not independent. The same would be true for any bivariate distribution supported on this same triangular region. ∎

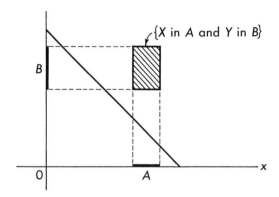

Figure 3-9 Support region for Example 3.7d

We define independence of three or more variables by a factorization condition extending (2):

DEFINITION Random variables $X_1, X_2, ..., X_n$ are *independent* if and

only if
$$f_{X_1,\ldots,X_n}(x_1,\ldots,x_n) = f_{X_1}(x_1) \cdots f_{X_n}(x_n) \quad \text{for all } (x_1,\ldots,x_n). \quad (5)$$

Events $E_1, \ldots, E_n$ are *independent* if and only if their indicator functions are independent random variables.

(In the first part of this definition, the "f" can mean either a p.f. or a p.d.f.) More generally, (5) defines independence of random vectors, when the X_i's are themselves taken to be vectors. [In checking (5) in a particular case, remember that the *support* must "factor"—not just the formula that defines f on the support.]

EXAMPLE 3.7e Suppose random variables X_i, $i = 1, \ldots, n$, are independent and that each has the p.d.f. $f_X(x) = 2x$, $0 < x < 1$. Their joint p.d.f. is then the product of the marginal p.d.f.'s:
$$f_{X_1,\ldots,X_n}(x_1,\ldots,x_n) = \prod_{i=1}^{n}(2x_i) = 2^n x_1 x_2 \cdots x_n, \quad 0 < x_i < 1.$$
(Observe that the joint p.d.f. is a function of n arguments. Even though the marginal density *functions* are the same, each one requires a distinct argument x_i.) ∎

When variables are independent, specifying their marginal distributions uniquely defines their joint distribution, and we'll often use the factorization condition (5), in what follows, to *construct* the joint distribution when we want a model to represent independence.

An important consequence of (5) is the following:

THEOREM 12 *The variables in any subset of a set of independent random variables are independent. Moreover, vectors consisting of distinct subsets of a set of independent random variables are independent.*

COROLLARY *If events $E_1, \ldots, E_n$ are independent, the events in any subset of them are independent.*

The theorem becomes clear upon calculating the marginal joint p.d.f.'s. For instance, if we integrate (or sum) over x_3 through x_n in (5), the l.h.s. becomes the joint p.d.f. of (X_1, X_2), and the r.h.s., the product of the two marginal p.d.f.'s. The corollary follows directly from the Theorem. For instance, suppose E, F, and G are independent events. This means that

$$P(EFG) = P(I_E = 1, I_F = 1, I_G = 1)$$
$$= f(1, 1, 1) = f_{I_E}(1) f_{I_F}(1) f_{I_G}(1) = P(E)P(F)P(G)). \quad (6)$$

So $P(EFG)$ must factor; but in addition, $P(EF)$, $P(FG)$, and $P(EG)$ must factor similarly—the three events are also independent in pairs. Indeed, as we'll see in Problems 3-65 and 3-66, the factorization (6) does not imply that the events are pairwise independent; and conversely, pairwise independence does not imply (6).

EXAMPLE 3.7f Suppose events E, F, G, and H are independent; are the events EF and $G \cup H$ independent? The answer is yes, a result of the strong factorization condition (for "all $\mathbf{x}$") in (5):

$$P[(EF)(G \cup H)] = P(EFG \cup EFH)$$
$$= P(E)P(F)P(G) + P(E)P(F)P(H) - P(E)P(F)P(G)P(H)$$
$$= P(EF)[P(G) + P(H) - P(GH)] = P(EF)P(G \cup H).$$

But if we knew *only* that $P(EFGH) = P(E)P(F)P(G)P(H)$, then EF and $G \cup H$ could be dependent. ∎

THEOREM 13 *If $\mathbf{X}_1, ..., \mathbf{X}_k$ are independent random vectors, the functions $g_1(\mathbf{X}_1), ..., g_k(\mathbf{X}_k)$ are independent.*

We have seen this to be so in the case of two random variables; the argument in this more general situation is quite analagous.

PROBLEMS

3-52. Check each of the following problems from the preceding problem set for independence of X and Y:
(a) Problem 3-41. (c) Problem 3-43.
(b) Problem 3-42. (d) Problem 3-44.

3-53. Check for independence in the bivariate distributions defined in
(a) Problem 3-16. (b) Problem 3-27.

3-54. Given that the random variables X_1, X_2, X_3 are i.i.d,
(a) determine the joint p.f. when the p.f. of X_i is $p^x(1-p)^{1-x}$, $x = 0$, 1, where p is a constant such that $0 < p < 1$.
(b) determine the joint p.d.f. when each X_i has the p.d.f. e^{-x} for $x > 0$.

3-55. The p.f. $f(x) = 1/6$, $x = 1, 2, 3, 4, 5, 6$, represents the number of points showing in the roll of a fair die. Let X_1 and X_2 denote the numbers of points showing in two successive rolls. Assuming independence, give the joint p.f. of (X_1, X_2).

3-56. Let X be uniformly distributed on an interval A of the x-axis and Y be uniformly distributed on an interval B of the y-axis. Describe the distribution of (X, Y) when X and Y are independent.

3-57. Referring to the joint distribution in Example 3.6c (page 71), determine joint probabilities for *independent* variables U and V whose distributions are those of X and Y, respectively.

3-58. Show that if X is discrete with equally likely values x_i, and Y is discrete with equally likely values y_j, then if X and Y are independent, the pairs (x_i, y_j) are equally likely.

3-59. For a uniform distribution of probability on a unit hemisphere, the joint c.d.f. of the spherical coordinates (Θ, Φ), from Problem 3-30, is

$$F(\theta, \phi) = \begin{cases} \dfrac{\phi}{2\pi}(1 - \cos\theta), & 0 < \phi < 2\pi, \ 0 < \theta < \pi/2, \\ \dfrac{\phi}{2\pi}, & 0 < \phi < 2\pi, \ \theta \geq \pi/2, \\ 1 - \cos\theta, & 0 < \theta < \pi/2, \ \phi \geq 2\pi. \end{cases}$$

(a) Find the joint p.d.f. and show that Φ and Θ are independent.
(b) Let (X, Y, Z) denote the rectangular coordinates of a point on that sample space. Show that X and Y are not independent. [Hint: Their distribution is just the projection in the xy-plane of the distribution on the hemisphere.]

3-60. In the toss of a fair die (as in Problem 2-32) show that the events "even" and "divisible by 3" are independent.

3-61. I pick a card at random from one deck of playing cards, put it in my right pocket, and then pick a card at random from another deck of cards and put it in my left pocket. Find the probability that both cards are red, given that
(a) the card in my left pocket is red.
(b) at least one of the cards I picked is red.
Assume that the selections are independent experiments, so that any event in the sample space of the first experiment is independent of any event in the sample space of the second.

3-62. Show the following, for independent events A, B, C:
(a) AB and C are independent.
(b) $A \cup B$ and C are independent.
(c) A^c, B, and C are independent.

3-63. Suppose that B and C are disjoint; show that if A is independent of B and independent of C, then A is independent of $B \cup C$.

3-64. Define the random variables "suit" and "denomination" on the sample space of 52 cards in a deck of playing cards. Show that these variables are independent when the 52 cards are equally likely.

3-65. Consider the sample space $\Omega = \{a, b, c, d, e\}$ and assign these corresponding probabilities: $1/8$, $3/16$, $3/16$, $3/16$, $5/16$. Let $E = \{a, b, c\}$, $F = \{a, b, d\}$, $G = \{a, c, d\}$. Show that these events are not pairwise independent, but that $P(EFG) = P(E)P(F)P(G)$.

3-66. Consider the sample space of these four points in $\Re^3$: $(1, 0, 0)$, $(0, 1, 0)$, $(0, 0, 1)$, $(1, 1, 1)$, and let these be equally likely. Define events E_i: the ith coordinate is 1. Show that the events E_1, E_2, E_3 are pairwise independent, but not independent.

3-67. In a sequence of independent tosses of a (fair) coin, let U and V denote the numbers of the trials in which the first heads appears and the second heads appears, respectively. Give the p.f. table.

3-68. Given that the p.d.f. of the n-vector $\mathbf{X}$ is $f(\mathbf{x}) = 1$ for $0 < x_i < 1$, $i = 1, \ldots, n$, show that the univariate marginal variables are independent.

3-69. Show that if the distribution of (X, Y) is circularly symmetric, i.e., that $f(x, y) = g(x^2 + y^2)$, then the variables $Z = \sqrt{X^2 + Y^2}$ and $T = \text{Arctan}(Y/X)$ are independent.

3.8 Exchangeable variables

We have seen some distributions with the property that the p.f. or the p.d.f. is a symmetric function of its arguments, that is, that

$$f(x_1, \ldots, x_n) = f(x_{i_1}, \ldots, x_{i_n}), \tag{1}$$

for every permutation $(i_1, \ldots, i_n)$ of the integers $(1, \ldots, n)$. Random variables whose p.f.'s or p.d.f.'s have this property are **exchangeable**. Because of the symmetry, all marginal distributions of a given dimension are identical: The distribution of X_i is the same for all i, the joint distribution of (X_i, X_j) is the same for all i and j, and so on.

A multivariate distribution important in sampling theory is the distribution of independent random variables $X_1, X_2, ..., X_n$ that have a common univariate distribution; we'll describe these as being **i.i.d** (for "independent and identically distributed"). They are exchangeable, since the joint p.d.f. is

$$f(x_1, ..., x_n) = \prod_{i=1}^{n} f(x_i),$$

which is clearly a symmetric function. But random variables can be exchangeable without being independent.

EXAMPLE 3.8a Example 3.7b dealt with these joint probabilities:

	1	2	3	4	
1	0	1/12	1/12	1/12	1/4
2	1/12	0	1/12	1/12	1/4
3	1/12	1/12	0	1/12	1/4
4	1/12	1/12	1/12	0	1/4
	1/4	1/4	1/4	1/4	1

This defines the joint distribution of the numbers on the first and second balls, respectively, drawn at random and without replacement from an urn with four balls numbered 1 to 4. As we pointed out in Example 3.7b, the variables are not independent. But since it doesn't matter which margin refers to X and which to Y, they *are* exchangeable. So the marginal distributions are identical. That Y has the same distribution as X seems surprising at first, until it is realized that the marginal distribution of Y involves probabilities unconditioned by any knowledge of X. When you don't know *which* ball turned up at the first selection, it's as though that ball is still available for the second. ∎

The experiment in this last example is a special case of an important, general sampling process. Consider a population of finite size N, to be sampled at random, one at a time without replacement, obtaining a sample of size n. When, at each selection, we observe a function $X(\omega)$ upon drawing population member ω, the result is a sequence $(X_1, ..., X_n)$ of exchangeable random variables. (If the sampling is done with replacement, of course, the observations are i.i.d.)

PROBLEMS

3-70. In each case determine whether the random variables X and Y are exchangeable:
(a) $f(x, y) = K \exp\{-\frac{1}{2}(x^2 + y^2)\}$ (Example 3.7b).
(b) $f(x, y) = 24xy$ for $x > 0, y > 0, x + y < 1$ (Example 3.6b, p.70).
(c) Example 3.5e (p. 68).

3-71. In each of the following problems, determine whether the distribution is that of exchangeable random variables:
(a) Problem 3-7 (p. 47)
(b) Problem 3-8 (p. 47)
(c) Problem 3-26 (p. 62)
(d) Problem 3-27 (p. 62)
(e) Problem 3-28 (p. 62)
(f) Problem 3-41 (p. 75)
(g) Problem 3-42 (p. 75).

3-72. Show that if a random vector $(X_1, ..., X_n)$ has a uniform distribution on the region bounded by $x_1 + \cdots + x_n \leq 1$ and $x_i > 0$, $i = 1, ..., n$, then the variables X_i are exchangeable but not independent.

4
Expectations

"Mathematical expectation" is a traditional term used to describe a kind of average amount one can expect from participating in a lottery or other game of chance with a monetary reward.

EXAMPLE 4a Suppose you will receive a dollar for each point showing after the roll of a fair die. How much would you expect to get? Of course, the outcome of the roll is random, so you can't say for sure. Thinking of probabilities as implied by the symmetry of the die, you might take the middle of the distribution, or 3.5, as the number of dollars to "expect." If you have your own personal assessment of the probabilities, you might be willing to pay an amount in the "middle" of your distribution to play the game.

Thinking of probabilities as long-run limits, you would perhaps want to know how much you'd get on average after a large number of rolls. If the die is fair, the law of large numbers suggests that in a long sequence of rolls, the six sides of the die will turn up about equally often. Suppose, in N tries, each side did turn up exactly $N/6$ times; your reward in dollars would be

$$1 \cdot \frac{N}{6} + 2 \cdot \frac{N}{6} + 3 \cdot \frac{N}{6} + 4 \cdot \frac{N}{6} + 5 \cdot \frac{N}{6} + 6 \cdot \frac{N}{6}$$

Dividing this by N, the number of trials, yields $3.50 per trial as the average amount per trial. This is your *mathematical expectation*. It is not an amount you can actually get on one trial, so is not to be "expected" in that sense. But if a gambling house charges a customer $3.50 for each play of this game, it should collect about as much as it pays out, and the game would appear to favor neither the player nor the house. (In real-life casinos, the charge for playing a particular game is more than its mathematical expectation, so that it can pay expenses and make a profit.) ∎

The amount won in such games is a random variable, and in this chapter we define the notion of average value for any random variable. Averages of random variables and functions of random variables are useful, in particular, for describing various aspects of a probability distribution.

In mathematical terms, an average or expectation or expected value is an integral with respect to a probability measure, and the appropriate general tool is Lebesgue integration. However, in dealing with random variables that are discrete or continuous, sums and Riemann integrals, respectively, are adequate for our purposes.

4.1 Discrete random variables

Let X denote a discrete random variable defined on Ω, with possible values $x_1, x_2, \ldots$ and corresponding probabilities $f(x_i)$.

DEFINITION The *expected value* (or *expectation* or *average value* or *mean value*) of a random variable X with p.f. f and possible values x_i, for $i = 1, 2, \ldots$, is

$$E(X) = \sum_{i=1}^{\infty} x_i f(x_i), \qquad (1)$$

provided this sum is absolutely convergent.

When the list of possible values is finite, the sum is a finite sum, and there is no question of convergence. Other notations for (1) are μ, or μ_X, or just EX (without parentheses) when the meaning is clear.

EXAMPLE 4.1a Let X denote the number of tosses of a fair coin required to get heads, with possible values $1, 2, 3, \ldots$. The probability that it takes k tosses is the probability of $k-1$ tails followed by heads, or $1/2^k$. The average number of required tosses is then

$$EX = \sum_{k=1}^{\infty} k 2^{-k} = \sum_{k=1}^{\infty} \left\{ \sum_{i=1}^{k} 1 \right\} 2^{-k}$$

$$= \sum_{i=1}^{\infty} \left\{ \sum_{k=i}^{\infty} 2^{-k} \right\} = \sum_{i=1}^{\infty} 2^{-(i-1)} = 2.$$

(The series are convergent, so we can interchange the order of summations.) ∎

4.1 Discrete random variables

If Ω itself is discrete, the p.f. is $f(x_i) = \sum P(\omega)$, where the sum extends over all ω such that $X(\omega) = x_i$. In such cases we can calculate the mean EX as a sum over Ω:

$$EX = \sum_i x_i f(x_i) = \sum_i x_i \left\{ \sum_{X(\omega)=x_i} P(\omega) \right\} = \sum_i \sum_{X(\omega)=x_i} X(\omega)P(\omega).$$

Since X partitions Ω, the last double sum accounts for values of XP at every $\omega \in \Omega$. Thus we have the following:

THEOREM 1 When X is discrete and EX exists,

$$EX = \sum_\Omega X(\omega)P(\omega). \qquad (2)$$

EXAMPLE 4.1b Consider a sequence of three coin tosses in which the eight possible outcomes are equally likely, and define $X(\omega)$ as the number of heads in the outcome ω. The distribution of X is shown in the following table, in which is included a column of the products that appear in the definition (1): *[probability]*

[X(ω) = the # of heads]

[possible Values]

x_i	$f(x_i)$	$x_i f(x_i)$
0	1/8	0
1	3/8	3/8
2	3/8	6/8
3	1/8	3/8

The mean value is the sum of the entries in the last column:

$$E(X) = \sum x f(x) = 3/2.$$

But we can also get this from the uniform distribution in Ω by using (2):

$$E(X) = \sum_\Omega X(\omega)P(\omega) = (0+1+1+1+2+2+2+3) \cdot \frac{1}{8} = \frac{12}{8}.$$

The sum is the same, because each of its terms is accounted for in just one of the terms of $\sum x f(x)$. ∎

The alternative formula (2) is useful in establishing important properties of expectations. In stating these we assume that mean values exist:

THEOREM 2 If X is a discrete random variable, the mean value of a function $Y = g(X)$ is

$$E(Y) = E[g(X)] = \sum_i g(x_i) P(X = x_i). \qquad (3)$$

[f(y) = Σ g(xᵢ) P(X = xᵢ)]

THEOREM 3 *Let X and Y be discrete random variables. Then*

(i) *For any constant a, $E(a) = a$ and $E(aX) = aE(X)$.* (4)
(ii) $E(X+Y) = E(X) + E(Y)$. (5)
(iii) *Given constants a, b, c,*

$$E(aX + bY + c) = aE(X) + bE(Y) + c. \quad (6)$$

Theorem 2 follows upon applying (2) to Y as a random variable defined on the probability space of X-values. For (i) of Theorem 3, we apply Theorem 2 [with "ω" $= x_i$, "$X(\omega)$" $= g(x_i)$, and "$P(\omega)$" $= f(x_i)$] to the functions $Y = a$ and $Y = aX$:

$$E(a) = \sum a f(x_i) = a \sum f(x_i) = a,$$

$$E(aX) = \sum a x_i f(x_i) = a \sum x_i f(x_i) = aEX.$$

For (ii) and (iii), which involve the two variables X and Y, we need to assume a joint distribution, defined either directly as a joint distribution in the xy-plane or as induced by $[X(\omega), Y(\omega)]$. [The former is a special case of the latter in which $\omega = (x, y)$ and $X(\omega) = x$, $Y(\omega) = y$.] Then

$$E(X+Y) = \sum [X(\omega) + Y(\omega)] P(\omega)$$

$$= \sum X(\omega) P(\omega) + \sum Y(\omega) P(\omega) = EX + EY,$$

and (iii) follows from (i) and (ii). It also follows, by induction, that expectations are additive not only for two summands but for any finite sum.

EXAMPLE 4.1c In Example 3.6a we defined X to be the number of points assigned to a playing card in a system of bidding for the game of bridge, with p.f.

$$f(x) = \begin{cases} 1/13, & x = 1, 2, 3, 4, \\ 9/13, & x = 0. \end{cases}$$

Suppose we deal all 52 cards, 13 each to four bridge players. Let Y denote the total number of points in one player's hand. To find EY we can express Y as the sum $X_1 + \cdots + X_{13}$ and find EY as the sum of the means of the X's. The X's are exchangeable, with common mean

$$E(X_i) = (1 + 2 + 3 + 4) \cdot \tfrac{1}{13} = \tfrac{10}{13},$$

so then $EY = \sum E(X_i) = 13 \cdot \tfrac{10}{13} = 10.$

Another approach, again using the additive property of expectations, is to let Y_i ($i = 1, 2, 3, 4$) denote the number of points in player i's hand, and realize that these Y's are exchangeable and have the same mean, EY. There are 40 points in the whole deck, so
$$Y_1 + Y_2 + Y_3 + Y_4 = 40.$$
Taking expectations on both sides we get $4EY = 40$, or $EY = 10$. ∎

The calculation of a mean as given by (1) is exactly the same as the calculation of the center of gravity of a system of point masses on a line—masses m_i at x_i, where $f(x_i) = m_i / \sum m_j$. The significance of this is that we can think of probability distributions as mass distributions and exploit the intuition we have for the latter. In particular, when you calculate a mean value, it should at least look reasonable as a balance point. And if a distribution is symmetrical about some value, that value is the mean or expected value, if indeed the expected value exists.

4.2 Continuous random variables

Suppose next that X is a continuous random variable with p.d.f. $f(x)$ and support (a, b). Divide the support interval into n subintervals of width $\Delta x = (b - a)/n$. Let x_i be a point on the ith subinterval (say, the left-hand endpoint) and define a discrete random variable X^* with values x_i and p.f.
$$f^*(x_i) = P(X^* = x_i) \doteq f(x_i) \Delta x.$$
The variable X^* is an approximation to X, with mean value
$$E(X^*) = \sum_{i=1}^{n} x_i f^*(x_i) = \sum_{i=1}^{n} x_i f(x_i) \Delta x.$$
As n becomes infinite, this sum converges to a definite integral:
$$\lim_{n \to \infty} E(X^*) = \int_a^b x f(x) d(x), \quad \text{continuous R.V.}$$
when $xf(x)$ is integrable. We take this integral as the definition of EX. When its support is not finite, we approximate X by truncation to a finite interval (a, b): $X_{a,b} = f(x) I_{(a,b)}$, where I denotes the indicator function. Taking the limit as the interval endpoints go to infinity independently, we define the mean to be

$$E(X) = \lim_{\substack{a \to -\infty \\ b \to \infty}} E(X_{a,b}) = \int_{-\infty}^{\infty} x f(x) dx,$$

provided the limit exists.

DEFINITION If X is a continuous random variable with p.d.f. $f(x)$, its mean or *expected value* is

$$E(X) = \int_{-\infty}^{\infty} x f(x) dx, \tag{1}$$

provided that, if improper, the integral is absolutely convergent.

If we had the luxury of a background in Stieltjes integrals, we could write expected values as

$$EX = \int_{-\infty}^{\infty} x \, dF(x). \tag{2}$$

This integral is evaluated as the integral (1) for continuous distributions and as the sum (1) in §4.1 for discrete distributions; and with these rules of evaluation understood, we can simplify some statements about means using (2) simply as a notation.

We often want to use the fact that when a distribution has a support which is less than the whole real line, we need only integrate over the support. For, an integral over a given range can be expressed as the sum of integrals over subintervals whose union is that range:

$$\int_{-\infty}^{\infty} x f(x) dx = \int_{-\infty}^{a} x f(x) dx + \int_{a}^{b} x f(x) dx + \int_{b}^{\infty} x f(x) dx.$$

If the support is (a, b), the first and last integrals on the right are 0.

As in the discrete case, the expected value (1) is a weighted "sum" of the values of X, where the weights are now the probability elements, $f(x)dx$. Moreover, the analogy with mass carries over: Finding the mean of a continuous probability distribution is just like finding the center of gravity of a continuous distribution of mass. In particular, if a distribution is symmetric about some value, the mean (if it exists) is that value:

DEFINITION A continuous distribution is *symmetric* about $x = a$ if for all x, $f(a + x) = f(a - x)$.

THEOREM 4 *If the distribution of the continuous variable X is symmetric about $x = a$, then $EX = a$.*

4.2 Continuous random variables

To show Theorem 4, we write the integral (1) as follows:

$$EX = \int_{-\infty}^{\infty} x\,f(x)dx = a + \int_{-\infty}^{a}(x-a)\,f(x)dx + \int_{a}^{\infty}(x-a)f(x)dx$$

$$= a - \int_{0}^{\infty} y\,f(a-y)\,dy + \int_{0}^{\infty} y\,f(a+y)\,dy.$$

(We made changes of variable $y = a - x$ and $y = x - a$, respectively, to get these last two integrals.) By the assumption of symmetry, the last two integrals cancel, leaving $EX = a$.

EXAMPLE 4.2a Let X have the p.d.f. $f(x) = \frac{1}{\pi}(1+x^2)^{-1}$, encountered earlier (in Problem 3-20). Since the product $xf(x)$ behaves like $1/x$ at $x = \pm\infty$, the integral (1) diverges at both ends. We say that the mean value does not exist, and that $E|X| = \infty$. The distribution is symmetric about $x = 0$, but the point of symmetry is not the mean value, since the mean value does not exist. [Had we given the definition (1) as a "Cauchy principal value" of the improper integral, it would yield 0 in this instance. However, it is the existence of (1) as given that will be important in understanding the behavior of sample averages.] ■

THEOREM 5 If $P(X \geq 0) = 1$, then $EX \geq 0$. Moreover, if $X \geq 0$ and $EX = 0$, then $P(X = 0) = 1$.

We'll not prove Theorem 5, but recall from calculus that the definite integral of a nonnegative integrand is nonnegative. (A rigorous proof requires a background in measure theory.)

A *conditional* distribution is a distribution and may have a mean value. Suppose the pair (X, Y) has a continuous distribution; then the **conditional mean** of X given $Y = y$ is

$$E(X \mid y) = \int_{-\infty}^{\infty} x\,f(x \mid y)dx, \tag{3}$$

The value of this conditional mean is a function of y, the **regression function of X on Y**. (We'll have much to do with this notion in Chapter 14.)

EXAMPLE 4.2b Suppose (X, Y) is uniformly distributed within the triangle in the first quadrant bounded by the line $x + y = 1$. According to Theorem 10 of Chapter 3 (page 72), the conditional distribution of X given $Y = y_0$ is *uniform* on its support interval:

$$f(x \mid y_0) = \frac{1}{1-y_0} \quad \text{for } 0 < x < 1-y_0.$$

The conditional mean is the balance point—the middle of the interval from 0 to $1-y_0$. Thus, $E(X \mid y) = \frac{1}{2}(1-y)$ for any y on the unit interval. This function of y is the regression function of X on Y. ■

We'll deal mostly with distributions that are either discrete or continuous; but a type of distribution that is neither entirely discrete nor entirely continuous is sometimes useful, namely, a mixture of a discrete and a continuous distribution. (See Example 3.2a and Problem 3-12.) Consider the c.d.f.

$$F(x) = \alpha F_1(x) + (1-\alpha) F_2(x), \quad 0 \le \alpha \le 1. \tag{4}$$

This was seen in Problem 3-15 to be a c.d.f. of a random variable X when F_1 and F_2 are c.d.f.'s of random variables X_1 and X_2. Then

$$EX = \alpha \int x\, dF_1 + (1-\alpha) \int x\, dF_2$$

$$= \alpha E(X_1) + (1-\alpha)E(X_2). \tag{5}$$

Suppose X has a mixture distribution of the form (4) where F_1 is continuous, with p.d.f. $f(x)$, and F_2 is discrete, with possible values x_i and p.f. $f(x_i)$. According to (5), the mean of X is

$$EX = \alpha \int_{-\infty}^{\infty} x f_1(x)\, dx + (1-\alpha) \sum_{i=1}^{\infty} x_i f_2(x_i). \tag{6}$$

EXAMPLE 4.2c Consider the c.d.f. shown in Figure 4-1, defining a mixture of a discrete and a continuous distribution:

$$F(x) = \begin{cases} \frac{1}{4}x, & 0 < x < 1, \\ \frac{1}{2}, & 1 \le x < 2, \\ \frac{1}{2} + \frac{1}{4}(x-2)^2, & 2 \le x < 3. \end{cases}$$

As defined by this c.d.f., half the mass is at the points $x=1$ and $x=3$, and the other half is spread out over the interval from 1 to 3. So the distribution is a mixture of a discrete distribution and a continuous distribution, with $\alpha = 1/2$ in (4). The discrete distribution has probabilities $1/2$ at $x=1$ and $x=3$, and the continuous distribution has the following p.d.f.:

$$f(x) = \begin{cases} \frac{1}{2}, & 0 < x < 1, \\ 0, & 1 < x < 2, \\ x - 2, & 2 < x < 3. \end{cases}$$

The mean of the discrete component is clearly $x = 2$, and the mean of the continuous component is

$$\int_0^1 x \frac{1}{2} dx + \int_2^3 x(x-2) \, dx = \frac{19}{12}.$$

Averaging these means with weights $1/2$ each we find $EX = 43/24$. ∎

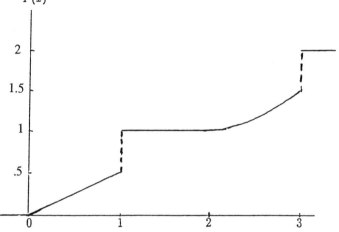

Figure 4-1. C.d.f. for Example 4.2c

PROBLEMS

4-1. Find the mean value of each of the following discrete variables:
(a) X = number of heads showing in three independent tosses of a coin (Problem 3-1): $f(x) = 1/8$ for $x = 0, 3$, and $f(x) = 3/8$ for $x = 1, 2$.
(b) Y = number of correctly identified cola brands (Problem 3-4):
$$f(3) = 1/6, \quad f(1) = 3/6, \quad f(0) = 2/6.$$
(c) Z = number of rotten eggs in a random selection of two from a dozen, two of which are rotten (Problem 3-6):
$$f(0) = 45/66, \quad f(1) = 20/66, \quad f(2) = 1/66.$$

4-2. Problem 2-39 dealt with trying to open a door when one has four keys that look alike, only one of which opens the door. Assuming we select keys at random without replacement, let W denote the number of tries required to get the door open. Find EW. (First find the p.f.)

4-3. Let X = total number of points showing on three dice. Find EX. [Hint: X is the sum of the numbers of points on the individual dice.]

4-4. Example 3.6a gave the following table of the joint p.f. of (X,Y):

		Y		
		1	2	3
	1	1/12	1/6	1/12
X	2	1/6	0	1/6
	3	0	1/3	0

(a) Find $E(X+Y)$. (b) Find $E(X \mid Y = 1)$.

4-5. Problem 3-41 gave the following table of the joint p.f. of (X,Y):

	0	1	2	3
0	1/27	3/27	3/27	1/27
1	3/27	6/27	3/27	0
2	3/27	3/27	0	0
3	1/27	0	0	0

(a) Find $E(X+Y)$. (b) Find $E(X \mid Y = 1)$.

4-6. Find $E(X \mid Y = k)$ when $f(x,y) = 1/6$ for $x = 1,2,3$, $y = 1,2,3$, and $x \neq y$.

4-7. Consider the sample space $\Omega = \{1, 2, 3, ...\}$, and the discrete distribution defined by
$$P(\omega) = \frac{6}{\pi^2 \omega^2}.$$
Let $X(\omega) = (-1)^{\omega+1}\omega$. Show that the series that we'd need to use to find EX is convergent but not absolutely convergent. (That is, mean is not defined.)

4-8. Find EX, when X has the p.d.f.
(a) $f(x) = \frac{1}{2}e^{-|x|}$.
(b) $f(x) = 1 - |x|$, $|x| < 1$.
(c) $f(x) = 6x(1-x)$, $0 < x < 1$.

4-9. Find the mean value of the distribution of the random variable U defined by the c.d.f. $F(u) = 1 - \cos u$, $0 < u < \pi/2$.

4-10. Given the joint p.d.f. $f(x,y) = e^{-x-y}$, $x > 0$, $y > 0$, find the conditional mean, $E(X \mid y)$.

4-11. Let (X, Y) have the p.d.f. $f(x, y) = 3(x + y)$ for $x > 0$, $y > 0$, $x + y < 1$. Find $E(Y \mid 1/2)$.

4-12. Consider the random pair (X, Y) such that $X \mid Y = 0$ is uniform on $(0, 1)$, $P(X = 1 \mid Y = 1) = 1$, and $Y = 0$ or 1 according as a coin falls tails or heads. Find EX.

4-13. Use the result of Problem 3-47 to show that if $\{E_i\}$ is a partition of the sample space, then

$$E(X) = \sum_i E(X \mid E_i) P(E_i). \tag{7}$$

4.3 Functions of random variables

The operation $E(\,\cdot\,)$ defined above for continuous and earlier for discrete variables is linear, essentially because integration and summaton are linear operations. Before stating this property more formally we consider more general functions of a random variable: $Y = g(X)$. Theorem 1 of §4.1, for discrete variables, suggests the correct formula for the expected value of a function of a continuous random variable, when that function is absolutely integrable with respect to the distribution of the variable:

THEOREM 6 *When X has p.d.f. $f(x)$, the mean value of $g(X)$ is*

$$E[g(X)] = \int_{-\infty}^{\infty} g(x)f(x)dx. \tag{1}$$

Moreover, if (X, Y) has a continuous bivariate distribution with p.d.f. $f(x, y)$, then

$$E[g(X, Y)] = \int_{-\infty}^{\infty} \int_{-\infty}^{\infty} g(x, y)f(x, y)\,dxdy. \tag{2}$$

More generally, if $\mathbf{X}$ has a continuous n-variate distribution with p.d.f. $f(\mathbf{x})$, then

$$E[g(\mathbf{X})] = \int_{\Re^n} g(\mathbf{x})f(\mathbf{x})d\mathbf{x}. \tag{3}$$

[Sometimes (1) is taken as the *definition* of $E[g(X)]$; doing this requires a theorem asserting the consistency of the definition, which amounts to our Theorem 6. For, with (1) as definition there would be two ways of calculating $E[g(X)]$: as an integral with respect to the density of $Y = g(X)$: $\int y f_Y(y) \, dy$, and as the integral (1). These must then be shown equivalent.]

Proofs of the results in Theorem 6 in any generality are beyond our scope. However, when g is strictly monotonic, a change of variable in the integral establishes (1): Let $Y = g(X)$, where g is strictly monotonically increasing, with inverse $x = g^{-1}(y)$. Then

$$\int_{-\infty}^{\infty} g(x) f_X(x) \, dx = \int_{-\infty}^{\infty} y f_X[g^{-1}(y)] \frac{d}{dy}[g^{-1}(y)] \, dy$$

$$= \int_{-\infty}^{\infty} y f_Y(y) \, dy. \qquad (4)$$

EXAMPLE 4.3a Let X be uniform on $(-1, 1)$, and define $Y = X^2$. Using (1), we find the mean value of Y to be

$$EY = \int_{-1}^{1} x^2 f(x) \, dx = \int_{-1}^{1} x^2 \left(\frac{1}{2}\right) dx = \frac{1}{3}.$$

If, instead, we exploit the symmetry about $x = 0$, to write the integral over $(-1, 1)$ as twice the integral over $(0, 1)$, we can make the change of variable $y = x^2$, $dy = 2x \, dx$, since the function x^2 is monotone increasing over the range of integration:

$$EY = 2 \int_0^1 x^2 \frac{1}{2} dx = \int_0^1 \frac{y}{2\sqrt{y}} dy = \frac{1}{3}.$$

The first calculation is simpler, and generally speaking, using (1) is simpler than first finding the p.d.f. of Y and integrating over y. ∎

Applied to the function $g(y) = f(x \mid y)$, Theorem 6 (for fixed x) gives an interpretation of (5) in §3.6: The unconditional p.d.f. of X is the mean, with respect to the distribution of Y, of the conditional p.d.f. of X given $Y = y$:

$$E[f(x \mid Y)] = \int_{-\infty}^{\infty} f(x \mid y) f_Y(y) \, dy = f_X(x). \qquad (5)$$

Applying Theorem 6 to the function $g(y) = E(X \mid y)$, we obtain an important and often useful result:

THEOREM 7 $E[E(X \mid Y)] = \int_{-\infty}^{\infty} E(X \mid y) f_Y(y) \, dy = EX. \qquad (6)$

4.3 Functions of random variables

We'll refer to (6) as the **iterated expectation** formula. Notice the inner "E" in the left-hand side of (6) is with respect to *conditional* distribution of X given $Y = y$, and the outer "E" is with respect to the distribution of Y.

EXAMPLE 4.3b Consider the joint density function $f(x, y) = 24xy$ for $x > 0, y > 0$, and $x + y < 1$. The *marginal* p.d.f. of X is

$$f_X(x) = 24x \int_0^{1-x} y\, dy = 12x(1-x)^2, \quad 0 < x < 1,$$

and its mean value is

$$EX = 12 \int_0^1 x^2(1-x)^2 = \tfrac{2}{5}.$$

discrete or continuous p.f.

Because X and Y are exchangeable, EY is also $2/5$. The *conditional* p.d.f. of Y, given $X = x$, is

$$f(y \mid x) = \frac{f(x, y)}{f_X(x)} = \frac{2y}{(1-x)^2}, \quad 0 < y < 1 - x.$$

Conditional p.d.f of Y, given X = x

The conditional mean of Y given X is then

$$E(Y \mid x) = \frac{1}{(1-x)^2} \int_0^{1-x} 2y^2\, dy = \tfrac{2}{3}(1-x).$$

The conditional mean of Y given X

To find EY we average this with respect to the distribution of X:

$$E[E(Y \mid X)] = E\left[\tfrac{2}{3}(1-X)\right] = \tfrac{2}{3}(1 - EX) = \tfrac{2}{5} = EY,$$

thus verifying (6).

To illustrate (2), we'll find $E(XY)$ by integrating the product with respect to the joint p.d.f. To evaluate the double integral as an iterated integral, we set the limits of integration according to the support of f, defined by $0 < x < 1 - y < 1$:

$$E(XY) = \int_0^1 \int_0^{1-y} xy(24xy)\, dx\, dy = \tfrac{2}{15}.$$

In case you might have thought that an average product is the product of averages, here it is not: $(EX)(EY) = 4/25 \neq E(XY)$. ∎

Applying (2) to the function $g(x, y) = a + bx + cy$ we can establish the linearity of the expectation operation:

$$E(a+bX+cY) = \int\int (a+bx+cy)f(x,y)\,dxdy$$
$$= a + b\int\int f(x,y)\,x\,dxdy + c\int\int f(x,y)\,y\,dxdy$$
$$= a + bE(X) + cE(Y).$$

[With this we institute a convenient convention: When integral signs appear without specific limits, it is to be understood that the integration is over the whole space (of appropriate dimension).] This establishes the following:

THEOREM 8 *If continuous variables X and Y have expected values, then (for given constants a, b, c),*

$$E(a + bX + cY) = a + b\,EX + c\,EY. \tag{7}$$

THEOREM 9 *If the continuous random variables $X_1, \ldots, X_n$ have expected values, then*

$$E\left\{\sum X_i\right\} = \sum E(X_i). \tag{8}$$

The additivity over finitely many terms asserted by Theorem 9 follows from (7) by induction.

COROLLARY 1 *If $X_1, \ldots, X_n$ are identically distributed with mean EX, then $E\{\sum X_i\} = n(EX)$.*

COROLLARY 2 *If $P(X \geq Y) = 1$, then $EX \geq EY$.*

Corollary 1 is simply a special case of Theorem 9. Corollary 2 follows from Theorem 8 and Theorem 5 (§4.2). The next theorem and corollaries are important properties of expectations related to the above Corollary 2:

THEOREM 10 *If $g(x) \leq h(x)$ for all x, then $E[g(X)] \leq E[h(X)]$.*

COROLLARY 1 $|EX| \leq E|X|.$

COROLLARY 2 *If $h(x) \geq 0$, then $E[h(X)] \geq 0$.*

Theorem 10 is a standard result for integrals, a consequence of Theorem 5 (§4.2). Corollary 2 follows upon setting $g = 0$. For Corollary 1, note that for all x, $-x \leq |x|$ and $x \leq |x|$. Applying the theorem yields

$-E(X) \leq E|X|$ and $EX \leq E|X|$.

In Example 4.3b we saw that the average of a product is not always the product of the averages. But it is in one important case, that of *independent* variables. When X and Y are independent, their joint p.d.f. factors: $f(x,y) = f_X(x)f_Y(y)$, so

$$E(XY) = \int\int xy f_X(x) f_Y(y) dx dy$$
$$= \int x f_X(x) dx \int y f_Y(y) dy = (EX)(EY).$$

However, the average product can equal the product of the averages even when the variables are *not* independent. For example, if (X,Y) is uniformly distributed on the unit disk, the average product is zero and the averages are zero. (See Problem 4-21.) But the variables cannot be independent—the support is not a product set.

When X and Y are independent, so are *functions* of these variables: $g(X)$ and $h(Y)$, as we showed in §3.7. As a result, we have the following:

THEOREM 11 *If X and Y are independent, then*

$$E[g(X)h(Y)] = E[g(X)]E[h(Y)]. \tag{9}$$

PROBLEMS

4-14. Given that $X \sim \mathcal{U}(-1, 1)$,
(a) calculate $E(X^2)$ in two ways.
(b) determine $E|X|$ by visualizing the distribution of $|X|$.

4-15. Calculate $E(X^2)$, where X has p.d.f. $2x$, $0 < x < 1$.

4-16. Given the p.f. $f_Y(k) = 2^{-k}$, $k = 1, 2, ...$, try to find $E(2^Y)$.

4-17. Find $E(e^{tX})$, as a function of t,
(a) when $X \sim \mathcal{U}(a,b)$.
(b) when X has p.d.f. e^{-x}, $x > 0$. [Give the restriction on t needed for the expectation to exist.]

4-18. Let θ have density $f_\theta(u) = \sin u$, $0 < u < \pi/2$.
(a) Calculate $E(\cos \theta)$ in two ways.
(b) Calculate $E(\cos^2 \theta)$.

4-19. Given this joint p.d.f. for the pair (X,Y): $f(x,y) = 6(1 - x - y)$, $0 < y < 1 - x < 1$, find $E(XY)$.

4-20. Suppose that (X,Y) is uniform on the triangle $0 < x < y < 1$.
(a) Find $E(X \mid Y = y)$, without a formal integration, by examining the

cross section at y.
(b) Use (a) to show that $EX = EY/2$.
(c) Calculate $E[(Y - X)^2]$.

4-21. Let (X, Y) be uniformly distributed on the unit disk: $x^2 + y^2 \le 1$. Show that $E(XY) = (EX)(EY)$ by calculating each side.

4-22. Show that if X and Y are independent, $E(X \mid Y = y) = EX$. Is the converse true? Either prove it or give a counterexample.

4-23. Let X, Y, and Z be i.i.d. (independent, identically distributed), each with the p.d.f. $e^{-\lambda}$ for $\lambda > 0$.
(a) Calculate $E(X + Y \mid Z = k)$.
(b) Calculate $E(Z \mid X + Y = k)$.
(c) Calculate by expanding the square: $E[(X + Y + Z)^2]$.
(d) Calculate $E(X \mid X + Y + Z = 1)$. [Hint: Exploit exchangeability.]

4-24. Let $Y = X_1 + X_2$, where X_1 and X_2 are independent, each with p.d.f. $e^{-\lambda}$, $\lambda > 0$. Find $E(e^{tY})$, using the result of Problem 4-17(b).

4-25. Given the joint p.d.f. $f_{X,Y}(x, y) = (4xy)^{-1/2}$, $0 < y < x < 1$, find $E(Y \mid x)$ and verify that $E[E(Y \mid X)] = EY$.

4-26. Show the following:
(a) If X and Y are i.i.d, then $E[(X - Y)^2] = 2[E(X^2) - (EX)^2]$.
(b) If $X_1, ..., X_n$ are exchangeable, then
$$E[(\sum X)^2] = n \sum E(X_1^2) + n(n-1)E(X_1 X_2).$$

4.4 Moments

The mean or expected value of a random variable describes an aspect of its distribution, namely, the "middle" or "center." The expected value of various functions of a random variable describe other aspects of the distribution. Particularly useful are the expected values of integer powers of random variables, as well as powers of deviations about a particular value. These are called *moments*:

DEFINITION The *k*th *moment* of X is $\mu'_k = E(X^k)$. The *k*th *central moment* is $\mu_k = E[(X - \mu)^k]$.

The exponent k is referred to as the *order* of the moment. A moment may fail to exist because the expectation that would define it does not exist. (Recall that an expected value EY "does not exist" if the integral defining $E|Y|$ does not converge.) When it exists, the first moment is the expected value of X itself: $\mu'_1 = \mu$. The first central moment is zero: $\mu_k = E(X - \mu) = EX - E\mu = \mu - \mu = 0$. The second central moment is an important characteristic of probability distributions which is the topic of the next section.

THEOREM 12 *If X has moments of order k, it has moments of all lower orders.*

This theorem is true for both moments μ'_k about zero and central moments μ_k. To see why, we observe first that if X has a kth moment (about 0), it has moments (about zero) of all lower orders. This follows from the fact that for $j < k$ and all x, $|x^j| \leq |x^k| + 1$, because then $E\{|X^j|\} \leq E\{|X^k|\} + 1$, a consequence Theorem 10 of §4.3. But central moments of order k are expressible in terms of moments about zero of order no larger than k, as is apparent upon expanding $(X - \mu)^k$ and averaging term by term. So the theorem is true for central moments.

The term "moment" is borrowed from mechanics, extending the analogy between probability and mass distributions. The first moment of a discrete mass distribution is the sum of products xm, for masses m at points x; for a continuous mass distribution with density $m(x)$ at x, the first moment is the integral of products $xm(x)$. Since we normalize probability to have total value 1, a probability moment corresponds to an *average* moment of a mass distribution.

The various moments describe different aspects of a distribution. The first moment—the mean—is an indication of the middle or center. "Absolute" moments $E\{|X - \mu|^k\}$ will tend to be large when there are large deviations $(X - \mu)$ from the center, so they tell us something about how mass or probability is spread out. If we know *all* of the moments, do we know all there is to know about a distribution? This question—the "moment problem"—has had considerable attention in classical mathematics, and the answer is that, subject to mild restrictions, the moments do indeed characterize a distribution.

In mechanics, the first two moments are the important ones. The same is true in statistical theory, for reasons that will emerge. The next section takes up the second central moment in some detail.

4.5 Variance

Averaging squared deviations from the mean of a distribution results in a measure of dispersion about the mean called the **variance** of the distribution, or of a random variable with that distribution:

DEFINITION The mean squared deviation of a random variable about its mean is its *variance*:

$$\text{var } X = \sigma^2 = E[(X - \mu)^2].$$

(As with the symbol μ, we may affix a subscript when needed for clarity, and write σ_X^2.) The variance is calculated as a sum or an integral, according as the distribution is discrete or continuous:

$$\text{var } X = \int (x - \mu)^2 dF(x) = \begin{cases} \sum (x - \mu)^2 f(x), & X \text{ discrete}, \\ \int (x - \mu)^2 f(x) dx, & X \text{ continuous}. \end{cases} \quad (1)$$

It is clear (Corollary 2 of Theorem 10 in §4.3) that a variance is always nonnegative. Moreover, if the variance of a discrete variable is zero, then every term in the sum (1) must be zero, which means that X can have only one value assigned a positive probability, $X = \mu$. We say then that the distribution is *singular* at that value. The same conclusion can be drawn when X is continuous (and even more generally), as will be shown in the next section.

Since the variance is an average *squared* deviation, its units are the square of the units of X: If X is a number of pounds, then X^2 is a number of "square pounds," which is a bit hard to appreciate intuitively. On the other hand, the square root of the variance has the same units as X itself, so it is easier to interpret.

DEFINITION The *standard deviation* of X (or of its distribution) is the positive square root of its variance:

$$\sigma = \sqrt{\text{var } X} = \sqrt{E[(X - \mu)^2]}. \quad (2)$$

This kind of average is referred to as a *root-mean-square* or *rms* average. It is thought of as a kind of "typical" deviation from the mean. We may use the abbreviation "s.d." for standard deviation and will sometimes write s.d.(X) for σ_X.

In measuring dispersion in a distribution, positive and negative deviations count alike, so it would be quite natural to take the average *absolute* deviation as a measure. This is called the **mean deviation**:

$$\text{m.a.d.} = E|X - \mu|. \quad (3)$$

Like the s.d., the mean (absolute) deviation has the units of X. It is never larger than the standard deviation:

$$\text{m.a.d.}(X) \leq \sigma_X. \tag{4}$$

This we can see as follows: The variance of any random variable is non-negative; in particular, for the variable $|X - \mu|$,

$$\text{var}\{|X - \mu|\} = E[(X - \mu)^2] - [E|X - \mu|]^2 \geq 0.$$

Transposing the second term and taking square roots yields (4).

EXAMPLE 4.5a Suppose X has a "triangular" density, as defined by the p.d.f. $1 - |x|$ for $|x| < 1$. Because the support is finite, the mean exists; because the p.d.f. is symmetrical about 0, the mean is $\mu = 0$. The variance is therefore the second moment about 0:

$$\text{var } X = E(X^2) = \int_{-1}^{1} x^2(1 - |x|)\,dx = 2\int_0^1 x^2(1 - x)\,dx = 1/6.$$

The mean deviation is found similarly:

$$\text{m.a.d.} = E|X - 0| = 2\int_0^1 x(1 - x)\,dx = 1/3.$$

The standard deviation is $\sigma = \sqrt{1/6} \doteq .408 > \text{m.a.d.}$ ∎

An alternative formula for the variance is often more convenient than (1), obtained by expanding the square in (1) and averaging term by term:

$$\text{var } X = E(X^2) - 2\mu E(X) + \mu^2 = E(X^2) - \mu^2. \tag{5}$$

A slightly more general relation is often useful:

THEOREM 13 (**Parallel axis theorem**): *For any constant a,*

$$E[(X - a)^2] = \sigma_X^2 + (a - \mu_X)^2. \tag{6}$$

COROLLARY *The variance is the smallest second moment.*

To show (6) we add and subtract μ inside $(X - a)$, regroup terms, expand the binomial, and average term by term:

$$\begin{aligned} E[(X - a)^2] &= E\{[(X - \mu) - (a - \mu)]^2\} \\ &= E[(X - \mu)^2] - 2(a - \mu)E(X - \mu) + E[(a - \mu)^2]. \end{aligned}$$

Since $E(X - \mu) = \mu - \mu = 0$, this establishes (6). (The term "parallel axis theorem" again borrows from terminology in mechanics.) From (6), it is clear that when $a = \mu_X$, that the right side is as small as it can be and that the left side is the variance. This is therefore the *smallest possible second moment*, as asserted by the corollary.

EXAMPLE 4.5b We'll use (5) first to calculate the variance of a discrete uniform distribution, in particular, of the number of points showing at the roll of a single die. The p.f. is $f(x) = 1/6$, $x = 1, 2, ..., 6$, and the mean is $\mu = 7/2$. Then,

$$E(X^2) = \tfrac{1}{6} \cdot (1^2 + 2^2 + 3^2 + 4^2 + 5^2 + 6^2) = \tfrac{91}{6},$$

and $\sigma^2 = 91/6 - (7/2)^2 = 35/12$. The s.d. is $\sigma \doteq 1.71$.

Next, suppose X is continuous and uniform on the interval $(0, b)$, with p.d.f. $f(x) = 1/b$, $0 < x < b$, and mean $b/2$. Then

$$E(X^2) = \int_0^b x^2 \cdot \tfrac{1}{b} dx = \tfrac{b^2}{3},$$

so var $X = \tfrac{b^2}{3} - (\tfrac{b}{2})^2 = \tfrac{1}{12} b^2$. ∎

We've seen that the mean value of a variable $Y = a + bX$ is that same linear transformation of the mean value of X. Consider next how the standard deviation transforms: Since it is a kind of average *deviation* from the mean, it should not depend on the reference origin. But it should involve the scale factor b, and this is the case:

$$\text{var}(a + bX) = E\{[a + bX - E(a + bX)]^2\}$$
$$= E\{[b(X - EX)]^2\} = b^2 \text{ var } X.$$

Taking square roots, we have

$$\sigma_{a + bX} = |b| \sigma_X. \qquad (7)$$

In particular, the s.d. of $-X$ is the same as the s.d. of $+X$, a fact that should be clear intuitively: The distribution of $-X$ is the mirror image of the distribution of $+X$, and the dispersion is the same.

A univariate *conditional* distribution, being a distribution, may have a variance. The variance of X given $Y = y$ is

$$\text{var}(X \mid y) = E[(X - E(X \mid y))^2 \mid Y = y] = E(X^2 \mid y) - \{E(X \mid y)\}^2.$$

This depends on the particular value of Y that is given—it's a function of y: $g(y) = \text{var}(X \mid y)$. If we use this to define a random variable, $g(Y)$, we might think at first that averaging with respect to the distribution of Y would produce the *unconditional* variance of X. But this is not correct, in general. To find the correct relation we first apply the parallel axis theorem (6) with $a = \mu_X$ to obtain

$$E[(X - \mu_X)^2 \mid Y = y] = \text{var}(X \mid y) + (\mu_X - \mu_{X \mid y})^2.$$

Each side is a function of y; replacing y by the random variable Y, we average with respect to the distribution of Y:

$$E[(X - \mu_X)^2] = E[\text{var}(X \mid Y)] + E[(\mu_{X \mid Y} - \mu_X)^2].$$

With $h(y) = \mu_{X \mid y}$, the last term is the variance of $h(Y)$. Hence,

THEOREM 14 *When X has second moments,*

$$\text{var } X = E[\text{var}(X \mid Y)] + \text{var}(\mu_{X \mid Y}). \qquad (8)$$

In words: The variance of X is the mean of the conditional variance (given $Y = y$) plus the variance of the conditional mean. Moreover, $\text{var } X = E[\text{var}(X \mid y)]$ only if $\mu_{X \mid y}$ is a constant (with probability 1).

EXAMPLE 4.5c In Example 4.2b we considered a uniform distribution within the triangle bounded by the line $x + y = 1$ and the coordinate axes. There we found $X \mid y$ to be uniform on $(0, 1 - y)$:

$$f(x \mid y) = \frac{1}{1-y}, \quad 0 < x < 1 - y,$$

and $E(X \mid y) = \frac{1}{2}(1-y)$. In Example 4.5b we found that the variance of a uniform distribution is $1/12$ of the square of the length of the support interval; so

$$\text{var}(X \mid y) = \frac{1}{12}(1-y)^2.$$

The marginal p.d.f. $f_Y(y)$, for $0 < y < 1$, is the area of the rectangular cross-section under the joint p.d.f. at y or $2(1-y)$ [height 2, times width $1-y$]. With y now replaced by the random variable Y, $\text{var}(X \mid Y)$ becomes a random variable, whose mean we find by integrating with respect to $f_Y(y)$:

$$E[\text{var}(X \mid Y)] = \frac{1}{12} E[(1-Y)^2] = \frac{1}{12} \int_0^1 (1-y)^2 \cdot 2(1-y)\, dy = \frac{1}{24}.$$

110   Chapter 4   Expectations

The joint p.d.f. is symmetrical in x and y, so X and Y are exchangeable, with equal means and variances: $\mu = 1/3$ and $\sigma^2 = 1/18$. The variance of the conditional mean [from (7)] is then

$$\operatorname{var}[E(X \mid Y)] = \operatorname{var}\left\{\tfrac{1}{2}(1 - Y)\right\} = \tfrac{1}{4}\operatorname{var} Y = \tfrac{1}{72},$$

and in this instance we can verify (8): $\tfrac{1}{18} = \tfrac{1}{24} + \tfrac{1}{72}$. ∎

4.6 Chebyshev and related inequalities (only good for proving shyfs)

The Chebyshev inequality is a useful theoretical tool, relating the variance of a distribution and the notion of dispersion as defined by probabilities of events. It is a consequence of a more basic inequality:

THEOREM 15 (Markov inequality)   *For any positive constant b and nonnegative function h for which the quantities are defined,*

$$P[h(X) \geq b] \leq \tfrac{1}{b} E[h(X)]. \tag{1}$$

To prove this, let $A = \{x\colon h(x) \geq b\}$. Using the result of Problem 4-13 we obtain

$$E[h(X)] = E[h(X) \mid A]\, P(A) + E[h(X) \mid A^c]\, P(A^c). \tag{2}$$

We now apply Corollary 2 of Theorem 9 (§4.3) to drop the second term and replace $E[h(X) \mid A]$ by b:

$$E[h(X)] \geq E[h(X) \mid A]\, P(A) \geq b\, P(A).$$

Division by b yields (1). Writing (1) with $h(x) = (x - \mu)^2$ and $\sqrt{b} = c$ we obtain the inequality usually referred to as Chebyshev's:

THEOREM 16 (Chebyshev inequality)   *For any constant $c > 0$, if X has mean μ and s.d. σ, then*

$$P(\,|X - \mu| \geq c) \leq \tfrac{\sigma^2}{c^2}. \tag{3}$$

Equivalent forms of the Chebyshev inequality (3) are useful:

$$P(|X - \mu| < c) \geq 1 - \tfrac{\sigma^2}{c^2}. \tag{4}$$

4.6 Chebyshev and related inequalities

$$P(|X - \mu| \geq k\sigma) \leq \frac{1}{k^2}, \text{ for } k > 0. \tag{5}$$

From (5) we see that the probability outside two standard deviations from the mean is less than 1/4, outside three s.d.'s from the mean, less than 1/9, and so on. For most of the distributions we'll be working with, these bounds are quite crude—the inequalities are far from equalities. However, as a general rule, (3) must include the possibility of an equality, as the next example shows.

EXAMPLE 4.6a Consider the discrete random variable X, defined by $P(X = 0) = 3/4$ and $P(X = a) = P(X = -a) = 1/8$. By symmetry, the mean is $\mu = 0$. The standard deviation is $\sigma = a/2$, so

$$P(|X| \geq 2\sigma) = P(|X| \geq a) = 1/4,$$

and (5) is an *equality*. ∎

Some additional relationships between standard deviation and probability are given in the following results:

THEOREM 17 *If $\sigma = 0$, then $P(X = \mu) = 1$.*

THEOREM 18 *If $P(|X - \mu| > K) = 0$, then $\sigma \leq K$.*

To show Theorem 17, we set $\sigma = 0$ and $c = 1/n$ in (4):

$$P\{|X - \mu| < 1/n\} = 1.$$

The limit as n becomes infinite of the left side is $P(X = \mu)$, as an application of Theorem 4(b) in §2.4 (page 28), and of course the limit on the right is 1. For Theorem 18 we define the event $A = \{x: (x - \mu)^2 \leq K^2\}$, and write the identity (2) with $h(x) = (x - \mu)^2$:

$$\sigma^2 = E[(X - \mu)^2 \mid A] P(A) + E[(X - \mu)^2 \mid A^c] P(A^c).$$

By assumption, $P(A^c) = 0$, so the second term on the right is 0. And on A, $(x - \mu)^2 \leq K^2$, so the first term does not exceed $K^2 P(A) \leq K^2$.

PROBLEMS

4-27. Find the variance for each of the following variables, whose means were found in Problem 4-1:
(a) X = number of heads in three independent coin tosses (Prob. 3-1):

$$f(0) = f(3) = 1/8, \quad f(1) = f(2) = 3/8.$$

(b) Y = number of correctly identified cola brands (Problem 3-4):
$$f(3) = 1/6, \quad f(1) = 3/6, \quad f(0) = 2/6.$$

(c) Z = number of rotten eggs in a random selection of two from a dozen, two of which are rotten (Problem 3-6):
$$f(0) = 45/66, \quad f(1) = 20/66, \quad f(2) = 1/66.$$

4-28. Given that X has the c.d.f. $F(x) = x^2$ for $0 < x < 1$, find σ_X.

4-29. Find the variance for the continuous variable Y with p.d.f.

(a) $f(y) = \frac{1}{2} e^{-|y|}$.

(b) $f(y) = 1 - |y|$, $|y| < 1$.

(c) $f(y) = 6y(1-y)$, $0 < y < 1$.

(The means were calculated in Problem 4-8.)

4-30. In Problem 3-5 you found the distribution of the number of points thrown with two dice, when the 36 outcomes are asssumed to be equally likely: $f(x) = \frac{1}{36}(6 - |7 - x|)$ for $x = 2, 3, \ldots, 12$. Find μ and σ^2.

4-31. In Problem 4-4 we assumed the discrete bivariate distribution given by the following table of probabilities $f(x, y)$:

	Y	1	2	3
	1	1/12	1/6	1/12
X	2	1/6	0	1/6
	3	0	1/3	0

(a) Find the means and variances of X and Y.
(b) Find the conditional variances: $\mathrm{var}(X \mid y)$, $x = 1, 2, 3$.
(c) Verify equation (8) of §4.5.

4-32. Consider a random variable X with mean μ and variance σ^2. Find the mean and variance of the **Z-score**: $Z = (X - \mu)/\sigma$. [This transformation converts X to a standard scale, marked off in the number of standard deviations to the right (for $Z > 0$) or left (for $Z < 0$) of the mean. Its mean and variance are thought of as "standard" parameters, and the transformation is called a *standardizing* transformation.]

4-33. For distributions with finite fourth moments,
(a) obtain the fourth central moment in terms of moments about 0.
(b) show: $\mu_3 = \mu_3' - 3\mu\sigma^2 - \mu^3$.

4-34. Find the variance of the distribution defined by the following c.d.f.: $F(x) = 1 - .8 e^{-x}$, $x \geq 0$.

4-35. Show that if X is continuous, the smallest absolute first moment is the one about any *median* of the distribution. [Hint: Write $E|X - a|$ as the sum of an integral over $(-\infty, a)$ and an integral over (a, ∞); recast this as

$$2aF(a) - a + \int_a^\infty x f(x)\, dx - \int_{-\infty}^a x f(x)\, dx.$$

Then minimize as usual by differentiating with respect to a.]

4-36. Show that $E(X^2) = 0$ implies $P(X = 0) = 1$.

4-37. Show the following:
(a) If $P(|X| > K) = 0$, then $E(X^2) \leq K^2$.
(b) If $\sigma_X > K > 0$, then $P(|X| > K) > 0$.

4-38. Show that $P(|X| > b)$ does not exceed either $E(X^2)/b^2$ or $E|X|/b$. Show also that for $f_X(x) = e^{-x}$, $x > 0$, one bound is better when $b = 3$ and the other is better when $b = \sqrt{2}$.

4-39. Given the distribution of Problem 4-25 [with $f(x, y) = (4xy)^{-1/2}$ for $0 < y < x < 1$], calculate var Y and verify the relation (8) of §4.5 (page 109), with roles of X and Y interchanged.

4.7 Covariance

Given a random vector (X, Y) and positive integers r and s, an average of the form $E[(X - a)^r (Y - b)^s]$ is a **mixed moment** (or *product moment*). We speak of the *order* of the moment as being the number of factors, $r + s$. The second order mixed central moment is important in probability and statistics; it is called the **covariance**:

$$\text{cov}(X, Y) = E[(X - \mu_X)(Y - \mu_Y)] = E(XY) - (EX)(EY). \quad (1)$$

In a sense, this generalizes the notion of variance, in that the covariance of a random variable with itself is its variance:

$$\text{cov}(X, X) = E[(X - \mu)^2] = E(X^2) - (EX)^2 = \sigma_x^2. \quad (2)$$

The covariance is related to association of the two variables in this sense: If X tends to be large when Y is large and small when Y is small (algebraically), the covariance will be positive; if X tends to be small when Y

is large and conversely, the covariance will be negative. The magnitude of the covariance corresponds to the strength of the tendency. This will be made more precise in the next section in terms of "correlation."

EXAMPLE 4.7a Consider the distribution of the pair (X, Y), defined by the joint density $f(x, y) = 8xy$, $0 \leq x \leq y \leq 1$. The marginal p.d.f.'s are

$$f_X(x) = \int_x^1 8xy\,dy = 4x(1 - x^2), \quad 0 \leq x \leq 1,$$

$$f_Y(y) = \int_0^y 8xy\,dx = 4y^3, \quad 0 \leq y \leq 1.$$

Using these we can calculate $EX = \frac{8}{15}$, $EY = \frac{4}{5}$. The covariance is then

$$\text{cov}(X, Y) = \int_0^1 \int_0^y (xy) 8xy\,dx\,dy - \left(\frac{8}{15}\right)\left(\frac{4}{5}\right) = \frac{4}{225}. \blacksquare$$

We saw earlier (Theorem 11 in §4.3) that when two variables are *independent*, the average of their product is the product of their averages: $E(XY) = (EX)(EY)$. That is, their covariance of X and Y is zero [from the definition (1)], which we'll find to be a very useful fact:

THEOREM 19 *When X and Y are independent, $\text{cov}(X, Y) = 0$.*

The covariance is a *bilinear* operation—linear in each argument. First, we observe that for arbitrary a and b we have

$$\text{cov}(aX + bY, U) = E[(aX + bY)U] - E(aX + bY)E(U)$$

$$= a E(XU) + b E(YU) - a(EX)(EU) - b(EY)(EU)$$

$$= a\,\text{cov}(X, U) + b\,\text{cov}(Y, U).$$

Extending this, by induction, or simply by similar applications of the distributive law for n terms, we obtain the following much used formula, which is valid when the second moments involved are finite:

THEOREM 20 *For arbitrary constants a_i, b_i,*

$$\text{cov}\left(\sum_1^m a_i X_i, \sum_1^n b_j Y_j\right) = \sum_1^m \sum_1^n a_i b_j \,\text{cov}(X_i, Y_j). \tag{3}$$

COROLLARY 1 $\operatorname{var}\left(\sum_{1}^{n} X_i\right) = \sum_{i=1}^{n} \sum_{j=1}^{n} \operatorname{cov}(X_i, X_j)$ (4)

$$= \sum_{i=1}^{n} \operatorname{var} X_i + 2 \sum_{i<j} \operatorname{cov}(X_i, X_j).$$

COROLLARY 2 *If $X_1, \ldots, X_n$ are independent, then*

$$\operatorname{var}\left(\sum_{1}^{n} a_i X_i\right) = \sum_{1}^{n} a_i^2 \operatorname{var} X_i. \qquad (5)$$

COROLLARY 3 *Given random variables X and Y,*

$$\operatorname{var}(aX \pm bY) = a^2 \operatorname{var} X + b^2 \operatorname{var} Y \pm 2ab \operatorname{cov}(X, Y).$$

COROLLARY 4 *If X and Y are independent, then*

$$\operatorname{var}(aX \pm bY) = a^2 \operatorname{var} X + b^2 \operatorname{var} Y.$$

(Of course, these various formulas hold only when the second moments involved exist.) Clearly, Corollaries 3 and 4 are special cases of Corollaries 1 and 2, respectively.

EXAMPLE 4.7b The resistance of a combination of resistors connected in series is the sum of the individual resistances. Actual resistances vary from one resistor to another. Suppose the actual resistance of a single resistor is a random variable whose mean is the nominal resistance and whose s.d. is five percent of the nominal resistance. Consider three such "five percent" resistors in series, with nominal resistances 400, 400, and 200 ohms, respectively. The nominal resistance of the combination is then 1000 ohms. If we assume the actual resistances are independent random variables, the s.d. of the sum is

$$\sqrt{20^2 + 20^2 + 10^2} = 30,$$

which is three percent of the nominal resistance. In this sense the precision of the combination is greater than that of the individual components. (We don't *add* the given tolerance figures to get the corresponding figure for the combination, since "chances are" that there tends to be some cancellation of positive and negative errors.) ■

EXAMPLE 4.7c Suppose a quantity is to be determined as the quotient of two measured quantities $A = B/C$, where B and C include random errors β and γ:

$$B = B_0 + \beta, \quad C = C_0 + \gamma.$$

Using the measured quantities to compute A involves an error:

$$A - A_0 = \frac{B_0 + \beta}{C_0 + \gamma} - \frac{B_0}{C_0} \doteq \frac{C_0 \beta - B_0 \gamma}{C_0^2}. \qquad (6)$$

The approximation (which is the differential of B/C) serves to linearize the expression for error. If the errors β and γ are assumed independent, we can use (4) and (5) to approximate the variance of the error in B/C:

$$\text{var } \alpha = \frac{1}{C_0^2} \text{var } \beta + \frac{B_0^2}{C_0^4} \text{var } \gamma. \quad \blacksquare$$

4.8 Correlation

We said in the preceding section that covariance is related to the degree of association of two random variables. As a potential measure of association, the covariance suffers from being dependent for its value on the units of measurement that happen to be employed. To avoid this problem, we define the *coefficient of linear correlation*, or **correlation coefficient**:

$$\rho_{X,Y} = \frac{\sigma_{X,Y}}{\sigma_X \sigma_Y}. \qquad (1)$$

Any linear change of scale introduces the same factor above and below the line, so cancellation leaves the correlation unchanged.

EXAMPLE 4.8a In Examples 3.7b and 3.8a we gave the distribution of (X, Y), the numbers on the first and second balls, respectively, when two balls are selected at random and without replacement from an urn with balls numbered 1 to 4. The probability table for the joint p.d.f. is as follows:

	1	2	3	4	
1	0	1/12	1/12	1/12	1/4
2	1/12	0	1/12	1/12	1/4
3	1/12	1/12	0	1/12	1/4
4	1/12	1/12	1/12	0	1/4
	1/4	1/4	1/4	1/4	1

We pointed out in Example 3.8a that the variables are exchangeable, so the means are equal: $EX = EY = 5/2$, and the variances are equal:

$\sigma_X^2 = \sigma_Y^2 = 5/4$. To find the covariance we need the average product:

$$E(XY) = \tfrac{1}{12}(1\cdot 2 + 1\cdot 3 + \cdots 4\cdot 2 + 4\cdot 3) = \tfrac{70}{12}.$$

The correlation coefficient is then

$$\rho = \frac{\tfrac{70}{12} - \tfrac{25}{4}}{\sqrt{\tfrac{5}{4}}\sqrt{\tfrac{5}{4}}} = -\tfrac{1}{3}.$$

The negative sign is to be expected: When you remove a ball with a small number, the next number drawn is expected to be somewhat larger than 5/2; and if you remove a large-numbered ball, the next is expected to be somewhat smaller than 5/2. ∎

A correlation is zero when the covariance is zero, and we say that in this case the variables are **uncorrelated**. In particular, when two variables are independent, they are uncorrelated. (The converse of this is false; see Problem 4-40.) When n variables are independent, they are pairwise independent and therefore pairwise uncorrelated.

It is important to know that the magnitude of a correlation is bounded by 1. This statement is a form of the *Schwarz inequality*, which we'll derive in the present setting. For random variables U and V and real constant k, consider the nonnegative variable $(U - kV)^2$. Its expectation is nonnegative:

$$0 \le E[(U - kV)^2] = k^2 E(V^2) - 2kE(UV) + E(U^2). \tag{2}$$

The function of k on the right is quadratic, and the fact that it is nonnegative implies that its discriminant is nonpositive:

$$[E(UV)]^2 - E(U^2)E(V^2) \le 0. \tag{3}$$

(From algebra: The *discriminant* of the quadratic function $ak^2 + 2bk + c$ is $b^2 - ac$.) Substituting $U = X - EX$ and $V = Y - EY$ in (3), we get this important bound for the magnitude of ρ:

$$\sigma_{X,Y}^2 \le \sigma_X^2 \sigma_Y^2, \quad \text{or} \quad |\rho_{X,Y}| \le 1. \tag{4}$$

The significance of $|\rho| = 1$ should be noted: In this case the discriminant [l.h.s. of (3)] is equal to zero, which means that the parabola representing the quadratic function in (2) touches the axis at some point—there is a double zero, say at k_0. This means that the inequality in (2) is an equality for that value of k, which in turn implies that $U = k_0 V$ with probability 1 (see Problem 4-36). So the distribution of

118   Chapter 4   Expectations

(X, Y) is singular—concentrated on a straight line. (The slope of the line is positive or negative according as ρ is positive or negative.) We think of $\rho = \pm 1$ as "perfect" linear correlation; knowing the value of one variable you essentially know the value of the other (which is then redundant). A continuous bivariate distribution with $\rho = \pm 1$ does not have a density in the usual sense; a discrete bivariate distribution with $\rho = \pm 1$ has a square probability table with nonnegative entries on the main diagonal and 0's off the diagonal (when the marginal values for both variables are written in natural order).

The class of random variables with finite second moments (pairwise) is a *vector space*, in which the inner product is the covariance, and the norm is the s.d.: $\| X \| = \sqrt{\text{cov}(X, X)} = \sigma_X$. Suppose we represent random variables as ordinary two-dimensional vectors (arrows from the origin), and look at the usual vector addition of a sum (Figure 4-2). The squared length of the sum is

$$\| X + Y \|^2 = \text{var}(X + Y) = \text{var } X + \text{var } Y + 2\,\text{cov}(X, Y)$$
$$= \| X \|^2 + \| Y \|^2 + 2\rho_{X,Y} \| X \| \, \| Y \|.$$

According to the law of cosines from trigonometry, $\rho_{X,Y}$ is the cosine of the angle θ between the vectors representing X and Y. The vectors are perpendicular when the variables are uncorrelated and are parallel when one variable is a (positive or negative) multiple of the other.

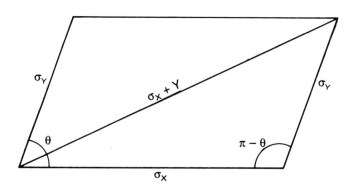

Figure 4-2. Vector representation of random variables

PROBLEMS

4-40. Let (X, Y) have the joint p.f. given by the following table:

X \ Y	0	1	2
1	1/4	0	1/4
3	1/12	1/3	1/12

(a) Find the means, the variances, and the covariance.
(b) Find the correlation coefficient.
(c) Find the covariance of X and Y given that $Y \neq 2$.
(d) Are X and Y independent?
(e) Are X and Y conditionally independent, given $Y \neq 1$?

4-41. Find the correlation coefficient for the bivariate distribution with constant density in the triangle of the first quadrant determined by the condition $0 < x < y < 1$. [As we'll see later, this is the joint distribution of the smaller and the larger of two independent observations on a uniform distribution on (0, 1). Can you derive it now?]

4-42. Consider again the distribution of Problems 4-39 and 4-25, with p.d.f. $f(x, y) = (4xy)^{-1/2}$ for $0 < y < x < 1$. [In those problems we found var $Y = 7/180$, $EY = 1/6$, $X \sim \mathcal{U}(0, 1)$.] Find the correlation coefficient.

4-43. A population includes M individuals of type 1, and $N - M$ of type 0. We select one at random and then another at random from those that remain. (Random sampling without replacement.) Let X and Y denote the type numbers of the individuals selected on the first and second draws, respectively. Obtain the table for $f(x, y)$, check for exchangeability, and calculate the covariance.

4-44. Suppose X, Y, and Z are i.i.d. with mean 0 and s.d. 1. Find
(a) $\text{var}(2X + 3Y - Z)$. (b) $\text{cov}(X - 2Y, 3X + Y + 2Z)$.

4-45. Given variables (X_1, X_2) with means $(2, -1)$, variances $(1, 9)$, and covariance -4, find the corresponding parameters of the transformed variables $Y_1 = X_1 + X_2$, $Y_2 = X_1 - X_2$.

4-46. The random vector (X, Y) has joint p.d.f. $f(x, y) = e^{-y}$, $0 < x < y$. Find $\text{cov}(X, Y)$. [You may find this integration formula useful:
$$\int_0^\infty x^k e^{-x} dx = k!.]$$

4-47. For the random orientation (Θ, Φ) considered in Example 2.6b, the joint p.d.f. was found in Problem 3-59 to be $f(\theta, \phi) = \frac{1}{2\pi} \sin \theta$ for $0 < \theta < \pi/2$ and $0 < \phi < 2\pi$. In that problem we also found that the corresponding rectangular coordinates $X = \cos \phi \sin \theta$, $Y = \sin \phi \sin \theta$ are not independent. Show that they are uncorrelated.

4-48. Use the Schwarz inequality to show: If $E(X^2)$ is finite, so is $E|X|$.

4-49. Show: $X+Y$ and $X-Y$ are uncorrelated if and only if $\sigma_X = \sigma_Y$.

4-50. Show: If $\rho^2 = 1$, then $\text{var}(X+Y) = (\sigma_X \pm \sigma_Y)^2$, according as $\rho = \pm 1$.

4-51. Suppose three gears have widths X, Y, and Z which are independent random variables with means .500, .300, and .700 inches and s.d.'s .001, .004, and .002 inches, respectively. They are to be assembled side by side on a single shaft. Find the mean and s.d. of the assembly width.

4-52. After a given period of operation, a guidance system has a position error of $\delta = 5\delta_a + 20\delta_g$ (nautical miles), where δ_a and δ_g are independent random errors in an accelerometer and a gyro (in ft/sec^2 and deg/hr, respectively). Determine the s.d. of the position error corresponding to errors in δ_a and δ_g with s.d.'s of 0.1 ft/sec^2 and .05 deg/hr.

4-53. Obtain an approximate expression for the variance of the error in calculating a product AB using measured values of A and B that involve independent errors α and β about the actual values A_0 and B_0.

4.9 Generating functions

The expected values of powers of X (moments) characterize its distribution, in many cases. The expected values of certain functions which involve all these powers, in series expansions, also characterize a distribution. For a function $g(x, t)$, the expected value $\int g(x, t) f(x)\,dx$ is a function of t, and as such is referred to as a *transform* of f. Transforms are valuable in dealing with sums of independent random variables. Their usefulness in probability stems from the familiar law of exponents: $a^{x+y} = a^x a^y$. The particular transforms we use are the moment generating function, the factorial moment generating function (including a special case, the probability generating function), and the characteristic function, for which "g" is (respectively) e^{tx}, t^x, e^{itx}.

DEFINITION The *moment generating function* (m.g.f.) of a random variable X is $\psi(t) = E(e^{tX})$, provided this expectation exists for t in a neighborhood of 0.

For discrete variables with p.f. f, the defining expectation is

4.9 Generating functions

$$\psi(t) = \sum_i e^{tx_i} f(x_i), \tag{1}$$

and for continuous variables with p.d.f. f, it is:

$$\psi(t) = \int e^{tx} f(x)\, dx. \tag{2}$$

Without worrying, for the moment, about the validity of interchanging integrals with sums and derivatives, we introduce the power series for the exponential function in (1) and (2):

$$\psi(t) = \int \sum_0^\infty \frac{(tx)^k}{k!} dF(x) = \sum_0^\infty \frac{t^k}{k!} \int x^k dF(x) = \sum_0^\infty E(X^k) \frac{t^k}{k!},$$

where we've used the integral with respect to dF to include both the discrete and the continuous case. Thus, the coefficient of $t^k/k!$ in the power series expansion of ψ is the kth moment. But we know from calculus that this coefficient is the kth derivative at 0:

$$E(X^k) = \psi^{(k)}(0) \tag{3}$$

We can also get (3) by formal differentiation under the integral sign:

$$\psi^{(k)}(t) = \frac{d^k}{dt^k} E(e^{tX}) = E(X^k e^{tX}),$$

setting $t = 0$ in this kth derivative. Thus, the m.g.f. "generates" the moments.

Of course, $\psi(0) = E(e^{0t}) = E(1) = 1$ for any distribution. But it can be shown that if $E(e^{tX})$ exists in a *neighborhood* of $t = 0$, then the distribution has moments of all orders, and the various interchanges are legitimate. Also, the m.g.f. then uniquely defines the distribution—different distributions cannot have the same m.g.f. We note that in the case of a distribution whose support is bounded, there is no question of convergence, so ψ exists; moreover, in this case the distribution is uniquely defined by the moments and hence by ψ.

EXAMPLE 4.9a Consider a random variable X with just two possible values, and corresponding probabilities $f(1) = p$, $f(0) = 1 - p$. For this distribution,

$$\psi(t) = E(e^{tx}) = pe^t + (1-p)e^0 = 1 + p(e^t - 1).$$

The derivatives are

$$\psi'(t) = \psi''(t) = \cdots = pe^t,$$

and $E(X^k) = \psi^{(k)}(0) = p$, for $k = 1, 2, \ldots$. Substituting the power series for e^t in $\psi(t)$, we have

$$\psi(t) = 1 + pt + p\frac{t^2}{2} + p\frac{t^3}{3!} \cdots,$$

and can read the moments as the coefficients of $t^k/k!$. These moments, of course, can be easily calculated directly. Indeed, when there are only the two values 0 and 1 for X, we have $X^k = X$, so $E(X^k) = EX = p$. ∎

EXAMPLE 4.9b Consider the p.d.f. $f(x) = e^{-x}$, $x > 0$. The m.g.f. is

$$\psi(t) = \int_0^\infty e^{tx} \cdot e^{-x} dx = \frac{1}{1-t}, \quad |t| < 1.$$

We could find the moments by differentiating and substituting $t = 0$, but in this case it is perhaps simpler to use the alternative tack—reading moments as coefficients in the power series expansion:

$$\psi(t) = \frac{1}{1-t} = 1 + t + t^2 + t^3 + \cdots = 1 + t + 2!\frac{t^2}{2!} + 3!\frac{t^3}{3!} + \cdots.$$

It is thus apparent that $E(X^k) = k!$, the coefficient of $t^k/k!$. ∎

THEOREM 21 *If $Y = a + bX$, and ψ_X exists, then*

$$\psi_Y(t) = e^{at}\psi_X(bt). \tag{4}$$

THEOREM 22 *If X and Y are independent and have m.g.f.'s, then the m.g.f. of their sum is the product of their m.g.f.'s:*

$$\psi_{X+Y}(t) = \psi_X(t) \cdot \psi_Y(t). \tag{5}$$

For Theorem 21, recall the law of exponents: $e^{ta + tbX} = e^{ta}e^{tbX}$. From this it follows that

$$E[e^{t(a + bX)}] = e^{at}E(e^{btX}).$$

Another application of the law of exponents yields Theorem 22:

$$E[e^{t(X + Y)}] = E[e^{tX}e^{tY}] = E(e^{tX})E(e^{tY}) = \psi_X(t)\psi_Y(t),$$

where the factorization of the average product as the product of the averages is another application of Theorem 11 in §4.3.

4.9 Generating functions

EXAMPLE 4.9c Let Y be a discrete random variable with three possible values and probabilities as follows: $f(0) = 1/4$, $f(1) = 1/2$, $f(2) = 1/4$. The m.g.f. is

$$\psi_Y(t) = e^0 \cdot \frac{1}{4} + e^t \cdot \frac{1}{2} + e^{2t} \cdot \frac{1}{4} = \left(\frac{e^t + 1}{2}\right)^2.$$

This happens to be the square of the m.g.f. of the distribution in Example 4.9a, with $p = 1/2$. In view of (5) and the fact that different distributions cannot have the same m.g.f., it follows that the distribution of the sum of two i.i.d. variables, each with the two-point distribution in Example 4.9a, is the distribution of the Y defined above. Of course, in this simple situation, we could reason from more basic principles: The four sequences 0 0, 0 1, 1 0, 1 1 are equally likely, and it is clear that the sum will be 0 with probability 1/4, 1 with probability 2/4, and 2 with probability 1/4—a replica of Y. ∎

Clearly, we would not have needed such a fancy tool as the m.g.f. to find the distribution of Y in this last example. However, finding the distribution of a sum directly is not always so easy:

THEOREM 23 *If X and Y are independent, continuous variables, the p.d.f. of their sum is*

$$f_{X+Y}(z) = \int_{-\infty}^{\infty} f_X(z-y) f_Y(y) \, dy. \qquad (6)$$

We'll derive this "convolution" formula for the p.d.f. of the sum using a standard technique—first finding the c.d.f. from its definition and then differentiating to obtain the p.d.f. Recall that the joint p.d.f. factors into the product of the marginal p.d.f.'s because of the assumed independence:

$$F_{X+Y}(z) = P(X+Y \leq z) = \int_{-\infty}^{\infty} \int_{-\infty}^{z-y} f_X(x) f_Y(y) dx \, dy$$

$$= \int_{-\infty}^{\infty} \left\{ \int_{-\infty}^{z-y} f_X(x) \, dx \right\} f_Y(y) \, dy$$

$$= \int_{-\infty}^{\infty} F_X(z-y) f_Y(y) dy.$$

[We've used (5) of §3.4 in expressing the probability as a double integral; the integration limits in the iterated integral are determined by the region where $x + y \leq z$.] Differentiating with respect to z under the integral sign yields (6).

The technique we used in Example 4.9c will be quite useful, in later chapters, for finding the distribution of a sum of independent random variables without having to deal with a convolution. This is important when there are more than two variables; an n-fold convolution is quite unpleasant to deal with directly.

Closely related to the m.g.f. is the **factorial moment generating function** (*f.m.g.f.*), defined as the following expected value, provided that it exists in a neighborhood of $t = 1$:

$$\eta_X(t) = E(t^X). \tag{7}$$

Observe that since $t^X = e^{(\log t)X}$, $\eta_X(t) = \psi_X(\log t)$, and the point $t = 1$ corresponds to a 0 in the argument of ψ. Since $t^{X+Y} = t^X t^Y$, it follows that when X and Y are *independent*, the f.m.g.f. of the sum is the product of the f.m.g.f.'s: $\eta_{X+Y}(t) = \eta_X(t)\eta_Y(t)$.

The f.m.g.f. also "generates" the moments of X, but in a roundabout way: If we differentiate with respect to t under the E (which is either an integration or summation), we obtain

$$\eta_X'(t) = E[Xt^{X-1}], \quad \eta_X''(t) = E[X(X-1)t^{X-2}], \text{ etc.}$$

Substituting $t = 1$ yields *factorial moments*:

$$\eta_X'(1) = E(X), \quad \eta_X''(1) = E[X(X-1)],$$

and so on. In general, we get the kth factorial moment from derivatives of η:

$$\eta_X^{(k)}(1) = E[(X)_k]. \tag{8}$$

The first factorial moment is the first moment. The second is

$$\eta_X''(1) = E[X(X-1)] = E(X^2) - EX,$$

so we can calculate the variance from the first and second factorial moments:

$$\text{var } X = \eta_X''(1) + EX - (EX)^2.$$

Which generating function should one use, ψ or η? The form of the p.f. or p.d.f. is sometimes such as to combine neatly with e^{tx}, and sometimes with t^x, in a function easily summed or integrated. So use the one that is easier in a particular case. (In some cases neither one is easy to use.)

EXAMPLE 4.9d The preceding example treated the discrete distribution with probabilities 1/4, 1/2, and 1/4 at $x = 0$, 1, and 2, respectively, in terms of the m.g.f. The *factorial* m.g.f. is

$$\eta(t) = t^0 \cdot \frac{1}{4} + t^1 \cdot \frac{1}{2} + t^2 \cdot \frac{1}{4} = \frac{1}{4}(1+t)^2.$$

This is the f.m.g.f. of the sum of independent observations from the distribution on 0 and 1 in Example 4.9a—it's the square of the f.m.g.f. of that distribution. The derivative is $\eta'(t) = (1+t)/2$, from which we get $E(X) = \eta'(1) = 1$. Differentiating once more we find $\eta''(1) = 1/2$, so var $X = 1/2 + 1 - 1 = 1/2$, which you can check directly. ■

When the value space X is $\{0, 1, 2, ...\}$, as in the preceding example, the f.m.g.f. plays an additional role. With the temporary notation $p_i = P(X = i)$, f.m.g.f. is

$$\eta(t) = p_0 + p_1 t + p_2 t^2 + p_3 t^3 + \cdots.$$

That is, the probabilities p_i are coefficients in the power series expansion of η. In this case, η is also called a **probability generating function**. It can be quite useful in finding probabilities for sums of integer-valued variables, as in the next example.

EXAMPLE 4.9e Let X denote the number of points showing at the roll of a die. If we assume equally likely outcomes, the f.m.g.f. is

$$\eta(t) = \frac{1}{6} \cdot (t + t^2 + t^3 + t^4 + t^5 + t^6) = \frac{t(1-t^6)}{6(1-t)}.$$

Suppose we roll the die three times in independent trials or, equivalently, roll three such dice together, and let Y denote the total number of points that show—the sum of the numbers of points on the three dice. The f.m.g.f. of Y is then the cube of the f.m.g.f. for a single die:

$$\eta_Y(t) = \frac{t^3(1-t^6)^3}{216(1-t)^3}. \tag{9}$$

We expand the numerator of (9) in a (finite) power series, using the binomial theorem:

$$t^3(1-t^6)^3 = \sum_{i=0}^{3} \binom{3}{i} t^3 (-t^6)^i.$$

We can write the factor $(1-t)^3$ in the denominator of (9) as a negative power of $(1-t)$ in the numerator. The terms in its power series expansion are formed like those in the expansion of a binomial with a positive exponent, and the resulting series converges when $|t| < 1$:

$$(1-t)^{-3} = \sum_0^\infty \binom{-3}{j}(-t)^j,$$

where

$$\binom{-n}{j} = \frac{(-n)_j}{j!} = \frac{(-n)(-n-1)(-n-2)\cdots(-n-j+1)}{j!}.$$

Substituting these series in (9) we obtain

$$\eta_Y(t) = \frac{1}{216} \sum_{i=0}^3 \sum_{j=0}^\infty (-1)^{i+j} \binom{3}{i}\binom{-3}{j} t^{6i+j+3}.$$

This is a power series in t, and the coefficient of t^k is $p_k = P(Y = k)$. For instance, suppose we want to find the probability of a total of 15. We then look for the terms in t^{15}—terms in which the exponent of t is $6i + j + 3 = 15$. There are three: one with $i = 0$ and $j = 12$, one with $i = 1$ and $j = 6$, and one with $i = 2$ and $j = 0$. So

$$p_{15} = \frac{1}{216} \cdot \left[\binom{3}{0}\binom{-3}{12} - \binom{3}{1}\binom{-3}{6} + \binom{3}{2}\right] = \frac{10}{216}. \blacksquare$$

PROBLEMS

4-54. Obtain the m.g.f. of a distribution which is singular at $x = b$: $f(b) = 1$, and $f(x) = 0$ for $x \neq b$. Use it to obtain the moments and the central moments.

4-55. Given that X has the m.g.f. $\psi(t) = (1 - t^2)^{-1}$, find its variance.

4-56. Find the m.g.f. of a random variable with a uniform density on $(0,1)$: $f(x) = 1$, $0 < x < 1$, and use it to find the mean and variance. [Hint: In your closed form expression for ψ, substitute the power series for e^t to obtain the power series for ψ (and read the moments from that).]

4-57. Find the m.g.f. of a random variable with uniform density on the interval (a, b). [This was done in Problem 4-17, but another way is to use the result of the preceding problem and a suitable linear transformation.]

4-58. Find the m.g.f. of the distribution whose density function is $f(x) = K \exp\left(-\frac{1}{2}x^2\right)$. [Hint: It isn't necessary to know the value of the constant K. Complete the square in the exponent and use the fact that the integral of $\exp(-\frac{1}{2}x^2)$ is the same as that of $\exp\{-\frac{1}{2}(x-a)^2\}$ for any constant a.]

4-59. Given that X has the distribution of the preceding problem, find the m.g.f. of the variable $Y = a + bX$.

4-60. A table of integrals (or successive integrations by parts) gives the following formula:
$$\int_0^\infty x^n e^{-x} dx = n!,$$
which implies that $f(x) = x^n e^{-x}/n!$ is a p.d.f. on $(0, \infty)$.
(a) Use the formula to find the corresponding m.g.f.
(b) Use the result of (a) to find the mean and variance.

4-61. Let $Y = X_1 + X_2$, where X_1 and X_2 are i.i.d. with the uniform distribution of Problem 4-56. Find the m.g.f. of Y and use it to obtain the mean and variance of Y. [Hint: Start with the power series expression for the m.g.f. of X and obtain the first few terms of its square.]

4-62. Let $Y = X_1 + X_2$, where X_1 and X_2 are i.i.d. with the distribution in Problem 4-58. Find the m.g.f. of Y and derive the mean and variance of Y from its m.g.f.

4-63. Is either of the following a m.g.f.? (Justify your answer.)
(a) $(1 + t^2)^{-1}$.
(b) $(1 - t^3)^{-1}$.

42-64. Given that $\psi_X(t) = (1 - 2t)^{-1/2}$, find μ_X and σ_X.

4-65. Find the m.g.f. of the distribution with p.d.f. $\frac{1}{2}e^{-|x|}$.

4-66. Find the distribution of X, given that its m.g.f. is
$$\psi(t) = 1 - p + \frac{p}{2}(e^t + e^{-t}).$$
[Hint: X is a discrete variable.]

4-67. Find the f.m.g.f. of the discrete variable with p.f. $f(k) = 1/2^{k+1}$ for $k = 0, 1, \ldots$ and use it to obtain the mean and variance.

4-68. Find K so that $f(n) = K/n!$ is a p.f. on $n = 0, 1, 2, \ldots$, find the corresponding f.m.g.f. η, and use η to find the mean and variance of the distribution.

4-69. Let $Y = X_1 + \cdots + X_n$, where the X_i's are i.i.d. with the distribution of Problem 4-67. Find the f.m.g.f. of Y.

4-70. Let $Y = X_1 + \cdots + X_n$, where the X_i's are i.i.d. with the distribution of Problem 4-64. Find the m.g.f. of Y and from it, var Y.

4-71. Find the probability of a total of 12 when three dice are rolled. (See Example 4.9e.)

4-72. Three numbers are drawn at random from the digits 0, 1, ..., 9 (with replacement). Find the probability that their sum is 10.

4.10 Characteristic functions

Use of the m.g.f. is limited to distributions for which moments of all orders exist and are finite. A transform which does not have this limitation is the **characteristic function**:

$$\phi(t) = E[\cos(tX)] + i\, E[\sin(tX)],$$

where $i = \sqrt{-1}$. Using the deMoivre formula: $\cos\theta + i\sin\theta = e^{i\theta}$, we write ϕ in the form

$$\phi(t) = E(e^{itX}) = \int e^{itx} dF(x). \tag{1}$$

[Interpret $\int () dF(x)$ as $\int () f(x)\, dx$ in the continuous, and as $\sum () f(x)$ in the discrete case.] The sine and cosine functions are bounded, so the integral converges for every F. Since ϕ is a complex-valued function, the calculation of characteristic functions can be difficult, often requiring contour integration.

EXAMPLE 4.10a The distribution defined by the p.d.f. e^{-x} for $x > 0$ is one for which we can calculate the characteristic function:

$$\phi(t) = \int_0^\infty e^{-x} \cos tx\, dx + i \int_0^\infty e^{-x} \sin tx\, dx$$

$$= \frac{1}{1+t^2} + i\frac{t}{1+t^2} = \frac{1}{1-it}.$$

(One can find the values of these definite integrals in any table of integrals, but the integration can be carried out by "integration by parts.") It is significant that you'll get the right answer if you simply overlook the fact that e^{itx} is complex-valued and use a standard formula in a formal integration:

$$\int_0^\infty e^{itx-x}dx = \int_0^\infty e^{-(1-it)x}dx = \frac{-1}{1-it}e^{-(1-it)x}\Big|_0^\infty = \frac{1}{1-it}. \blacksquare$$

What's really going on in this last example is that when the m.g.f. exists, then $\phi(t) = \psi(it)$, as can be shown using complex variable theory. Ineed, when ψ exists, this is the way to find ϕ: Calculate $\psi(t)$ and then replace t by it to get $\phi(t)$.

The usefulness of the characteristic function depends on uniqueness, continuity, and expansion theorems that we'll not prove:

THEOREM 24 (Inversion formula) *If $x-h$ and $x+h$ are points of continuity of a c.d.f F_X, then the increment in F_X over the interval $(x-h, x+h)$ is given in terms of ϕ_X by*

$$F_X(x+h) - F_X(x-h) = \lim_{T\to\infty} \frac{1}{\pi}\int_{-T}^T \frac{\sin ht}{t}e^{-itx}\phi_X(t)\,dt.$$

THEOREM 25 (Uniqueness theorem) *Corresponding to each characteristic function ϕ there is a unique distribution function having that ϕ as its characteristic function.*

THEOREM 26 (Continuity theorem) *If distribution functions $\{F_n\}$ converge pointwise to a distribution function F, the corresponding characteristic functions $\{\phi_n\}$ converge to the characteristic function of F. Conversely, given a sequence of characteristic functions $\{\phi_n\}$ converging to a function ϕ which is continuous at $t=0$, then ϕ is a characteristic function, and the corresponding c.d.f.'s $\{F_n\}$ converge to the c.d.f. determined by ϕ.*

Theorem 25 is a consequence of Theorem 24. For, the recovery of F with the inversion formula uniquely defines it, except at points of discontinuity of F, where uniqueness is the result of the requirement of right-continuity for c.d.f.'s. Using Theorems 25 and 26 we can get limiting distributions by going to transforms, passing to the limit, and then going back to c.d.f.'s

Because of the relation between ϕ and ψ, one can use ϕ to generate moments—even when ψ fails to exist. It can be shown that if a distribution has moments up to order k, then for $j \le k$ we can find μ_j as $i^{-j}\phi^{(j)}(0)$.

EXAMPLE 4.10b We've seen that the p.d.f. $f(x) = [\pi(1+x^2)]^{-1}$ has no moments of finite (integral) order. Its characteristic function is

$$\phi(t) = \frac{1}{\pi}\int \frac{\cos tx}{1+x^2}\,dx + \frac{i}{\pi}\int \frac{\sin tx}{1+x^2}\,dx = e^{-|t|}.$$

(The integrals can be found in a table of integrals; they are evaluated using contour integration.) This function is not differentiable at $t = 0$, a fact that corresponds to the nonexistence of moments. ∎

In the case of continuous distributions, the inversion formula yields a formula for the p.d.f. in terms of the characteristic function:

$$f(x) = \lim_{h\to 0} \frac{F(x+h) - F(x-h)}{2h}$$

$$= \lim_{h\to 0} \lim_{T\to\infty} \frac{1}{2\pi} \int_{-T}^{T} \frac{\sin ht}{ht} e^{-itx} \phi(t)\,dt.$$

When limits and integrals can be interchanged, we have

$$f(x) = \lim_{T\to\infty} \frac{1}{2\pi} \int_{-T}^{T} e^{-itx} \phi(t)\,dt. \tag{2}$$

The function ϕ is called the *Fourier transform* of f, and (2) gives f as the *inverse* Fourier transform of ϕ.

EXAMPLE 4.10c Consider once more (from Example 4.10b) the characteristic function $\phi(t) = e^{-|t|}$. Using the inversion formula (2), we find

$$f(x) = \lim_{T\to\infty} \frac{1}{2\pi} \int_{-T}^{T} \exp\{-|t| - itx\}\,dx$$

$$= \lim_{T\to\infty} \frac{1}{\pi} \int_{0}^{T} e^{-t}\cos tx\,dx = \frac{1/\pi}{1+x^2}.$$

[We've exploited the fact that the first integrand is an even function plus i times an odd function; the integral of the odd function is zero, and for the even function we can integrate over $(0, T)$ and multiply by 2.] So we recover the f which led to the ϕ in the first place: The inverse transform of the transform brings us back to the original p.d.f. ∎

The next theorem gives a Taylor's expansion in which, because of the special nature of a characteristic function, the error term is of smaller order than can generally be claimed in such expansions. In stating this theorem we use the notation $o(x)$ to denote any function of x with the property that $o(x)/x$ tends to 0 with x. [The symbol $o(x)$ is read "a function of smaller order than x."] Examples of such functions are x^3, $\sin^2 x$, $e^x - 1 - x$, etc.

THEOREM 27 (Expansion with remainder) *If X has a finite kth absolute moment, the characteristic function can be expressed as follows:*

$$\phi(t) = 1 + E(X)(it) + E(X^2)\frac{(it)^2}{2!} + \cdots + E(X^k)\frac{(it)^k}{k!} + o(t^k).$$

4.11 Multivariate generating functions

Multivariate analogs of univariate generating functions and characteristic functions can be defined and serve similar purposes. There are similar uniqueness theorems, and we'll give only some formal definitions and relations.

The bivariate m.g.f. is $\psi(s, t) = E(e^{sX + tY})$, and the bivariate characteristic function is

$$\phi(s, t) = E[e^{i(sX + tY)}]. \tag{1}$$

The m.g.f.'s of the marginal distributions are obtained from ϕ or ψ by setting the argument corresponding to the unwanted variable equal to 0. Thus, $\psi_X(s) = \psi(s, 0)$, and $\psi_Y(t) = \psi(0, t)$. The following theorem gives us a way of characterizing independence in terms of characteristic functions (or in terms of m.g.f.'s, when these exist):

THEOREM 28 *If X and Y ar independent, their joint characteristic function (or m.g.f.) factors into the product of the marginal characteristic functions (m.g.f.'s). Conversely, if a joint characteristic function factors into the product of a function of s and a function of t: $\phi_{X,Y}(s, t) = g(s)h(t)$, then X and Y are independent. Moreover, if one of these factors is a marginal characteristic function, so is the other.*

To prove this we note that setting $s = 0$ produces $\phi_Y(t)$, which is then proportional to $h(t)$, and similarly, $\phi_X(s)$ is proportional to $g(s)$. From this it is clear [since $\phi(0, 0) = 1$] that $\phi_{X,Y}(s, t) = \phi_X(s)\phi_Y(t)$. With these marginal functions written as integrals, we have

$$\phi_{X,Y}(s, t) = \int e^{isx} f_X(x)\, dx \int e^{ity} f_Y(y)\, dy$$

$$= \int\int e^{i(sx+ty)} f_X(x) f_Y(y)\,dx\,dy.$$

This tells us that ϕ is the characteristic function of the product $f_X f_Y$; and because there is only one p.d.f. with ϕ as its characteristic function, it follows that $f_{X,Y} = f_X f_Y$, which shows the asserted independence. (Of course an analogous result holds for moment generating functions, when these exist.)

We define the m.g.f. of the random vector $\mathbf{X} = (X_1, ..., X_n)$ in similar fashion:

$$\psi(t_1, ..., t_n) = E[\exp(t_1 X_1 + \cdots + t_n X_n)].$$

In vector and matrix notation, we write this as

$$\psi(\mathbf{t}) = E(e^{\mathbf{t}'\mathbf{X}}),$$

where the prime denotes transpose. Again, setting some arguments t_i equal to zero produces the m.g.f. of the variables corresponding to the remaining arguments.

Moments of a multivariate distribution, marginal and mixed, of various orders, are "generated" as partial derivatives of corresponding orders. For instance, the product moment $E(XY)$ is the second order mixed partial derivative of the m.g.f. evaluated at $s = t = 0$.

EXAMPLE 4.11a Consider a "random sample of size n," to be defined later as a sequence of n i.i.d. observations on a random variable X (the "population"). When studying sample variability we'll be dealing with the deviations $X_i - \bar{X}$ about the average $\bar{X} = \frac{1}{n}\sum X_i$. Here we'll find the joint m.g.f. of these deviations in terms of the population m.g.f., ψ:

$$\psi_{X_1-\bar{X},...,X_n-\bar{X}}(t) = E\left[\exp\left\{\sum_1^n t_i(X_i - \bar{X})\right\}\right].$$

Exploiting the fact that the sum of the deviations of any set of numbers about their mean is 0, we rewrite the sum in the exponent as follows:

$$\sum_1^n t_i(X_i - \bar{X}) = \sum_1^n (t_i - \bar{t})(X_i - \bar{X}) = \sum_1^n (t_i - \bar{t}) X_i.$$

Since the X's are independent, the joint m.g.f. we're after is

$$E\left[\exp\left\{\sum_1^n (t_i - \bar{t}) X_i\right\}\right] = [\psi(t_i - \bar{t})]^n. \blacksquare$$

As in the case of two variables, when random variables $X_1, ..., X_n$ are *independent*, their m.g.f. factors into the product of the univariate m.g.f.'s. For example, if X, Y, Z are independent, then

$$\psi_{X,Y,Z}(s,t,u) = E(e^{sX}e^{tY}e^{uZ}) = \psi_X(s)\psi_Y(t)\psi_Z(u).$$

Conversely, if the joint m.g.f. completely factors as a function of s times a function of t times a function of u, then X, Y, and Z are independent, by the uniqueness property. In this extension of Theorem 28, either or both X and Y can be vector variables.

PROBLEMS

4-73. Obtain a formula for the characteristic function of $Y = a + bX$ in terms of the characteristic function of X.

4-74. Find the characteristic function of a random variable whose p.d.f. is constant on $(0, 1)$.

4-75. Find the characteristic function of the random variable X whose p.d.f. is that of Problem 4-58: $K \exp\{-x^2/2\}$.

4-76. Consider the p.d.f.

$$f(x) = \frac{a/\pi}{(x-m)^2 + a^2}.$$

Find the characteristic function, using the result of Example 4.10b. [Hint: Make the change of variable $u = (x-m)/a$ in the integral that defines ϕ.]

4-77. Let $Y = X_1 + \cdots + X_n$, where the X_i's are i.i.d. Find the characteristic function of the average, $\bar{X} = Y/n$, in terms of the common characteristic function of the X_i.

4-78. Use the preceding result to find the characteristic function of $\frac{1}{n}(X_1 + \cdots + X_n)$ in the particular cases where the common p.d.f. is
(a) e^{-x}, $x > 0$ (as in Example 4.10a).
(b) $K \exp(-x^2/2)$ (as in Problem 4-58).
(c) $[\pi(1+x^2)]^{-1}$ (as in Example 4.10b).

4-79. Find the bivariate p.d.f. of the distribution whose joint m.g.f. is $[(1-s)(1-t)]^{-1}$.

4-80. Given the bivariate m.g.f. $\psi_{X,Y}(s,t) = \exp(-s^2 + st - t^2)$, find the m.g.f.'s of the marginal distributions. (Are X and Y independent?)

4-81. Find the bivariate m.g.f. of the random pair (X, Y) with joint p.d.f. $f(x, y) = e^{-y}$, $0 < x < y$, and from it find the marginal m.g.f.'s.

4-82. Obtain the joint m.g.f. of $U = aX + bY$ and $V = cX + dY$ in terms of the m.g.f. of (X, Y).

5
Limit Theorems

The theory of "large" sample statistical inference depends on certain mathematical results concerning the limiting behavior of sequences of random variables and probability distributions. Here we present some important results together with some proofs and outlines of proofs.[1]

5.1 Convergence

Consider a probability space Ω, an infinite sequence of random variables $\{Y_n(\omega)\}$, and the corresponding sequence of c.d.f.'s, $\{F_n\}$. Random variables and c.d.f.'s are functions, so limits of such sequences are limits of sequences of functions. Convergence of sequences of functions is defined in various ways:

1. *Pointwise convergence*: Functions $\{F_n\}$ converge to F pointwise when, for each x, the sequence of numbers $\{F_n(x)\}$ converges to the number $F(x)$.

2. *Convergence in distribution*: The sequence of random variables $\{Y_n\}$ converges in distribution to Y with c.d.f. F when the sequence of c.d.f.'s, $\{F_n\}$ converges pointwise to F at each continuity point of F.

3. *Convergence in quadratic mean*: The sequence of random variables $\{Y_n\}$ converges to Y in quadratic mean if and only the mean squared difference tends to 0:

$$\lim_{n \to \infty} E[(Y_n - Y)^2] = 0. \qquad (1)$$

---

[1] More complete treatments are found in H. Cramér, *Mathematical Methods of Statistics*, Princeton U. Press, Princeton, N. J., (1946), or W. Feller, *An Introduction to Probability Theory and Its Applications*, Vols. I & II, New York: John Wiley & Sons.

(In mathematics this is also known as L_2-convergence.)

4. *Convergence in probability*: The sequence of random variables $\{Y_n\}$ converges to Y in probability if and only if, for any $\epsilon > 0$,

$$\lim_{n \to \infty} P(|Y_n - Y| \geq \epsilon) = 0. \qquad (2)$$

(This is also referred to as convergence in *measure* or *weak convergence*.) We write $Y_n \xrightarrow{p} Y$ to indicate this type of convergence.

5. *Almost sure convergence*: The sequence of random variables $\{Y_n\}$ converges to Y almost surely if and only if

$$P(\lim_{n \to \infty} Y_n = Y) = 1. \qquad (3)$$

(This type of convergence is also known as *strong convergence* or *convergence with probability 1*.)

THEOREM 1 *If Y_n converges to Y in quadratic mean, then $Y_n \xrightarrow{p} Y$.*

This is a consequence of the Markov inequality [Theorem 15 of §4.6, page 110], with $h(x) = x^2$, $b = \epsilon^2$ (for given $\epsilon > 0$), and $X = |Y_n - Y|$:

$$P(|Y_n - Y| > \epsilon) \leq \frac{1}{\epsilon^2} E[(Y_n - Y)^2].$$

Clearly, if the right side tends to 0, so does the left side.

THEOREM 2 *If Y_n converges to a constant k in distribution, then Y_n converges to k in probability, and conversely.*

To show this, suppose $\epsilon > 0$ is given. It is easily verified that

$$P(|Y_n - k| \geq 2\epsilon) \leq [1 - F_n(k + \epsilon)] + F_n(k - \epsilon).$$

Then, if $Y_n \to k$ in distribution, the right side tends to 0, which implies that the left side tends to zero. Thus, $Y_n \to k$. To see the converse, again let $\epsilon > 0$ be given, and observe that

$$P(|Y_n - k| \geq \epsilon) = [1 - F_n(k + \epsilon)] + F_n(k - \epsilon) + P(Y_n = k + \epsilon).$$

If $Y_n \to k$ in probability, the left side tends to zero, as does then the right side. So, being nonnegative, each term on the right tends to zero. This means that $F_n(x)$ converges to 1 at any $x > k$ and to 0 at any $x < k$. So, except for the discontinuity point at k, the limiting function is the c.d.f. of a singular distribution at k.

THEOREM 3 *If $X_n \xrightarrow{p} k$ (a constant), and if $g(x)$ is continuous at $x = k$, then $g(X_n) \xrightarrow{p} g(k)$.*

Given $\epsilon > 0$, choose δ so that $|x - k| < \delta$ implies $|g(x) - g(k)| < \epsilon$. This can be done because g is continuous at k. Then [Theorem 2, p.27]
$$P(|X_n - k| < \delta) \leq P(|g(X_n) - g(k)| < \epsilon).$$
As $n \to \infty$, the left-hand side tends to 1, as must therefore the right-hand side, which proves the theorem.

THEOREM 4 *Almost sure convergence implies convergence in probability.*

THEOREM 5 (Slutsky) *If the sequence $\{X_n\}$ tends to a in probability, and the sequence of $\{Y_n\}$ tends to Y in distribution, then the sequence of products $\{X_n Y_n\}$ tends to aY in distribution.*

Proof of Theorems 4 and 5 are beyond our scope. (See texts in advanced probability.)

5.2 Law of large numbers

The frequency interpretation of probability is based on a phenomenon that is observed in long sequences of independent trials of an experiment of chance: The relative frequency of occurrence of an event tends to stabilize at a limiting value, and in cases where an event E has *a priori* probability p, this long-run limit is near p. Thus, if Y_n denotes the number of occurrences of an event E in n trials, it is observed that, in some sense,
$$\lim_{n \to \infty} \frac{Y_n}{n} = P(E). \tag{1}$$
This kind of conclusion actually follows from our axioms of probability and definition of independent trials; theorems stating such results are *laws of large numbers*.

Consider a sequence of random variables $X_1, X_2, \ldots$ which are identically distributed, with the property that the variables in any finite subsequence are independent. (We say they are "i.i.d.") The *weak law of large numbers* is the following:

THEOREM 6 (Khintchine) *If the common distribution of the i.i.d. variables X_i has a finite first moment μ, then the sequence of averages*

$$Y_n = \tfrac{1}{n}(X_1 + \cdots + X_n)$$

converges to μ in probability.

Proving that $Y_n \xrightarrow{P} \mu$ is relatively easy when the common distribution has finite *second* moments (see Problem 5-2). We'll give just the essence of the proof in the more general case, which requires complex variable theory for rigorization. [As before, let $o(x)$ denote any function of x with the property that $o(x)/x$ tends to 0 with x.]

LEMMA $\quad \lim\limits_{x \to 0} [1 + ax + o(x)]^{1/x} = e^a$.

Although we need this for complex a, we'll show it only for real a: Let $y = ax + o(x) = x[a + o(1)]$. Then

$$[1 + ax + o(x)]^{1/x} = [(1+y)^{1/y}]^{a + o(1)}$$

$$= [(1+y)^{1/y}]^a \cdot [(1+y)^{1/y}]^{o(1)}.$$

As x tends to 0, so does y, and the first factor on the right tends to e^a. The second factor tends to 1, because its logarithm tends to 0.

Returning to Khintchine's theorem, we'll show that the characteristic function of the average Y_n tends to the characteristic function of the singular distribution at μ. Let ϕ denote the characteristic function of the common distribution of the X_i, which we can write (from Theorem 27 of §4.10, page 131) as

$$\phi(t) = 1 + i\mu t + o(t).$$

The characteristic function of $nY_n = \sum\limits_{1}^{n} X_i$ is $[\phi(t)]^n$, so

$$\phi_{Y_n}(t) = [\phi(t/n)]^n = [1 + i\mu t/n + o(t/n)]^n.$$

As n becomes infinite, $x = t/n$ tends to 0, and application of the above lemma (for complex $a = i\mu$) shows that the characteristic function of Y_n converges to the characteristic function of the distribution which concentrates probability 1 at μ. The continuity theorem for characteristic functions (Theorem 26 of §4.10) then asserts that convergence in distribution follows, and convergence in probability is the result of Theorem 2 of the preceding section. (An alternative demonstration, depending on l'Hospital's rule for complex functions, is suggested in Problem 5-4.)

Convergence of relative frequencies to probabilities is obtained as a special case of Khintchine's theorem: Let $X = 1$ or 0 according as an event E does or does not occur. The p.f. of X is given by $f(0) = P(E^c)$ and $f(1) = P(E) = p$. The sum $X_1 + \cdots + X_n$ is the number of times E occurs in n trials, so the Y_n of Khintchine's theorem is the relative frequency of success—which then converges to $EX = p$. We note that in this case, X has moments of *all* orders, so the proof of convergence does not require the mathematics of complex variable theory—the method of Problem 5-2 is adequate.

Remark: Even though the average Y_n of the first n X's converges to $\mu = EX$, the *sum* is not necessarily close to $n\mu$ for large n. The s.d. of the sum nY_n is $\sqrt{n}\,\sigma$, which can be quite large.

5.3 The central limit theorem

We've seen that Y_n, the arithmetic average of the first n of a sequence of i.i.d. observations tends in distribution to the constant μ as n becomes infinite. We now want to study the *shape* of the distribution, as n becomes infinite, under the assumption that the second moments of the common distribution are finite. For this, we center the distribution at zero and stretch the scale so as to see the shape more clearly, using the standardizing transformation

$$Z_n = \frac{Y_n - \mu}{\sigma/\sqrt{n}} = \frac{nY_n - n\mu}{\sqrt{n}\,\sigma}. \tag{1}$$

The characteristic function of $X_i - \mu$, from Theorem 27 of §4.10, is

$$\phi_{X_i - \mu}(t) = 1 - \frac{\sigma^2 t^2}{2} + o(t^2),$$

since $E(X_i - \mu) = 0$ and $\text{var}(X_i - \mu) = \sigma^2$. Let $U_i = \dfrac{X_i - \mu}{\sqrt{n}\,\sigma}$. Then

$$\phi_{U_i}(t) = \phi_{X_i - \mu}\left(\frac{t}{\sqrt{n}\,\sigma}\right) = 1 - \frac{t^2}{2n} + o\!\left(\frac{t^2}{n}\right).$$

The characteristic function of $Z_n = \sum_1^n U_i$ is the nth power:

$$\phi_{Z_n}(t) = \left[1 - \frac{t^2}{2n} + o\!\left(\frac{t^2}{n}\right)\right]^n.$$

From the lemma of the preceding section, the limit as $n \to \infty$ is $e^{t^2/2}$, the characteristic function of the distribution defined by the density

140   Chapter 5   Limit Theorems

$f(z) \propto e^{-t^2/2}$. [See Problems 4-58 and 4-75.] Again using the continuity theorem for characteristic functions, we find this p.d.f. as that of the limiting distribution of Z_n. This completes an outline of the proof of the following important result:

THEOREM 7 (Central Limit Theorem)   Let $X_1, X_2, \ldots$, be i.i.d. with mean μ and s.d. σ.   Let $S_n = X_1 + X_2 + \cdots X_n$.   Then

$$\lim_{n \to \infty} P\left\{\frac{S_n - n\mu}{\sqrt{n}\,\sigma} \le z\right\} = \Phi(z) = \frac{1}{\sqrt{2\pi}} \int_{-\infty}^{z} e^{-u^2/2} du. \qquad (2)$$

The special case of this theorem in which the common distribution of the X's is a two-point distribution is due to the mathematicians deMoivre [1733, for the case of $f(0) = f(1) = 1/2$] and Laplace [1812, for more general case: $f(1) = p$]. The remarkable feature of the theorem is that the limiting distribution is independent of the nature of the common distribution of the X's! That distribution can be discrete, asymmetric, multimodal, or whatever, provided only that it has a second moment.

The practical import of the central limit theorem lies in the fact that for large but finite n, the c.d.f. of the sum S_n is close to the function Φ defined by (2). Substituting $z = (y - n\mu)/(\sqrt{n}\,\sigma)$ in (2) we obtain this approximation:

$$P(X_1 + \cdots + X_n \le y) \doteq \Phi\left(\frac{y - n\mu}{\sqrt{n}\,\sigma}\right). \qquad (3)$$

For values of Φ (calculated by numerical integrations) see Table I of the Appendix.

What "large" and "close" mean is not easy to specify as a value of n and a degree of closeness. It depends on the nature of the common distribution of the X's. The experience we'll gain as we proceed will afford some guidance in this regard.

EXAMPLE 5.3a   Suppose we can assume that the weights of persons using a certain elevator are independent observations (X_i's) from a population X with a finite second moment. (In practice, possible weights are surely limited, so the second moment will exist.) If the population mean and s.d. are $\mu = 160$ lb and $\sigma = 20$ lb, what is the probability that 18 persons would have a combined weight exceeding the posted load limit of 3000 lb?

Denoting the total weight by W and applying (3), we have

$$P(W > 3000) = 1 - P(X_1 + \cdots X_{18} < 3000)$$

$$= 1 - \Phi\left(\frac{3000 - 2880}{\sqrt{18 \times 20}}\right) = 1 - \Phi(\sqrt{2}) \doteq .079,$$

where $\Phi(\sqrt{2}) \doteq .921$ is found in Table I. ∎

The central limit theorem is so-called because it is central to much of the theory of statistical inference. However, the theorem as we have given it is not the most general theorem of this nature. The conclusion of the theorem is in fact valid under weaker assumptions. For instance, it is not necessary that the terms in the sum be identically distributed, provided that their third moments satisfy a certain mild condition. Also, the assumption of independence can be relaxed somewhat. (See texts in advanced probability theory.)

PROBLEMS

5-1. Show that Y_n converges in quadratic mean to a constant k if and only if both $\lim_{n \to \infty} E(Y_n) = k$ and $\lim_{n \to \infty} \text{var } Y_n = 0$.

5-2. Deduce the weak law of large numbers for the case in which the population variance is finite by showing that the arithmetic average Y_n converges in quadratic mean to the population mean.

5-3. Show that if $X, X_1, X_2, \ldots$ is an i.i.d. sequence with $E(X^k) < \infty$, then $\frac{1}{n}\sum_{1}^{n} X_i^k$ converges in probability to $E(X^k)$.

5-4. Given that the first moment $\mu = -i\phi'(0)$ exists, use l'Hospital's rule to show that $[\phi(t/n)]^n \to e^{i\mu t}$ as $n \to \infty$, assuming that the usual rules of calculus hold for the functions are complex-valued. [Hint: Take the limit of the logarithm divided by t.]

5-5. Show that if $X_n \xrightarrow{p} a$ and $Y_n \xrightarrow{p} b$, then $X_n + Y_n \xrightarrow{p} a + b$. [Hint: Use the "triangle inequality": $|U + V| \leq |U| + |V|$ with $U = X_n - a$ and $V = Y_n - b$.]

5-6. Booklets are packaged in bundles of 100 by weighing them. Suppose that the weight of each booklet is a random variable with mean 1 oz and standard deviation .02 oz. Find the probability that a bundle of exactly 100 booklets would be counted as containing 101 booklets or more—that is, that the bundle weighs more than 100.5 oz.

5-7. The weight of a certain food package is a random variable with $\mu = 4.0$ oz and $\sigma = 0.3$ oz. Find the probability that a carton of 24 packages weighs less than 94 oz.

5-8. An elevator is designed for a maximum capacity of 20 persons and a maximum weight of 3,000 lb. Suppose that the weights of those who ride this elevator have a standard deviation of 20 lb. Find the mean weight that would result in a probability of .05 that 20 persons would exceed the 3,000 lb. limit.

5-9. Suppose we add n real numbers, each rounded to the nearest integer, where the round-off error is a random variable uniformly distributed on the interval $(-.5, .5)$. Find the probability that the error in the sum is bounded by $\sqrt{n}/2$ in magnitude when n is large, assuming independent errors.

5-10. Show (as asserted on page 138) that $[(1+y)^{1/y}]^{o(1)}$ tends to 1 as y tends to 0. [Here "$o(1)$" denotes a function that tends to 0.]

5-11. Assuming finite second moments, so that $\phi''(0) = -\sigma^2$, and assuming that the rules of calculus apply (as in Problem 5-4), show:

$$\left\{\phi\left(\frac{t}{\sqrt{n}\,\sigma}\right)\right\}^n \to e^{-t^2/2} \text{ as } n \to \infty.$$

(Cf. the proof of the central limit theorem.)

6
Some Parametric Families

In the preceding chapters we have encountered distributions that can be thought of as special cases of general, important classes of models in which the individual members of a class are indexed by one or more real parameters. In this chapter we'll study several such classes or families in some detail, using the various tools we now have at our disposal. Some are classes of discrete distributions and others, continuous. Some involve a single parameter and others, two or three parameters. In later chapters we'll encounter other families of distributions, ones that we'll use in analyzing the processes of inference. The families to be considered in this chapter are rather basic, often used as population models, representing variables we observe and measure.

Like mathematical models generally, the various distributions we study are idealizations, involving assumptions that cannot be verified with certainty. Indeed, one goal of statistical inference is to be able to conclude, with some (but limited) assurance, that a model describes "reality" closely enough for practical purposes.

6.1 Bernoulli trials

A *Bernoulli experiment* is an experiment of chance with just two possible outcomes: Either an event A happens or it does not. For example, a tossed coin falls either heads (A) or tails (A^c); an individual drawn at random is either male (A) or female (A^c); an experiment either succeeds (A) or fails (A^c); a product is good (A) or defective (A^c); an inoculation takes (A) or does not take (A^c); and so on. Because of the variety of situations in which such phenomena occur, we find it convenient to refer to A simply as "success," and to A^c as "failure."

The indicator function of A defines a **Bernoulli random variable**:

Chapter 6 Some Parametric Families

$$X(\omega) = \begin{cases} 1 & \text{if } \omega \in A, \\ 0 & \text{if } \omega \in A^c. \end{cases} \tag{1}$$

Its distribution is called a *Bernoulli distribution*. The p.f. has just two values: $f(1) = P(A) = p$, $f(0) = 1 - p$, so the distribution is characterized by the single parameter $p = P(A)$. It is common practice to denote $1 - p$ by q, so that $p + q = 1$. The p.f. can be given by the single formula

$$f(x \mid p) = p^x q^{1-x}, \quad x = 0, 1. \tag{2}$$

When a variable X has a Bernoulli distribution with parameter p, we'll write $X \sim \text{Ber}(p)$, or "X is $\text{Ber}(p)$."

EXAMPLE 6.1a A ball is selected at random from a container with 4 black and 5 red balls. The outcome is either *red* or *black*; introducing the Bernoulli coding $red = 1$ and $black = 0$ defines a random variable X which is $\text{Ber}(p)$, where $p = P(\text{red}) = 5/9$, and $q = P(\text{black}) = 4/9$.

If a second ball is selected without replacement of the first, we can define the Bernoulli variable Y, defined as 1 or 0 according as the second ball is red or black. We have seen in earlier examples that X and Y are exchangeable, so they are identically distributed. Of course, they are not independent. However, if the second ball had been selected only after replacement and mixing, the indicator variables for the first and second selections would be i.i.d. ∎

With the coding adopted in (1), we have defined X as a *numerical* variable, one that has moments. In particular, the mean is

$$EX = 1 \cdot p + 0 \cdot q = p. \tag{3}$$

And because $X = X^k$ for both of the possible values of X, the kth moment is $E(X^k) = EX = p$. So the variance of a Bernoulli variable is

$$\text{var } X = p - p^2 = p(1 - p) = pq. \tag{4}$$

The factorial moment generating function of a Bernoulli distribution is

$$\eta(t) = E(t^X) = t^0 q + t^1 p = pt + q. \tag{5}$$

Sampling a Bernoulli population at random with replacement and mixing between selections generates a sequence of independent, Bernoulli trials. We'll refer to such a sequence as a **Bernoulli process**. (A more standard use of this term refers to the sequence of partial sums of the sequences of 0's and 1's.) Such a sequence can also arise in observing repeated trials of an experiment with two possible outcomes where there is not a tangible "population" that is being sampled. For example, in repeated spins of a pointer with a scale from 0 to 1 around the circle traced by its tip, it may be that the model of a uniform p.d.f. on (0, 1) describes each single spin, and that successive spins can be considered independent—if, for instance, at each trial the pointer turns many revolutions before coming to rest. If we define $X = 0$ or 1 according as the outcome of a spin is greater or less than .3, then $X \sim \text{Ber}(.3)$, and the successive spins generate a Bernoulli process.

Some important characteristics of a Bernoulli process should be noted. Because the trials are i.i.d., the outlook for the future, as seen from some point in the process, is the same for all such vantage points—and no matter what may have happened up to that point. Indeed, looking forward *or* looking backward in a Bernoulli sequence, from any point, one sees the same random process. Moreover, the probability characteristics of any finite portion of the sequence of specified length are independent of where in the sequence that portion begins.

In the next two sections we study the distribution of the number of successes in a fixed number of independent, identical Bernoulli trials, and the distribution of the number of trials needed to obtain a specified number of successes. Then in §6.4 we study the distribution of the number of successes when the sampling is without replacement from a finite Bernoulli population, in which case the trials are identically distributed but not independent.

6.2 The binomial distribution

Let S_n denote the number of "successes" in a sequence of n independent trials of a Bernoulli experiment with parameter p:

$$S_n = X_1 + \cdots X_n, \tag{1}$$

where $X_i \sim \text{Ber}(p)$. Each X is 1 or 0 according as the experiment results in "success" or a "failure," so the sum S_n, the sum of 0's and 1's, is the number of 1's—the number of successes in the n trials. Since the factorial m.g.f. of each X_i [from (5) of the preceding section] is $\eta(t) = pt + q$, it follows that the f.m.g.f. of S_n is

$$\eta_{S_n}(t) = (pt + q)^n = \sum_{i=0}^{n} \binom{n}{i}(pt)^i q^{n-i} = \sum_{i=0}^{n} \left\{ \binom{n}{i} p^i q^{n-i} \right\} t^i. \tag{2}$$

The coefficient of t^k in this (finite) power series expansion of the f.m.f.g. is the probability that S_n has the value k. This defines the probability function for the **binomial distribution**, given by the *binomial formula*:

$$P(S_n = k) = \binom{n}{k} p^k q^{n-k}. \tag{3}$$

A random variable with this p.f. is a *binomial random variable*, and we write $S_n \sim \text{Bin}(n, p)$. Substituting $t = 1$ in (2) shows that the probabilities (3) sum to 1, as they must.

The binomial formula (3) can be reasoned somewhat more directly: The probability of any particular sequence of outcomes with k successes and $n - k$ failures is $p^k q^{n-k}$. The probability of exactly k successes is then the sum of these quantities over all sequences with that many successes, which is the number of such sequences, $\binom{n}{k}$, multiplied by the probability of each. And this produces the binomial formula (3).

The mean and variance of a binomial variable can be obtained from the f.m.g.f., but it is even easier to get them from its structure as the sum of i.i.d. Bernoulli variables. Using (3) and (4) of the preceding section we find

$$E(S_n) = E(X_1 + \cdots + X_n) = EX_1 + \cdots + EX_n = np, \tag{4}$$

and

$$\text{var } S_n = \text{var}(X_1 + \cdots + X_n) = \text{var } X_1 + \cdots + \text{var } X_n = npq. \tag{5}$$

EXAMPLE 6.2a Suppose a treatment is "successful" 90 percent of the time. What is the probability that it will be successful in the cases of at least eight of the next ten patients?

We can calculate this if we make enough assumptions: Suppose the results on those ten patients are independent, and that the probability of success is .90 for each one. Then Y, the number of patients for which it is successful is binomial with parameter $p = .90$. The probability of at least eight successes is the probability of 8 or 9 or 10 successes:

$$P(Y = 8, 9, 10) = \binom{10}{8}.9^8 .1^2 + \binom{10}{9}.9^9 .1^1 + \binom{10}{10}.9^{10}.1^0 \doteq .93.$$

The mean and variance, from (4) and (5), are $E(Y) = np = 10 \times .9 = 9$ and $\sigma_Y = \sqrt{npq} \doteq .95$. ∎

The binomial distribution has an important reproductive property: If $X \sim \text{Bin}(m, p)$, it can be thought of as the number of successes in m independent trials; and if $Y \sim \text{Bin}(n, p)$, it can be thought of as the number of successes in n trials. If X and Y are independent, the sum can be interpreted as the number of successes in $m + n$ independent trials—the n following the m, say. Then $X + Y \sim \text{Bin}(m + n, p)$. By induction we obtain the following:

THEOREM 1 *If the random variables* $Y_1, \ldots, Y_r$ *are independent, and* $Y_i \sim \text{Bin}(n_i, p)$, *then*

$$Y_1 + \cdots + Y_r \sim \text{Bin}(n_1 + \cdots + n_r, p).$$

(A proof using generating functions is left as an exercise—Problem 6-8.)

For inferences about Bernoulli populations, we'll be needing to evaluate cumulative binomial probabilities for large n. Because this can be quite tedious, an approximation based on the central limit theorem is very useful. We have seen that when $Y \sim \text{Bin}(n, p)$, it can be expressed as the sum of n independent Bernoulli variables X_i. Then, from (3) of §5.3 (page 140) we obtain

$$P(Y \leq y) \doteq \Phi\left(\frac{y - np}{\sqrt{npq}}\right). \tag{6}$$

EXAMPLE 6.2b Suppose $Y \sim \text{Bin}(8, .5)$. Then $E(Y) = 4$, $\sigma_Y^2 = 2$, and the c.d.f. is

$$P(Y \leq y) = \sum_{k=0}^{y} \binom{8}{k} (.5)^k (.5)^{n-k}.$$

The approximating normal c.d.f. is

$$P(Y \leq y) \doteq \Phi\left(\frac{y - 4}{\sqrt{2}}\right).$$

The graphs of these two c.d.f.'s are shown in Figure 6-1. Probabilities for $y = 5$ are marked on the figure, showing that the normal approximation can be appreciably in error for this small an n. ∎

Figure 6-1 shows that for $n = 8$ the normal approximation (6) is not very good (at the integer values of Y), but it also suggests a way of improving the approximation. The normal c.d.f. nearly bisects the stair treads of the step function, and at an integer k, the normal c.d.f. is closer to the desired $F(k)$ at $x = k + .5$:

$$F(k) = P(Y \leq k) \doteq \Phi\left(\frac{k + .5 - np}{\sqrt{npq}}\right). \tag{7}$$

With the extra .5, a *continuity correction,* the approximation is usually much improved. It gives reasonable accuracy even when $n = 8$, provided p is neither close to 0 nor close to 1, in which cases the binomial distribution is quite skewed. For a given p, how large n has to be depends on the desired accuracy. Various rules of thumb are used: One rule says that npq should be at least 5 or 10; another says that n should be at least 10 times q/p or p/q, whichever is larger.

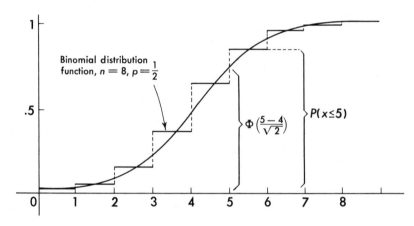

Figure 6-1 Binomial and approximating normal c.d.f.'s

EXAMPLE 6.2c For an example of the degree of accuracy of a normal approximation consider $Y \sim \text{Bin}(20, .15)$. Exact and approximate values of the c.d.f. are given in the following table, for $k \leq 6$.

k	$F(k)$	$\Phi\left(\dfrac{k + .5 - 3}{\sqrt{2.55}}\right)$
0	.039	.059
1	.176	.174
2	.405	.377
3	.648	.623
4	.830	.826
5	.933	.941
6	.978	.970

The approximation isn't all that bad, even though both rules of thumb are violated: $npq = 2.55$, and $q/p = 17/3$. ∎

When n is large, but not large enough that, say, npq is at least 5, either p or q is small, and the distribution is skewed. An approximation for this case will be taken up in §6.5.

PROBLEMS

6-1. For any random variable X, consider the event $A = [X > 1]$. Give

6-2. A person with no clairvoyant powers attempts to predict the fall of a coin in each of several successive trials. Find the probability of
 (a) four correct predictions in four trials.
 (b) at most eight correct predictions in ten trials.
 (c) at least three correct predictions in twelve trials.
 (d) three correct predictions in the next three trials, given that the three preceding predictions were correct.

6-3. The reliability of a device is defined as the probability that it functions correctly. Suppose the reliability of a certain cut-off device is 11/12. Find the reliability of a system consisting of four such cut-off devices so arranged that any one of them operating properly would provide the cut-off function.

6-4. Find $P(X = 7)$, given $X \sim \text{Bin}(n, p)$, with $EX = 3$ and var $X = 2$.

6-5. Fifty Cadillac owners are each given blindfolded rides in a Cadillac and in a Lincoln, to test the "ride." Each person is required to state a preference for one or the other. Suppose the two cars "ride" equally well.
 (a) What is the distribution of the number of subjects who state a preference for the Lincoln?
 (b) Find the mean and s.d. of the distribution in (a).
 (c) Find the probability (using a hand calculator with a factorial key and a y^x key) that exactly 25 of the subjects state a preference for the Lincoln.

6-6. Given that 5 percent of the articles produced in a certain process are defective, find
 (a) the mean and standard deviation of X, the number of defectives in a lot of 100 articles.
 (b) the probability of at most three defectives among 50 articles.
[Assume that successively produced articles are independent Bernoulli trials, in which $p = P(\text{defective}) = .05$.]

6-7. Derive a formula for the factorial moments of a binomial distribution from the f.m.g.f. and use the result to find the mean and variance.

6-8. Use generating functions to show that the sum of finitely many independent, binomial variables with the same p is binomial with that p.

6-9. Evaluate each of the following expressions:

(a) $\sum_{k=0}^{10} k \binom{10}{10-k}(.5)^{10}$.

(b) $\sum_{k=0}^{20} k^2 \binom{20}{k}(.3)^k(.7)^{20-k}$.

150 Chapter 6 Some Parametric Families

6-10. Let S_n denote the sum of the first n of a sequence of i.i.d. Bernoulli variables, $0 < p < 1$. Show that S_n converges in probability to the common probability p of success. Apply this result to show that in a sequence of independent observations on a random variable X, the relative frequency of occurrences of the event $[X \leq x]$ converges to $F_X(x)$, for each x.

6-11. A coin is tossed 100 times. Find an approximate probability that the number of heads in the 100 tosses is
(a) less than 50.
(b) exactly 50.
(c) at least 40 but at most 60.

6-12. Suppose $X \sim \text{Bin}(10, .1)$.
(a) Draw a graph of the c.d.f. of X and superimpose a graph of the approximating normal c.d.f.
(b) Compare exact binomial probabilities with normal approximations.

6-13. Find the probability that at least 28 of 72 rolls of a pair of dice result in a total of 5 or fewer.

6.3 Hypergeometric distributions

Consider a finite population in which individual members are of one of two types, and code these 1 ("success") and 0 ("failure"). When sampling is done at random with replacement and mixing between selections, a sequence of trials is an i.i.d. sequence in which the common distribution is Bernoulli. When sampling is done at random *without* replacement, the individual selections are again Bernoulli trials, as we explain in the next paragraph; but they are not independent.

Denote the population size by N and the number of type 1 individuals by M. Consider a sequence of n random selections (made without replacement): $X_1, X_2, ..., X_n$. Each of these variables is $\text{Ber}(p)$, where $p = M/N$ and $q = 1 - p$. This fact is obvious for X_1; and for the subsequent trials it is a special case of a general result mentioned at the end of §3.8 (page 86). The trials are exchangeable, so X_i has the same distribution as X_1. [Also, that $X_i \sim \text{Ber}(p)$ was demonstrated in the present context following Example 2.7c (page 36).]

As in the case of independent trials, we are again interested in

$$S_n = \sum_{i=1}^{n} X_i, \tag{1}$$

the number of individuals of type 1 in this sample of size n. Just before

Example 2.7c we showed that with sampling as described, the various possible combinations of n individuals are equally likely. So to find the probability of a specified number of "successes" we need only count the number of combinations with that many successes, and divide by the total number of possible combinations:

$$f_{S_n}(k \mid N, M, n) = P(S_n = k) = \frac{\binom{M}{k}\binom{N-M}{n-k}}{\binom{N}{n}}, \quad k = 0, 1, 2, ..., n. \quad (2)$$

This is the **hypergeometric** p.f. When X has the distribution defined by (2), we may write $X \sim \mathrm{Hyper}(N, M, n)$.

In applying (2) it is to be understood that $\binom{n}{r} = 0$ when $r > n$. It will then give the correct probability, for example, of the event $S_n = 5$, when $M = 3$: If only three type 1 individuals are in the population you can't get five in the sample, so the probability is zero.

We obtain the first two moments of S_n by exploiting its structure as given by (1). For the first moment, we find

$$E(S_n) = E\left(\sum_{i=1}^{n} X_i\right) = \sum_{i=1}^{n} E(X_i) = np. \quad \text{Mean} \quad (3)$$

This is the same formula as for sampling *with* replacement, since expectations are additive whether or not summands are independent. To get the variance we use Corollary 1 of Theorem 20 in §4.7 (page 114): Variance

$$\mathrm{var}\, S_n = \mathrm{var}\left(\sum_{i=1}^{n} X_i\right) = \sum_{i=1}^{n} \mathrm{var}\, X_i + \sum_{i \neq j} \mathrm{cov}(X_i, X_j). \quad (4)$$

The X's are exchangeable, so $\mathrm{var}\, X_i = \mathrm{var}\, X_1$. Moreover, the joint distribution of any two X's is the same as that of the first two, so that (in particular) $\mathrm{cov}(X_i, X_j) = \mathrm{cov}(X_1, X_2)$. Since there are $n(n-1)$ pairs (i, j) in which $i \neq j$, we can rewrite (4) as

$$\mathrm{var}\, S_n = n\, \mathrm{var}\, X_1 + n(n-1)\, \mathrm{cov}(X_1, X_2). \quad (5)$$

In the first term on the right, $\mathrm{var}\, X_1 = pq$; for the second term we need the covariance. We can find this most easily by realizing that when $n = N$, the number of "successes" in the sample is always the same—no variability:

$$\mathrm{var}\, S_N = 0 = N\, \mathrm{var}\, X_1 + N(N-1)\, \mathrm{cov}(X_1, X_2).$$

Solving for the covariance and substituting in (4) we obtain

152   Chapter 6   Some Parametric Families

$$\text{var } S_n = npq \cdot \frac{N-n}{N-1}. \tag{6}$$

The factor multiplying npq in (6) is a "finite population correction factor." Its value is 1 when $N = n$ and tends to 1 as N becomes infinite. (When $N \gg n$, it is close to 1.)

The p.f. (2) does not combine easily with the e^{tk} or t^k in the sum that defines a generating function, but we can find the factorial moments directly. It is convenient to allow indices to range from 0 to N, with the understanding (as before) that $\binom{a}{b} = 0$ when $a < 0$ or when $b > a > 0$:

$$E[(S_n)_r] = \sum_{k=r}^{N} (k)_r f(k \mid N, M, n) = \sum_{k=r}^{N} (k)_r \frac{\binom{M}{k}\binom{N-M}{n-k}}{\binom{N}{n}}.$$

Using the identity

$$(k)_r \binom{M}{k} = (M)_r \binom{M-r}{k-r}$$

we obtain

$$\sum_{k=r}^{N} (k)_r \binom{M}{k}\binom{N-M}{n-k} = \sum_{k=r}^{N} (M)_r \binom{M-r}{k-r}\binom{N-M}{n-k}$$

$$= (M)_r \sum_{j=0}^{N} \binom{M-r}{j}\binom{(N-r)-(M-r)}{(n-r)-j}$$

$$= (M)_r \binom{N-r}{n-r} \sum_{j=0}^{N} f(j \mid N-r, M-r, n-r)$$

$$= (M)_r \binom{N-r}{n-r} = \frac{(M)_r (n)_r}{(N)_r} \binom{N}{n}.$$

So, for $r \leq N$, the rth factorial moment of a hypergeometric variable is

$$E[(S_n)_r] = \frac{(M)_r (n)_r}{(N)_r}. \tag{7}$$

In particular, $r = 1$ yields $EX = np$, as found in (3).

EXAMPLE 6.3a   Let X denote the number of defective articles in a random selection of three from a lot which includes two defective and eight good articles. The p.f. of X is

$$f(k \mid 10, 2, 3) = \frac{\binom{2}{k}\binom{8}{3-k}}{\binom{10}{3}}, \quad k = 0, 1, 2.$$

These probabilities are 7/15, 7/15, 1/15 for 0, 1, 2, respectively, so

$$EX = 0 \cdot \frac{7}{15} + 1 \cdot \frac{7}{15} + 2 \cdot \frac{1}{15} = \frac{3}{5} = np,$$

and

$$EX^2 = 0^2 \cdot \frac{7}{15} + 1^2 \cdot \frac{7}{15} + 2^2 \cdot \frac{1}{15} = \frac{11}{15}.$$

Then

$$\text{var } X = E(X^2) - (EX)^2 = \frac{11}{15} - \frac{9}{25} = \frac{28}{75}$$

$$= 3 \times .2 \times .8 \times \frac{10-3}{10-1} = npq\frac{N-n}{N-1}. \blacksquare$$

We pointed out that the hypergeometric variance (6) tends to the binomial variance npq as the population size N becomes infinite, with the sample size n fixed. We can say more: The hypergeometric p.f. (2) tends to the binomial p.f. as $N \to \infty$ with $M/N = p$ (fixed). The reason for this is that when N is large compared to the sample size n, the ratio of 1's and 0's is practically the same after as before some individuals are removed. Formally,

$$\frac{\binom{M}{k}\binom{N-M}{n-k}}{\binom{N}{n}} = \frac{(M)_k}{k!} \cdot \frac{(N-M)_{n-k}}{(n-k)!} \cdot \frac{n!}{(N)_k} = \binom{n}{k} \cdot \frac{(M)_k(N-M)_{n-k}}{(N)_n}$$

$$= \binom{n}{k} \cdot \frac{M}{N} \cdot \frac{M-1}{N-1} \cdots \frac{M-k+1}{N-k+1} \cdot \frac{N-M}{N-k} \cdots \frac{N-M-n+k+1}{N-n+1}$$

$$\to \binom{n}{k} \cdot p \cdot p \cdots p \cdot q \cdots q = \binom{n}{k} p^k q^{n-k}, \quad \text{as } N \to \infty.$$

As is typical of limit theorems, the practical importance of this result lies in the fact that for large N, a hypergeometric probability can be approximated by an easier-to-calculate binomial probability. A rule of thumb is not easy to give, partly because it depends on how close an approximation is required. But as a very rough guide, the population size should perhaps be at least 10 or 20 times the sample size. The following example gives some hint of the kind of accuracy that is typical.

EXAMPLE 6.3b Consider the case $n = 5$ and $p = .5$. The following table gives the hypergeometric p.f. for $k = 0$ to 5, for several population sizes, as well as the binomial approximation, corresponding to $N = \infty$. The latter is fairly close even for $N = 100$.

k	N = 50	N = 100	N = 200	N = ∞
0	.0003	.0006	.0008	.0010
1	.0050	.0072	.0085	.0098
2	.0316	.0380	.0410	.0439
3	.1076	.1131	.1153	.1172
4	.2181	.2114	.2082	.2051
5	.2748	.2539	.2525	.2461 ■

6.4 Negative binomial distributions

Suppose, instead of sampling a Bernoulli experiment a fixed, preset number of times, we sample only until a specified number of successes are obtained. The required number of trials is then random, and it is this random variable upon which we now focus.

Consider first the case of sampling until a success is *first* observed. Let X denote the number of trials required to achieve this first success. The possible values of X are 1, 2, 3, ... The event $X = k$ is observed if and only if the sequence of results consists of $k - 1$ failures followed by a success:

$$f_X(k \mid p) = P(X = k) = q^{k-1}p, \quad k = 1, 2, 3, \ldots \qquad (1)$$

This p.f. defines the family of **geometric** distributions, and when X has the p.f. (1) we write $X \sim \text{Geo}(p)$. The factorial m.g.f. of a geometric distribution is

$$\eta_X(t) = E(t^X) = \sum_{k=1}^{\infty} t^k q^{k-1} p = \frac{pt}{1 - qt}. \qquad (2)$$

From this we can find the first factorial moment of the distribution, which is the mean: $E(X) = \eta'(1) = 1/p$. This result is easy to remember and rather intuitive: For instance, if $p = 1/7$, there is one chance in seven of a success, and on average it takes seven trials to get a success.

To find the variance we differentiate the f.m.g.f. once more:

$$\eta_Y''(t) = 2pq(1 - qt)^{-3},$$

so the second factorial moment is

$$E[X(X - 1)] = \eta_Y''(1) = \frac{2q}{p^2}.$$

From this we find

$$\operatorname{var} X = E[X(X-1)] + EX - (EX)^2$$
$$= \frac{2q}{p^2} + \frac{1}{p} - \left(\frac{1}{p}\right)^2 = \frac{q}{p^2}. \qquad (3)$$

Sometimes it is more convenient to focus on the number of failures prior to the first success: $Y = X - 1$. The p.f. is

$$f_Y(k) = q^k p, \quad k = 0, 1, 2, \ldots \qquad (4)$$

The distribution of Y could also be called geometric (and sometimes is), but to avoid confusion we'll use the term "geometric" and the abbreviation Geo(p) only to refer to the distribution with p.f. (1). The mean value of Y is

$$EY = EX - 1 = \frac{1}{p} - 1 = \frac{q}{p}. \qquad (5)$$

Of course, since $Y = X - 1$, $\operatorname{var} Y = \operatorname{var} X = q/p^2$.

EXAMPLE 6.4a One approach to testing the validity of an assumption is to make the assumption and see how its various consequences compare with what one observes. Suppose a basketball player misses 20 percent of his free-throws. We interpret this to mean that the probability of a miss is .20. If we assume that successive free-throws are independent, how many tries will result in a basket before a miss?

The number of tries is a random variable whose distribution is Geo(.2). (Here we interpret the "success" of the above discussion to be a *miss*.) The average number of tries to a miss, including the one on which the basket is missed, is $1/p = 5$. The average number of baskets made before the first miss is $q/p = .8/.2 = 4$. ∎

Next let U denote the number of trials required to obtain the rth success, starting from any point in a Bernoulli process. To find the p.f. we observe that the event $[U = k]$ occurs if and only if there are $r - 1$ successes in the first $k - 1$ trials followed by a success:

$$f_U(k \mid p) = P(U = k \mid p)$$
$$= P(r - 1 \text{ successes in } k - 1 \text{ trials}) \cdot P(\text{success})$$
$$= \binom{k-1}{r-1} p^{r-1} (1-p)^{k-r} p.$$

Thus,

$$f_U(k \mid p) = \binom{k-1}{r-1} p^r (1-p)^{k-r}, \quad k = r, r+1, r+2, \ldots \tag{6}$$

This is the p.f. of a **negative binomial** distribution, and we may write $U \sim \text{Negbin}(r, p)$. [The slight ambiguity of "geometric" carries over—we could have defined the negative binomial as the distribution of $U - r$, the number of failures prior to the rth success.]

The mean and variance of U are easy to find from the fact that U is the sum of the number of trials needed to get the first success plus the number needed to get the second, and so on: $U = X_1 + \cdots + X_r$, where each X_i is $\text{Geo}(p)$. So,

$$E(U) = r/p, \quad \text{var } U = rq/p^2. \tag{7}$$

We could have derived the p.f. (6) using the factorial m.g.f., which in this case is also a probability generating function. The f.m.g.f. of X_i, given by (2), is

$$\eta_{Y_i}(t) = pt(1 - qt)^{-1}, \quad i = 1, 2, \ldots, r.$$

The f.m.g.f. of U is then the rth power,

$$\eta_U(t) = [pt(1-qt)^{-1}]^r = p^r t^r (1-qt)^{-r}. \tag{8}$$

As we saw in Example 4.9e, the power series for a negative power of a binomial is developed just as it is for a positive power:

$$(1 - qt)^{-r} = 1^{-r} + (-r) 1^{-r-1}(-qt)^1$$

$$+ \frac{(-r)(-r-1)}{2} 1^{-r-2}(-qt)^2 + \cdots = \sum_{j=0}^{\infty} \binom{-r}{j}(-qt)^j,$$

where

$$\binom{-r}{j} = \frac{(-r)(-r-1)\cdots(-r-j+1)}{j!}$$

$$= (-1)^j \cdot \frac{(r+j-1)\cdots(r+1)r}{j!} = (-1)^j \binom{r+j-1}{j}.$$

So we can write the power series for the f.m.g.f. of U as

$$\eta_U(t) = \sum_{j=0}^{\infty} \left\{ \binom{r+j-1}{j} p^r q^j \right\} t^{r+j},$$

a convergent series for $|t| < 1$. And the coefficient of t^k is $P(U = k)$;

this is the quantity in braces with $j = k - r$:

$$f_U(k \mid p) = \binom{k-1}{k-r} p^r q^{k-r}, \quad k = r,\ r+1,\ r+2, ...,$$

which agrees with (6), since $\binom{k-1}{k-r} = \binom{k-1}{r-1}$.

EXAMPLE 6.4b Example 6.4a dealt with a Bernoulli process—a sequence of free-throw attempts where at each try the probability of a basket is $p = .8$. The number of tries it takes to get the first basket, counted from any starting point in the sequence, is Geo(.8). The number of tries required to get the *fourth* basket is Negbin(4, .8), with p.f.

$$f(k \mid .8, 4) = \binom{k-1}{3} .8^4 . 2^{k-4}, \quad k = 4,\ 5,\ 6, ...$$

The probability that it takes at least six tries (say) to get that fourth basket (or that it takes at least two misses before the fourth basket) is 1 minus the probability that it takes at most five tries:

$$1 - (.8)^3 \left\{ \binom{3}{3}(.8) + \binom{4}{3} .2 \right\} \doteq .181.$$

The mean number of tries required to obtain four baskets is

$$EU = \frac{r}{p} = \frac{4}{.8} = 5.$$

The s.d. of the number required [from (7)] is

$$\sqrt{\operatorname{var} U} = \frac{\sqrt{rq}}{p} = \frac{\sqrt{4 \times .2}}{.8} \doteq 1.12. \blacksquare$$

If sampling is from a finite population *without* replacement, the number of individual selections it takes to obtain a given number of successes is not negative binomial. Some call it "negative hypergeometric." The following example shows how to find the p.f. in a particular case.

EXAMPLE 6.4c Suppose we draw items one at a time, at random and without replacement, from a lot of 20 items, 5 of which are defective. The *first* defective is obtained as the kth selection if the first $k-1$ selections are all good items *and* the kth is defective:

$$P(X_1 = k) = \frac{\binom{15}{k-1}}{\binom{20}{k-1}} \cdot \frac{5}{20 - (k-1)}, \quad k = 1,\ 2,\ ...,\ 16. \blacksquare$$

158 Chapter 6 Some Parametric Families

More generally, we can find the p.f. of U, the number of the selection on which the rth item of a given type is located, patterning the reasoning in the case of independent trials. Thus, for a population of size N which includes M of type 1 and $N - M$ of type 0, and sampling at random without replacement, the p.f. of the number of selections it takes to get r of type 1, for $k = r, r+1, \ldots, N - M + r$, is

$$P(U = k) = \frac{\binom{M}{r-1}\binom{N-M}{k-r}}{\binom{N}{k-1}} \cdot \frac{M-(r-1)}{N-(k-1)}. \quad \blacksquare$$

PROBLEMS

6-14. Let X denote the number of white chips among five chips drawn at random from a bowl containing four white and six red chips.
(a) Find the p.f. of X.
(b) Calculate the mean and variance using the p.f. and check these against the formulas [(3) and (6) in §6.3].

6-15. Five articles are drawn at random, without replacement, from a lot of 100 articles, of which ten are defective. Find the (exact) probability that there are two defectives among the five and compare this with what the answer would be if the selections had been made with replacement and mixing.

6-16. In the game of bridge, a "hand" is a selection of 13 cards from the 52 cards in the deck. Find
(a) the expected number of face cards in a hand.
(b) the expected number of aces in a hand.
[The deck includes four aces and 12 face cards.]

6-17. Suppose 45 percent of a population of 10,000 voters favor a particular proposition. Find the probability that among 10 voters chosen at random (no replacement) six or more favor the proposition.

6-18. Twenty cups of coffee are brewed, 15 in the usual way and 5 from instant coffee. After tasting all 20 cups, a taster selects five and declares them to the ones made from instant coffee. Find the probability that if the taster's selection is random, exactly k of those selected are made from instant.

6-19. Evaluate: $\sum_{1}^{4} k \binom{4}{k}\binom{8}{6-k}$, without resorting to the "brute force" method of evaluating each term in the sum.

6-20. Three balls are selected at random from an urn containing three

6.4 Negative binomial distributions

white, five red, and two black balls. Let X denote the number of red, and Y the number of white balls among those selected and find
(a) $P(X = 1, Y = 1)$.
(b) the marginal distributions of X and Y.
(c) $E(X + Y)$.

6-21. Given that 55 percent of the voters in a city of 50,000 voters will vote for candidate A, find the probability that less than a majority of a random selection of 100 voters vote for A.

6-22. Calculate the variance of a hypergeometric distribution [(6) in §6.3],
(a) using (4) of §6.3 and $\text{cov}(X_1, X_2)$ from Problem 4-43.
(b) using the factorial moments, (7) of §6.3.

6-23. Given independent variables X_1 and X_2, where $X_1 \sim \text{Bin}(n_1, ,p)$ and $X_2 \sim \text{Bin}(n_2, ,p)$, find the conditional distribution of X_1 given $X_1 + X_2 = m$, where $0 \leq m \leq n_1 + n_2$.

6-24. A pair of ordinary, six-sided (fair) dice is rolled in a sequence of independent trials.
(a) Find the probability that it takes at least five rolls to get a double.
(b) Find the mean number of rolls required to get a double.
(c) Find the mean number of rolls required to show a 7 or an 11.

6-25. People with a particular blood type constitute five percent of a certain population. Assume that the population is large enough that successive selections of a person are essentially i.i.d. trials.
(a) Find the expected number of tests necessary to locate three people with that blood type.
(b) Find the probability that it takes at least eight tests to locate two people with that blood type.

6-26. Find the most probable value(s) of the random variable X, defined as the number of failures prior to the fifth success in a sequence of i.i.d. Bernoulli trials with probability success $p = .5$.

6-27. Evaluate each of the following sums:

(a) $\sum_{0}^{\infty} \binom{4+k}{k} 2^{-k}$.

(b) $\sum_{0}^{\infty} k \binom{5+k}{k} (.3)^6 (.7)^k$.

6-28. Derive an expression for the kth factorial moment of a geometric distribution by calculating the kth derivative of the f.m.g.f.

6-29. Use generating functions to show that the sum of finitely many independent, negative binomial variables with the same p is again negative binomial. Interpret the result in terms of a sequence of

Bernoulli trials.

6.5 The Poisson process

An experiment that unfolds with time, with an outcome which at each instant t is a random variable X_t, is called a *random process* or *stochastic process*. When t is discrete, as in the case of a Bernoulli process (§6.1), the process is termed a *time series*. In this section we define a random process which may be thought of as a continuous-time version or analog of a Bernoulli process with a small p. As in the Bernoulli case, we'll be interested in a happening or event of some sort ("success"), in how many of these events there are in a specified interval of time, and in how long it is until the first event or until the rth event (after some point in time).

Examples of "events" for which our model has been found useful include arrivals at a service counter, industrial accidents, clicks or counts recorded by a Geiger counter in the presence of a radioactive substance, breakdowns of equipment (with failures corrected as they occur), and, substituting distance for time, flaws in a long manufactured tape or wire. Because of these varied kinds of application, we'll refer to the happening we're counting simply as an *event*, for purposes of our general discussion.

Since the process to be defined is an analog of the Bernoulli process, we recall some characteristics of the latter: When p is small, successes occur rather infrequently, albeit sporadically, and rarely occur in consecutive trials. The number of successes in a given number of trials has a distribution which does not depend on where the trials begin in the sequence. And the numbers of successes in nonoverlapping finite sequences of trials are independent (because the trials are independent). Moreover, although we did not go into this earlier, the probability of a success in a short sequence is approximately proportional to its length: $P(1 \text{ success in } k \text{ trials}) = kpq^{k-1}$, which is nearly proportional to k when k is not large (so that q^{k-1} is still near 1).

With these properties of a Bernoulli process in mind, we base our definition of a **Poisson process** on the following postulates. As a temporary notation, let $P_n(t)$ denote the probability of n events in an interval of length t. We'll derive a sequence of differential equations for the P_n and solve them to get the p.f. of a Poisson random variable X, the number of events in an interval of given length t:

$$P_n(t) = P(X = n \mid \lambda) = f_X(n \mid \lambda t).$$

6.5 The Poisson process   161

POISSON POSTULATES:

1. *The numbers of events occurring in nonoverlapping time intervals are independent.*
2. *The probability structure is time-invariant.*
3. *The probability of exactly one event in an infinitesimal interval h is approximately proportional to its length. That is, for some λ,*

$$P_1(h) = \lambda h + o(h). \tag{1}$$

4. *The probability of finding more than one event in an infinitesimal interval h is of smaller order than the probability of exactly one:*

$$\sum_{n>1} P_n(h) = o(h).$$

[The notation "$o(x)$" was defined in §4.10 (page 131).] The second postulate implies, in particular, that the notation $P_n(t)$ is justified, since this probability does not depend on when the interval of length t begins. And postulates 3 and 4 imply $P_0(h) = 1 - \lambda h + o(h)$.

We show now that the function $P_n(t)$ is implied by the postulates. Consider the interval $(0, t+h)$. (Because of time-invariance, we may as well start our clock at $t = 0$.) There are exactly n events in that interval if and only if there are n in the interval $(0, t)$ and none in $(t, t+h)$, or $n-1$ in $(0, t)$ and one in $(t, t+h)$, ..., or none in $(0, t)$ and n in $(t, t+h)$. Events relating to the nonoverlapping intervals $(0, t)$ and $(t, t+h)$ are independent, so when $n > 0$,

$$P_n(t+h) = P_n(t) \cdot P_0(h) + P_{n-1}(t) \cdot P_1(h) + \cdots + P_0(t) \cdot P_n(h)$$

$$= P_n(t)[1 - \lambda h + o(h)] + P_{n-1}(t)[\lambda h + o(h)] + o(h).$$

Subtracting $P_n(t)$ from both sides and dividing by h, we obtain the increment quotient:

$$\frac{P_n(t+h) - P_n(t)}{h} = -\lambda P_n(t) + \lambda P_{n-1}(t) + \frac{o(h)}{h}.$$

Passing to the limit as $h \to 0$, we obtain a differential equation for $P_n(t)$, when $n \geq 1$. Problem 6-42 asks you to derive the equation for the case $n = 0$, thus completing this system of differential equations:

$$\begin{cases} P_0'(t) = -\lambda P_0(t), \\ P_n'(t) = -\lambda P_n(t) + \lambda P_{n-1}(t), \quad n \geq 1. \end{cases} \tag{2}$$

Together with the obvious initial condition $P_0(0) = 1$, the first equation

162   Chapter 6 Some Parametric Families

leads to $P_0(t) = e^{-\lambda t}$. Now substitute this in the second equation with $n = 2$, and solve for P_1 with the initial condition $P_1(0) = 0$; substitute the result in the second equation with $n = 3$; and so on, to obtain the general formula for P_n:

$$P_n(t) = \frac{e^{-\lambda t}(\lambda t)^n}{n!}. \tag{3}$$

[You can verify that the function (3) does indeed satisfy the system (2).] Problem 6-46 outlines a derivation of (3) by taking the limit of approximations to the Poisson process by a discrete time Bernoulli process.

Returning to more standard notation, we define the **Poisson** family of distributions by the p.f.

$$f(n \mid \mu) = \frac{e^{-\mu}\mu^n}{n!}, \quad n = 0, 1, 2, \ldots . \tag{4}$$

(Problem 6-40 asks you to verify that these probabilities sum to 1.) A random variable X with the p.f. (3) is said to have a *Poisson distribution*, and we write $X \sim \text{Poi}(\mu)$. Table II gives cumulative Poisson probabilities $P(X \leq n)$ for selected values of μ.

Since λ and t appear in (2) only in the combination λt, we have set $\mu = \lambda t$, using the notation "μ" deliberately because this parameter is in fact the mean of the distribution:

$$EX = \sum_{n=0}^{\infty} n \left(\frac{e^{-\mu}\mu^n}{n!} \right) = \mu e^{-\mu} \sum_{n=1}^{\infty} \frac{\mu^{n-1}}{(n-1)!} = \mu. \tag{5}$$

The parameter λ appearing in the Poisson postulates is now seen to be the expected number of events in a *unit* interval.

The form of the Poisson p.f. combines easily with s^n to give us the factorial m.g.f. of a Poisson variable X:

$$\eta_X(s) = E(s^X) = \sum_{n=0}^{\infty} s^n \left(\frac{e^{-\mu}(\mu)^n}{n!} \right) = e^{-\mu} \sum_{n=0}^{\infty} \frac{(s\mu)^n}{n!} = e^{(s-1)\mu}. \tag{6}$$

(We have used the "dummy" variable s in the f.m.f.g. rather than the earlier t, so as not to confuse it with the t we are using as the time variable in a Poisson process.) Successive differentiation and substitution of $s = 1$ verifies (5) and shows the variance to be equal to the mean: $\sigma_X^2 = \mu$.

EXAMPLE 6.5a Customers enter a waiting line at random, at a rate of four per minute. Assuming that the arrival process is a Poisson process, we can find the probability that at least one customer arrives in any given half-minute period: The rate per half-minute is $\lambda t = 4 \cdot (1/2) = 2$,

so
$$P(\text{at least one}) = 1 - P(\text{none}) = 1 - f(0 \mid 2) = e^{-2}2^0/0! \doteq .135.$$

Similarly, the probability of at most ten arrivals in a three-minute period, with $\lambda t = 12$, is

$$\sum_{n=0}^{10} e^{-12}12^n/n! = .347,$$

according to Table II. ∎

We have introduced the Poisson process as a continuous-time analog of a Bernoulli process, and the Poisson distribution is the analog of the binomial distribution—the distribution of the number of "events" in an interval of specified length. Indeed, the limit of the binomial p.f., as n becomes infinite with np fixed, is a Poisson p.f., as we show next. Let $\lambda = np$, and rearrange the binomial p.f. as follows:

$$\binom{n}{k}p^k(1-p)^{n-k} = \frac{n(n-1)\cdots(n-k+1)}{k(k-1)\cdots 3\cdot 2\cdot 1} \cdot \left(\frac{\lambda}{n}\right)^k(1-p)^{-k}\left(1-\frac{\lambda}{n}\right)^n$$

$$= \frac{n}{n} \cdot \frac{n-1}{n} \cdots \frac{n-k+1}{n} \cdot \frac{\lambda^k}{k!}(1-p)^{-k}\left(1-\frac{\lambda}{n}\right)^n.$$

As n becomes infinite and p tends to zero, with $\lambda = np$ fixed, the first k factors tend to 1, the next factor is unchanged, the next tends to 1, and with $x = 2/n$ or $n = -\lambda/x$, the last factor is

$$(1+x)^n = [(1+x)^{1/x}]^{-\lambda} \to e^{-\lambda}.$$

Thus, with $np = \lambda$ (fixed),

$$\lim_{\substack{n\to\infty \\ p\to 0}} \binom{n}{k}p^k(1-p)^{n-k} = \frac{\lambda^k}{k!}e^{-\lambda}. \tag{7}$$

This gives an approximation to a binomial probability with large n and small p as a Poisson probability with $\mu = np$:

$$\binom{n}{k}p^k(1-p)^{n-k} \doteq \frac{(np)^k}{k!}e^{-np}. \tag{8}$$

EXAMPLE 6.5b Consider (as in Problem 6-25) a population in which five percent have a particular blood type. What is the probability that in a random selection of 100 individuals (no replacement), at least four have that blood type?

Again considering that the population is much larger than 100, we can consider the number X of individuals with the given blood type as binomially distributed: $X \sim \text{Bin}(100, .05)$. Then $np = 5$, and it follows from (8) that $X \sim \text{Poi}(5)$. So

$$P(X \geq 4) = 1 - P(X \leq 3) = 1 - e^{-5}(1 + 5 + 5^2/2 + 5^3/3!) \doteq 1 - .265.$$

$[P(X \leq 3)$ can also be found in Table II, with $m = 5$, $c = 3$.] ∎

The indexing parameter t in a Poisson process, as described above, is a linear measure—a real variable. However, in the Poisson postulates, one can substitute an area element dA or a volume element dV for "h," location-invariant for time-invariant, and nonoverlapping regions for nonoverlapping intervals. The "rate" parameter λ is the average number of events per unit area or per unit volume, as the case may be. Thus, the Poisson distribution has been used for the number of bacterial growths in a Petri dish, the number of fish in a given volume of water, etc. (See Problem 6-34.)

Since it is derived, in a sense, from the binomial distribution, one could naturally expect that the Poisson distribution would enjoy a reproductive property similar to that of the binomial:

THEOREM 2 *If $X_1, X_2, \ldots, X_n$ are independent random variables, with $X_i \sim Poi(\lambda_i)$, then*

$$\sum X_i \sim \text{Poi}\left(\sum \lambda_i\right).$$

Problem 6-43 calls for a proof of this theorem using generating functions. Together with the central limit theorem, the theorem provides a way to approximate Poisson probabilities when λ is large. (Table II ends at $\lambda = 15$ because, for most purposes, the normal approximation is adequate when λ is larger than 15.)

EXAMPLE 6.5c Suppose $X \sim \text{Poi}(100)$, and we want $P(X \leq 80)$. To find this we think of X as the sum of 100 i.i.d. variables, each distributed as $\text{Poi}(1)$. Then by the central limit theorem, the distribution of the Z-score $Z = \frac{X - 100}{10}$ is approximately that given in Table I. The Z-score for $X = 80$ is -2, so $\Phi(-2) \doteq .023$ is the desired probability. ∎

PROBLEMS

6-30. Telephone calls are placed through a certain exchange at a rate of four per minute. Assuming the arrival process is Poisson, find the probability of
 (a) at most six calls in one minute.
 (b) three or more calls in a 15-second period.

(c) no calls in a half-minute period.

6-31. Suppose X is Poisson with variance 1. Find $P(X = 2)$.

6-32. Defects in a certain manufactured tape occur on the average of one per 2,000 feet. Assuming a Poisson distribution for the number of defects in a given length of tape, find the probability of
 (a) at most two defects in a 2,400 ft roll.
 (b) no defects in a 1000 ft roll.
 (c) no more than one defect in each 1,200 ft roll in a box of five.

6-33. A Geiger counter records on the average of 40 counts per minute when in the neighborhood of a certain radioactive substance. Assuming a Poisson process, find the probability of
 (a) exactly two counts in a 6-second period.
 (b) exactly k counts in a T-second period.
 (c) no counts in a 5-second period.

6-34. Flaws in the plating of large sheets of metal occur at random, on the average of one flaw per 50 square feet. Assuming that the distribution of flaws is described by the Poisson distribution, find the probability
 (a) that a sheet 5 ft by 8 ft has no flaws.
 (b) that a sheet 10 ft by 10 ft has at most two flaws.

6-35. Bolts produced by a certain machine are occasionally defective—an average of one defective bolt per 200. They are packaged in boxes of 50, and there are 100 boxes in a carton. Find the probability
 (a) that a box has at most one defective bolt.
 (b) that none of the boxes in a carton has more than one defective.

6-36. Evaluate: $\sum_{k=0}^{\infty} \dfrac{(k^2 - k)\, 3^k}{k!}$.

6-37. Given that $X \sim \text{Bin}(50, .03)$, find $P(X \geq 4)$
 (a) using a Poisson approximation.
 (b) using a hand calculator to find individual binomial probabilities.

6-38. In a random selection of 200 individuals from a population of 100,000, two percent of whom are tune to a classical music radio station, what is the probability that at least two of the 200 are tuned to that station? [Compare binomial and Poisson approximations.]

6-39. Use appropriate approximations and tables to evaluate each of the following:
 (a) $P(X \leq 3)$, where $X \sim \text{Bin}(100, .05)$.
 (b) $P(X > 50)$, where $X \sim \text{Poi}(64)$.

(c) $P(X \leq 5)$, where $X \sim$ Hyper(50000, 3000, 40).

6-40. Show that the Poisson probabilities $e^{-\mu}\mu^n/n!$ sum to 1, $n = 0, 1,$ ... (as of course they must).

6-41. Show that the Poisson p.f. [(2) of §6.5] satisfies the Poisson postulates.

6-42. Derive the differential equation for $P_0(t)$ given as part of (2) in §6.5.

6-43. Use generating functions to show the reproductive property of the Poisson distribution (Theorem 2 of the preceding section).

6-44. Men arrive at a service counter at the rate of six per hour, women at a rate of 12 per hour, and children at the rate of 12 per hour. Assuming that the arrival processes are independent and each is Poisson, find the probability that at least two customers arrive in a five-minute period.

6-45. For a particular Poisson arrival process with rate parameter λ, let X_1 denote the number of arrivals in the interval $(0, t_1)$ and X_2, the number in (t_1, t_2). Find and identify the conditional distribution of X_1 given $X_1 + X_2 = m$.

6-46. Consider a process in which the Poisson postulates are satisfied, and let X denote the number of events in the interval $(0, T)$. Divide the interval into n equal subintervals of length $h = T/n$, and consider each subinterval as a "trial." With $Y_i =$ number of events in the ith subinterval, show that Y_i is approximately Ber($\lambda T/n$). Use (7) to deduce the Poisson formula (3).

6.6 Exponential and related distributions

In a Poisson process, the time T, measured from any given point until the next "event" is a random variable of interest, called the *waiting time*. Its distribution is implied by the Poisson postulates, as follows: The **tail-probability** function of the distribution is

$$R(t) = P(T > t). \tag{1}$$

This is complementary to the c.d.f.: if you know R you know $F = 1 - R$. But $T > t$ if and only if there are no "events" in the interval $(0, t)$. So, if f denotes the Poisson p.f. with parameter λt,

$$F_T(t) = 1 - R(t) = 1 - f(0 \mid \lambda) = 1 - e^{-\lambda t}, \quad t > 0. \tag{2}$$

6.6 Exponential and related distributions

The corresponding p.d.f. is

$$f_T(t) = F'_T(t) = \lambda e^{-\lambda t}, \quad t > 0. \tag{3}$$

The waiting time T is said to have an **exponential** distribution, and we write $T \sim \text{Exp}(\lambda)$. The p.d.f. and c.d.f. are shown in Figure 6-2 for $\lambda = 1$.

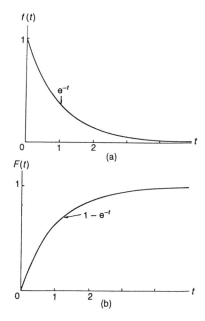

Figure 6-2 Exponential density and c.d.f.

The m.g.f. of the exponential distribution is

$$\psi(s) = E(e^{sT}) = \bigg|_{t=0}^{\infty} e^{st}(\lambda e^{-\lambda t})\, dt = \frac{\lambda}{\lambda - s} = \frac{1}{1 - s/\lambda}, \tag{4}$$

for $s < \lambda$. Expanding in powers of s, we get

$$\psi(s) = \sum_{k=0}^{\infty} \left(\frac{s}{\lambda}\right)^k = \sum_{k=0}^{\infty} \left(\frac{k!}{\lambda^k}\right) \cdot \frac{s^k}{k!}.$$

The coefficient of $s^k/k!$ is the kth moment:

$$E(T^k) = k!/\lambda^k, \quad k = 0, 1, 2, \ldots. \tag{5}$$

In particular,

$$ET = 1/\lambda, \quad E(T^2) = 2/\lambda^2, \quad \text{and} \quad \text{var } T = 1/\lambda^2. \tag{6}$$

Thus, the mean equals the standard deviation.

The exponential distribution is "memoryless." In an application in which the "event" is a customer arrival, this means that how long one must wait for the next arrival is a random variable which does not depend on how long one may have been waiting already: The future is independent of the past, in that the future unfoldment of the process, as seen from any point in time, does not depend on what has gone on before. To show this memoryless property, we find the tail-probability of the conditional distribution:

$$P(T > r+t \mid T > r) = \frac{P(T > r+t)}{P(T > r)} = \frac{e^{-\lambda(r+t)}}{e^{-\lambda r}} = e^{-\lambda t}, \quad t > 0,$$

which is the same as the tail-probability of the *un*conditional distribution. This is not unexpected, inasmuch as the Bernoulli process has this property.

Again, suppose you have been watching an arrival process from time 0 to time t and have counted k arrivals. What is then the conditional distribution of the time to the next arrival after t? To find this we exploit the property of the Poisson process that the numbers of arrivals in non-overlapping intervals are independent variables: Let T denote the time from point t to the next arrival. Then

$$P[T > r \mid k \text{ arr. in } (0, t)] = \frac{P[k \text{ arr. in } (0, t) \text{ and } 0 \text{ arr. in } (t, t+r)]}{P[k \text{ arr. in } (0, t)]}$$

$$= P[0 \text{ arr. in } (t, t+r)] = e^{-\lambda r}, \quad r > 0,$$

so the conditional distribution of T is independent of the condition—it is still Exp(λ): The future looks the same no matter when you start watching the process, and no matter how much you may have learned about previous arrivals.

In an application in which the "waiting time" is the time to failure of a system, the distribution of future life is the same at $t = 0$ as at any later point in time. This would mean that the exponential distribution does not describe a system whose failure is caused by wearing out. Rather, it may be appropriate when failures are caused by shocks occurring at random times (as with surges of voltage or hazards on a road). In particular, it does not describe the distribution of human life length.

EXAMPLE 6.6a Consider a Poisson arrival process with rate parameter $\lambda = 4$/min, as in Example 6.5a. The time to the next arrival after $t = 0$ is Exp(4), with mean $1/\lambda = 1/4$ min and standard deviation $1/\lambda = 1/4$ min. If after two minutes, say, we restart the stop watch, the time to the

6.6 Exponential and related distributions

next arrival after that point is again Exp(4)—independent of what you learned about arrivals during the first two minutes! ∎

Next we consider the time to the rth arrival. We denote by T_1 the time to the first arrival, by T_2 the time from the first to the second arrival, by T_3 the time from the second to the third arrival, and so on. These are successive *inter-arrival times*.

THEOREM 3 *The times T_i in a sequence of r successive inter-arrival times are independent random variables.*

We show this by finding the joint p.d.f. of the T_i, using a differential method—calculating the probability of the "parallelepiped" defined by increments $dt_1, ..., dt_r$ and reading the coefficient of the differential element $dt_1 \cdots dt_r$ as the joint p.d.f. This is the probability of the intersection of the events $[t_i < T_i < t_i + dt_i]$, $i = 1, ..., r$. These r inequalities are all satisfied if and only if there is exactly one arrival in each interval $(t_i, t_i + dt_i)$ and no other arrivals between 0 and Σt_i (see Figure 6-3). The probability of an arrival in an interval of length dt_i is approximately λdt_i, and the probability of no arrivals in an interval of approximate length t_i is $\exp(-\lambda t_i)$. So

$$P(t_1 < T_1 < t_1 + dt_1, ..., t_r < T_r < t_r + dt_r)$$
$$\doteq \prod_{i=1}^{r} (\lambda t_i) \cdot \prod_{i=1}^{r} e^{-\lambda t_i} = \{\lambda^r \exp(-\lambda \Sigma t_i)\} dt_1 \cdots dt_r.$$

The coefficient (in braces) is the joint p.d.f., and it is clearly the product of the univariate exponential p.d.f.'s of the individual T_i. This means that those times are independent, as claimed.

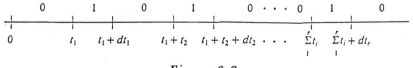

Figure 6-3

We consider next the time T to the rth arrival, which is the sum of the first r interarrival times:

$$T = T_1 + T_2 + \cdots + T_r, \tag{7}$$

where the variables T_i are i.i.d., $T_i \sim \text{Exp}(\lambda)$. Because the summands are independent and identically distributed, the m.g.f. of the sum is the rth power of the m.g.f. of T_1:

$$\psi_T(s) = [\psi_{T_1}(s)]^r = (1 - s/\lambda)^{-r}. \tag{8}$$

In Problem 4-60 we encountered this m.g.f. for $\lambda = 1$, derived from the p.d.f.

$$f(y) = \frac{y^{r-1} e^{-y}}{(r-1)!}.$$

Because of the uniqueness property of m.g.f.'s, this must be the p.d.f. of T in (7) when $\lambda = 1$. For general λ, we make the change of variable $T = Y/\lambda$ to obtain the following p.d.f. for the distribution defined by (8):

$$f(t \mid r, \lambda) = \frac{\lambda^r t^{r-1} e^{-\lambda t}}{(r-1)!}, \quad t > 0. \tag{9}$$

Problem 6-56 obtains a formula for the corresponding c.d.f. in terms of Poisson probabilities.

We can get a general formula for the moments of the distribution (9) by expanding the m.g.f. (8) in a power series (a negative binomial expansion):

$$(1 - s/\lambda)^{-r} = \sum_{k=0}^{\infty} \binom{r+k-1}{k} \left(\frac{s}{\lambda}\right)^k.$$

The kth moment is the coefficient of $s^k/k!$:

$$E(T^k) = \binom{r+k-1}{k} \frac{k!}{\lambda^k}. \tag{10}$$

The mean and variance are implied by (10), but it's perhaps simpler and more insightful to obtain them from the structure of T in (7), as the sum of i.i.d. exponential variables. The mean is

$$E(T) = E(T_1 + \cdots + T_r) = \frac{r}{\lambda}, \tag{11}$$

and the variance,

$$\operatorname{var} T = \operatorname{var}(T_1 + \cdots + T_r) = \frac{r}{\lambda^2}. \tag{12}$$

EXAMPLE 6.6b Continuing the setting of Example 6.6a, consider again the Poisson arrival process with rate parameter $\lambda = 4/\min$. The time to the eighth arrival after any given point in time is Gam(8, 4), with mean value $\mu = 8/4 = 2$ min, and variance $\sigma^2 = 8/16 = 1/2$ min^2, or $\sigma \doteq .71$ min. ∎

In a process which records the occurrences of system failure, the tail-probability function (1) is called the *reliability function*. Its logarithmic derivative is called the *hazard function*:

$$h(x) = -\frac{d}{dx}[\log R(x)] = -\frac{R'(x)}{R(x)} = \frac{f(x)}{R(x)}. \qquad (13)$$

This is interpreted as the rate of "dying" among many such systems, relative to the number still operating. For a Poisson process, the hazard is constant: $h(x) = \lambda$. (See also Problem 6-51.) But for systems that wear out, a constant hazard is not appropriate. When the hazard is proportional to a power of x:

$$h(x) = \frac{\alpha x^{\alpha-1}}{\beta^\alpha}, \quad x > 0,\ \alpha > 0,\ \beta > 0, \qquad (14)$$

the population p.d.f. is

$$f(x \mid \alpha,\ \beta) = \frac{\alpha x^{\alpha-1}}{\beta^\alpha} e^{-(x/\beta)^\alpha}, \quad x > 0, \qquad (15)$$

(see Problem 6-52). This defines the family of *Weibull distributions*. When hazard is exponential:

$$h(x) = ae^{bx}, \quad x > 0, \qquad (16)$$

the p.d.f. is

$$f(x) = a\,\exp\!\left[bx - \frac{a}{b}(e^{bx} - 1)\right], \quad x > 0, \qquad (17)$$

defining the *Gompertz distribution*.

PROBLEMS

6-47. Find the median of the exponential distribution, $\text{Exp}(\lambda)$.

6-48. Customers join a waiting line according to a Poisson law, with an average time between arrivals of two minutes.
(a) Find the probability that six minutes will elapse with no customers.
(b) Find the probability of at most two customers in a 6-minute period.
(c) Find the probability that the next customer (after a given point in time) arrives within the next two minutes.
(d) How long is the average wait for the third customer from now?
(e) On average, how long has it been (assuming you are not told how long when you arrive on the scene) since the most recent customer?

6-49. The time to failure of a unit is $T \sim \text{Exp}(\lambda)$, with $\lambda = .4/\text{day}$. If the unit is immediately reset and put in operation after each failure, find
(a) the mean time to the third failure.

172   Chapter 6   Some Parametric Families

(b) the expected number of failures in 10 days.
(c) the probability of no failures in a day.
(d) the probability of at most three failures in a week.

6-50. Suppose a system consists of four components in "series," that is, so interconnected that the system fails if any component fails. Assuming independence of times to failure of the components, find the distribution of the time to failure of the system if each unit has an exponential life distribution with mean time to failure of 5 hours.

6-51. Show that if the hazard function (13) is constant, the life distribution is exponential.

6-52. Obtain the p.d.f. (15) assuming a hazard function given by (14).

6-53. Obtain the p.d.f. (17) assuming a hazard function given by (16).

6-54. Each of two units has an exponential distribution of time to failure with mean one hour. If these are placed in a system such that there is system failure if and only if *both* units fail, find the expected time to system failure.

6-55. Show that if successive times to failure are independent exponential variables with parameter λ, then for a small time increment h,
(a) $P(\text{one or more failures in } h) = \lambda h + o(h)$.
(b) $P(\text{more than one failure in } h) = o(h)$.

6-56. Show that the distribution function of S_r, the time to the rth "failure" in a Poisson process, can be expressed as follows:

$$F_{S_r}(x) = 1 - \sum_{k=0}^{r-1} \frac{(\lambda x)^k}{k!} e^{-\lambda x}, \quad x > 0.$$

Do this in two ways:
(a) Differentiate this expression for the c.d.f. to obtain the p.d.f. (9) of §6.6 (and check to see that the additive constant is correct).
(b) Calculate $P(S_r \leq x)$ in terms of a Poisson random variable.

6-57. Use (7), (11), and (12) to obtain a simpler approximate distribution for the S_r of the preceding problem when r is large, and use this to find the probability, in the setting of Problem 6-48, that the time to the arrival of the 20th customer is at least 35 minutes.

6.7 Gamma and beta distributions

The following integral, depending on the value of a real parameter α, is encountered frequently in applied mathematics:

$$\int_0^\infty u^{\alpha-1} e^{-u}\, du = \Gamma(\alpha). \tag{1}$$

The integral is an "improper" definite integral, which can be shown to be convergent for $\alpha > 0$. As a function of α it is called the **gamma function**. It satisfies the following recurrence relation:

$$\Gamma(\alpha+1) = \alpha\, \Gamma(\alpha). \tag{2}$$

(See Problem 6-61.) Clearly, $\Gamma(1) = 1$, and by induction on the integer n, it follows that $\Gamma(n) = (n-1)!$. We may think of $\Gamma(\alpha)$ as a generalized factorial function.

Making the change of variable $u = \lambda x$ in the integral (1), we obtain another formula for the gamma function:

$$\Gamma(\alpha) = \lambda^\alpha \int_0^\infty x^{\alpha-1} e^{-\lambda x}\, dx, \tag{3}$$

which implies this convenient integral formula:

$$\int_0^\infty x^{\alpha-1} e^{-\lambda x}\, dx = \frac{\Gamma(\alpha)}{\lambda^\alpha}. \tag{4}$$

And the change of variable $u = z^2/2$ in (1) yields yet another useful integration formula—one we'll need in the next section:

$$\int_0^\infty z^{2\alpha-1} e^{-z^2/2}\, dz = 2^{\alpha-1} \Gamma(\alpha). \tag{5}$$

Multiplying (4) through by $\lambda^\alpha / \Gamma(\alpha)$, we see that

$$\int_0^\infty \left\{ \frac{\lambda^\alpha}{\Gamma(\alpha)} x^{\alpha-1} e^{-\lambda x} \right\} dx = 1.$$

The integrand here is nonnegative; and because its integral over $(0, \infty)$ is 1, it defines a probability density function, the p.d.f. of the **gamma distribution**:

$$f(x \mid \alpha, \lambda) = \frac{\lambda^\alpha}{\Gamma(\alpha)} x^{\alpha-1} e^{-\lambda x}, \quad x > 0. \tag{6}$$

When X has this distribution, we use the notation $X \sim \text{Gam}(\alpha, \lambda)$.

In §6.6 we found the distribution of the time T to the rth event in a Poisson process with rate parameter λ. The p.d.f. given by (9) in §6.6 is a special case of (6), in which α is the integer r: $T \sim \text{Gam}(r, \lambda)$.

The moments and m.g.f. of a gamma distribution can be evaluated with the aid of gamma functions. Thus the mean is

$$E(X) = \frac{\lambda^\alpha}{\Gamma(\alpha)} \int_0^\infty x \, x^{\alpha-1} e^{-\lambda x} \, dx = \frac{1}{\lambda \Gamma(\alpha)} \int_0^\infty y^{(\alpha+1)-1} e^{-y} \, dy$$

$$= \frac{\Gamma(\alpha+1)}{\lambda \Gamma(\alpha)} = \frac{\alpha}{\lambda}. \tag{7}$$

A similar calculation yields $E(X^2) = \frac{(\alpha+1)\alpha}{\lambda^2}$, from which we obtain

$$\text{var } X = \alpha/\lambda^2. \tag{8}$$

The moment generating function is

$$\psi_X(t) = E(e^{tX}) = \frac{\lambda^\alpha}{\Gamma(\alpha)} \int_0^\infty x^{\alpha-1} e^{-(\lambda-t)x} \, dx = \frac{1}{(1-t/\lambda)^\alpha}. \tag{9}$$

Another useful family of distributions is the *beta* family, which we now introduce by evaluating the product of two gamma functions, each expressed using (5) in the form

$$\Gamma(\alpha) = \frac{1}{2^{\alpha-1}} \int_0^\infty z^{2\alpha-1} e^{-z^2/2} \, dz. \tag{10}$$

In writing integrals for $\Gamma(s)$ and $\Gamma(t)$ we use different dummy variables:

$$\Gamma(s)\Gamma(t) = \int_0^\infty x^{2s-1} e^{-\frac{1}{2}x^2} \, dx \cdot \int_0^\infty y^{2t-1} e^{-\frac{1}{2}y^2} \, dy$$

$$= \int_0^\infty \int_0^\infty x^{2s-1} y^{2t-1} e^{-\frac{1}{2}(x^2+y^2)} \, dx \, dy.$$

In this integral over the first quadrant, we change to polar coordinates: $x = r \cos \theta$, $y = r \sin \theta$. The element of area $dx \, dy$ becomes $r \, dr \, d\theta$; the product $\Gamma(s)\Gamma(t)$ is then

$$\frac{1}{2^{s+t-2}} \int_0^{\pi/2} \left(\int_0^\infty r^{2(s+t-1)} e^{-\frac{1}{2}r^2} r \, dr \right) \cos^{2s-1}\theta \sin^{2t-1}\theta \, d\theta.$$

The r-integral, which does not involve θ, can be moved outside the θ-integral. With the change of variable $u = r^2/2$, $du = r \, dr$, we obtain

$$\Gamma(s)\Gamma(t) = \int_0^\infty u^{s+t-1} e^{-u} \, du \int_0^{\pi/2} 2 \cos^{2s-1}\theta \sin^{2r-1}\theta \, d\theta.$$

The u-integral we recognize as $\Gamma(s+t)$, and the θ-integral depends on the parameters s and t. Thus,

$$\Gamma(s)\Gamma(t) = \Gamma(s+t)\,B(s,t),$$

where $B(s, t)$ denotes the **beta function**:

$$B(s,t) = \frac{\Gamma(s)\Gamma(t)}{\Gamma(s+t)} = 2\int_0^{\pi/2} \cos^{2s-1}\theta \, \sin^{2t-1}\theta \, d\theta. \tag{11}$$

Notice that if s and t are interchanged, the value of $B(s, t)$ is unchanged:

$$B(s,t) = B(t,s). \tag{12}$$

The beta function is easily evaluated at $s = t = \frac{1}{2}$, which in turn implies the value of $\Gamma(\frac{1}{2})$:

$$B(\tfrac{1}{2},\tfrac{1}{2}) = \frac{\Gamma(\tfrac{1}{2})\Gamma(\tfrac{1}{2})}{\Gamma(1)} = 2\int_0^{\pi/2} d\theta = \pi.$$

Since $\Gamma(1) = 1$, we see that $\Gamma(\tfrac{1}{2}) = \sqrt{\pi}$. From this and the recurrence relation (2) we can evaluate the gamma function at points midway between successive integers:

$$\Gamma(k+\tfrac{1}{2}) = (k-\tfrac{1}{2})\Gamma(k-\tfrac{1}{2}) = (k-\tfrac{1}{2})(k-\tfrac{3}{2})\cdots\tfrac{5}{2}\cdot\tfrac{3}{2}\cdot\tfrac{1}{2}\sqrt{\pi}. \tag{13}$$

The beta function has other forms, obtained by changes of variable in the integral. Substituting $x = \cos^2\theta$ and $1 - x = \sin^2\theta$ in (11), we obtain

$$B(s,t) = \int_0^1 x^{s-1}(1-x)^{t-1}\,dx. \tag{14}$$

Substituting $x = \frac{u}{1+u}$, $1 - x = \frac{1}{1+u}$, and $dx = \frac{du}{(1+u)^2}$ in (11) yields yet another expression for the beta function:

$$B(s,t) = \int_0^\infty \frac{u^{s-1}}{(1+u)^{s+t}}\,du. \tag{15}$$

The integration formulas (11), (14), and (15) give rise to useful distributions on $(0, \pi/2)$, $(0, 1)$, and $(0, \infty)$, respectively. The **beta distribution** is defined by the p.d.f.

$$f(x\mid s,t) = \frac{1}{B(s,t)}x^{s-1}(1-x)^{t-1}, \quad 0 < x < 1. \tag{16}$$

176 Chapter 6 Some Parametric Families

(See also Problem 6-65.) Several beta densities are shown in Figure 6-4.

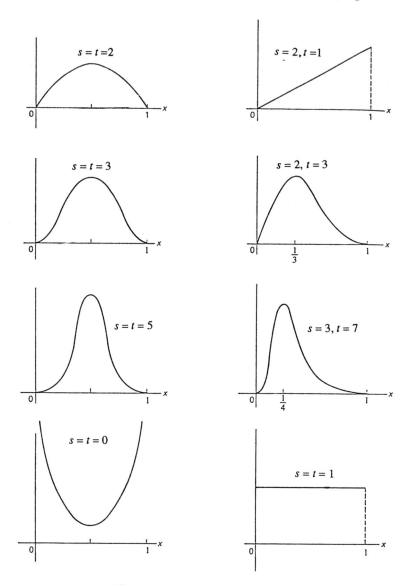

Figure 6-4. Some beta densities

The mean and variance of a beta distribution are

↑ mean

$$EX = \frac{s}{s+t}, \quad \text{var } X = \frac{st}{(s+t+1)(s+t)^2}. \tag{17}$$

Problem 6-60 asks you to obtain the formula for the mean; to get the variance, we first find

$$E(X^2) = \frac{1}{B(s,t)} \int_0^1 x^{s+1}(1-x)^{t-1}\,dx$$

$$= \frac{B(s+2,t)}{B(s,t)} = \frac{s(s+1)}{(s+t+1)(s+t)}$$

and then subtract $(EX)^2$.

PROBLEMS

6-58. Evaluate the following:

(a) $\Gamma(6)$.

(b) $\Gamma(\frac{11}{2})$.

(c) $B(2, \frac{3}{2})$.

(d) $\int_0^1 (-\log u)^{3/2} u\,du$.

(e) $\int_0^\infty e^{-3x} x^4\,dx$.

(f) $\int_0^1 x^5(1-x)^9\,dx$.

(g) $\int_0^{\pi/2} \cos^4\theta \sin^6\theta\,d\theta$.

(h) $\int_0^\infty \frac{u^2\,du}{(1+u)^5}$.

6-59. Derive (8), the formula for the variance of $\text{Gam}(\alpha, \lambda)$.

6-60. Verify the formula given in (17) for the mean of a beta distribution, by recognizing the integral expression for the mean as involving a beta function.

6-61. Use an integration by parts to establish the recurrence relation (2).

6-62. Use formula (10) for the gamma function to define a family of distributions on $(0, \infty)$ and calculate the expected value and mean square when $\alpha = 3$.

6-63. Show that the sum of independent gamma variables which have a common λ has a gamma distribution.

6-64. Find $E(1/T)$, where $T \sim \text{Gam}(\alpha, \lambda)$

6-65. Use formula (15) for the beta function to define a family of distributions on the interval $(0, \infty)$ and calculate the mean.

6-66. Given that W has the distribution defined in the preceding problem, show that $1/W$ has a distribution in the same family.

6-67. Show that $\Gamma(\alpha)$ is concave up (on $\alpha > 0$) by showing that its second derivative is always positive.

6-68. Show that $\Gamma(\alpha) \to \infty$ as $\alpha \to 0$. [Hint: Look at (2).]

6.8 Normal distributions

The limiting distribution of a standardized sum of i.i.d. variables, given by the central limit theorem (§5.3), is called **standard normal**. Its p.d.f. is

$$f(z) = \frac{1}{\sqrt{2\pi}} \exp\left(-\frac{1}{2}z^2\right). \tag{1}$$

(The constant $\sqrt{2\pi}$, which makes the area under the p.d.f. equal to 1, will be derived below.) The corresponding c.d.f. is

$$\Phi(z) = \frac{1}{\sqrt{2\pi}} \int_{-\infty}^{z} e^{-\frac{1}{2}u^2} du \tag{2}$$

Values of this function are given in Table I. (It is not an "elementary" function in the traditional classification.) The c.d.f. and p.d.f. are shown in Figure 6-5.

The standard normal distribution is clearly symmetric about $z = 0$, so the area to the left of $-z$ is the same as the area to the right of z:

$$\Phi(-z) = 1 - \Phi(z). \tag{3}$$

This symmetry implies that if Z is standard normal, so also is the variable $-Z$ (Problem 6-70).

Moments of all orders exist, because the negative exponential in the p.d.f. (1) dominates any power of z and guarantees convergence of the defining integrals. Because of the symmetry about 0, all odd-order moments of a standard normal variable are 0, including (in particular) the mean. The m.g.f. was obtained in Problem 4-58:

$$\psi_Z(t) = e^{t^2/2} = \sum_0^\infty \frac{(t^2/2)^k}{k!} = \sum_0^\infty \frac{(2k)!}{2^k k!} \cdot \frac{t^{2k}}{(2k!)}. \tag{4}$$

The terms in odd powers of t are missing, confirming the fact that all odd-order moments are 0. The moments of even order: $\mu_{2k} = E(Z^{2k})$, ($n = 2k$) appear in the series expansion for ψ_Z as the coefficients of $t^{2k}/(2k)!$:

$$\mu_{2k} = \frac{(2k)!}{2^k k!} = (2k-1)(2k-3)\cdots 5\cdot 3\cdot 1, \text{ for } k = 1, 2, \ldots \quad (5)$$

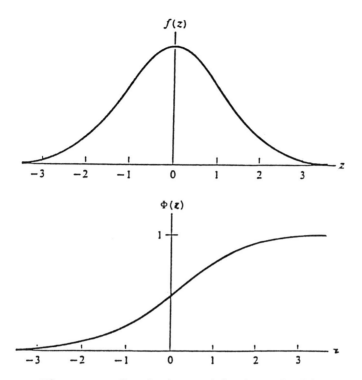

Figure 6-5. Standard normal density and c.d.f.

The moments μ_{2k} can also be expressed in terms of the gamma function, using (5) of the preceding section:

$$E(Z^{2k}) = \frac{1}{\sqrt{2\pi}} \int_{-\infty}^{\infty} z^{2k} e^{-z^2/2}\, dz = \sqrt{\frac{2}{\pi}} \int_{0}^{\infty} z^{2k} e^{-z^2/2}\, dz = \frac{2^k}{\sqrt{\pi}} \Gamma(k + \tfrac{1}{2}).$$

With $k = 0$, this reduces to $E(1) = 1$, which verifies that the constant $1/\sqrt{2\pi}$ is the correct constant in (1).

More generally, a random variable X is said to be **normal**, or normally distributed, if the Z-score

$$Z = \frac{X - \mu_X}{\sigma_X} \quad (6)$$

is standard normal. If X is normally distributed, it can be expressed as

$$X = \mu_X + \sigma_X Z, \quad (7)$$

where Z is standard normal. We'll write $X \sim \mathcal{N}(\mu, \sigma^2)$ to mean that X is normal with mean μ and variance σ^2.

THEOREM 3 *If $X \sim \mathcal{N}(\mu_X, \sigma_X^2)$, and Y is a linear function of X: $Y = a + bX$, then $Y \sim \mathcal{N}(a + b\mu_X, b^2\sigma_X^2)$.*

The formulas for mean and variance of $a + bX$ follow from properties of the mean and variance (Chapter 4). To show that $Y = a + bX$ is normal we form the Z-score:

$$Z_Y = \frac{a + bX - (a + b\mu_X)}{|b|\sigma_X} = \frac{b}{|b|} Z_X.$$

Since both Z and $-Z$ are standard normal, Y is normal, as claimed.

When $X \sim \mathcal{N}(\mu, \sigma^2)$, we can use (7) to find probabilities for X in the standard normal table. The c.d.f. of X is

$$F(x) = P(X \leq x) = P\left(Z \leq \frac{x-\mu}{\sigma}\right) = \Phi\left(\frac{x-\mu}{\sigma}\right). \tag{8}$$

Thus, to find the area to the left of x, we enter the standard normal table at the corresponding standard or Z-score.

EXAMPLE 6.8a Suppose $X \sim \mathcal{N}(10, 4)$. To illustrate use of a standard normal table we use Table I to find the following probability:

$$P(|X - 9| > 1) = 1 - P(|X - 9| < 1) = 1 - P(8 < X < 10)$$

$$= 1 - [F_X(10) - F_X(8)] = 1 - \Phi\left(\frac{10 - 10}{2}\right) + \Phi\left(\frac{8 - 10}{2}\right)$$

$$= 1 - .5 + .1587 = .6587. \blacksquare$$

Since a general normal X is a linear function of the standard normal Z, we can find the p.d.f. of X from the p.d.f. of Z using a transformation formula from Chapter 3 [(2) of §3.5, page 64]:

$$f_{\mu + \sigma Z}(x) = \frac{1}{\sigma} f_Z\left(\frac{x - \mu}{\sigma}\right).$$

Substituting f_Z from (1), we obtain

$$f_X(x) = \frac{1}{\sqrt{2\pi}\,\sigma} \exp\left[-\frac{1}{2\sigma^2}(x - \mu)^2\right]. \tag{9}$$

The fact that the integral of a p.d.f. is 1, implies the following useful integration formula:

$$\int_{-\infty}^{\infty} e^{-c(x-\mu)^2} dx = \sqrt{\frac{\pi}{c}}. \tag{10}$$

The exponent in the normal density (9) is a *quadratic* function with a negative leading term; and an exponential function of such a quadratic function serves to define a normal p.d.f., as we illustate in the next example.

EXAMPLE 6.8b Consider the function $e^{-3x^2 + 6x}$. This is nonnegative, and with a suitable constant multiplier to make the area under the graph equal to 1, it defines a normal p.d.f. We see this by completing the square in the exponent:

$$-3x^2 + 6x = -3(x-1)^2 + 3 = -\frac{(x-1)^2}{1/3} + 3.$$

Comparing this with the exponent in (9) we see that it is the exponent in a normal p.d.f. with $\mu = 1$ and $2\sigma^2 = 1/3$. The p.d.f. is therefore

$$f(x) = \frac{1}{\sqrt{2\pi(1/6)}} \exp\left\{-\frac{(x-1)^2}{2(1/6)}\right\}. \blacksquare$$

When $X \sim \mathcal{N}(\mu, \sigma^2)$, its moments are easily obtained from the corresponding moments of the standard normal Z. The central moments of odd order are clearly zero, in view of the symmetry about the mean. Moments of order $n = 2k$, for k a positive integer, are obtained from the corresponding moments of Z, given by (5):

$$E[(X-\mu)^{2k}] = \sigma^{2k} E(Z^{2k}) = (2k-1)(2k-3)\cdots 5\cdot 3\cdot 1\, \sigma^{2k}. \tag{11}$$

The moment generating function of X is obtained from (4) using (7):

$$\psi_X(t) = E[e^{tX}] = E[e^{t(\sigma Z + \mu)}] = e^{t\mu}\psi_Z(\sigma t) = \exp(t\mu + \tfrac{1}{2}\sigma^2 t^2). \tag{12}$$

Substitution of it for t yeilds the characteristic function:

$$\phi_X(t) = \exp(it\mu - \tfrac{1}{2}\sigma^2 t^2). \tag{13}$$

Using this (or the m.g.f.) we can show the following important result:

THEOREM 4 *If $X_1, X_2, ..., X_n$ are i.i.d. $\mathcal{N}(\mu, \sigma^2)$, then*

182 Chapter 6 Some Parametric Families

$$\sum_{i=1}^{n} X_i \sim \mathcal{N}(n\mu, n\sigma^2). \tag{14}$$

To establish this theorem we find the characteristic function of the sum S_n:

$$\phi_{S_n}(t) = [\phi_X(t)]^n = [\exp(it\mu - \tfrac{1}{2}\sigma^2 t^2)]^n = \exp[(n\mu)it - \tfrac{1}{2}(n\sigma^2)t^2],$$

which is the characteristic function of $\mathcal{N}(n\mu, n\sigma^2)$. The uniqueness theorem for characteristic functions [§4.10, page 129] establishes (14).

6.9 Chi-square distributions

Closely related to the normal is the chi-square family of distributions. The m.g.f. function of the square of a standard normal Z is

$$\psi_{Z^2}(t) = E(e^{tZ^2}) = \frac{1}{\sqrt{2\pi}} \int_{-\infty}^{\infty} e^{tz^2 - z^2/2}\, dz = \frac{1}{\sqrt{1-2t}}, \tag{1}$$

obtained using the integration formula (10) of the preceding section. When $Z_1, \ldots, Z_k$ are i.i.d. standard normal, we denote the sum of their squares by χ^2:

$$\chi^2 = Z_1^2 + \cdots + Z_k^2.$$

Because the Z's are independent, so are their squares. The m.g.f. of χ^2 is therefore the kth power of the m.g.f. of a single term:

$$\psi_{\chi^2}(t) = \frac{1}{(1-2t)^{k/2}} \tag{2}$$

In view of (6) of §6.7, we see that the chi-square family is a subfamily of the gamma family in which $\alpha = k$ and $\lambda = \tfrac{1}{2}$.

DEFINITION The *chi-square distribution with k degrees of freedom* is the distribution of the sum of squares of k independent, standard normal variables. When χ^2 has this distribution, we write $\chi^2 \sim \text{chi}^2(k)$.

The p.d.f. of $\text{chi}^2(k)$ is a *gamma* density. Thus, from (6) of §6.7, with $\alpha = k/2$ and $\lambda = 1/2$, we see that the p.d.f. is

$$f_{\chi^2(k)}(u \mid k) = \frac{1}{2^{k/2}\Gamma(k/2)} u^{k/2 - 1} e^{-u/2}, \quad u > 0.$$

Figure 6-6 shows the graphs of the chi-square p.d.f. for several values of the parameter k. The c.d.f. is not an elementary function, but of course, as an integral, it can be evaluated numerically. Table V gives the c.d.f. in terms of distribution percentiles as well as right-tail areas—probabilities of the form $P(\chi^2 > K)$.

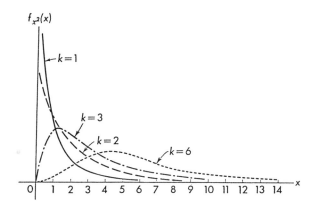

Figure 6-6. Some chi-square p.d.f.'s

The mean and variance of a chi-square variable can be calculated from its m.g.f., but they follow more simply from its structure:

$$E(\chi^2) = E(Z_1^2 + \cdots + Z_k^2) = k\, E(Z^2) = k, \tag{3}$$

and

$$\operatorname{var} \chi^2 = \operatorname{var}(Z_1^2 + \cdots + Z_k^2) = k \operatorname{var} Z = 2k. \tag{4}$$

The structure of χ^2 as a sum is also useful in establishing the reproductive property stated in the next theorem:

THEOREM 5 *If $Y_i \sim chi^2(k_i)$, $i = 1, \ldots, n$, are independent, their sum has a chi-square distribution: $\sum Y_i \sim chi^2(\sum k_i)$.*

This is clear since each Y_i is the sum of squares of k_i independent, standard normal variables, so that $\sum Y_i$ is the sum of squares of $\sum k_i$ independent, standard normal variables. The result can also be obtained using m.g.f.'s [Problem 6-79(b)], a tool that we use in deriving the following result which we'll have several occasions to appeal to in Chapters 8 and 14:

THEOREM 6 *If $X = U + V$, where U and V are independent, and* where $X \sim chi^2(m)$ and $U \sim chi^2(k)$, then $V \sim chi^2(m-k)$.

The proof is as follows: The m.g.f. of X is the product of the m.g.f.'s of U and V. Thus,

$$\psi_X(t) = (1-2t)^{-m/2} = \psi_U(t) \cdot \psi_V(t) = (1-2t)^{-k/2} \cdot \psi_V(t),$$

whence

$$\psi_V(t) = (1-2t)^{-(m-k)/2}.$$

This is the m.g.f. of $chi^2(m-k)$, so (by the uniqueness property of m.g.f.'s) the distribution of V is as asserted in the theorem. Yet another consequence of the structure of χ^2 is the following:

THEOREM 7 *If $U \sim chi^2(k)$, then for large k, $U \approx \mathcal{N}(k, 2k)$.*

Since (by Theorem 5) a variable which is $chi^2(k)$ is the sum of k i.i.d. variables whose common distribution is $chi^2(1)$, the theorem follows from the central limit theorem. This means that we can get approximate chi-square probabilities using the normal table, as we illustrate in the next example.

EXAMPLE 6.9a To approximate the 95th percentile of $chi^2(50)$, we can use the standard method—the mean (50) plus the 95th percentile of Z times the standard deviation ($\sqrt{100}$):

$$\chi^2_{.95} \doteq 50 + 1.645 \times 10 = 66.45.$$

It has been found that a somewhat better approximation is obtained using the variable

$$Y = \sqrt{2\chi^2} - \sqrt{2k-1},$$

which is also approximately standard normal, so that

$$\chi^2_p \doteq \tfrac{1}{2}(z_p + \sqrt{2k-1})^2.$$

Thus, in the present case,

$$\chi^2_{.95} \doteq \tfrac{1}{2}(1.645 + \sqrt{100-1})^2 = 67.2.$$

This is closer to the correct value, which is 67.5. ∎

PROBLEMS

6-69. Given $X \sim N(10, 4)$, find
(a) $P(X < 11.5)$.
(b) $P(|X - 10| > 3)$.
(c) $E(X^2)$.
(d) $E[(X - 10)^4]$.
(e) $E(X^3)$.
(f) the first and third quartiles.

6-70. Show that if X is normal with zero mean, so is $-X$.

6-71. Use characteristic functions to prove Theorem 3.

6-72. Suppose a certain test score is normally distributed. One student gets a score of 24.5, at the 75th percentile, and a second student, a score of 19, at the 40th percentile. Find the mean and standard deviation of the distribution.

6-73. Determine the p.d.f. and the expected value of $Y = |X - \mu|$, where $X \sim N(\mu, \sigma^2)$.

6-74. Find the density function of of $V = Z^2$, where $Z \sim N(0, 1)$, by first finding the c.d.f., $P(Z^2 \leq v)$, and differentiating with respect to v. (Do you recognize the distribution?)

6-75. Suppose X and Y are independent, standard normal variables.
(a) Find the joint p.d.f. of $|X|$ and Y. [See Problem 6-73.]
(b) Find the p.d.f. of $U = Y/|X|$. [Hint: First find the c.d.f of U; that is, find $P(Y/|X| \leq u)$.]

6-76. Obtain an expression for the even order moments of a standard normal distribution in terms of a gamma function.

6-77. Show that when the normal variables $X_1,..., X_n$ are independent, every linear combination $\sum a_i X_i$ is normal.

6-78. Show that if $X \sim U(0, 1)$, the variable $-2 \log X$ has a chi-square distribution with 2 degrees of freedom.

6-79. Use the chi-square m.g.f.
(a) to derive the formulas for mean and variance.
(b) to show the reproductive property in Theorem 5.

6-80. Using a normal approximation, find approximate values of
(a) the 80th percentile of $chi^2(60)$.
(b) $P(Y > 60)$, when $Y \sim chi^2(50)$.

6-81. Show that the variable $Y = \sqrt{2Y^2} - \sqrt{2k - 1}$ (from Example 6.9a) is approximately standard normal.

6-82. Let $Y = 2\lambda \sum X_i$, where $X_1, X_2, ..., X_n$ are i.i.d. $Exp(\lambda)$. Show that $Y \sim Chi^2(2n)$.

6-83. When $Y \sim \mathcal{N}(\mu, \sigma^2)$, the variable $X = e^Y$ is said to be *lognormal*, since $\log X$ is normal. (The lognormal distribution find applications in physics, biology, economics, and various other fields.)
  (a) Find the density of X.
  (b) Find the mean and variance of X in terms of the parameters of Y.
  (c) Show that if X is lognormal, so is X^α for any real α.

6-84. Suppose X, Y, and Z are i.i.d., each $\mathcal{N}(0, \sigma^2)$.
  (a) Find the p.d.f., mean, and variance of $R = (X^2 + Y^2)^{1/2}$. (The variable R is said to have a *Rayleigh* distribution.)
  (b) Find the p.d.f. of $T = (X^2 + Y^2 + Z^2)^{1/2}$. [This defines the *Maxwell* distribution.]

6.10 Multinomial distributions

In §6.2 we obtained the *bi*nomial distribution as the distribution of the number of "successes" in n independent trials of a Bernoulli experiment—sampling a "population" whose members are in one of two categories. We now generalize the situation to one in which population members are in one of k categories, $A_1, A_2, \ldots, A_k$. Let p_i denote the population proportion of individuals in category A_i, or the probability that, at a given trial, the outcome is A_i. These probabilities or proportions must satisfy
$$p_1 + p_2 + \cdots + p_k = 1.$$
Now define X_i as the indicator variable for the event that the outcome is A_i—a Bernoulli variable with parameter p_i. These k X's define a generalized Bernoulli trial; consider their joint m.g.f.:
$$\psi_\mathbf{X}(\mathbf{t}) = E(e^{\mathbf{X}'\mathbf{t}}) = E[\exp(X_1 t_1 + \cdots + X_k t_k)].$$
Since one and only one of the X's is 1 at any given trial, the possible values of the exponent are $t_1, \ldots, t_k$, with corresponding probabilities $p_1, \ldots, p_k$. So
$$\psi_\mathbf{X}(\mathbf{t}) = p_1 e^{t_1} + \cdots + p_k e^{t_k}. \tag{1}$$
Because $\sum X_i = 1$, the distribution is really $k - 1$ dimensional—we could get by with just $k - 1$ X's, as we did in the case $k = 2$, where the single variable X_1 is Bernoulli.

Now consider n independent trials of the above experiment, and let Y_i denote the number of times the category A_i occurs in the n trials:

6.10 Multinomial distributions

$$Y_i = \sum X_i.$$

The vector $\mathbf{Y} = \sum \mathbf{X}$ is then the vector of category frequencies. As with the X's, $k-1$ of the Y's define the kth: $Y_k = n - (Y_1 + \cdots + Y_{k-1})$. So we can think of the distribution either as k-dimensional, but restricted to a hyperplane, or as $k-1$ dimensional. We refer to it as **k-nomial**. When $k = 2$, the distribution is binomial; when $k = 3$, it is trinomial; etc. In general, we refer to the distributions as **multinomial**.

We'll obtain the joint p.d.f. using the kind of reasoning that leads to the binomial formula [(3) in §6.2]: In a sequence of the results of n independent trials, the probability of finding exactly y_i A_i's ($i = 1, 2, ..., k$) in some particular order is the product

$$p_1^{y_1} p_2^{y_2} \cdots p_k^{y_k}. \tag{2}$$

The probability of exactly y_i A_i's ($i = 1, 2, ..., k$) *without* regard to order is the product of (2) and the number of distinct arrangements of the A_i's:

$$P(Y_1 = y_1, ..., Y_k = y_k) = \binom{n}{y_1, ..., y_k} p_1^{y_1} p_2^{y_2} \cdots p_k^{y_k}, \tag{3}$$

where each y_i is an integer from 0 to n, and $\sum y_i = n$. This is the p.f. of the multinomial distribution.

The m.g.f. of a k-nomial distribution, defined by n independent trials of $\mathbf{X}$, is the nth power of the m.g.f. (1):

$$\psi_\mathbf{Y}(\mathbf{t}) = [\psi_\mathbf{X}(\mathbf{t})]^n = (p_1 e^{t_1} + \cdots + p_k e^{t_k})^n.$$

The distribution of $\mathbf{Y}$ is restricted to the hyperplane $y_1 + \cdots + y_k = n$, so one of the Y's is redundant—expressible in terms of the others. The joint m.g.f. of the first $k-1$ Y's is obtained by setting $t_k = 0$:

$$(p_1 e^{t_1} + \cdots + p_{k-1} e^{t_{k-1}} + p_k)^n. \tag{4}$$

Marginal distributions are obtained by setting appropriate t_i's equal to 0, which results in a m.g.f. of the same type as (4). For instance, when $k = 5$, the joint m.g.f. of (Y_1, Y_2) is

$$[p_1 e^{t_1} + p_2 e^{t_2} + (p_3 + p_4 + p_5)]^n,$$

the m.g.f. of a trinomial distribution. [This is perhaps more directly reasoned by simply observing that focusing on Y_1 and Y_2 amounts to considering just the three categories: A_1, A_2, and "none of the above," with probabilities p_1, p_2, and $1 - p_1 - p_2$, respectively.] Similar reasoning yields the following:

THEOREM 8 *If the distribution of $(Y_1 + \cdots + Y_k)$ is multinomial, the marginal distributions of any subset of variables Y_i is multinomial.*

EXAMPLE 6.10a Suppose we roll eight dice—or a single die eight times in succession. Let Y_i denote the number of times the face numbered i appears. There are six categories for each die, so (in the above notation) $n = 8$, $k = 6$, $p_i = 1/6$. Then, for instance,

$$P[\mathbf{Y} = (1,2,1,3,0,1)] = \binom{8}{1,2,1,3,0,1}\left(\frac{1}{6}\right)^8 = \frac{3360}{1679616} \doteq .002.$$

The p.f. of the marginal distribution of (Y_1, Y_2) is trinomial:

$$f(y_1, y_2) = P(Y_1 = y_1, Y_2 = y_2)$$

$$= \frac{8!}{y_1! y_2! (8 - y_1 - y_2)!} \left(\frac{1}{6}\right)^{y_1 + y_2} \left(\frac{4}{6}\right)^{8 - y_1 - y_2}. \blacksquare$$

EXAMPLE 6.10b In recording the value of a continuous random variable X, with p.d.f. $f(x)$, it is necessary to round off—a process that effectively partitions the value space into mutually disjoint intervals $I_1, \ldots, I_k$. In making n independent observations on X, one usually records only the frequency Y_i of each I_i. $Y_i \sim \text{Bin}(n, p_i)$, where

$$p_i = P(I_i) = \int_{I_i} f(x)\,dx,$$

and the vector of k interval frequencies is multinomial. $\blacksquare$

6.11 Exponential families

Many of the distribution families we've encountered so far are special cases of general families called *exponential*. This is of more than merely academic interest: When we establish a result for the general family, it applies automatically to all subfamilies.

The **one-parameter exponential family** is defined by the following:

$$f(x; \theta) = B(\theta) h(x) \exp[Q(\theta) R(x)], \tag{1}$$

where "f" can be either the p.d.f. of a continuous distribution or the p.f. of a discrete distribution. Table 6-1 lists some one-parameter families by name with corresponding identifications of the functions B, Q, R, and h.

6.11 Exponential families

Table 6-1

Name	p.f. or p.d.f.	$B(\theta)$	$Q(\theta)$	$R(x)$	$h(x)$
Ber(θ)	$\theta^x(1-\theta)^{1-x}$, $x=0,1$	$1-\theta$	$\log\dfrac{\theta}{1-\theta}$	x	1
Bin(n,θ)	$\binom{n}{x}\theta^x(1-\theta)^{n-x}$, $x=0,\ldots,n$	$(1-\theta)^n$	$\log\dfrac{\theta}{1-\theta}$	x	$\binom{n}{x}$
Geo(θ)	$\theta(1-\theta)^{x-1}$, $x=1,2,\ldots$	$\dfrac{\theta}{1-\theta}$	$\log(1-\theta)$	x	1
Negbin(r,θ)	$\binom{x-1}{r-1}\theta^r(1-\theta)^{(x-r)}$, $x=r,r+1,\ldots$	$\dfrac{\theta^r}{(1-\theta)^r}$	$\log(1-\theta)$	x	$\binom{x-1}{r-1}$
Poi(θ)	$\dfrac{\theta^x}{x!}e^{-\theta}$, $x=0,1,\ldots$	$e^{-\theta}$	$\log\theta$	x	$1/x!$
Exp(θ)	$\theta e^{-\theta x}$, $x>0$	θ	$-\theta$	x	1
$\mathcal{N}(0,\theta)$	$\dfrac{1}{\sqrt{2\pi\theta}}e^{-x^2/2\theta}$	$(2\pi\theta)^{-1/2}$	$-(2\theta)^{-1}$	x^2	1
$\mathcal{N}(\theta,1)$	$\dfrac{1}{\sqrt{2\pi}}e^{-(x-\theta)^2/2}$	$(2\pi)^{-1/2}e^{-\theta^2/2}$	θ	x	$e^{-x^2/2}$
Gam(α,θ)	$\dfrac{\theta^\alpha}{\Gamma(\alpha)}x^{\alpha-1}e^{-\theta x}$, $x>0$ (α given)	θ^α	$-\theta$	x	$\dfrac{x^{\alpha-1}}{\Gamma(\alpha)}$

The Cauchy distribution and the hypergeometric distribution are not included in the list; they are not in the exponential family.

A multiparameter exponential family is defined by a p.d.f. or p.f. of the following form, in which $\boldsymbol{\theta} = (\theta_1, ..., \theta_k)$:

$$f(x; \boldsymbol{\theta}) = B(\boldsymbol{\theta})\, h(x) \exp[Q_1(\boldsymbol{\theta})R_1(x) + \cdots + Q_k(\boldsymbol{\theta})R_k(x)]. \qquad (2)$$

The general normal distribution, with two parameter (μ, σ^2), belongs to this family with $k = 2$, as we explain in the next example.

EXAMPLE 6.11a The p.d.f. of the general normal distribution is

$$f(x; \mu, \sigma^2) = \frac{1}{\sqrt{2\pi\sigma^2}} \exp\left\{-\frac{(x-\mu)^2}{2\sigma^2}\right\}$$

$$= \frac{1}{\sqrt{2\pi\sigma^2}} \exp\left(-\frac{\mu^2}{2\sigma^2}\right) \exp\left\{-\frac{x^2}{2\sigma^2} + \frac{x\mu}{\sigma^2}\right\}.$$

Then, with

$$B(\mu, \sigma^2) = \frac{1}{\sqrt{2\pi\sigma^2}} \exp\left(-\frac{\mu^2}{2\sigma^2}\right), \quad Q_1(\mu, \sigma^2) = -\frac{1}{2\sigma^2}, \quad R_1(x) = x^2,$$

and

$$Q_2(\mu, \sigma^2) = \frac{\mu}{\sigma^2}, \quad R_2(x) = x,$$

we see that f is in the two-parameter exponential family. ∎

A multivariate p.d.f. for the random vector $\mathbf{X}$ is said to belong to the multivariate exponential family if it is of the form (2) in which the single argument x is replaced by the vector argument $\mathbf{x}$. If the p.d.f. of X is of the form (1): $f(x; \theta) = C(\theta)g(x)\exp[Q(\theta)S(x)]$, then the joint p.d.f. of n independent copies of X is

$$f(\mathbf{x}; \theta) = \prod f(x_i; \theta) = [C(\theta)]^n \prod g(x_i) \exp\left\{Q(\theta) \sum S(x_i)\right\},$$

which is in the multivariate exponential family.

6.12 The Cauchy family

In the limit theorems of Chapter 5, we assumed existence of at least the first moment, and most of the distributions we have encountered so far have finite first moments. A notable exception is the distribution defined in Examples 4.2a and 4.10b and used in Problems 4-76 and 78(c).

The family of **Cauchy distributions** is the family generated by linear transformations on the variable of Example 4.10b, with p.d.f.

$$f(x) = \frac{1/\pi}{1+x^2}. \tag{1}$$

The corresponding characteristic function, from Example 4.10b, is

$$\phi(t) = e^{-|t|}. \tag{2}$$

A Cauchy distribution has no moments of positive integral order. However, it is apparent from symmetry that the *median* of the distribution is 0.

We can find the c.d.f. by integrating the p.d.f.:

$$F(x) = \int_{-\infty}^{x} \frac{1/\pi}{1+y^2} dy = \tfrac{1}{\pi}[\text{Arctan } x] + \tfrac{1}{2}.$$

From this we can find the quartiles. Thus, the third quartile Q_3 is a number such that $F(Q_3) = 3/4$, or $\text{Arctan}(Q_3) = \pi/4$. So $Q_3 = 1$, and by symmetry the first quartile is $Q_1 = -1$. We'll refer to a Cauchy variable with these parameters as *standard* Cauchy.

We say that any distribution with p.d.f. reducible to (1) by a linear transformation is a Cauchy distribution. Thus, if Y has the p.d.f.

$$f_Y(y \mid \alpha, \theta) = \frac{\alpha/\pi}{\alpha^2 + (y-\theta)^2},$$

the variable $X = (Y - \theta)/\alpha$ has the p.d.f.

$$f_X(x) = \alpha f_Y(\theta + \alpha x) = \frac{1/\pi}{1+x^2},$$

which is (1). So Y is Cauchy.

Since the Cauchy distribution is a counterexample to limit theorems with even weaker moment conditions than those we've given, one might think it unusual to encounter a random variable with a Cauchy distribution. But, as we see in the corollary of next theorem, the ratio of two independent, standard normal variables is Cauchy:

THEOREM 9 *Suppose (X, Y) has a circularly symmetric, continuous, bivariate distribution, with p.d.f.*

$$f(x, y) = g(x^2 + y^2),$$

where g is a nonnegative function on $(0, \infty)$. Then $U = Y/X$ has a standard Cauchy distribution.

To prove Theorem 9, we first observe that because f is a p.d.f.,

$$1 = \int\int f(x,y)\,dx\,dy = \int_0^{2\pi}\int_0^\infty g(r^2)\,r\,dr\,d\theta = 2\pi\int_0^\infty r\,g(r^2)\,dr.$$

With this we can calculate the c.d.f. of the ratio $U = Y/X$. The event $[Y/X \leq u]$ defines two sectors of the plane (centered at the origin), with central angle $\phi = \text{Arctan}\,u + \pi/2$. (See Figure 6-7.) But for any such sector R_ϕ with central angle ϕ, $P(R_\phi)$ is proportional to ϕ:

$$P(R_\phi) = \int\int g(x^2+y^2)\,dx\,dy = \int_0^\infty \left\{\int_a^{a+\phi} d\theta\right\} r\,g(r^2)\,dr = K\phi,$$

Hence

$$F_U(u) = P(Y/X \leq u) = \frac{1}{\pi}\left(\text{Arctan}\,u + \frac{\pi}{2}\right),$$

which is the c.d.f. of the standard Cauchy distribution.

COROLLARY *The distribution of the ratio of two independent, standard normal variables is standard Cauchy.*

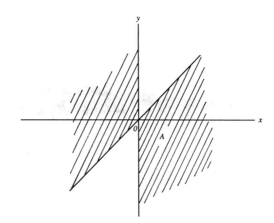

Figure 6-7

PROBLEMS

6-85. Consider a sample of n individuals drawn at random from a very large population whose responses to a survey question are either "yes", "no", or "no opinion." Give the joint distribution of the sample frequencies of these three categories in terms of the corresponding population proportions.

6-86. Let X have a triangular density: $f(x) = 1 - |x|$, $-1 < x < 1$, and consider six independent observations on X. If we divide the support into four equal class intervals, what is the probability that there will be one observation in each of the outside intervals and two each in the two inside intervals?

6-87. Suppose $X \sim N(0, 1)$, and consider the eight intervals determined by the values $x = 0$, ± 1, ± 2, , ± 3. Give the distribution of the frequencies of occurrence of these intervals in n independent observations on X.

6-88. A group of students consists of five each of freshmen, sophomores, juniors, and seniors. Find the probability that a committee of eight chosen at random (without replacement) from the 20 students represents the four classes equally.

6-89. A four-sided die is tossed eight times. Find the probability that each side turns up exactly twice. (Cf. the preceding problem.)

6-90. Given that $(X_1, ..., X_k)$ has a multinomial distribution with parameters $(p_1, ..., p_k)$, use the marginal m.g.f. of X_1 and X_2 to obtain $\text{cov}(X_1, X_2) = -np_1p_2$.

6-91. Show that the p.d.f. (9) of §6.6, for given r, is in the exponential family.

6-92. Show that the family of densities with p.d.f.'s

$$f(x \mid r, s) \propto x^{r-1}(1-x)^{s-1}, \quad 0 < x < 1,$$

where $r > 0$, $s > 0$, is in the two-parameter exponential family. (The constant of proportionality will involve the parameters r and s.)

6-93. Show that the Rayleigh distribution with parameter θ (Problem 6-84) is in the one-parameter exponential family.

6-94. Let X_1 and X_2 be independent and identically distributed with a discrete p.f. in the exponential family (1) of §6.11. Show that the distribution of the sum $R(X_1) + R(X_2)$ is in the exponential family.

7
Sampling and Reduction of Data

We'll refer to the way things are, to the true state of affairs—in particular, with regard to some phenomenon of interest—as the *state of nature*. When the phenomenon involves "chance," we use a probability model for it. However, the appropriate model in any given situation is usually not known. In some situations, the model may be one of a family of models indexed by a parameter θ, say, either by necessity or by assumption. It is then the value of θ that is unknown. For instance, if the variable of interest can take on one of only two possible values, the model to use is a Bernoulli distribution; but the precise value of the Bernoulli parameter p is usually unknown.

It seems reasonable to think that one can learn something about the model for an experiment of chance—about the "laws" of nature—by doing the experiment one or more times. The results constitute a **sample**, and obtaining them is called **sampling**. We'll see that when the process of sampling is suitably structured, we can indeed make sensible (yet often risky) inferences about the model. **Statistical inference** is the drawing of conclusions about a model on the basis of data—on statistical evidence.

7.1 Random sampling

The object of an investigation is sometimes a group of individuals or "population," or more specifically, the characteristics or responses of the individuals in a population. In such cases a sample is a subset of the population; and if the sample is to have an assessable chance of reflecting population characteristics, it should be drawn at random. Selecting at random from the population is an "experiment." The randomness of selection implies a probability model for the population, or for the experiment of selecting from it.

When we measure or observe some characteristic X of the individual selected, we create a random variable on the sample space of the population. We may then transfer our attention and our modeling to the population of values of X, and refer to the distribution of X as the population distribution—the "state of nature," with respect to the problem at hand.

EXAMPLE 7.1a Suppose the student population of a certain university includes 24 percent freshmen, 20 percent sophomores, 18 percent juniors, 15 percent seniors, and 23 percent graduate students. Selecting a student at random assigns equal probabilities to all students. Observing X, the class of the student selected, defines a discrete random variable with this distribution:

Class	Fr	So	Jr	Sr	Gr
Probability	.24	.20	.18	.15	.23

This is the population distribution, and it is the distribution of any single observation X. ∎

The simplest form of random sampling from a finite population proceeds in one of the following two ways:

(1) Objects or individuals are drawn at random, one at a time, and a particular characteristic of interest is determined for each one drawn. After each observation and before the next selection, the object just drawn is replaced and the population thoroughly mixed.

(2) Objects are drawn and observed as in (1), except that they are *not* replaced.

We'll find that analysis is simpler under (1). This is because the model for any given observation is not affected by the results of other observations. It is precisely to achieve this independence that individuals selected are replaced and the population mixed. Because the population is restored to its original state, the model for subsequent selections is the same at any point, no matter what one may know about earlier selections. With this kind of sampling, observations are i.i.d., and their common distribution is that of the population.

Method (2) is usually more practical and, for a given sample size, more informative. We have already seen in special cases that with this method, results are somewhat less variable—more precise. On the other hand, we also saw that when the population size is much greater than the sample size: $N \gg n$, there is little difference between sampling with replacement and sampling without replacement. In some cases, even though some individual population members are "real" enough, the population is conceptual when we take it to be infinite. Thus, we may want to consider the population of all conceivable males, or the population of all items of a particular type that will ever or could ever be produced. For these, $N = \infty$—replacement is not an issue.

In another kind of situation we observe a quantity X that is generated by an experiment of chance, but not by selection from a real population. The toss of a coin is such an experiment. We again speak of the distribution of X as a population distribution, and of observing X as sampling that distribution. Since there is not a finite, real population, there is no question of "with" or "without" replacement when more than one observation is made. Sometimes it can be safely assumed that successive observations are independent. When that is the case, the population distribution remains the same at each observation, and the sampling process is mathematically the same as (1) above:

(3) Observations are obtained as the result of repeated, independent trials of an experiment, under conditions that are identical (with respect to those factors that can be controlled).

We'll refer to samples obtained by methods (1) and (3) as **random samples**, and those obtained by method (2) as **simple random samples**.

We denote successive observations on X by $X_1, X_2, \ldots$, and for much of what is to follow we assume that there is a fixed, preset sample size which we denote by n. If the population is described by a p.d.f. (or p.f.) f, this is the model for each observation.

DEFINITION A *random sample* of observations on X is a set of i.i.d. random variables $X_1, \ldots, X_n$, each with the distribution of X.

Because the observations in a random sample are independent, their joint p.d.f. (or p.f.) is the product of their p.d.f.'s (or p.f.'s), and these functions are the same for all observation—the population f:

$$f_X(\mathbf{x}) = \prod_{i=1}^{n} f(x_i). \tag{1}$$

This is defined on the n-dimensional sample space of the observation vector, and of course requires n arguments! The dummy variable x_i is associated with X_i, the ith sample observation.

EXAMPLE 7.1b Measurement errors often follow a normal distribution, at least approximately. In n successive measurements, the joint distribution of the errors X_i, according to (1), is the product of normal densities:

$$f(x_1, \ldots, x_n) = (2\pi\sigma^2)^{-n/2} \exp\left\{\frac{-1}{2\sigma^2} \sum_{i=1}^{n} (x_i - \mu)^2\right\}.$$

This is an n-variate p.d.f., describing a distribution in $\mathcal{R}^n$, the space of all possible samples of size n. Note in particular that there are n arguments x_i on the left, and the *same* n arguments are on the right. ∎

Strictly speaking, the question of whether or not an actual sampling process is faithfully represented by a particular sampling model is unanswerable. In practice, one can only make every effort to see that the sampling process conforms to the conditions imposed by the model.

7.2 Describing samples

We introduce here a useful but artificial probability distribution associated with a sample, which is to be thought of as an empirical version of the population distribution:

DEFINITION Given a sample $\mathbf{X} = (X_1, \ldots, X_n)$, the *sample distribution* is the discrete distribution that assigns probability $1/n$ to each sample value X_i.

When the population variable X is discrete, with possible values $x_1, \ldots, x_n$, assigning probability $1/n$ to each observation in the sample implies a probability f_j/n for the value x_j, where f_j is the frequency of occurrence of x_j in the sample. Samples with repeated values are often given by means of a frequency table, as in the next example.

EXAMPLE 7.2a Given a random sample of n students from the student population of Example 7.1a, it is customary to summarize the sample by giving the sample frequency of each category. For instance, in a random sample of 50 students, we might find these frequencies:

Class	Fr	So	Jr	Sr	Gr
Probability	10	12	8	7	13

The sample distribution is defined by the relative frequencies:

Class	Fr	So	Jr	Sr	Gr
Frequency	.20	.24	.16	.14	.26

∎

When X is *numerical*, a sample distribution is conveniently described by a distribution function:

7.2 Describing samples

DEFINITION The *sample distribution function* is the c.d.f. of the sample distribution, which jumps an amount $1/n$ at each sample value:

$$F_n(x) = \frac{\#\,(\text{observations} \leq x)}{n}. \tag{1}$$

If a particular value occurs k times in a data set, that value has probability k/n, and the d.f. jumps an amount k/n there. Ideally, when X is continuous, single values would not occur more than once in a finite sample. But in practice they can and do occur more than once, because of the round-off necessary in the recording of data.

The sample d.f. is a statistic—a reduction of the data, defined by the values where the jumps occur and the amounts of the jumps. The sample d.f. is an empirical version of the population c.d.f., and would be expected to resemble that ideal, especially when n is large. Indeed, according to the law of large numbers (§5.2), we know that for each x,

$$F_n(x) \xrightarrow{p} F(x),$$

where $F(x)$ is the population c.d.f. For, $F_n(x)$ is the relative frequency of the event $[X \leq x]$ and $F(x)$ is its probability. Actually, one can make a stronger claim: The sample d.f. converges to $F(x)$ *uniformly in* x, with probability 1.[1]

EXAMPLE 7.2b Lengths of remission (in weeks) for 21 leukemia patients receiving a drug treatment (6-MP) were observed, as follows:

10, 7, 32, 21, 22, 6, 16, 34, 32, 25,
11, 20, 19, 6, 17, 35, 6, 13, 9, 6, 10

Figure 7-1 shows the sample d.f. ∎

Like probability distributions, data sets are often more readily understood (especially large ones) in terms of various descriptive quantities. We have used the term "parameter" for numerical quantities that describe a population or the distribution of a random variable. And although the term "sample parameter" may be logically correct, we use the traditional term **statistic** for quantities that describe a sample. Such quantities are of course calculated from the observations in the sample.

DEFINITION A *statistic* is a function of the observations in a sample.

---

[1] C. R. Rao, *Linear Statistical Inference and Its Applications*, 2nd Ed., New York: John Wiley & Sons, Inc., 1973, page 42.

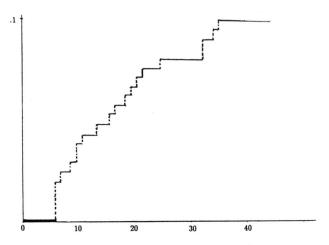

Figure 7-1 Sample d.f. for Example 7.2b

A statistic can be a single real-valued function, $Y = t(X_1, ..., X_n)$. Such a function is a random variable, depending as it does on the sample observations, which are random variables. But to be a *statistic*, Y cannot depend on anything other than the sample observations.

Sometimes we'll consider several descriptive measures, and so define a vector-valued statistic. Unless the statistic consists of the whole sample itself, calculating a statistic *reduces* the data to a single number, or to a vector of numbers. (In §7.8 we address the question of whether one loses any pertinent sample information by such reduction.)

EXAMPLE 7.2c In the setting of Example 7.2a, the vector of frequencies of the various classes, $(f_F, f_{So}, f_J, f_{Sr}, f_G)$, is a statistic, depending as it does on the particular sample that results from the sampling process, and of course it is a reduction of the raw data. On the other hand, the vector (10, 12, 8, 7, 13) is not variable; rather, it is simply a particular realization of the random vector of frequencies. ∎

Data from continuous variables are often recorded as tallies in a scheme of, say, k class intervals and rounded to the class interval midpoints. The result is a **frequency distribution**, with possible values x_i (those midpoints) and corresponding frequencies f_i. A frequency distribution is commonly represented graphically as a **histogram**. This is a plot in which there is a bar above each class interval whose height is proportional to the class frequency—if the class intervals are of equal width. If the class intervals are not of the same width, the heights of the bars are adjusted so that the *areas* of the bars are proportional to frequency. The area above any interval is then (at least approximately) the relative frequency of that interval. Thus, a histogram is a kind of "density function" of the sample distribution.

EXAMPLE 7.2d The frequency distribution in the following table summarizes the hemoglobin levels of 50 cancer patients at a veterans' hospital (in gm/100 ml).

Class interval	Midpoint	Frequency
9.95-10.85	10.4	1
10.85-11.75	11.3	6
11.75-12.65	12.2	7
12.65-13.55	13.1	12
13.55-14.45	14.0	12
14.45-15.35	14.9	9
15.35-16.25	15.8	3
	Total:	50

The corresponding histogram is shown in Figure 7-2. This frequency distribution has resulted from rounding the hemoglobin levels that constitute the raw data, so it only approximately represents the sample. Indeed, other choices of a system of class-intervals would result (from the same raw data) in a somewhat different frequency distribution and histogram. ∎

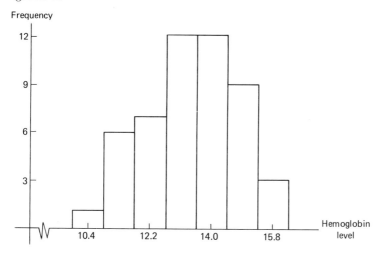

Figure 7-2. Histogram for Example 7.2d

A preliminary reduction of the data in a sample from a numerical population is given by the **order statistic**:

$$\mathbf{Y} = (X_{(1)}, \ldots, X_{(n)}), \qquad (2)$$

where $Y_i = X_{(i)}$ denotes the ith smallest sample observation. (There will be some ambiguity in the case of a discrete variable, because of the possibility of repeated values.) The sample d.f. is determined by the order statistic (and vice versa), so it too is a statistic.

We have worked with various descriptive parameters of a probability distribution, and when we apply these to the sample d.f., they define sample statistics. Thus, moments and quantiles (percentiles, quartiles, median, etc.) are automatically defined for samples, as the corresponding parameters of the sample d.f.

Because sample distributions are discrete, sample quantiles are often not uniquely defined. For instance, when the sample size is odd: $n = 2k - 1$, the median is the middle observation, $X_{(k)}$; but when the sample size is even: $n = 2k$, there is not a single middle observation, and any number between $X_{(k)}$ and $X_{(k+1)}$ is a median. It is customary to take the average of these two middle observations as "the" median.

DEFINITION The **sample median** is

$$\tilde{X} = \begin{cases} X_{(k)}, & n = 2k - 1, \\ \frac{1}{2}[X_{(k)} + X_{(k+1)}], & n = 2k. \end{cases}$$

The quartiles Q_1 and Q_3 are the 25th and 75th percentiles, respectively, with some convention for definiteness when these are not unique. For instance, Q_1 and Q_3 might be defined as the medians of the lower and upper halves, respectively, of the ordered data, where in the case of odd n, each "half" would include the middle observation.

Some other statistics based on the ordered observations are these:

The **range**: $R = X_{(n)} - X_{(1)}$,

The **interquartile range**: $\text{IQR} = Q_3 - Q_1$,

The **midrange**: $\frac{1}{2}[X_{(1)} + X_{(n)}]$.

The midrange is a measure of the middle of a sample. The range and interquartile range are measures of dispersion or spread. The range is easily calculated for small samples, and as such it has been found useful in industrial settings, in the construction of charts for quality control.

The five numbers $(X_{(1)}, Q_1, \tilde{X}, Q_3, X_{(n)})$ constitute a crude but handy summary of the information in a sample from a continuous population. They are the basis of a **box plot**, to be illustrated in the next example.

EXAMPLE 7.2e The 21 observations in Example 7.2b are repeated here, but arranged in numerical order—in the order statistic:

6, 6, 6, 6, 7, 9, 10, 10, 11, 13, 16, 17, 19, 20, 21, 22, 25, 32, 32, 34, 35.

The median is $\tilde{X} = 16$, the middle observation (11th from either end), and the quartiles are $Q_1 = 9$, $Q_3 = 22$. Thus, IQR = 13, and the midrange is 41/2. The five-number summary is thus (6, 9, 16, 22, 35). The *box-plot*, based on these five numbers, is shown in Figure 7-3. Vertical lines are drawn at the quartiles to define the box. Even in this crude plot, pertinent features of the data set stand out; for instance, a skewing to the right is evident. ∎

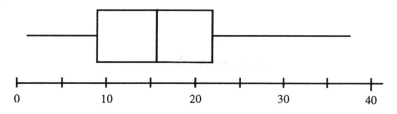

Figure 7-3. Box plot for Example 7.2e

As in the case of numerical populations, the most used moments of a sample distribution are the mean and the variance. The first moment is the sum of sample values each multiplied by the "probability" $1/n$:

$$\bar{X} = \frac{1}{n} \sum_{i=1}^{n} X_i \qquad (3)$$

This is the **sample mean**. Being defined as the mean of a distribution, it has the properties of a population mean:

THEOREM 1 *If $Y_i = a + bX_i$ for each i, then $\bar{Y} = a + b\bar{X}$.*

THEOREM 2 *The sum of the deviations of a set of numbers about their mean value is always 0:*

$$\sum_{i=1}^{n} (X_i - \bar{X}) = 0.$$

When data are summarized in a frequency distribution, with k possible values x_j and corresponding frequencies f_j, for $j = 1, ..., k$, the sum (3) is found by multiplying the possible values by corresponding frequencies and summing. Thus, in this case

$$\bar{X} = \frac{1}{n} \sum_{j=1}^{k} f_j x_j, \qquad (4)$$

where $f_1 + \cdots + f_k = n$. When a frequency distribution is obtained by rounding the observations on a continuous variable, the mean as given by (4) is only approximately equal to the mean of the raw data. But the approximation is usually fairly good.

The *variance* of the sample distribution is the average squared deviation about the mean, where to "average" now means to sum and divide by n. This variance V is given by the equivalent formulas

$$V = \frac{1}{n} \sum_{i=1}^{n} (X_i - \bar{X})^2 = \frac{1}{n} \sum_{i=1}^{n} X_i^2 - \bar{X}^2. \qquad (5)$$

The next two theorems follow from general theorems about variances:

THEOREM 3 (Parallel axis theorem): *For any constant a,*

$$\frac{1}{n} \sum_{i=1}^{n} (X_i - a)^2 = V + (a - \bar{X})^2. \qquad (6)$$

Because it is a variance—the variance of a probability distribution, V is the *smallest* of all second sample moments. This is again evident from (6), where it is clear that the right-hand side is smallest when $a = \bar{X}$.

Although (5) could be taken as the "sample variance," it is customary, instead, to define **sample variance** as

$$S^2 = \frac{1}{n-1} \sum_{i=1}^{n} (X_i - \bar{X})^2 = \frac{\sum X_i^2 - n\bar{X}^2}{n-1} = \frac{n}{n-1} V. \qquad (7)$$

We then define the **sample standard deviation** S as the positive square root of this version of variance. (Using the denominator $n - 1$ makes various formulas and tables somewhat more convenient.)

For data summarized in a frequency distribution, with possible values x_j and corresponding frequencies f_j, the sums in (5) and (7) are calculated by weighting squares involving possible values with the corresponding frequency:

$$S = \sqrt{\frac{\sum f_i x_i^2 - n\bar{X}^2}{n-1}}. \tag{8}$$

[On a hand calculator with statistical functions you may see two keys for the standard deviation, perhaps marked σ_{n-1} and σ_n, or sx and σx; these provide the option of calculating the s.d. as either (8) or as $\sqrt{V}$, respectively.]

EXAMPLE 7.2f Calculations with the data in Example 7.2e yield these statistics: $\sum X = 357$ and $\sum X^2 = 8049$. Substituting them in (3) and (7), we find $\bar{X} = 17$ and $S = 9.95$. As when calculating the mean and standard deviation of a population, one should check them against a plot of the data to see that $\bar{X}$ and S appear reasonable as the balance point and "typical" deviation from the balance point, respectively. ∎

THEOREM 4 *If $Y_i = a + bX_i$ for each i, then*

$$V_Y = b^2 V_X, \text{ and } S_Y = |b| S_X. \tag{9}$$

Relations (9) follow from corresponding properties of probability distributions.

PROBLEMS

7-1. Write the joint p.d.f. or p.f. of the observations in a random sample of size n from
(a) $\mathcal{U}(a, b)$.
(b) $\mathcal{N}(\mu, \sigma^2)$.
(c) Ber(p).
(d) Poi(m).
(e) Geo(p): $f(x \mid p) = (1-p)^x p$, $x = 0, 1, 2, ...$.
(f) Cauchy(θ): $f(x|\theta) = \dfrac{1/\pi}{1 + (x - \theta)^2}$.

7-2. Give the joint p.f. of $(X_1, ..., X_n)$, obtained by sampling one at a time without replacement from a Bernoulli population of size N which includes M items coded 1 and $N - M$ items coded 0. [Hint: See Problem 2-42.]

7-3. Determine the joint p.f. of $(Y_1, ..., Y_k)$, where Y_i is the frequency of A_i in a random sample of a categorical variable with "values" A_i and corresponding probabilities $p_i = P(A_i)$, $i = 1, ..., k$.

7-4. The following are 30 test scores:
51, 52, 52, 58, 59, 59, 61, 62, 63, 69, 72, 74, 76, 80, 80
80, 81, 81, 82, 83, 83, 84, 86, 87, 87, 88, 88, 89, 90, 93.
(a) Plot the sample d.f.

(b) Find the mean and standard deviation.
(c) Find the median and midrange of the scores.
(d) Find the sample quartiles.
(e) Find the range and interquartile range.
(f) Construct a box-plot.
(g) Round the scores to the nearest multiple of 5 and calculate the mean and standard deviation of the rounded data. [Cf. (b).]

7-5. Given $\sum X^2 = 198$, $S^2 = 12$, and $n = 10$, find two possible values of $\bar{X}$.

7-6. Given $\bar{X} = 5$, $n = 10$, and $S_X = 2$,
(a) find $\sum X$ and $\sum X^2$.
(b) find $\bar{Y}$ and S_Y, where $Y_i = 3 + 4X_i$.

7-7. Define the statistic $M = \frac{1}{n}\sum |X_i - \bar{X}|$, the *mean deviation* of the sample distribution. (See §4.5.)
(a) Show that $M < S$ if $n > 1$.
(b) Find the mean deviation for the data in Problem 7-4.

7-8. For the hemoglobin data referred to in Example 7.2d, use the frequency distribution given in that example to find
(a) the mean.
(b) the standard deviation.
(c) the mean deviation (about the mean).
(d) the sample median (approximate).

7-9. Show that the second moment V of the sample $(X_1, ..., X_n)$ can be calculated as follows:
$$V = \frac{1}{2n^2} \sum_i \sum_j (X_i - X_j)^2.$$

7-10. Use the method of calculus for minimizing a function of a to show that $\sum (X_i - a)^2$ has its minimum value when $a = \bar{X}$.

7-11. Show that $\sum |X_i - a|$ has its minimum value at $a = \tilde{X}$, a median of the sample distribution. [Hint: Differentiate the sum, using
$$\frac{d|x|}{dx} = \frac{d\sqrt{x^2}}{dx} = \frac{x}{|x|} = \text{sign of } x,$$
and examine the value of the sum as a increases through the sample values.]

7.3 Sampling distributions

We pointed out that a statistic $T = t(X_1, ..., X_n)$ calculated from a random sample is a random variable. Its value varies from sample to sample, and this variation is described by its distribution, referred to as its **sampling distribution**. This distribution is implied by the assumed population distribution—the distribution of X. When the population distribution depends on a parameter θ, the sampling distribution of any statistic T depends on θ as well.

EXAMPLE 7.3a Let Y denote the number of successes in n independent observations on Ber(p). This sample sum is a statistic. Its sampling distribution was studied in §6.2: $Y \sim \text{Bin}(n, p)$. ∎

If a population is finite and sampling is done at random, we can (in principle) find the sampling distribution of a statistic by calculating its value for every possible sample. The samples are equally likely, so the probability of a particular value of the statistic is just the number of samples in which it has that value divided by the number of possible samples.

EXAMPLE 7.3b Consider a bowl with seven tickets numbered 1 to 7 as a population of size $N = 7$, and random selections of size $n = 3$. The statistic U defined as the absolute difference between the largest and smallest number in the sample is a random variable. We list here all of the 35 selections of three, along with the value of U for each of these samples:

X	U	X	U	X	U	X	U	X	U
123	2	136	5	167	6	247	5	356	3
124	3	137	6	234	2	256	4	357	4
125	4	145	4	235	3	257	5	367	4
126	5	146	5	236	4	267	5	456	2
127	6	147	6	237	5	345	2	457	3
134	3	156	5	245	3	346	3	467	3
135	4	157	6	246	4	347	4	567	2

The 35 samples are equally likely, so we can find the probabilities for the various values of U by counting. They are as follows:

u	2	3	4	5	6
$f(u)$	5/35	8/35	9/35	8/35	5/35

This table defines the sampling distribution of U. ∎

In some cases, a mathematical derivation of the sampling distribution of a particular statistic T is intractable, and we then settle for an approximate distribution, obtaining it by computer simulation: A computer is programmed to generate a string of (pseudo) random digits, taking n sequences of k of these digits to create a random sample from $\mathcal{U}(0, 1)$. These n observations are suitably transformed (for instance, by using the result in Problem 3-35) to constitute a random sample from the specified population distribution. The statistic T is then calculated for this sample. Repeating the process hundreds or thousands of times produces that many random samples and a value of T from each. If one could get infinitely many observations in this way, their distribution would be the desired sampling distribution. With a large but finite number of observations on T, their distribution will be a good approximation to the sampling distribution of T.

EXAMPLE 7.3c A computer was used to generate 400 samples of size $n = 5$ from $U(0, 1)$ and calculate the sample range R for each. The distribution of the 400 sample ranges is shown as a histogram in Figure 7-4 together with the actual sampling distribution found mathematically. (The latter will be derived in §7.6.) ∎

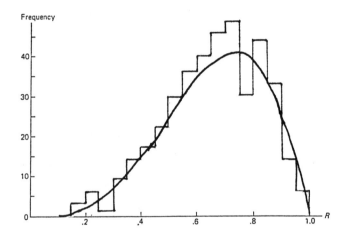

Figure 7-4. Empirical sampling distribution of R when $n = 5$

7.4 Sampling distributions of sample moments

In principle, the distribution of the mean of a random sample is derivable using either the moment generating function or (when this fails to exist) the characteristic function. The latter (from Problem 4-77) is

$$\phi_{\bar{X}}(t) = \left[\phi_X\!\left(\frac{t}{n}\right)\right]^n. \tag{1}$$

7.4 Sampling distributions of sample moments

This characterizes the distribution of $\bar{X}$, and the inversion formula (Theorem 24 of §4.10, page 129) gives the c.d.f. as an integral formula. If the characteristic function or the m.g.f. is one we recognize, it may be one for which we know the corresponding p.d.f., as in the following example.

EXAMPLE 7.4a Consider a random sample of size n from Exp(λ). We found the m.g.f. of the sample sum in §6.6 [(8), page 170],

$$\psi_{\Sigma X}(t) = (1 - t/\lambda)^{-n},$$

and recognized this as the m.g.f. of a distribution with p.d.f.

$$f_{\Sigma X}(y) = \frac{\lambda^n y^{n-1} e^{-\lambda y}}{(n-1)!}, \quad y > 0.$$

We then get the p.d.f. of $\bar{X} = \frac{1}{n}\sum X$ by applying (2) of §3.5 (page 64):

$$f_{\bar{X}}(u) = \frac{(n\lambda)^n u^{n-1} e^{-\lambda n u}}{(n-1)!}. \blacksquare$$

The important moments of the sample mean are its mean and variance. For a random sample from a population with mean μ and variance σ^2, these moments are easy to obtain without resorting to the m.g.f., simply by using properties of moments of sums of random variables:

THEOREM 5 If $(X_1, \ldots, X_n)$ is a random sample from a population with mean μ and variance σ^2, then

$$E(\bar{X}) = \mu, \quad \operatorname{var} \bar{X} = \frac{\sigma^2}{n}. \tag{2}$$

THEOREM 6 If $X_1, \ldots, X_n$ are obtained by sampling at random without replacement from a population of size N with mean μ and variance σ^2, then

$$E(\bar{X}) = \mu, \quad \operatorname{var} \bar{X} = \frac{\sigma^2}{n} \cdot \frac{N-n}{N-1}. \tag{3}$$

To show $E(\bar{X}) = \mu$, we use the linearity of expectations and the fact that (in both cases) observations are identically distributed, so that $EX_i = \mu$ for all i:

$$E(\bar{X}) = \frac{1}{n} \sum_{i=1}^{n} EX_i = \mu.$$

When observations are identically distributed, $\text{var} X_i = \sigma^2$; and if they are independent, the variance of their sum is the sum of their variances:

$$\text{var}\bar{X} = \frac{1}{n^2}\sum_{i=1}^{n}\text{var}X_i = \frac{\sigma^2}{n}.$$

What remains, for Theorem 6, is to obtain the formula for variance when the observations are merely exchangeable. This was essentially done in §6.3, where we found [as equations (5) and (6), page 151]

$$\text{var}\left(\sum X_i\right) = n\,\text{var}X_1 + n(n-1)\text{cov}(X_1, X_2),$$

and (with $n = N$)

$$0 = N\,\text{var}X_1 + N(N-1)\text{cov}(X_1, X_2).$$

Substituting the covariance from the second into the first of these last two equations yields the formula for variance in (3).

The *large sample* distribution for the sample mean follows from the central limit theorem [Theorem 5, §5.3], which asserts, when the population has a finite variance, that the standard score

$$Z = \frac{\bar{X} - \mu}{\sigma/\sqrt{n}}$$

converges in distribution to a standard normal variable. Thus, approximate probabilities for $\bar{X}$ can be found using the standard normal table.

We take up next the sampling distribution of the sample variance, but first consider the second sample moment about the sample mean, the statistic V as defined by (5) in §7.2:

THEOREM 7 *For random samples of size n from a population with mean μ and finite variance σ^2,*

$$E(V) = \sigma^2\left(\frac{n-1}{n}\right), \qquad (4)$$

and

$$\text{var}\,V = \frac{\mu_4 - \mu_2^2}{n} - \frac{2(\mu_4 - 2\mu_2^2)}{n^2} + \frac{\mu_4 - 3\mu_2^2}{n^3}. \qquad (5)$$

COROLLARY *For random samples of size n, $E(S^2) = \sigma^2$.*

To obtain (4), we turn to the parallel axis theorem [Theorem 3, page 204 with $u = \mu$] and average term by term:

$$E\left(\frac{1}{n}\sum (X_i - \mu)^2\right) = E\left(\frac{1}{n}\sum (X_i - \bar{X})^2\right) + E[(\bar{X} - \mu)^2]$$

or

$$\sigma^2 = E(V) + \frac{\sigma^2}{n}.$$

The corollary then follows upon setting $S^2 = \frac{n}{n-1}V$. The derivation of (5), although straightforward, is extremely tedious and will be omitted.[2]

Because the mean and variance are by far the most important sample moments, and because a study of higher order sample moments is quite laborious, we leave this study to the references[3] and simply quote the following result:

THEOREM 8 *Sample moments are asymptotically normal and tend in probability to the corresponding population moments, when these exist.*

COROLLARY 1 *If $\sigma < \infty$, and S_n is the standard deviation of a random sample, then $S_n \xrightarrow{p} \sigma$.*

COROLLARY 2 *If $S_n \xrightarrow{p} \sigma$, then $1/S_n \xrightarrow{p} 1/\sigma$.*

COROLLARY 3 *For random samples from a population with mean μ and finite variance σ^2, the "standard" score*

$$Z = \frac{\bar{X} - \mu}{S/\sqrt{n}}$$

converges in distribution to $\mathcal{N}(0, 1)$.

Corollaries 1 and 2 follow from Theorem 8 above and from Theorem 3 of Chapter 5 (page 137). Corollary 3 follows from Corollary 2, with the aid of Theorem 4 of Chapter 5.

7.5 Sampling normal populations

A normal population is completely defined by its mean μ and variance σ^2. In §7.8 we'll see that everything a random sample has to tell us about these parameters is summarized in the corresponding sample moments, namely, in the sample mean and sample variance. From

---

[2] See H. Cramér, *Mathematical Methods of Statistics*, Princeton, N. J., Princeton University Press, 1946, page 348.

[3] Ibid., page 365.

Theorem 4 of §6.8 we obtain the following important result:

THEOREM 9 *For a random sample of size n from $\mathcal{N}(\mu, \sigma^2)$,*
$$\bar{X} \sim \mathcal{N}(\mu, \sigma^2/n).$$

In this section we derive the distribution of the sample variance and show that the sample mean and variance are actually independent statistics.

THEOREM 10 *For a random sample $(X_1, ..., X_n)$ from $\mathcal{N}(\mu, \sigma^2)$, the statistic $\bar{X}$ and the vector $(X_1 - \bar{X}, ..., X_n - \bar{X})$ are independent.*

COROLLARY *The mean $\bar{X}$ and the sample variance S^2 of a random sample from a normal population are independent.*

To prove Theorem 10, we write out the joint m.g.f. of the $n+1$ quantities involved:

$$\psi(t, t_1, ..., t_n) = E[\exp\{t\bar{X} + t_1(X_1 - \bar{X}) + \cdots + t_n(X_n - \bar{X})\}].$$

The quantity in braces in the exponent can be rewritten as follows:

$$t\bar{X} + \sum t_i(X_i - \bar{X}) = \sum \left[\frac{t}{n} + (t_i - \bar{t})\right] X_i = \sum a_i X_i.$$

This is a linear combination of independent normal variables, in which the coefficients are

$$a_i = \frac{t}{n} + (t_i - \bar{t}),$$

where

$$\sum a_i = t, \text{ and } \sum a_i^2 = \frac{t^2}{n} + \sum (t_i - \bar{t})^2.$$

Thus,

$$\psi(t, t_1, ..., t_n) = E\left[\exp\left(\sum a_i X_i\right)\right] = \prod \psi_{X_i}(a_i)$$

$$= \prod \exp\left(\mu a_i + \frac{\sigma^2}{2} a_i^2\right) = \exp\left(\mu \sum a_i + \frac{\sigma^2}{2} \sum a_i^2\right)$$

$$= \exp\left(\mu t + \frac{\sigma^2 t^2}{2n}\right) \cdot \exp\left(\frac{\sigma^2}{2} \sum (t_i - \bar{t})^2\right).$$

The first factor in the exponent of this last expression is the m.g.f. of $\bar{X}$, and the second factor does not involve t. An obvious extension of

7.5 Sampling normal populations

Theorem 28 of §4.11 (page 131) shows the asserted independence. It also shows that the second factor is the n-variate m.g.f. of the vector of deviations: $(X_1 - \bar{X}, ..., X_n - \bar{X})$. The corollary then follows because the sample variance is a function of those deviations.

THEOREM 11 *If S^2 is the variance of a random sample from $N(\mu, \sigma^2)$, then*

$$(n-1)S^2/\sigma^2 \sim \text{chi}^2(n-1).$$

COROLLARY *For random samples of size n from $N(\mu, \sigma^2)$,*

$$\text{var } S^2 = \frac{2\sigma^4}{n-1}. \tag{1}$$

To establish Theorem 11 we use a form of the parallel axis theorem:

$$\sum \left(\frac{X_i - \mu}{\sigma}\right)^2 = \frac{(n-1)S^2}{\sigma^2} + \frac{(\bar{X} - \mu)^2}{\sigma^2/n}.$$

The term on the left is the sum of n squares of independent standard normal variables, so it is $\text{chi}^2(n)$. The second term on the right is the square of a single standard normal variable and is therefore $\text{chi}^2(1)$. The conclusion of Theorem 11 now follows from Theorem 6 of Chapter 6 (page 184) because the independence of the two terms on the right, as guaranteed by the corollary to Theorem 10. The variance of S^2 as given in the above corollary follows from the fact that the variance of a chi-square distribution is twice the number of degrees of freedom.

PROBLEMS

7-12. Consider a simple random sample of size $n = 2$ (sampling at random without replacement) from a population of four chips numbered 1 through 4.
(a) Find the joint distribution of the smaller and the larger of the two sample observations. (Give it as a table of joint probabilities.)
(b) Find the distribution of the sample range, using (a).
(c) Find the distribution of the sample mean.

7-13. Find the probability that the mean of a random sample of size 100 from a normal population with $\sigma = 4$ differs from the population mean by more than .8.

7-14. Find the mean and variance of S^2, the variance of a random sample from $U(0, 1)$.

7-15. Consider a random sample from Exp(2). Find $P(\bar{X} > .6)$,
(a) when $n = 10$. [Hint: You may find Problem 6-56 useful.]
(b) when $n = 100$.

7-16. Two independent random samples are taken from a population with mean 80 and s.d. 6. The sample sizes are 100 and 150. Let D denote the difference between the sample means. Find
(a) $E(D)$. (b) var D. (c) $P(|D| > 2)$.

7-17. Let $Y_n = \overline{X^k} = \frac{1}{n} \sum X_i^k$, the sample kth moment about 0.
(a) Find the mean, variance, and asymptotic distribution of Y_n in terms of population moments.
(b) Show that Y_n converges in probability to the corresponding population moment.

7-18. Show that the mean of a random sample from Cauchy(θ) [see Problem 7-1(f)] has the same distribution as the population.

7-19. Consider a random sample of size n without replacement from a population of size N. Use Theorem 3 (page 204) and Theorem 6 (page 209) to show that
$$E(S^2) = \frac{N}{N-1} \sigma^2.$$

7-20. Use the Schwarz inequality [(3), §4.8, page 117] to show: $ES \leq \sigma$.

7-21. Let $\bar{X}$ denote the mean of a random sample of size 10 from $N(30, 25)$.
(a) Find $P(\bar{X} < 28)$. (b) Find $P(S^2 > 40)$.

7-22. Two independent random samples are taken from $N(80, 36)$. The sample sizes are 10 and 15. Let D denote the difference between the sample means. Find $P(|D| > 2)$.

7-23. Consider a random sample from a normal population. Show:
(a) The sample mean deviation, $\frac{1}{n} \sum |X_i - \bar{X}|$, and the sample mean are independent statistics.
(b) The sample mean and the sample range are independent statistics.

7-24. In the case of normal populations, show how the formulas for mean and variance of S^2 follow from Theorem 11.

7-25. Show that for random samples from a normal population, the expected value of the quantity $(\bar{X} - \mu)/S$ is 0. [Hint: The mean of a quotient is *not* necessarily the quotient of the means. Write the quotient as a product and exploit independence of the sample mean and sample variance.]

7.6 Sampling distributions of order statistics

We have encountered various statistics calculated as or from certain of the ordered observations in a sample: median, range, midrange, and quartiles. Each function of the order statistic has a sampling distribution, which we now derive, for the case of continuous populations.

The c.d.f. of $X_{(n)}$, the *largest* of n observations in a random sample from a population with c.d.f. $F(x)$, is

$$F_{X_{(n)}}(v) = P(X_{(n)} \leq v) = P(X_1 \leq v, \, X_2 \leq v, \, ..., \, X_n \leq v) = [F(v)]^n.$$

Differentiating, we obtain the p.d.f. in terms of f, the population p.d.f.:

$$f_{X_{(n)}}(v) = n \, [F(v)]^{n-1} f(v). \tag{1}$$

For the *smallest* of the n observations, a symmetrical argument yields

$$1 - F_{X_{(1)}}(u) = [1 - F(u)]^n,$$

and thus

$$f_{X_{(1)}}(u) = n \, [1 - F(u)]^{n-1} f(u). \tag{2}$$

EXAMPLE 7.6a Suppose n units with exponential lifetime distributions are operated simultaneously and independently. Denote their times to failure by T_i, and suppose the distribution of each is Exp(λ). The time to the first failure among the n units is the *minimum*, $T_{(1)}$. From (2), the p.d.f. of the distribution of $T_{(1)}$ is

$$f_{T_{(1)}}(u) = n \, [e^{-\lambda u}]^{n-1} \lambda e^{-\lambda u} = n\lambda e^{-n\lambda u}, \quad u > 0,$$

which is the p.d.f. of Exp($n\lambda$), with mean value $1/(n\lambda)$.

By similar reasoning we can find the p.d.f. of $Y = T_{(2)} - T_{(1)}$, the time from the first failure to the second failure. At the point of the first failure, and with no memory on the part of those still operating, the time from then until the second failure is the minimum of the remaining $n - 1$ unit lifetimes, so $Y \sim \text{Exp}[(n-1)\lambda]$. Using this, we can find the mean time to the second failure:

$$E(T_{(2)}) = E[(T_{(2)} - T_{(1)}) + T_{(1)}] = \frac{1}{(n-1)\lambda} + \frac{1}{n\lambda}. \quad \blacksquare$$

For $u < v$, the *joint* c.d.f. of $(X_{(1)}, X_{(n)})$, from Problem 7-31, is

$$F(u, v) = F^n(v) - [F(v) - F(u)]^n.$$

Differentiation with respect to u and then v yields the p.d.f:

$$f_{X_{(1)}, X_{(n)}}(u, v) = n(n-1)[F(v) - F(u)]^{n-2} f(u)f(v), \quad u < v. \quad (3)$$

Using this one can calculate the sampling distribution of functions of the smallest and largest observations, such as the range and midrange.

EXAMPLE 7.6b For a uniform population on (0, 1), the joint p.d.f. of the smallest and largest observations in a random sample is

$$f(u, v) = n(n-1)(v-u)^{n-2}, \quad 0 < u < v < 1, \quad (4)$$

obtained by substituting $F(x) = x$ and $f(x) = 1$ in (3). Integrating out u produces the p.d.f. of the largest observation:

$$f_{X_{(n)}}(v) = n(n-1) \int_0^v (v-u)^{n-2} du = nv^{n-1}, \quad 0 < v < 1,$$

which agrees with (1).

The c.d.f. of the sample range, $R = X_{(n)} - X_{(1)}$, is $P(R \leq r)$, given by the integral of $f(u, v)$ [from (4)] over that portion of its support where $v - u \leq r$, for $0 < r < 1$. It is somewhat simpler to integrate over the complement of that region (see Figure 7-5):

$$1 - F_R(r) = n(n-1) \int_r^1 \int_0^{v-r} (v-u)^{n-2} du\, dv = 1 - r^n - n(1-r)r^{n-1}.$$

The density function of the range is therefore

$$f_R(r) = \frac{d}{dr} F_R(r) = n(n-1)(1-r)r^{n-2}, \quad 0 < r < 1. \blacksquare$$

The distribution of $X_{(k)}$, the kth smallest observation in a random sample of size n is defined by its c.d.f. :

$$P\{X_{(k)} \leq y\} = P(\text{at least } k\ X_i\text{'s} \leq y) = \sum_{j=k}^n \binom{n}{j} [F(y)]^j [1 - F(y)]^{n-j}.$$

Differentiation produces the p.d.f., but it's a little easier to find the p.d.f. by using a differential approach, as follows: We calculate the probability that $X_{(k)}$ is in a small increment $(y, y + dy)$. This probability element will be proportional to dy, and the multiplier of dy is the p.d.f. In the present context, the increment dy defines three intervals; these are as follows, shown with the approximate probability of each and the number of observations in each:

7.6 Sampling distributions of order statistics

Interval	Approx. probability	Frequency
$(-\infty, y)$	$F(y)$	U
$(y, y+dy)$	$f(y)dy$	V
$(y+dy, \infty)$	$1 - F(y)$	W

Then, for small dy, (U, V, W) is approximately multinomial, and

$$P(y < X_{(k)} < y+dy) \doteq P(U = k-1, V = 1, W = n-k)$$

$$\doteq \binom{n}{k-1, 1, n-k}[F(y)]^{k-1}[f(y)dy]^1[1-Fy)]^{n-k}.$$

The p.d.f. of $X_{(k)}$ is the coefficient of dy:

$$f_{X_{(k)}}(y) = \binom{n}{k-1, 1, n-k}[F(y)]^{k-1}[1-F(y)]^{n-k}f(y). \quad (5)$$

EXAMPLE 7.6c The p.d.f. of the kth smallest observation in a random sample from $\mathcal{U}(0,1)$ is given by (5) with $F(y) = y$ and $f(y) = 1$, for $0 < y < 1$:

$$f_{X_{(k)}}(y) = \frac{n!}{(k-1)!(n-k)!}y^{k-1}(1-y)^{n-k}, \quad 0 < y < 1.$$

This is a beta density: $X_{(k)} \sim \text{Beta}(k, n-k+1)$. The mean and variance of $X_{(k)}$, from (17) of §6.7 with $s = k$ and $t = n - k + 1$, are

$$E(X_{(k)}) = \frac{k}{n+1}, \quad \text{var } X_{(k)} = \frac{k(n-k+1)}{(n+2)(n+1)^2}. \blacksquare \quad (6)$$

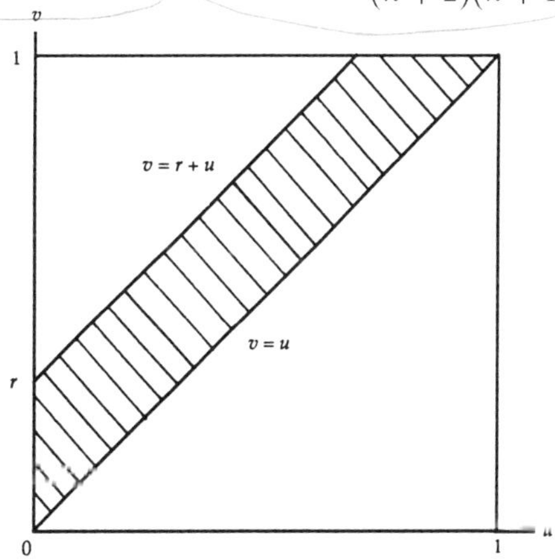

Figure 7-5

An important order statistic is the sample median. When n is odd: $n = 2k - 1$, the sample median is the middle observation, or $X_{(k)}$. And unless np is an integer, the 100pth percentile of the sample distribution is the kth smallest observation, where $k = [np] + 1$, that is, 1 plus the largest integer less than np. It can be shown[4] that the *asymptotic* distribution of the 100pth percentile is normal with mean x_p, the corresponding population percentile, and variance

$$\frac{1}{[f(x_p)]^2} \frac{p(1-p)}{n}, \tag{7}$$

when the population p.d.f. f has a continuous derivative near x_p.

EXAMPLE 7.6d The sample median $\widetilde{X}$ is the sample 50th percentile. Its asymptotic distribution is normal with $E(\widetilde{X}) = x_{.5}$ and, from (6), var $\widetilde{X} = [4nf^2(x_{.5})]^{-1}$. If the population is $N(\mu, \sigma^2)$, the population median is μ, and the density at $x = \mu$ is $f(\mu) = (2\pi\sigma^2)^{-1/2}$; so in this case, $\widetilde{X} \approx N(\mu, \pi\sigma^2/2n)$. It is of interest to note that $\bar{X} \approx N(\mu, \sigma^2/n)$, somewhat narrower than the asymptotic distribution of the median. ∎

PROBLEMS

7-26. For a random sample of size n from $U(0, 1)$,
(a) find the mean and variance of the smallest observation.
(b) find the mean and variance of the largest observation using the results in (a) and exploiting symmetry.

7-27. Consider n identical units with i.i.d. times to failure, $T_i \sim \text{Exp}(\lambda)$.
(a) Note that when connected in series (Problem 6-50), the time to system failure is the *minimum* of the unit times to failure, $T_{(1)}$. Find the mean time to failure, $ET_{(1)}$.
(b) Note that when connected in parallel (Problem 6-54), the time to system failure is the *maximum*, $T_{(n)}$. Find $ET_{(n)}$ when $n = 3$ by integrating with respect to the p.d.f. (1).
(c) Extend the technique of Example 7.6a to find a formula for $ET_{(n)}$, and observe that when $n = 3$ it produces the answer to (b).

7-28. Let Y denote the median of a random sample from $U(0, 1)$. For simplicity, assume that n is odd: $n = 2k - 1$, and find
(a) the p.d.f. of Y. (b) the mean and variance of Y.

---

[4]See H. Cramér, *Mathematical Methods of Statistics*, Princeton, N. J., Princeton University Press, 1946, page 367ff. Also, E. L. Lehmann, *Theory of Point Estimation*, New York, John Wiley & Sons (1983), page 354.

7.6 Sampling distributions of order statistics

7-29. For a random sample of size n from $\mathcal{U}(a, b)$, use a linear transformation
 (a) to find the mean and variance of the largest observation.
 (b) to find the density function of the sample range. (See Example 7.6b.)

7-30. Show that if a population is symmetric about $x = a$, the distribution of the median of a random sample is symmetric about $x = a$. (Assume n is odd.)

7-31. Consider a random sample from a continuous population with c.d.f. F. With $U = X_{(1)}$ and $V = X_{(n)}$, show that the joint c.d.f. of (U, V) can be expressed in terms of the population c.d.f. as follows:

$$F(u, v) = F^n(v) - [F(v) - F(u)]^n,$$

as asserted just after Example 7.6a.

7-32. Consider a random sample of size n from a population with c.d.f. F and p.d.f. f.
 (a) Show that the p.d.f. of the sample range R is given by

$$f_R(r) = n(n-1) \int_{-\infty}^{\infty} [F(r+u) - F(u)]^{n-2} f(r+u) f(u)\, du, \quad r > 0.$$

 (b) Find the p.d.f. of R for samples of size $n = 5$ from $\mathcal{U}(0, 1)$.
 (c) Show that in the case of $\mathcal{N}(\mu, \sigma^2)$, the distribution of R/σ depends only on the sample size n—not on the population parameters. [Hint: Let $Z_i = (X_i - \mu)/\sigma$ and find R_Z, the range of $(Z_1, ..., Z_n)$, in terms of R_X, the range of $(X_1, ..., X_n)$.]

7-33. Find the p.d.f. of the sample midrange: $M = \frac{1}{2}[X_{(1)} + X_{(n)}]$,
 (a) as an integral involving the population c.d.f. and p.d.f., for random samples of size n from an arbitrary continuous population.
 (b) for a random sample from $\mathcal{U}(0, 1)$. [Hint: When $x < 1/2$, the product $f(2x - s)f(s)$ is 1 if $0 < s < 2x$ (and 0 otherwise); when $x > 1/2$, the product is 1 if $2x - 1 < s < 1$ (and 0 otherwise).]

7-34. For random samples of size n from a population with c.d.f. F,
 (a) show that $Y_{(k)} = F(X_{(k)})$ is the kth smallest observation in a random sample from $\mathcal{U}(0, 1)$. [See Theorem 9, Chapter 3, page 66.]
 (b) find the mean area under the p.d.f. of X between $X_{(k)}$ and $X_{(k+1)}$, for a random sample of size n. [Hint: Use (5) in §7.6.]

7-35. Use the differential method to find the joint p.d.f. of the jth smallest and kth smallest observations in a random sample in terms of the population c.d.f. and p.d.f.

7-36. For random samples from a Cauchy population with location parameter θ, find the asymptotic distributions of the sample median and quartiles.

7-37. Find the asymptotic distributions of the median and quartiles of a random sample from $\text{Exp}(\lambda)$.

7.7 Likelihood

When the "correct" model for a population variable X is unknown, one often assumes that it is one of a family of models, $\{f(x \mid \theta)\}$, where θ is either a numerical or vector parameter—or simply a label for one of a collection of possible models. The essence of a statistical problem is to infer θ from "data," or observations on X. Because of the stochastic nature of the problem, investigators are usually willing to settle for a probability distribution for θ, one that embodies personal beliefs about θ in view of the data. In discrete cases, what they want is $P(\text{model} \mid \text{data})$. But so far, in our study of probability distributions, our focus has been on quantities of the form $P(\text{data} \mid \text{model})$. Clearly, we need a way to interchange roles of the data and the model, and for this the obvious tool is Bayes' theorem:

$$P(\text{model} \mid \text{data}) = \frac{P(\text{data} \mid \text{model}) \cdot P(\text{model})}{P(\text{data})}.$$

For given data, the denominator is a constant, and we write

$$P(\text{model} \mid \text{data}) \propto P(\text{data} \mid \text{model}) \cdot P(\text{model}). \tag{1}$$

Whether a "model" can be assigned a probability—whether θ can be considered to be a random variable—is a controversial question. Regarding θ as a random variable does not mean to suggest that nature plays games of chance to determine its value; rather, it is simply that we use a probability distribution to describe our uncertainty about something whose precise value we don't know. Those who are comfortable with this approach refer to the left-hand side of (1) as a *posterior* probability and the second factor on the right, as the *prior* probability. For given data, we call $P(\text{data} \mid \text{model})$ the **likelihood** of the model; and we write

$$\text{posterior} \propto \text{likelihood} \times \text{prior}, \tag{2}$$

as a schematic that shows how to use the data to convert a prior distribution for θ to a posterior distribution for θ. In this section we focus on the likelihood, which is the contribution of the data in this conversion.

Roughly speaking, given some data, the likelihood of a model is the probability of the data under that model. Likelihood plays an important role in the theory of statistical inference even for those who question the validity of the Bayesian approach. For example, suppose there are two models that are possible, as the one that could have produced some observed data. To many, it is intuitively logical that the model under which the observed data have a higher probability is to be preferred as the explanation of how the data came to be.

EXAMPLE 7.7a I have two pairs of dice, one (pair 1) of the ordinary kind, and another (pair 2) of which one die has nothing but 5's on its six faces and the other, four 2's and two 6's. Suppose I roll one set of dice and get a seven; which set did I use?

We know that $P(\text{seven} \mid \text{pair 1}) = 1/6$, and $P(\text{seven} \mid \text{pair 2}) = 4/6$. These are the *likelihoods* of pair 1 and pair 2, respectively. The observed "seven" has a higher probability with pair 2 than with pair 1, and of these two possible models we tend to prefer thinking that pair 2 was the one I rolled—it is four times as "likely" as pair 1. Notice that $1/6$ and $4/6$ don't add up to 1; indeed, there is no reason to add these probabilities, since they are probabilities in different models. ■

Since the probability of any given data depends on the model we assume for it, we call $P(\text{data} \mid \theta)$ the **likelihood function**, a function of θ. But we need some flexibility in this terminology. We'll be using likelihoods only in comparisons and in forming ratios, as in the above example. So two functions that are the same except for a constant factor (one not involving θ) will be considered equivalent. Thus, "the" likelihood function is actually an equivalence class, a class of functions that are identical except for constant factors. When the population variable is *discrete* and the data are given as a vector $\mathbf{X}$, the likelihood function is

$$L(\theta) = P(\mathbf{X} = \mathbf{x} \mid \theta) = f_\mathbf{X}(\mathbf{x} \mid \theta). \tag{3}$$

EXAMPLE 7.7b Consider a random selection of size $n = 2$ from a lot of size $N = 6$ which includes M that are defective. Let Y denote the number of defective items in the sample. Suppose we find $Y = 1$. In the model implied by the phrase "random selection," the probability of this outcome is

$$L(M) = P(Y = 1 \mid M) = \frac{\binom{M}{1}\binom{6-M}{1}}{\binom{6}{2}}.$$

In comparing likelihoods of different values of M, we can ignore the denominator and write
$$L(M) = M(6 - M).$$
Then $\dfrac{L(3)}{L(4)} \doteq 1.125$ and $\dfrac{L(3)}{L(2)} \doteq 1.125$, so $M = 3$ is preferred to either 2 or 4 as being a more likely explanation of the sample result. Indeed, $M = 3$ is the *most* likely value of M when $Y = 1$. In the columns of the following table we show the distributions of Y for the various possible values of M. The probabilities in the *columns* add up to 1 in each case. The *rows*, on the other hand, define the likelihood functions for $Y = 0$, for $Y = 1$, and for $Y = 2$.

		\multicolumn{7}{c}{M}
		0	1	2	3	4	5	6
Y	0	$\frac{15}{15}$	$\frac{10}{15}$	$\frac{6}{15}$	$\frac{3}{15}$	$\frac{1}{15}$	0	0
	1	0	$\frac{5}{15}$	$\frac{8}{15}$	$\frac{9}{15}$	$\frac{8}{15}$	$\frac{5}{15}$	0
	2	0	0	$\frac{1}{15}$	$\frac{3}{15}$	$\frac{6}{15}$	$\frac{10}{15}$	$\frac{15}{15}$

From this it is easy to find the "most likely" value of M for each Y. ∎

When the population variable is continuous, the data vector has zero probability. Rather than compare probabilities, we compare probability elements, as in the ratio
$$\frac{f(\mathbf{x} \mid \theta_1)\,d\mathbf{x}}{f(\mathbf{x} \mid \theta_2)\,d\mathbf{x}}.$$
Since the differentials cancel, we take the joint p.d.f., or any function proportional thereto (for given $\mathbf{x}$), as the likelihood function. The proportionality "constant" can depend on $\mathbf{x}$.

DEFINITION Given $\mathbf{X} = \mathbf{x}$, the *likelihood function* is any function of θ proportional to the joint p.f. (discrete case) or the joint p.d.f. (continuous case) of $\mathbf{X}$:
$$L(\theta) \propto f(\mathbf{x} \mid \theta). \tag{4}$$

("The" likelihood function is thus an equivalence class of functions.)

If $\mathbf{X} = (X_1, \ldots, X_n)$ is a random sample from $f(x \mid \theta)$, the likelihood function (4) is a product:
$$L(\theta) \propto \prod_{i=1}^{n} f(x_i \mid \theta). \tag{5}$$

EXAMPLE 7.7c Consider a random sample of size 10 from Ber(p), that is, a sequence of ten independent Bernoulli trials with $p = P(X = 1)$. The population p.f. is

$$f(x \mid p) = p^x(1-p)^{1-x}, \quad x = 0, 1.$$

If the observed sample sequence is (1, 0, 1, 1, 0, 1, 1, 1, 1, 0), the likelihood function (5) is

$$L(p) = \prod_{i=1}^{10} f(x_i \mid p) = p^{\Sigma x_i}(1-p)^{10 - \Sigma x_i} = p^7(1-p)^3, \quad 0 \le p \le 1.$$

This likelihood function is pictured in Figure 7-6. ∎

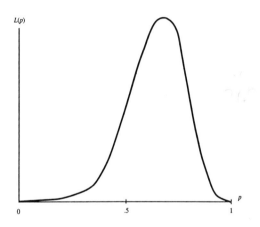

Figure 7-6 Likelihood function for Example 7.7c.

EXAMPLE 7.7d The joint p.d.f. of n independent observations on $N(\mu, 1)$, considered as a function of μ for fixed x's, is the likelihood function:

$$L(\mu) = \prod_{i=1}^{n} \frac{1}{\sqrt{2\pi}} e^{-(x_i - \mu)^2/2} \propto e^{-\Sigma(x_i - \mu)^2/2}.$$

To see the nature of dependence on μ more clearly we rewrite the sum in the exponent as

$$\sum_{i=1}^{n}(x_i - \mu)^2 = \sum_{i=1}^{n} x_i^2 - 2\mu n \bar{x} + n\mu^2$$

$$= n(\mu^2 - 2\mu\bar{x} + \bar{x}^2) + \text{terms free of } \mu.$$

Thus, the likelihood function (when we omit some factors not involving μ) is

$$L(\mu) = e^{-\frac{n}{2}(\mu - \bar{x})^2}. \tag{6}$$

The graph of $L(\mu)$, shown in Figure 7-7, is similar to a normal curve, centered at $\mu = \bar{x}$. ∎

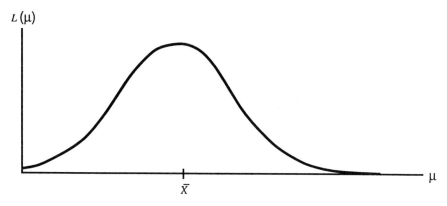

Figure 7-7 Likelihood function for Example 7.7d.

The likelihood function L depends on the data. So in Example 7.7c, a different number of successes would produce a different L. In Example 7.7d, we did not assume a particular data vector, but gave L as depending on a generic data vector. When we treat the data vector as a random variable, the likelihood function is a function-valued function of the data, $L_X(\theta)$, so we can think of it as a *statistic*. The phrase "likelihood function" can refer either to this statistic or to a particular realization, an $L(\theta)$ calculated from a particular set of data.

The intuitive notion that relative likelihood is a useful criterion for use in comparing two possible states of nature, as explanations of what has been observed, does indeed lead to statistical procedures that are "good" from many points of view, as we shall see. It is not so clear that when we consider more than two θ's, the one with the largest likelihood is the "best" model, in some sense. But we state this as a principle—not as a principle that can be proved, but as one that might be followed with the hope that it would lead to good procedures:

7.7 Likelihood

MAXIMUM LIKELIHOOD PRINCIPLE *A statistical inference or procedure should be consistent with the assumption that the best explanation of a set of data is provided by $\hat\theta$, a value of θ that maximizes the likelihood function, $L(\theta)$.*

There can be more than one θ that maximizes $L(\theta)$, but any $\hat\theta$ that does so will be called a **maximum likelihood state**. When θ is a single real parameter, it is typical that L is continuously differentiable, so that L' vanishes at $\hat\theta$, and (since $\log L$ has a maximum value whenever L does) $\hat\theta$ satisfies the **likelihood equation**:

$$\frac{\partial}{\partial \theta} \log L(\theta) = 0. \tag{7}$$

In such cases, to find a maximum likelihood state, we solve the likelihood equation and examine its roots as possible maximizing values of θ.

In some instances, however, it will happen that the maximum of L occurs at a point where the derivative is not zero or does not exist, and one must be alert to such possibilities.

EXAMPLE 7.7e In the context of Example 7.7c [random sample of size ten from $\text{Ber}(p)$], suppose we found ten successes in the ten trials. The likelihood function is

$$L(p) = p^{10}, \quad 0 \le p \le 1,$$

and the derivative, $10p^9$, vanishes at $p = 0$—hardly a plausible explanation for ten successes. Of course, the *maximum* of L occurs at the other endpoint of the interval $[0, 1]$, namely, $p = 1$. (See Figure 7-8.) This is the (unique) maximum likelihood state. ∎

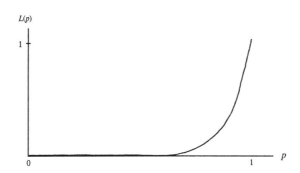

Figure 7-8 Likelihood function for Example 7.7e.

EXAMPLE 7.7f Suppose $X \sim \mathcal{U}(0, \theta)$, and consider a random sample of two: (X_1, X_2), with the outcome $X_1 = .4$ and $X_2 = 1.5$. Each p.d.f. is either 0 (when $x > \theta$) or $1/\theta$ (when $0 < x < \theta$). The likelihood function is then the product of two $1/\theta$'s (when θ exceeds both x_i's), or 0:

$$L(\theta) = \begin{cases} 1/\theta^2, & \text{if } \theta > 1.5, \\ 0, & \text{if } \theta < 1.5. \end{cases}$$

The graph is shown in Figure 7-9. Clearly, no value of θ *less* than the larger of the two observations is a possible explanation. And of the values of θ that are larger than the largest observation, the closer to that largest observation, the more likely the θ. ■

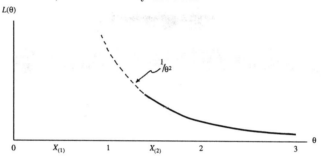

Figure 7-9 Likelihood function for Example 7.7f.

Since parametrization of a family of possible models can often be done in more than one way, we should check to see that different parametrizations could not lead to different maximum likelihood models. Suppose we have a family $\{f(x \mid \theta)\}$, and introduce a new parameter $\xi = g(\theta)$. If these are to define the same family, the relationship must be one-one, so that g has an inverse: $\theta = g^{-1}(\xi)$. In terms of ξ, the likelihood function is

$$\tilde{L}(\xi) = f(x \mid g^{-1}(\xi)) = L(g^{-1}(\xi)).$$

If $\hat{\theta}$ maximizes L and $\hat{\theta} = g(\hat{\xi})$, then for any $\xi = g(\theta)$,

$$\tilde{L}(\xi) = L(g^{-1}(\xi)) = L(\theta) \le L(\hat{\theta}) = L(g^{-1}(\hat{\xi})) = \tilde{L}(\hat{\xi}).$$

Thus, $\hat{\xi}$ maximizes $\tilde{L}$, so a model that maximizes the likelihood is the same no matter which of the equivalent parametrizations is used.

PROBLEMS

7-38. Give the likelihood function for a random sample of size n from each of the following populations:

(a) $N(\mu, \sigma^2)$.   (b) $Exp(\lambda)$.   (c) $Poi(\lambda)$.   (d) $Geo(p)$.

7-39. Find the likelihood function $L(\theta)$, given a random sample of size n from a population with p.d.f. $f(x \mid \theta) \propto [1 + (x-\theta)^2]^{-1}$ (a member of the Cauchy location family).

7-40. A discrete variable Z has just three possible values. Consider the three competing models indexed by θ_i defined in the following probability table:

	θ_1	θ_2	θ_3
z_1	.4	.6	.2
z_2	.2	.3	.1
z_3	.4	.1	.7

(a) Give $L(\theta)$ for each z_i. [Do the likelihoods sum to 1?]
(b) Show that the likelihood functions for $Z = z_1$ and $Z = z_2$ are the same.

7-41. In the setting of Example 7.7b (page 221), suppose $N = 12$, $n = 4$, and the sample sequence is (1, 0, 1, 1). Find the maximum likelihood value of M (an integer).

7-42. Find the maximum likelihood state for each of the distribution families in Problem 7-38.

7-43. Give the likelihood function for a random sample of size n from a member of the exponential family: $f(x \mid \theta) = B(\theta) h(x) \exp[Q(\theta) R(x)]$.

7-44. Find the θ that maximizes the likelihood for a random sample of size n from $f(x \mid \theta) = \frac{1}{\theta} e^{-x/\theta}$, $x > 0$. Does this define a model different from the one found in Problem 7-38(b)?

7-45. Consider a random sample from $U(\theta - \frac{1}{2}, \theta + \frac{1}{2})$. Find the likelihood function and determine all maximum likelihood states.

7-46. Experimenter A observes Poisson process with rate parameter λ and records t_0 as the elapsed time until the kth event occurs, where k was a pre-set number. Experimenter B observes the same process, counting the number of events in a pre-set time t_0, and records k events in that time period. Show that their likelihood functions are the same.

7-47. Suppose that in 12 trials of a Bernoulli experiment with parameter p there are five successes. Show that the likelihood function is the same, whether the sampling was stopped according to a specified number of trials or according to a specified number of successes.

7.8 Sufficiency

In §7.2 we suggested the possibility that reducing sample data to one or just a few numbers might lose information about the appropriate population model. The notion of *sufficiency* of a statistic deals with this question. Our treatment of this topic will not be consistently rigorous in its derivations, since in continuous cases it will involve conditional distributions. We first make the notion of data "reduction" more precise.

A statistic $T = t(\mathbf{X})$ partitions the sample space. (Recall the discussion of partitions in §1.4.) We'll term statistics T and U *equivalent* if they define the same partition; when this is the case, they are in one-one correspondence: $T = s(U)$, and $U = s^{-1}(T)$. Moreover, any assignment of a number to each partition set of T such that no two partition sets get the same number defines a statistic equivalent to T.

A partition Π_1 is said to be a **reduction** of partition Π_2 if and only if each partition set of Π_1 is precisely the union of partition sets of Π_2. And when this is the case, a statistic T_1 that defines Π_1 is a function of any statistic T_2 that defines Π_2: $T_1 = g(T_2)$, and we say that T_1 is a reduction of T_2. Unless Π_1 and Π_2 are exactly the same partition, the function g does not have a unique inverse, in which case we may say that Π_1 (or a statistic T_1 that defines it) is a *proper* reduction of Π_2 (or of T_2).

EXAMPLE 7.8a Consider a sample space with six elements: {a, b, c, d, e, f}. The partition {a, b, c}, {d, e}, {f} is defined by the random variable $T = t(\omega)$, where

$$t(a) = t(b) = t(c) = 1, \quad t(d) = t(e) = 2, \quad t(f) = 3.$$

Any other function of ω with distinct values on these three partition sets is equivalent to T. But consider $U = (T - 2)^2$. This has the value 0 on {d, e} and the value 1 on {a, b, c, f}. These two sets constitute a partition which is a proper reduction of the original one, and U is a proper reduction of T.

The partition $\Pi = [\{a, b\}, \{c, d\}, \{e, f\}]$ is *not* a reduction of either of the above partitions, nor is either of them a reduction of Π. ∎

EXAMPLE 7.8b For a sample of size $n = 3$ from a numerical population, $(\sum X_i, \sum X_i^2)$ is a statistic. Its partition sets are sets of points (x_1, x_2, x_3) which satisfy

$$\begin{cases} x_1 + x_2 + x_3 = a \\ x_1^2 + x_2^2 + x_3^2 = b, \end{cases}$$

for specified a and $b \geq 0$. These sets are circles—the intersections of planes with spheres centered at the origin. The union of these circles is the whole sample space; and the same partition is defined by the pair $(\bar{X}, S^2)$, since these two statistics can be calculated from the sample sum and sample sum of squares, and conversely. ∎

The notion of "sufficiency" of a statistic for a family of models $\{f(x \mid \theta)\}$ can be described roughly as follows: Reducing a set of data to a sufficient statistic does not lose anything in the data that would help us in choosing a model from among that family; so, for making inferences, it is sufficient to know only the value of that statistic, as opposed to knowing the whole sample in detail. (This somewhat vague claim will be shown more concretely in the problems of testing and estimation, taken up in the next two chapters.)

EXAMPLE 7.8c Consider a random sample of size $n = 3$ from $\text{Ber}(p)$. The joint p.f. of the three observations is

$$P(\mathbf{X} = \mathbf{x}) = P(X_1 = x_1, X_2 = x_2, X_3 = x_3) = p^k(1-p)^{3-k},$$

where each x_i is 0 or 1, and $k = x_1 + x_2 + x_3$. Now define the statistic $Y = \sum X_i$. This is the number of successes in the three independent trials, so $Y \sim \text{Bin}(3, p)$:

$$P(Y = k) = \binom{3}{k} p^k (1-p)^{3-k}, \quad k = 0, 1, 2, 3.$$

The *conditional* distribution in the sample space, given $Y = k$, is given by

$$P(\mathbf{X} = \mathbf{x} \mid Y = k) = \frac{P(\mathbf{X} = \mathbf{x} \text{ and } Y = k)}{P(Y = k)}.$$

We illustrate this calculation with $k = 2$. If $\sum x_i \neq 2$, the numerator is 0. If $\sum x_i = 2$, the sample is one of these: $(0, 1, 1)$, $(1, 0, 1)$, $(1, 1, 0)$, and the event $[Y = 2]$ in the numerator is redundant. So for each of these three sample points we find

$$P(\mathbf{X} = \mathbf{x} \mid Y = 2) = \frac{p^2(1-p)^1}{3p^2(1-p)^1} = \frac{1}{3}.$$

That is, the three sample points with $Y = 2$ are equally likely—no matter what the value of p may be! Similar calculations show:

$P(\mathbf{X} = \mathbf{x} \mid Y = 1) = 1/3$ for $\mathbf{x} = (0, 0, 1), (0, 1, 0), (1, 0, 0)$,

$P(\mathbf{X} = \mathbf{x} \mid Y = 0) = 1$ for $\mathbf{x} = (0, 0, 0)$,

$$P(\mathbf{X} = \mathbf{x} \mid Y = 3) = 1 \quad \text{for } \mathbf{x} = (1, 1, 1),$$

all of which are independent of p.

We can think of a particular sample $\mathbf{x}$ as being generated in two stages: First, sample $\text{Bin}(3, p)$ to get a value of Y, say $Y = k$. Then, if there is more than one sample point with $Y = k$, sample the conditional distribution given $Y = k$ to pin down which sample point it was that led to $Y = k$. The first step, sampling $\text{Bin}(3, p)$, clearly involves p; but the second does not! That is, no information about p could possibly be gained by learning which sample point (among those for which $Y = k$) produced the value $Y = k$. In this sense, it is sufficient to know the value of Y, and we say that Y is *sufficient*—sufficient, that is, for any inference to be made about the parameter p.

If we know Y, we also know the relative frequency of 1's, Y/n. These statistics define the same partition of the sample space, namely:

$$\{000\}, \{001, 010, 100\}, \{011, 101, 110\}, \{111\}.$$

It is really this partition that is sufficient for p, and any statistic that defines this same partition is sufficient for p. ∎

DEFINITION: A statistic $T = t(\mathbf{X})$ is *sufficient* for a family of distributions $\{f(x \mid \theta)\}$ if and only if the conditional distribution of $\mathbf{X}$ given the value of T is the same for all θ.

If a statistic U defines the same partition as T, then a conditional distribution given the value of U is exactly the same as the conditional distribution given the value of T. If one is sufficient, so is the other. It is essentially the partition that is sufficient: It is sufficient, for inference about θ, to know which partition set the data point is in. Determining which point within that partition set has occurred tells us nothing further about θ.

The conditional distributions referred to in the above definition of sufficiency are not always easy to calculate, especially in the case of continuous variables. But there is an equivalent characterization of sufficiency that is easier to apply. Before stating it, we give an example to illustrate using the definition to show that a statistic is *not* sufficient.

EXAMPLE 7.8d Let (X, Y) denote a random sample of size $n = 2$ from $\text{Geo}(p)$. The p.f. of the joint distribution is

$$f(i, j) = p^2(1 - p)^{i + j - 2}, \quad i, j = 1, 2, \ldots.$$

Consider the statistic $T = X - Y$, and the particular value 0:

$$P(T=0) = P(X=Y) = \sum_{i=1}^{\infty} f(i, i) = p^2 \sum_{1}^{\infty}\left[(1-p)^2\right]^{i-1} = \frac{p}{2-p}.$$

Then

$$P[(i, j) \mid T = 0] = \frac{P[(i, j) \text{ and } X = Y]}{P(X = Y)}.$$

When $i = j$, the numerator is just $f(i, i) = p^2(1-p)^{2i-2}$, so

$$P[(i, j) \mid T = 0] = \frac{p^2(1-p)^{2i-2}}{p/(2-p)} = p(2-p)(1-p)^{2i-2}.$$

This conditional probability *does* depend on p, which means that the statistic $T = X - Y$ is not sufficient. ■

We now state a useful criterion, a necessary and sufficient condition for a statistic to be sufficient for θ. Although we give it as a theorem, a correct statement would have to include certain regularity conditions without which the conclusion may not be valid.

THEOREM 12 (Factorization criterion) *A statistic $T = t(\mathbf{X})$ is sufficient for the family $\{f(x \mid \theta)\}$ if and only if, for some functions g and h,*

$$f(\mathbf{x} \mid \theta) = g(t(\mathbf{x}), \theta) h(\mathbf{x}). \tag{1}$$

That the criterion is valid is hard to prove in any generality. We'll show that it holds for discrete distributions. Various theorems have been given, for different kinds of regularity conditions, by Fisher, Neyman, Halmos and Savage, and Bahadur.[5] These cover all the cases that will arise here, including the cases in which the support of the population distribution depends on the unknown parameter.

Suppose first that (1) holds and let $T = t(\mathbf{X})$. The p.f. $f_T(t_0)$ is obtained by summing (1) over sample points $\mathbf{x}$ for which $t(\mathbf{x}) = t_0$:

$$f_T(t_0 \mid \theta) = \sum_{t(\mathbf{x}) = t_0} g[t(\mathbf{x}), \theta] h(\mathbf{x}) = g(t_0, \theta) \sum_{t(\mathbf{x}) = t_0} h(\mathbf{x}).$$

The conditional p.f. of the sample observations given $T = t_0$ is

$$P(\mathbf{X} = \mathbf{x} \mid T = t_0) = \frac{P(\mathbf{X} = \mathbf{x} \text{ and } T = t_0)}{P(T = t_0)}.$$

If $t(\mathbf{x}) \neq t_0$, this is 0, independent of θ. If $t(\mathbf{x}) = t_0$, the condition

---

[5] Lehmann, E. L., *Testing Statistical Hypotheses*, 2nd ed. Pacific Grove, Cal.: Wadsworth & Brooks/Cole, 1986, page 55.

$[T = t_0]$ in the event in the numerator is redundant, so the intersection is just the event $\mathbf{X} = \mathbf{x}$. The conditional p.f. is therefore

$$\frac{P(\mathbf{X} = \mathbf{x})}{P(T = t_0)} = \frac{g(t_0, \theta) h(\mathbf{x})}{g(t_0, \theta) \sum_{t(\mathbf{x}) = t_0} h(\mathbf{x})},$$

which is (also) independent of θ, so T is sufficient.

Conversely, suppose the conditional p.f. given $T = t$ does not depend on the parameter θ; denote it by c:

$$P(\mathbf{X} = \mathbf{x} \mid T = t) = c(\mathbf{x}, t).$$

[This is defined if the event $[T = t]$ has positive probability. However, if $P(T = t_0 \mid \theta) = 0$, then $f(\mathbf{x} \mid \theta) = 0$ at any $\mathbf{x}$ where $t(\mathbf{x}) = t_0$, and the factorization (1) is trivial.] For a given sample point $\mathbf{x}$, let $t(\mathbf{x}) = t^*$. Then

$$f(\mathbf{x} \mid \theta) = P(\mathbf{X} = \mathbf{x} \mid \theta) = P[\mathbf{X} = \mathbf{x}, t(\mathbf{X}) = t^* \mid \theta]$$

$$= P(\mathbf{X} = \mathbf{x} \mid T = t^*) \cdot P[t(\mathbf{X}) = t^* \mid \theta]$$

$$= c(\mathbf{x}, t^*) f_T(t^* \mid \theta).$$

This is the factorization (1), since the second factor depends on the data only through t^*, the value of T at $\mathbf{x}$.

COROLLARY 1 *A statistic $T = t(\mathbf{X})$ is sufficient for θ if and only if the likelihood function is of the form $L(\theta) = g(t(\boldsymbol{x}), \theta)$.*

COROLLARY 2 *For random samples, the order statistic is sufficient.*

Corollary 1 is equivalent to the factorization criterion because the factor $h(\mathbf{x})$ in that criterion does not depend on θ. (The product of the two factors *is* the p.f. or p.d.f.) Indeed, considered as a statistic, the random function $L(\theta)$ is sufficient. Corollary 2 follows from the fact that for random samples $L(\theta)$, as a product, does not depend on the order of the factors. It can be calculated if we know the order statistic and θ.

To use the factorization criterion, Corollary 1 tells us that we need only examine the likelihood function to see whether it can be calculated from the parameter and some function of the data. If so, that function of the data is sufficient for the parameter.

EXAMPLE 7.8. Let $\mathbf{X}$ be a random sample from $Geo(p)$. The joint p.f. of the X's is

$$f(\mathbf{x} \mid p) = \prod_1^n p(1-p)^{x_i} = p^n(1-p)^{\Sigma x_i}.$$

This is the likelihood function, and it depends on the x_i's only through the value of their sum. So $\sum X_i$ is a sufficient statistic. ∎

EXAMPLE 7.8f Suppose the admissible states of nature are just two in number, θ_0 and θ_1. Let f_0 and f_1 denote the corresponding density or probability functions, and define the *likelihood ratio* statistic

$$\Lambda = \lambda(\mathbf{X}) = \frac{L(\theta_0)}{L(\theta_1)} = \frac{f_0(\mathbf{X})}{f_1(\mathbf{X})},$$

and the function

$$g(\lambda, \theta_0) = \begin{cases} \sqrt{\lambda}, & \theta = \theta_0, \\ 1/\sqrt{\lambda}, & \theta = \theta_1. \end{cases}$$

Upon setting $h = \sqrt{f_0 f_1}$ we find that then $f_0(\mathbf{x}) = \sqrt{\lambda(\mathbf{x})}\, h(\mathbf{x})$, and $f_1(\mathbf{x}) = \sqrt{1/\lambda(\mathbf{x})}\, h(\mathbf{x})$; so $f(\mathbf{X} \mid \theta) = g(\Lambda, \theta) h(\mathbf{X})$. Thus, the likelihood ratio Λ is sufficient. ∎

EXAMPLE 7.8g For a random sample of size n from $\mathcal{U}(0, \theta)$, the joint p.d.f. of the X's (as in Example 7.7f) is

$$f(\mathbf{x} \mid \theta) = (1/\theta)^n \quad \text{if } 0 < x_i < \theta \text{ for } i = 1, 2, \ldots, n,$$

and zero elsewhere. Since θ exceeds all of the observations if and only if it exceeds the largest x_i, the likelihood function can be written

$$L(\theta) = \begin{cases} 1/\theta^n, & \theta > X_{(n)}, \\ 0, & \theta \leq X_{(n)}. \end{cases}$$

Since this depends on the data only through the value of $X_{(n)}$, we conclude that this statistic is sufficient for θ. ∎

The factorization criterion, or the equivalent formulation in the corollary, is not a good tool for showing that a particular statistic is *not* sufficient, since it is not easy to see that a likelihood can *not* be expressed as a function of θ and only that statistic. The criterion does permit an easy proof of the following proposition.

THEOREM 13 *If a sufficient T is a function of U, then U is sufficient. And if T is a function of a U which is not sufficient, neither is T.*

234 Chapter 7 Sampling and Reduction of Data

To show this we note that if T is sufficient for θ and $T = c(U)$, then

$$f(\mathbf{x} \mid \theta) = g(T, \theta) h(\mathbf{x}) = g[c(U), \theta] h(\mathbf{x}) = \tilde{g}(U, \theta) h(\mathbf{x}),$$

which implies the sufficiency of U. The second statement in the theorem is essentially a restatement of the first.

PROBLEMS

7-48. Determine the partition of the plane defined by the order statistic T for a sample of size 2. Determine also the partition sets defined by the sample sum Y. Is either a reduction of the other? And is either statistic a function of the other?

7-49. Consider a sample space of three elements, $\{a, b, c\}$, with three possible distributions, θ_1: $(.4, .2, .4)$, θ_2: $(.6, .3, .1)$, and θ_3: $(.2, .1, .7)$. The sets $A = \{c\}$, $B = \{a, b\}$ constitute a partition.
(a) Find the conditional distribution in the sample space given A and the conditional distribution given B.
(b) Construct a statistic for the partition $\{A, B\}$. Is it sufficient?
(c) Let $t(a) = 0$ and $t(b) = t(c) = 1$. Is $T = t(\omega)$ sufficient?

7-50. Describe the partitions defined by $\bar{X}$ and by $\bar{X}^2$, where $\bar{X}$ is the mean of a sample of size 3. Which is a reduction of the other?

7-51. Three chips are drawn at random, without replacement, from a bowl containing five chips numbered 1 through 5. Let X_i denote the number on the ith chip selected.
(a) Count and describe the partition sets defined by the order statistic.
(b) List the ten possible order statistics with the probability of each.
(c) Find the partition of the ten vectors in (b) defined by the further reduction
  (i) $(X_{(1)}, X_{(3)})$.   (iii) $X_{(3)}$.
  (ii) $X_{(1)}$.   (iv) $X_{(3)} - X_{(1)}$.

7-52. Consider random selections of chips, one chip at a time and without replacement, from a bowl of M blue and $N - M$ white chips. In Problem 2-42 we found that the probability of a particular sample sequence is

$$f(\mathbf{x} \mid M) = \frac{\binom{M}{k}\binom{N-M}{n-k}}{\binom{n}{k}\binom{N}{n}},$$

where $k = \sum x_i$. Find the conditional distributions in the sample space given the sample sum. Conclusion?

7-53. For a random sample of size n from a Poisson population, find the conditional distribution in the sample space given the sample sum, and thus show that the sample sum is sufficient.

7-54. Let (X_1, X_2) be a random sample from a uniform discrete distribution on the integers from 1 to N. Find the conditional probabilities in the sample space, given $X_1 + X_2 = 4$ (say), and conclude from your result that the sum is not sufficient for the parameter N.

7-55. Use the factorization criterion to find, in each case, a sufficient one-dimensional statistic based on a random sample of size n from a population with the given p.f. or p.d.f.:
 (a) Ber(p). (b) Geo(p). (c) Exp(λ).

7-56. Use the factorization criterion to find a sufficient statistic for M in Problem 7-52.

7-57. Use the factorization criterion to find a sufficient statistic for the Rayleigh family, with p.d.f.
$$f(z \mid \sigma^2) = \frac{z}{\sigma^2} \exp\left(-\frac{z^2}{2\sigma^2}\right), \quad z > 0.$$

7-58. Let $X \sim \mathcal{U}(\theta - \frac{1}{2}, \theta + \frac{1}{2})$. Show that for a random sample $\mathbf{X}$ of size n, the likelihood function depends on $\mathbf{X}$ only through the value of $(X_{(1)}, X_{(n)})$. Assuming that the factorization criterion applies in this case, what conclusion can be drawn? [See Problem 7-45, page 227.]

7-59. Find a sufficient statistic for (μ, σ), where $\sigma > 0$, as a function of a random sample from
$$f(x \mid \mu, \sigma) = \frac{1}{\sigma} e^{-(x-\mu)/\sigma}, \quad x > \mu.$$

7.9 Minimal sufficiency

As suggested in the discussion of Example 7.8c, inferential procedures can be based on a sufficient statistic without losing any pertinent sample information. (This will be explained in later chapters, for certain types of inference.) Obvious questions arise: Is there always a sufficient statistic? The answer to this is yes, since the sample itself is trivially sufficient. More importantly, is there a sufficient statistic of low dimension? Also, how do we find a such a statistic? And when we find one, is it the simplest one, or can it be reduced further without sacrificing sufficiency?

DEFINITION A statistic is *minimal sufficient* if it is sufficient and if no proper reduction of it is sufficient.

Whether there is such a statistic, whether it is unique, and how to find it are answered when we give a method (due to Lehmann and Scheffé[6]) for constructing a minimal sufficient partition.

For random samples of size n, we define an *equivalence relation* in the n-space of sample points as follows:

$$\mathbf{x} \simeq \mathbf{y} \text{ if and only if } f(\mathbf{y} \mid \theta) = k(\mathbf{y}, \mathbf{x}) f(\mathbf{x} \mid \theta). \tag{1}$$

That is, points $\mathbf{x}$ and $\mathbf{y}$ are equivalent when the ratio of densities at $\mathbf{x}$ and $\mathbf{y}$ is independent of θ. Then for each point $\mathbf{x}$, define the set D as the set of points equivalent to it:

$$D(\mathbf{x}) = \{\mathbf{y} \colon \mathbf{y} \simeq \mathbf{x}\}. \tag{2}$$

The points where f vanishes for all θ are in one of these sets, call it D_0. Every point $\mathbf{x}$ is in some D, namely $D(\mathbf{x})$, and there is no overlapping of sets D, so they constitute a partition of the sample space. [If $\mathbf{x}$ is in both $D(\mathbf{y})$ and $D(\mathbf{z})$, then $\mathbf{y}$ and $\mathbf{z}$ are equivalent, and $D(\mathbf{y}) = D(\mathbf{z})$.] It is this partition which is minimal sufficient. A rigorous proof is beyond our scope, requiring measure theory, but the basic idea is as follows.

First we argue sufficiency: Choose for each D a representative point, x_D. Let $G(\mathbf{x})$ denote the mapping from $\mathbf{x}$ to $\mathbf{x}_D$ via $D(\mathbf{x})$. Then $G(\mathbf{X})$ is a statistic and defines the partition. For any partition set D except D_0, and for $\mathbf{x}$ in D,

$$f(\mathbf{x} \mid \theta) = k(\mathbf{x}, \mathbf{x}_D) f(\mathbf{x}_D \mid \theta) = k[\mathbf{x}, G(\mathbf{x})] f[G(\mathbf{x}) \mid \theta].$$

This gives us the factorization showing sufficiency, when we set $h(\mathbf{x}) = 0$ for $\mathbf{x} \in D_0$ and $h(\mathbf{x}) = k[\mathbf{x}, G(\mathbf{x})]$ elsewhere.

To see the minimality of $G(\mathbf{X})$ and the corresponding partition, consider any other sufficient statistic $t(\mathbf{X})$ and corresponding partition sets E. Minimality follows if we can show that each E is a subset of some D, so that D is a reduction of E. Let $\mathbf{x}$ and $\mathbf{y}$ be points in an E, which means that $t(\mathbf{x}) = t(\mathbf{y})$. Because $t(\mathbf{X})$ is sufficient,

$$f(\mathbf{x} \mid \theta) = r(\mathbf{x}) s[t(\mathbf{x}), \theta] = r(\mathbf{x}) s[t(\mathbf{y}), \theta], \tag{3}$$

where $r(\mathbf{x})$ does not depend on θ, and

$$f(\mathbf{y} \mid \theta) = r(\mathbf{y}) s[t(\mathbf{y}), \theta]. \tag{4}$$

The ratio of (3) and (4) does not involve θ, so $\mathbf{x}$ and $\mathbf{y}$ are equivalent in the sense (1), used to define sets D. Thus, they are in the same equiva-

---

[6] E. Lehmann and H. Scheffé, "Completeness, similar regions, and unbiased estimation," *Sankhya* 10 (1950) page 327ff.

lence set D, and $E \subset D$. [If $r(\mathbf{y}) = 0$, then for all θ, $f(\mathbf{y} \mid \theta) = 0$, and the totality of all such points has probability 0.] Thus we have the following informal result, giving us a method for finding a minimal sufficient statistic:

THEOREM 14 *A statistic $T = t(\mathbf{x})$ which determines the partition defined by the equivalence relation (1) is minimal sufficient for θ. Moreover, T is a function of every other sufficient statistic.*

EXAMPLE 7.9a Consider a random sample $\mathbf{X}$ from $\mathcal{N}(\mu, \sigma^2)$. The ratio of the joint p.d.f. at $\mathbf{x}$ to its value at $\mathbf{y}$ is

$$\frac{f(\mathbf{x} \mid \mu, \sigma^2)}{f(\mathbf{y} \mid \mu, \sigma^2)} = \exp\left\{-\frac{1}{2\sigma^2}\left[\sum x_i^2 - \sum y_i^2 - 2\mu\left(\sum x_i - \sum y_i\right)\right]\right\}.$$

This is independent of the two parameters if and only if it happens that both $\sum x_i^2 = \sum y_i^2$ and $\sum x_i = \sum y_i$. Thus, $\mathbf{x}$ and $\mathbf{y}$ are in the same partition set of the minimal sufficient statistic if and only if their coordinates have the same sum and the same sum of squares. So the pair of statistics $(\sum X_i, \sum X_i^2)$ is minimal sufficient for (μ, σ^2). Moreover, since these uniquely determine $\bar{X}$ and S^2, the sample mean and variance are (jointly) minimal sufficient. ∎

EXAMPLE 7.9b Let $\mathbf{X}$ be a random sample from a Cauchy population with location parameter θ: $f(x \mid \theta) \propto [1 + (x - \theta)^2]^{-1}$. Two sample points $\mathbf{x}$ and $\mathbf{y}$ are in the same partition set of the minimal sufficient partition if and only if the ratio

$$\frac{f(\mathbf{x} \mid \theta)}{f(\mathbf{y} \mid \theta)} = \prod_{j=1}^{n} \frac{1 + (y_j - \theta)^2}{1 + (x_j - \theta)^2}$$

is independent of θ. The numerator and denominator are both polynomials of degree $2n$ in θ, and if their ratio is to be independent of θ, the polynomials must be identical (their leading coefficients are equal). This means that their zeros agree. These are easy to find, since the polynomials are factored: The zeros of the numerator are $y_j \pm i$, and the zeros of the denominator are $x_j \pm i$ ($j = 1, ..., n$). These two sets of $2n$ numbers will consist of the same $2n$ numbers if and only if the components of $\mathbf{x}$ are a permutation of the components of $\mathbf{y}$. A partition set of the minimal sufficient partition must then consist of the $n!$ permutations of n numbers. So the order statistic, which defines this partition, is minimal sufficient. ∎

THEOREM 15 *The likelihood function, as a random function calculated from a sample* **X**, *is a minimal sufficient statistic.*

This can be seen as a result of the factorization criterion and the fact that a sufficient statistic which is a function of every sufficient statistic is minimal sufficient. It also follows from the fact that the likelihood function determined by a particular result **x** is the same as that determined by another result **y** if and only (1) holds.

THEOREM 16 *For a random sample from the exponential family:*

$$f(x \mid \theta) = B(\theta) \exp[Q(\theta) R(x)] h(x),$$

(a) the statistic $T = \sum R(x_i)$ *is minimal sufficient, and (b) the distribution of* T *is in the exponential family.*

For (a) we note that the p.d.f. of n independent observations from a member of this family is

$$f(\mathbf{x} \mid \theta) = B^n(\theta) \exp\left[Q(\theta) \sum R(x_i)\right] \prod h(x_i).$$

Two sample points **x** and **y** are in the same partition set of the minimal sufficient partition if and only if the ratio of p.d.f.'s at those two points,

$$\frac{f(\mathbf{x} \mid \theta)}{f(\mathbf{y} \mid \theta)} = \exp\left[Q(\theta)\{\sum R(x_i) - \sum R(y_i)\}\right] \prod \frac{h(x_i)}{h(y_i)},$$

does not involve θ. This is the case if and only if $\sum R(x_i) = \sum R(y_i)$. So a partition set of the minimal sufficient statistic is one whose every point **x** yields the same value of $\sum R(x_i)$. This value *identifies* the various partition sets, so the statistic $\sum R(X_i)$ is minimal sufficient.

We'll show Theorem 16(b) only in the discrete case: The p.f. is

$$f_T(t \mid \theta) = P[\sum R(X_i) = t]$$

$$= \sum_{\sum R(x_i) = t} \left\{ B^n(\theta) \exp\left[Q(\theta) \sum R(x_i)\right] \prod h(x_i) \right\}.$$

The factor involving $\sum R(x_i)$ has the same value at each term of the sum, so it factors out along with the $B^n(\theta)$, and we have

$$f_T(t \mid \theta) = b(\theta) \exp[t Q(\theta)] H(t),$$

where $b(\theta) = B^n(\theta)$, and

$$H(t) = \sum_{\Sigma R(x_i) = t} \left(\prod h(x_i)\right).$$

Clearly, then f_T is in the exponential family.

The exponential family includes most of the special one-parameter families we've encountered, and we observe that for this single-parameter family, the minimal sufficient statistic is one-dimensional—a single real-valued function of the data, $T = t(\mathbf{X})$. There is also an exponential family indexed by a multi-dimensional parameter, $\boldsymbol{\theta} = (\theta_1, ..., \theta_k)$. The p.d.f. is

$$f(x \mid \boldsymbol{\theta}) = B(\boldsymbol{\theta})h(x) \exp[Q_1(\boldsymbol{\theta})R_1(x) + \cdots + Q_k(\boldsymbol{\theta})R_k(x)]. \quad (5)$$

For a random sample $\mathbf{X}$, the minimal sufficient statistic is k-dimensional:

$$\left[\sum R_1(X_i), \ldots, \sum R_k(X_i)\right].$$

It is interesting that under certain assumptions of regularity, and if the support of X is independent of $\boldsymbol{\theta}$, then if there is a minimal sufficient statistic of the same dimension as that of $\boldsymbol{\theta}$, the population distribution must belong to the exponential family (5).[7]

PROBLEMS

7-60. Find a minimal sufficient statistic based on a random sample from each of the following populations:
(a) Ber(p).
(b) Normal with known variance.
(c) Poi(λ).
(d) Exp(λ).
(e) Normal with known mean.
(f) Geo(p).

7-61. Find a minimal sufficient statistic based on a random sample from each of the following populations:

(a) Rayleigh: $f(z \mid \theta) = \frac{z}{\theta} \exp\left(-\frac{z^2}{2\theta}\right)$, $z > 0$.

(b) Maxwell: $f(u \mid \theta) = \frac{2u^2}{\pi \theta^3} \exp\left(-\frac{u^2}{2\theta^2}\right)$, $u > 0$.

7-62. Determine a minimal sufficient statistic for sampling without replacement from a finite Bernoulli population.

---

[7] See E. W. Barankin and A. P. Maitra, "Generalization of the Fisher-Darmois-Koopman-Pitman theorem on sufficient statistics," *Sankhya A*, 25 (1963), pp. 217-244, and references contained therein.

7-63. Find the partition that is minimal sufficient for the family of three distributions of a random variable with three possible values (a, b, c):

$$\theta_1 = (.4, .2, .4), \quad \theta_2 = (.6, .3, .1), \quad \theta_3 = (.2, .1, .7).$$

(This family was encountered in Problem 7-49.)

7-64. Find a minimal sufficient statistic for a random sample from a distribution on a finite number of possible values: $a_1, ..., a_k$, with corresponding probabilities $p_1, ..., p_k$.

7-65. Show that the likelihood ratio, shown in Example 7.8f to be sufficient for a family of just two states, is actually minimal sufficient.

7-66. Find a minimal sufficient statistic for a random sample from $\mathcal{N}(\theta, \theta^2)$.

7-67. Find a minimal sufficient statistic for a random sample of size n from a population with c.d.f. $F(x \mid \theta) = x^\alpha$, $0 < x < 1$, where $\alpha > 0$.

7.10 Information in a sample

We consider random samples from $f(x \mid \theta)$. We'll assume that θ is a real parameter, although the notion of "information" we're about to define can be extended to the case of a vector parameter.[8]

The likelihood function depends on the observed data, and here it will be helpful to indicate this dependence, using the notation $L_\mathbf{X}(\theta)$ for the likelihood of θ calculated for an observed $\mathbf{X}$. If $\mathbf{X}$ and $\mathbf{Y}$ are independent vectors,

$$L_{\mathbf{X},\mathbf{Y}}(\theta) = f_\mathbf{X}(\mathbf{x} \mid \theta) f_\mathbf{Y}(\mathbf{y} \mid \theta) = L_\mathbf{X}(\theta) L_\mathbf{Y}(\theta).$$

Taking logarithms, we find that the logarithm of the likelihood is additive over independent pieces of data:

$$\log L_{\mathbf{X},\mathbf{Y}}(\theta) = \log L_\mathbf{X}(\theta) + \log L_\mathbf{Y}(\theta). \tag{1}$$

The proportionality constant in a likelihood becomes an additive constant for the log-likelihood, and when we differentiate with respect to θ, that constant disappears.

In what follows a prime denotes differentiation with respect to θ. We'll consider the likelihood function as a random function, which it becomes when we replace $\mathbf{x}$ by $\mathbf{X}$.

---

[8] See E. L. Lehmann, *Theory of Point Estimation*, New York, John Wiley & Sons (1983), pages 123ff.

DEFINITION The derivative of the log-likelihood is the *score* function:

$$V(\theta) = \frac{\partial}{\partial \theta} \log L(\theta) = \frac{L'(\theta)}{L(\theta)} = \frac{f'(\mathbf{X} \mid \theta)}{f(\mathbf{X} \mid \theta)}, \tag{2}$$

where $L \neq 0$. (Let $V = 0$ when $L = 0$.)

Like L, the score function V is a random function of θ; and because knowing one of these we know the other, V is minimal sufficient. We now give some facts about V stated as "theorems," although these statements will omit the regularity conditions needed for mathematical rigor.

THEOREM 17 $E[V(\theta)] = 0$ *for each* θ.

For this to hold, we require sufficient regularity to allow interchange of integral and derivative, or summation and derivative. In the continuous case,

$$E(V) = \int \frac{f'(\mathbf{x} \mid \theta)}{f(\mathbf{x} \mid \theta)} f(\mathbf{x} \mid \theta) d\mathbf{x} = \frac{d}{d\theta} \int f(\mathbf{x} \mid \theta) d\mathbf{x} = \frac{d}{d\theta} 1 = 0.$$

(Realize that the integrals are n-dimensional, and $d\mathbf{x} = dx_1 \cdots dx_n$. The integration is over the whole of $\real^n$, minus the set of points where $f = 0$.)

EXAMPLE 7.10a Consider $X \sim \mathcal{U}(0, \theta)$. For a random sample of size n, the likelihood function is $1/\theta^n$, for $\theta > X_{(n)}$. Then $L'/L = -n/\theta$, and the theorem does not hold. When the population support depends on the parameter of interest, the situation requires special attention. Some theorems work, others don't. ■

DEFINITION The **information** in a sample $\mathbf{X}$ about the parameter θ is the variance of the score function:

$$I_{\mathbf{X}}(\theta) = \text{var } V = E\left\{ \left[\frac{\partial}{\partial \theta} \log f(\mathbf{X} \mid \theta)\right]^2 \right\}. \tag{3}$$

$= E V^2$

(The notation could be misleading. The "E" averages over the distribution of the data $\mathbf{X}$, so the information is not a function of $\mathbf{X}$. However, it does depend on the model f—on the experiment that results in $\mathbf{X}$. This may be different for another experiment with outcome $\mathbf{Y}$.)

THEOREM 18 *Information is additive over independent experiments: If $\mathbf{X}$ and $\mathbf{Y}$ are independent, then*

$$I_{\mathbf{X}+\mathbf{Y}}(\theta) = I_{\mathbf{X}}(\theta) + I_{\mathbf{Y}}(\theta). \tag{4}$$

242   Chapter 7 Sampling and Reduction of Data

Theorem 18 becomes clear when we observe that the score function (2) is additive because of (1) and the fact that differentiation is additive. For, the variance is additive over independent scores. Then, because the n observations in a random sample are independent, we have this important special case:

COROLLARY   *The information in a random sample of size n is n times the information in a single observation:*

$$I_{\mathbf{X}}(\theta) = nI_X(\theta). \tag{5}$$

Sometimes it is easier to calculate the information using an alternative formula, given by the following:

THEOREM 19   *The information in $\mathbf{X}$ about θ can be calculated as*

$$I_{\mathbf{X}}(\theta) = -E\left[\frac{\partial^2}{\partial \theta^2}\log f(\mathbf{X}\mid\theta)\right] = -E\left(\frac{\partial V}{\partial \theta}\right). \tag{6}$$

To show Theorem 19, we again need to interchange operations freely, so we require sufficient regularity to justify this. First, observe that

$$\frac{\partial V}{\partial \theta} = \frac{\partial^2}{\partial \theta^2}\log f = \frac{\partial}{\partial \theta}\left(\frac{f'}{f}\right) = \frac{ff'' - (f')^2}{f^2} = \frac{f''}{f} - \left(\frac{f'}{f}\right)^2.$$

Then, because

$$E\left(\frac{f''}{f}\right) = \int \frac{f''}{f} f \, d\mathbf{x} = \int f'' \, d\mathbf{x} = \frac{d^2}{d\theta^2}\int f \, d\mathbf{x} = 0,$$

the asserted equality follows when we take expectations:

$$-E\left(\frac{\partial V}{\partial \theta}\right) = -E\left[\frac{f''}{f} - \left(\frac{f'}{f}\right)^2\right] = E\left[\left(\frac{f'}{f}\right)^2\right] = \text{var } V = I_{\mathbf{X}}(\theta).$$

(Recall that the variance of V is the expected square, since $EV = 0$.)

EXAMPLE 7.10b   Suppose $X \sim \mathcal{N}(0, \theta)$. The p.d.f. is

$$f(x\mid\theta) = \frac{1}{\sqrt{2\pi\theta}} e^{-x^2/(2\theta)}.$$

To find the information in a single observation we first take the logarithm:

$$\log f(x \mid \theta) = -\tfrac{1}{2}\log(2\pi) - \tfrac{1}{2}\log\theta - \frac{x^2}{2\theta}.$$

We then differentiate this with respect to θ, replacing x with the *random variable* X to obtain the score function:

$$V = \frac{\partial}{\partial\theta}\log f = -\frac{1}{2\theta} + \frac{X^2}{2\theta^2}.$$

(Note that $EV = 0$, as promised by Theorem 17.) The information is the variance of this random variable:

$$I_X(\theta) = \operatorname{var} V = \frac{\operatorname{var}(X^2)}{4\theta^4} = \frac{1}{2\theta^2}.$$

Alternatively, we could find $I_X(\theta)$ [using (6)] by differentiating V again with respect to θ:

$$\frac{\partial V}{\partial\theta} = \frac{\partial}{\partial\theta}\left(-\frac{1}{2\theta} + \frac{X^2}{2\theta^2}\right) = \frac{1}{2\theta^2} - \frac{X^2}{\theta^3},$$

and averaging:

$$I_X(\theta) = -E\left(\frac{\partial V}{\partial\theta}\right) = E\left(\frac{X^2}{\theta^3} - \frac{1}{2\theta^2}\right) = \frac{1}{2\theta^2}.$$

From the corollary to Theorem 18 we calculate the information in a random sample of size n by multiplying the information in one observation by n:

$$I_{\mathbf{X}}(\theta) = n I_X(\theta) = \frac{n}{2\theta^2}. \blacksquare \qquad (7)$$

THEOREM 20 *The information provided by a sufficient statistic T is the same as that in the sample:*

$$I_T(\theta) = I_{\mathbf{X}}(\theta). \qquad (8)$$

We show this only in the discrete case. When $T = t(\mathbf{X})$ is sufficient, the factorization criterion says that $\mathbf{X}$ has a joint p.d.f. of the form

$$f(\mathbf{X} \mid \theta) = g[t(\mathbf{X}), \theta]\, h(\mathbf{X}).$$

The p.f. of T is the probability of $[T = t]$, found by summing the joint p.d.f. over those points $\mathbf{x}$ in the sample space for which $t(\mathbf{x}) = t$:

$$f_T(t \mid \theta) = \sum_{t(\mathbf{x}) = t} g[t(\mathbf{X}), \theta]\, h(\mathbf{X}) = g(t, \theta) \sum_{t(\mathbf{x}) = t} h(\mathbf{X}).$$

Since the last sum does not depend on θ, the likelihood function (when we observe $T = t$) is $L_t(\theta) = g(t, \theta)$. But upon observing $\mathbf{x} = (x_1, \ldots, x_n)$

and finding that $t(\mathbf{x}) = t$, we obtain the *same* likelihood:
$$L_{\mathbf{x}}(\theta) = g(t, \theta)h(\mathbf{x}).$$
So the information will be the same in each case, as claimed.

EXAMPLE 7.10c We illustrate Theorem 20 in the case of $X \sim \text{Exp}(\lambda)$ and a random sample of size n. The statistic $T = \sum X$ is sufficient for λ, and for a single observation on X, the score function is
$$\frac{\partial}{\partial \lambda} \log(\lambda e^{-\lambda x}) = \frac{\partial}{\partial \lambda}[\log \lambda - \lambda X] = \frac{1}{\lambda} - X.$$
So for one and n observations, the informations are
$$I_X(\lambda) = E\left\{-\frac{\partial^2}{\partial \lambda^2} \log f\right\} = \frac{1}{\lambda^2}, \text{ and } I_{\mathbf{X}}(\lambda) = \frac{n}{\lambda^2},$$
respectively. The p.d.f. of the sufficient statistic T [from (9), §6.6] is
$$f_T(t \mid \theta) \propto \lambda^n t^{n-1} e^{-\lambda t},$$
and the corresponding V, for a single observation on T, is
$$V = \frac{\partial}{\partial \lambda}(-n \log \lambda - \lambda T) = \frac{n}{\lambda} - T.$$
The negative of its derivative gives the information in T:
$$I_T(\lambda) = \frac{n}{\lambda^2},$$
the same as the information in the sample. ∎

It is also true that the information in a statistic that is *not* sufficient is *less* than the information in the sample:

THEOREM 21: *If $\mathbf{X}$ is a random sample from $f(x \mid \theta)$, and $T = t(\mathbf{X})$ is any statistic, then $I_T(\theta) \leq I_{\mathbf{X}}(\theta)$, and the equality holds if and only if T is sufficient for θ.*

We can prove this only in the discrete case, using the following:

LEMMA *When X is discrete with p.f. $f(x \mid \theta)$, $\mathbf{X}$ is a random sample, and $T = t(\mathbf{X})$, a statistic with p.f. $g(t \mid \theta)$, then*
$$E\left\{\frac{\partial}{\partial \theta} \log f(\mathbf{X} \mid \theta) \mid T = t\right\} = \frac{\partial}{\partial \theta} \log g(t \mid \theta).$$

The p.f. of T is derived from that of $\mathbf{X}$:
$$g(t \mid \theta) = \sum_{t(\mathbf{x}) = t} f(\mathbf{x} \mid \theta).$$

And from this we find
$$f_{\mathbf{X} \mid T = t}(\mathbf{x} \mid \theta) = \frac{f(\mathbf{x} \mid \theta)}{g(t \mid \theta)} \quad \text{if } t(\mathbf{x}) = t$$

(and 0, otherwise). With prime denoting derivative with respect to θ,
$$E\left(\frac{f'}{f} \,\Big|\, T = t\right) = \sum \frac{f'}{f} \cdot f_{\mathbf{X} \mid T = t} = \sum \frac{f'}{g(t)} = \frac{g'}{g} = \frac{\partial}{\partial \theta} \log g(t \mid \theta),$$

as asserted by the lemma. Problem 7-72 calls for a completion of the proof of Theorem 21 in the discrete case.

PROBLEMS

7-68. Find the information in a random sample of size n from
(a) $\mathcal{N}(\mu, 1)$.
(b) Ber(p).
(c) Poi(λ).
(d) Geo(p).

7-69. For random samples from Geo(p), show that $I_T(p)$, the information in the sample sum: $T = \sum X_i$, is the same as the information in all the sample observations: $I_{\mathbf{X}}(p)$,
(a) by evaluating each.
(b) by using a general theorem.

7-70. Consider a population defined by the p.d.f. $f(x \mid \theta)$, and a 1-1 transformation of the parameter: $\xi = h(\theta)$. For a single observation X, find the relationship between $I(\theta)$ and $\tilde{I}(\xi)$.

7-71. Consider the family of normal distributions with mean 0, indexed by the parameter σ, the standard deviation. Find $I(\sigma)$, the information about σ in a random sample of size n.

7-72. In the setting of Theorem 21 and the lemma which follows it,
(a) use the iterated expectation formula to show the following:
$$E\left\{\frac{\partial}{\partial \theta} \log f(\mathbf{X} \mid \theta) \frac{\partial}{\partial \theta} \log g[t(\mathbf{X}) \mid \theta]\right\} = I_T(t).$$

(b) use the result in (a) and the lemma, together with the fact that
$$E\left\{\left[\frac{\partial}{\partial \theta} \log f(\mathbf{X} \mid \theta) - \frac{\partial}{\partial \theta} \log g[t(\mathbf{X}) \mid \theta]\right]^2\right\} \geq 0,$$

246   Chapter 7 Sampling and Reduction of Data

to show Theorem 21 in the discrete case.

7-73. Let X have a distribution in the exponential family:
$$\log f(x \mid \theta) = \log B(\theta) + \log h(x) + Q(\theta)R(x).$$

(a) Show that $E[R(X)] = -\dfrac{B'}{BQ'}$ and $\operatorname{var} R(X) = \dfrac{I_X(\theta)}{Q'^2}$.

(b) Obtain a formula for $I_X(\theta)$ in terms of B, Q, and their derivatives.

7.11 Prior to posterior

In §7.7 we broached the possibility of using a probability distribution to represent the state of uncertainty about θ. And we gave the basic formula (Bayes" theorem) for using data to convert a prior distribution to a posterior distribution:

$$P(\text{model} \mid \text{data}) \propto P(\text{data} \mid \text{model}) \cdot P(\text{model}). \tag{9}$$

For given data, $P(\text{model} \mid \text{model})$ is the *likelihood* of the model. In words, we remember the relation this way:

$$\text{posterior} \propto \text{likelihood} \times \text{prior}. \tag{10}$$

(In any particular instance, a "posterior" and "prior" will be given by either a density or a probability function.) Thus, using (2) one can update a prior distribution in light of the information provided by a set of data. Also, this new, posterior distribution will serve as a prior for further updating of one's beliefs about θ, in light of the information in any new data set. (See Problem 7-80.)

The posterior distribution is conditional, deriving from the joint distribution of data D and the parameter Θ. Their joint p.f. or p.d.f. is

$$f_{D,\Theta}(d, \theta) = f(d \mid \theta) g(\theta),$$

where g is the prior p.f. or p.d.f., and $f(d \mid \theta)$ is the p.f. or p.d.f. of the data *given* the parameter value θ. The posterior p.f. or p.d.f. is then

$$h(\theta \mid d) = \dfrac{f(d \mid \theta) g(\theta)}{f_D(d)},$$

where f_D denotes the unconditional p.f. or p.d.f. of the data:

$$f_D(d) = E_g[f(d \mid \Theta)] = \begin{cases} \int f(d \mid \theta) g(\theta)\, d\theta, & (\Theta \text{ continuous}), \\ \sum f(d \mid \theta_i) g(\theta_i) & (\Theta \text{ discrete}), \end{cases}$$

7.11 Prior to posterior

EXAMPLE 7.11a Suppose we have prior probability $g_0 = .8$ for $\mathcal{N}(0, 1)$ and $g_1 = .2$ for $\mathcal{N}(1, 1)$, and observe $\bar{X} = .6$ in a random sample of size $n = 10$. The likelihood function [from Example 7.7d, page 223] is

$$L(\mu) = e^{-(\mu - \bar{X})^2},$$

so $L(0) = e^{-1.8} = .165$, and $L(1) = e^{-.8} = .449$. Then

$$g_0 L(0) + g_1 L(1) = .8 \times .165 + .2 \times .449 = .222.$$

and

$$h_0 = \frac{.8 \times .165}{.222} = .595, \quad h_1 = 1 - h_0 = .405.$$

With a value of the sample mean closer to $\mu = 1$, our new probability distribution for μ puts more weight on $\mu = 1$ than does the prior. ∎

EXAMPLE 7.11b Consider a random sample of size n from $\text{Ber}(p)$, and a prior p.d.f. for the parameter p which is $\text{Beta}(r, s)$:

$$g(p) \propto p^{r-1}(1-p)^{s-1}.$$

The likelihood function, in terms of Y, the number of successes, is

$$L(p) = p^Y (1-p)^{n-Y},$$

so the posterior p.d.f., from (2), is

$$h(p \mid Y) \propto L(p)g(p) \propto p^{r+Y-1}(1-p)^{n+s-Y-1}.$$

This is a beta density: $p \sim \text{Beta}(r + Y, n + s - Y)$.

To take a specific example, suppose $n = 10$, $Y = 7$, and we have a prior which is $\text{Beta}(3, 3)$. The posterior is then $\text{Beta}(10, 6)$, which skews the distribution to the right and sharpens it, as shown in Figure 7-10. ∎

In this last example we observe that the posterior is a member of the same distribution family as the prior. When this is the case, we say that the family is conjugate to $\{f(x \mid \theta)\}$. It can be shown that when there is a sufficient statistic of fixed dimension for any sample size (as in the case of the exponential family), it is possible[9] to construct a conjugate family of distributions for the parameter. The problems give several

---

[9] Ferguson, T. S., *Mathematical Statistics*, New York: Academic Press, Inc., 1967, page 163, or M. H. DeGroot, *Optimal Statistical Decisions*, New York: McGraw-Hill Book Co., 1970.

more examples of conjugate families.

Although finding a posterior from a prior distribution is relatively straightforward, an obvious question is where does one get the prior? It can be shown that persons who behave "rationally," according to a reasonable definition of the term, do indeed have prior distributions. The elicitation of a prior distribution is a problem that continues to be the subject of research. In simple situations one may elicit a prior by making comparisons with experiments in which probabilities are generally agreed upon. For instance, for the p in a Bernoulli experiment one might ask, "would you prefer to receive \$10 if this experiment succeeds, or to receive \$10 if this coin I toss turns up heads?" If the coin toss is preferred, the person's personal probability for success is less than $1/2$.

An unresolved question is how best to represent a state of complete ignorance about a parameter. Bayes (1763) considered the notion that ignorance could be represented by a uniform distribution, and Laplace"s apparent espousal of the idea led to so many controversial applications that the Bayesian approach fell into general disfavor. What is particularly awkward about using a uniform distribution is that one who is ignorant of θ is equally ignorant of θ^2 or any other function of θ. But when Θ is uniform, the distribution of $g(\Theta)$ is generally *not* uniform.

Another difficulty with using a uniform prior to represent ignorance is that the parameter space we consider is sometimes infinite, and a constant density is not possible. Yet, if we think of $g(\theta)$ as simply a weighting function, using a constant for g in a formal calculation of h often yields a proper posterior. This posterior is sometimes the limit of posteriors obtained from a suitably chosen sequence of proper priors.

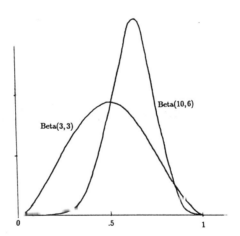

Figure 7-10. Prior and posterior for Example 7.11b

EXAMPLE 7.11c When $X \sim \text{Exp}(\lambda)$, the likelihood function is $\lambda^n e^{-\lambda Y}$, where $Y = \sum X_i$. If we adopt the prior $\lambda \sim \text{Exp}(\beta)$, the posterior is

$$h(\lambda \mid Y, \beta) \propto \lambda^n e^{-\lambda(\beta + Y)}, \quad \lambda > 0.$$

As β tends to 0, the prior density tends to a constant, an improper priod, but the posterior tends to a proper posterior:

$$h(\lambda \mid Y) = \frac{1}{n!} Y^{n+1} \lambda^n e^{-\lambda Y}, \quad \lambda > 0.$$

And this is what results if one applies (3) formally, with $g(\lambda) = 1$. ∎

PROBLEMS

7-74. Consider two models for x, a random variable with two values:

x	$f_0(x)$	$f_1(x)$
a	.5	.3
b	.5	.7

Find the posterior distributions given $X = a$ and given $X = b$, assuming prior probabilities $g_0 = .4$ and $g_1 = .6$.

7-75. Find a conjugate family for the case of random samples form a normal population with unknown mean and known variance $\sigma^2 = 1$.

7-76. Find a conjugate family for the case of random samples from a Poisson population with parameter λ: $f(x \mid \lambda) = \lambda^x e^{-\lambda}/x!$, $x = 0, 1, \ldots$.

7-77. Find a conjugate family for the case of random samples form an exponential population with parameter θ: $f(x \mid \theta) = \theta e^{-\theta x}$, $x > 0$.

7-78. Find a conjugate family for the case of random samples from a geometric distribution: $f(x \mid p) = p(1-p)^x$, $x = 0, 1, 2, \ldots$.

7-79. For the population distributions and prior probabilities in Problem 7-74, find the posterior distribution given that two independent observations result in (x_1, x_2). Compare this result with what you get using the posterior from Problem 7-74, given x_1, as a prior for combining with a second observation x_2.

7-80. Let (X, Y) have a joint p.d.f. $f(x, y) = f_1(x \mid \theta) f_2(y \mid x, \theta)$. Show that the posterior, assuming a prior $g(\theta)$, is the same as would be obtained using the posterior given X as a prior with the new data Y.

7-81. A lot of N articles contains M defectives. One article is drawn at random and found to be defective. Asusuming a $\text{Bin}(N, p)$ prior for M,

250 Chapter 7 Sampling and Reduction of Data

find the posterior distribution of M.

7-82. Given a random sample from $\mathcal{N}(\mu, 1)$, find the formal posterior distribution for the mean μ assuming a uniform (improper) prior for μ.

7-83. For the situation of the preceding problem, find the posterior for μ assuming the prior p.d.f. $g(\mu \propto e^{-\tau \mu^2/2}$ and examine the limiting posterior as $\tau \to 0$. (Compare with the answer to the preceding problem.)

7-84. Show the following:
(a) If the prior distribution is singular, concentrated on a single θ, then the posterior is the same as the prior. (That is, no finite amount of data can convince a statistican whose mind is already made up.)
(b) If the distribution of the data is independent of the state of nature, the posterior is the same as the prior.
(c) If T is sufficient for $f(x \mid \theta)$, the posterior depends on the sample $\mathbf{X}$ only through the value of T.

7.12 Principles of inference

Statistical inference—drawing consclusions about the general from the particular—is inherently imperfect. About all one can do is formulate some guiding "principles" and see where they lead. We have already mentioned or implied some principles that have adherents, but no principle seems to be universally accepted. Indeed, much of what is common statistical practice violates one or more of the principles we now state somewhat more precisely.

The *sufficiency principle* says roughly that inferences should depend on a sufficient statistic. More formally:

SUFFICIENCY PRINCIPLE *If the statistic $t(\mathbf{X})$ is sufficient for the family $\{f(x \mid \theta)\}$, and $t(\mathbf{x}_1) = t(\mathbf{x}_2)$, then the same inference should be drawn whether one observes $\mathbf{x}_1$ or $\mathbf{x}_2$.*

We have seen that the likelihood function $L(\theta \mid \mathbf{X})$ is minimal sufficient, and is therefore a function of every other sufficient statistic, suggesting that if we know less than the likelihood function, we are not using all the information about θ that is available in the sample. On the other hand, if we know the likelihood function, there is nothing more in the sample about θ. This suggests the likelihood principle. Before stating it we recall that the likelihood function (as we have defined it) is an equivalence class of functions that are identical except possibly for a factor independent of θ.

LIKELIHOOD PRINCIPLE *If an experiment $\mathcal{E}$ results in $\mathbf{X}$ with p.f. or p.d.f. $f(\mathbf{x} \mid \theta)$, whether one observes $\mathbf{x}_1$ or $\mathbf{x}_2$, the same inference should be drawn if the likelihoods are the same—that is, if*

$$L(\theta \mid \mathbf{x}_1) = L_2(\theta \mid \mathbf{x}_2).$$

As stated, the likelihood principle deals with only one experiment. We have seen instances where different experiments result in the same likelihood. (See Problems 7-45, 7-46.) Another version of the principle says that even in such cases we ought to draw the same inferences or come to the same conclusions.

STRONG LIKELIHOOD PRINCIPLE *If an experiment $\mathcal{E}$ results in $\mathbf{X}$ with p.f. or p.d.f. $f(\mathbf{x} \mid \theta)$ and a second experiment $\mathcal{F}$ results in $\mathbf{Y}$ with p.f. or p.d.f. $g(\mathbf{y} \mid \theta)$ involving the same parameter θ, then, whether one observes $\mathbf{x}$ with experiment $\mathcal{E}$ or $\mathbf{y}$ with experiment $\mathcal{F}$, the same inference should be drawn if the likelihood functions are the same: $L(\theta \mid \mathbf{x}) = L(\theta \mid \mathbf{y})$.*

A principle that we have not encountered is that of conditionality, which may seem almost more compelling than the likelihood principle.

EXAMPLE 7.12a Suppose we want to know about the rate parameter of a Poisson arrival process. Which experiment should we do: (a) observe the process for a fixed time, say one hour, and count the number of arrivals in that one-hour period, or (b) observe the process until a specified number of events have occurred, say 10, and record how long it takes? We may toss a coin to decide which one to do; suppose the coin-toss tells us to do the second experiment, and we find that it takes 40 minutes until the 10th arrival.

Now, suppose we never thought of doing experiment (a) at all, and simply did experiment (b) with the same result—40 minutes until the 10th trial. Is the information any different? ∎

CONDITIONALITY PRINCIPLE *Suppose an experiment $\mathcal{E}$ results in $\mathbf{X}$ with p.f. or p.d.f. $f(\mathbf{x} \mid \theta)$ and a second experiment $\mathcal{F}$ results in $\mathbf{Y}$ with p.f. or p.d.f. $g(\mathbf{y} \mid \theta)$ involving the same parameter θ. Whether we simply carry out $\mathcal{F}$, or happen to choose $\mathcal{F}$ in a random selection from $\mathcal{E}$ and $\mathcal{F}$, the inference should be the same, depending only on the outcome of $\mathcal{F}$.*

In 1962, A. Birnbaum showed that the conditionality principle together with the sufficiency principle implied the strong likelihood principle!

The likelihood principle does not say *how* we should use the likelihood function. Many people do concede that between two possible models, the one with the larger likelihood is the better explanation of observed data, as we pointed out in §7.7. But they may not agree that the largest of more than two models is the *best* explanation and thus accept the maximum likelihood principle stated in §7.7. Yet, this principle is the basis of some commonly used procedures for the problems of estimation and testing that we take up in the next couple of chapters.

Finally, we mention briefly that there is a principle of invariance[10] which implies that, ordinarily, our procedures and conclusions should not depend on the particular units of measurement we happen to use in obtaining data.

---

[10] See for instance R. L. Berger and G. Casella, *Statistical Inference*, (1990) Pacific Grove, CA: Brooks/Cole Pub. Co., p. 274, or J. Berger and R. Wolpert, The Likelihood Principle, Inst. of Math. Stat. Lecture Notes—Monograph Series, Hayward, CA: IMS (1984).

8
Estimation

In this chapter we study the use of sample information to estimate one or more population parameters. It is natural to expect that if a sample is at all representative of the population, one can use it to make an estimate that is better than a sheer guess.

The set of possible parameter values in a particular problem is the *parameter space*. For a single real parameter it is a subset of the real line. The *data space* is the set of possible values of the sample observation vector, $\mathbf{X} = (X_1, ..., X_n)$. When the parameter-value announced as an estimate is calculated from the sample data, it is a *function* of $\mathbf{X}$—a mapping from the data space to the parameter space.

DEFINITION An **estimator** is a function on the data space with values in the parameter space: $T = t(\mathbf{X})$.

Since an estimator is calculated from sample data alone, it is a statistic. If the sample is obtained by random sampling, an estimator is a *random variable*. Its value varies from one sample to another according to its sampling distribution, which is derived from the population distribution.

Because of sampling variability, sample information is (ordinarily) not totally reliable, and parameter estimates based on sample information are typically in error. The error in any particular estimate is unknown, depending as it does on the true parameter value. (If we knew that true value, we wouldn't be estimating it.) About all we can do is base our judgment of an estimator's performance on how likely it is to come close.

Some statistics are more successful than others in estimating a parameter. Sometimes it is intuitively clear what statistic to try as an estimator. For instance, the sample mean is a rather natural choice to use in estimating a population mean. Whatever statistic we try, we need to assess its performance as an estimator, and this requires a criterion for

8.1 Mean squared error

In estimating a single real parameter to have the value T, the *error of estimation* is defined as $T - \theta$. According to a version of the Chebyshev inequality [obtained by setting $h(T) = (T - \theta)^2$ in Theorem 15 of §4.6, page 110], we know that for every $\epsilon > 0$,

$$P(\,|T - \theta| > \epsilon\,) \le \frac{1}{\epsilon^2} E[(T - \theta)^2].$$

This shows that if the mean of the squared error is small, the chance that T will be far from θ in any particular instance is small.

DEFINITION The **mean squared error** of T as an estimator for the parameter θ is

$$\text{m.s.e.}(T) = E[(T - \theta)^2] = \text{var}\,T + (ET - \theta)^2. \tag{1}$$

The **bias** in T as an estimator of θ is

$$b_T(\theta) = ET - \theta. \tag{2}$$

An estimate is said to be **unbiased** when $b_T(\theta) = 0$, or $ET = \theta$.

It is clear from (1) that when we use T to estimate θ, the mean squared error is small if and only if T has *both* a small bias and a small variance. (In particular, unbiasedness alone is not enough.) When T is unbiased, its m.s.e. is just its variance.

EXAMPLE 8.1a From §7.4 (Theorems 5 and 6) we know that when sampling is random, $E(\bar{X}) = \mu$. So the sample mean is unbiased as an estimator of the population mean. And then m.s.e.$(\bar{X}) = \text{var}\,\bar{X}$. So if the observations are independent, m.s.e.$(\bar{X}) = \sigma^2/n$.

The statistic $T = X_1$ (the first observation in the sample) is also unbiased, because $E(X_1) = \mu$. However, its mean squared error is var $X = \sigma^2$, which is larger than that of $\bar{X}$ when $n > 1$. So unbiasedness alone is not enough. On the other hand, the estimator $T^* = 17$ has zero variance, but its mean squared error is the square of the bias, which will usually be large. ∎

If an estimator has a mean *proportional* to the parameter being estimated: $ET = k\theta$, it is easy to construct an unbiased estimator: T/k. (To do so may or may not serve a useful purpose.)

EXAMPLE 8.1b A natural estimator for the population variance σ^2 is V, the second moment of the sample distribution. We saw in §7.4 (Theorem 7, page 210) that V has a negative bias:

$$b_V(\sigma^2) = E(V) - \sigma^2 = -\frac{1}{n}\sigma^2.$$

However, since $E(V) = \frac{n-1}{n}\sigma^2$, we can construct an unbiased estimator by dividing V by $\frac{n-1}{n}$. The result is S^2, the statistic termed "sample variance" in Chapter 7. So S^2 is an unbiased estimator of σ^2. But it is not always better than V—at least, not in terms of mean squared error. For, consider the case of a normal population, where (from §7.5, Theorem 11, page 213)

$$\text{m.s.e.}(S^2) = \text{var } S^2 = \frac{2\sigma^4}{n-1}.$$

The variance of V is easy to get from the variance of S^2 (for $n > 1$):

$$\text{var } V = \left(\frac{n-1}{n}\right)^2 \text{var } S^2 = \left(\frac{n-1}{n}\right)^2 \frac{2\sigma^4}{n-1} = \frac{2\sigma^4(n-1)}{n^2}.$$

And with this we can calculate the mean squared error:

$$\text{m.s.e.}(V) = \text{var } V + b_V^2(\sigma^2) = \frac{\sigma^4}{n^2}[2(n-1)+1] = \frac{2n-1}{n^2}\sigma^4 < \text{m.s.e.}(S^2).$$

So in this case, and as measured by mean squared error, the biased V is better than the unbiased S^2. ∎

EXAMPLE 8.1c Consider **X**, a random sample of size n from $\mathcal{N}(\mu, \sigma^2)$, and let R denote the sample range, $X_{(n)} - X_{(1)}$. Define W as the range of the corresponding standardized X's: $(Z_1, ..., Z_n)$. Since standardizing is a monotone transformation, the largest Z comes from the largest X and the smallest Z from the smallest X. Hence,

$$W = Z_{(n)} - Z_{(1)} = \frac{X_{(n)} - \mu}{\sigma} - \frac{X_{(1)} - \mu}{\sigma} = \frac{R}{\sigma}.$$

Since the Z's are standard normal, the distribution of W does not involve the parameters μ and σ. In particular, its moments depend only on the sample size:

$$E(W) = a_n, \quad \text{var } W = b_n^2.$$

Then $E(R) = a_n \sigma$, a quantity proportional to σ. Thus, the statistic R/a_n is an unbiased estimator for σ, with variance

$$\text{var}\left(\frac{R}{a_n}\right) = \frac{\sigma^2 \text{ var } W}{a_n^2} = \frac{b_n^2}{a_n^2} \sigma^2.$$

The constants a_n and b_n are given in Table XIII for certain small sample sizes. These are calculated by numerical integrations, using the p.d.f. of W, from Problem 7-32:

$$f_W(w) = \frac{n(n-1)}{2\pi} \int_{-\infty}^{\infty} [\Phi(w+y) - \Phi(y)]^{n-2} e^{-[(w+y)^2 + y^2]} dy. \blacksquare$$

To see that requiring unbiasedness of an estimator would be overly restrictive (aside from being quite arbitrary), consider the next example, in which there is only one unbiased estimator. Indeed, it can also happen that there are *no* unbiased estimators!

EXAMPLE 8.1d Consider a Poisson arrival process with rate parameter λ and define $\theta = e^{-2\lambda}$. (This is the probability of two successive intervals of time with no "arrivals.") An estimator of θ based on a single observation X is a function on its possible values: $T = t(n)$, $n = 0, 1, 2, \ldots$. We look for coefficients $t_n = t(n)$ such that

$$E(T) = \sum_{n=0}^{\infty} t_n \frac{\lambda^n}{n!} e^{-\lambda} = e^{-2\lambda}.$$

Multiplying through by e^λ, we have

$$t_0 + t_1 \lambda + t_2 \frac{\lambda^2}{2!} + \cdots = e^{-\lambda} = 1 - \lambda + \frac{\lambda^2}{2!} - \cdots.$$

Power series expansions are unique, so the coefficients must match, and the only unbiased estimator is therefore: $T = (-1)^X$. With a value not possible for θ, this is not a useful "estimator" for θ. $\blacksquare$

The mean squared error usually depends on one or more unknown parameters, and it may not be possible to find an estimator with a mean squared error which is uniformly smallest. We'll see in §8.2 that if one admits only *unbiased* estimators, there is a lower bound for m.s.e.'s. However, the restriction to the class of unbiased estimators is rather artificial and may rule out some very good estimators, as we saw in Example 8.1b.

In the case of large samples, one can replace any unknown parameters that may appear in the m.s.e. by good sample estimates without introducing an appreciable error in the m.s.e. When this is done, and when the bias in the estimator is small, the square root of the m.s.e. is called the *standard error* (s.e.) of the estimate. It is an approximation to the standard deviation of the error, a kind of "typical" error, fairly intuitive as a measure of reliability of the estimator, especially when the estimator is asymptotically normal, as is often the case.

EXAMPLE 8.1e The mean of a random sample from a population with unknown mean and unknown variance σ^2 converges in probability to the population mean μ. (This is the "law of large numbers" of §5.2.) So it would seem to be a good large sample estimator for μ. It is unbiased, and its standard error can be approximated by replacing σ in the formula for its standard deviation by the sample standard deviation:

$$\text{s.e.}(\bar{X}) = \frac{S}{\sqrt{n}}. \qquad (3)$$

This is commonly referred to as the *standard error of the mean*. ∎

Mean squared error is a commonly used criterion for estimators, but it is not the only possibility. Other functions of the absolute error might be used, such as the kth absolute moment about the parameter: $E[|T - \theta|^k]$. Of these, the one we use ($k = 2$) is the easiest to handle mathematically, although this is not really a good excuse for using it. On the other hand, suppose we compare estimators based on the mean of $g(|T - \theta|)$, where g is an arbitrary *increasing* function such that $g(0) = 0$. Consider two normally distributed, unbiased estimators of θ, T and T', such that T is better than T' in terms of mean squared error: $\text{var}\, T' > \text{var}\, T$. Then

$$Eg(|T - \theta|) = \int_{-\infty}^{\infty} g(|t - \theta|)\, dF_T(t) = \int_0^1 g(|F_T^{-1}(u) - \theta|)\, du. \qquad (4)$$

Similarly,

$$Eg(|T' - \theta|) = \int_0^1 g(|F_{T'}^{-1}(u) - \theta|)\, du. \qquad (5)$$

Both F_T and $F_{T'}$ are normal c.d.f.'s with $F_{T'}^{-1}(\tfrac{1}{2}) = F_T^{-1}(\tfrac{1}{2}) = \theta$, and since T has the smaller variance, then (See Figure 8-1) for $u > \tfrac{1}{2}$

$$F_{T'}^{-1}(u) > F_T^{-1}(u) > \theta.$$

But $g(v)$ is an increasing function, so for $u > \frac{1}{2}$,

$$g(|F_{T'}^{-1}(u) - \theta|) > g(|F_T^{-1}(u) - \theta|). \qquad (6)$$

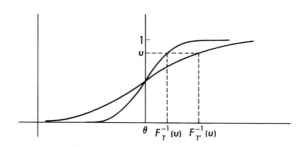

Figure 8-1

Similarly, when $u < \frac{1}{2}$, we find $F_{T'}^{-1}(u) < F_T^{-1}(u) < \theta$, or

$$F_T^{-1}(u) - \theta < F_{T'}^{-1}(u) - \theta < 0,$$

so (6) holds for *all* values of u. Therefore,

$$\int_0^1 g(|F_{T'}^{-1}(u) - \theta|) \, du \geq \int_0^1 g(|F_T^{-1}(u) - \theta|) \, du. \qquad (7)$$

In view of (4) and (5), then, it follows from (7) that

$$Eg(|T - \theta|) \leq Eg(|T' - \theta|).$$

Thus, T is at least as good as T' in terms of the measure defined by g. For normally distributed, unbiased estimators, we come to the same conclusion in using $g(|T - \theta|)$ as a measure of success (using mean absolute error, for instance) as in using mean squared error, $E[(T - \theta)^2]$.

PROBLEMS

8-1. Determine a condition on coefficients a_i so that the linear combination $T = \sum a_i X_i$ is an unbiased estimate of EX, and find its variance, assuming a random sample of size n from a population with finite variance σ^2.

8-2. Show that if T is an unbiased estimate of θ, then $aT + b$ is an unbiased estimate of $a\theta + b$.

8-3. Show that if T is an unbiased estimate of θ,
  (a) T^2 is biased as an estimator of θ^2.
  (b) $\sqrt{T}$ is biased as an estimator for $\sqrt{\theta}$.

8-4. Let Y denote the number of successes in n independent Bernoulli trials with parameter p, and define $T_1 = Y/n$ and $T_2 = (Y+1)/(n+2)$. Find and compare the m.s.e.'s of T_1 and T_2 when $n = 4$ and when $n = 8$. (In each case, sketch the m.s.e.'s on common axes.)

8-5. In Problem 7-29 we found the mean and variance of the largest observation in a random sample from $\mathcal{U}(0, \theta)$ to be
$$E[X_{(n)}] = \frac{n}{n+1}\theta, \quad \operatorname{var} X_{(n)} = \frac{n\theta^2}{(n+1)^2(n+2)}.$$
  (a) Find a multiple of $\bar{X}$ that is unbiased as an estimator of θ, and find its mean squared error.
  (b) Find a multiple of $X_{(n)}$ that is unbiased as an estimator of θ, and find its mean squared error. [Compare with the m.s.e. in (a).]

8-6. Find an unbiased estimator for the parameter $\xi = e^{-\lambda} = P(X = 0)$, based on a single observation from Poi(λ): $f(x \mid \lambda) = \lambda^x e^{-\lambda}/x!$, $x = 0, 1, 2, \ldots$. [Hint: A function $T = t(X)$ is a sequence, $t_n = t(n)$, as in Example 8.1d, so that $ET = \sum t_n f(n \mid \lambda)$.]

8-7. Show that the estimator $K\sum(X_i - \bar{X})^2$ of the variance of a normal population has the smallest mean squared error when $K = 1/(n+1)$.

8-8. For large random samples, find the standard error of estimate
  (a) when using $\bar{X}$ to estimate the parameter p, when $X \sim$ Ber(p).
  (b) when using $\bar{X}$ to estimate the parameter θ, when $X \sim$ Exp($1/\theta$).

8-9. In studying the effectiveness of a treatment, it is common practice to obtain a random sample of responses of treated individuals and an independent random sample of responses of untreated individuals. (The latter is referred to as the *control group*.) To estimate the mean difference in responses $\mu_T - \mu_C$, one might use the corresponding difference in sample means, $\bar{X}_T - \bar{X}_C$. For large random samples of sizes n_T and n_C, obtain an expression for the standard error estimate of the population mean difference as a function of the sample variances.

8-10. If $X \sim$ Exp(λ), then $EX = 1/\lambda$. So a natural candidate for estimating λ from a random sample of size n is $\hat{\lambda} = 1/\bar{X}$.
  (a) Calculate $E(1/\bar{X})$ when $n > 1$.
  (b) From the result of (a) find an unbiased estimate of the parameter λ and calculate its m.s.e.
  (c) Show that the multiple of $1/\bar{X}$ with the *smallest* m.s.e. in estimating

λ is $(n-2)/\sum X_i$.

8-11. For a single observation from Geo(p), show (by deriving it) that there is only one unbiased estimator for p, and that it is a rather silly estimator.

8-12. Consider a Bernoulli process with parameter p.
(a) Show that if we fix the number of trials, the sample proportion of successes is the only unbiased estimator of p which is a function of the number of successes (the minimal sufficient statistic).
(b) Suppose sample until we get m successes ($m > 1$, fixed). Let N denote the number of trials required. Show that the estimator $T = \frac{m-1}{N-1}$ is unbiased and is the only function of N that is unbiased in estimating p.

[For m successes in N trials, the likelihood function is the same with both stopping rules, but the (unique) unbiased estimates are different! (See Problem 7-47.) Unbiasedness is not a proper criterion for those who feel that when the likelihoods are the same, the statistical inference should be the same.]

8.2 Efficiency

Suppose there are competing estimators $T_1 = t_1(\mathbf{X})$ and $T_2 = t_2(\mathbf{X})$ for a parameter θ, and that m.s.e.(T_1) < m.s.e.(T_2). We say then that T_1 makes "more efficient" use of the data $\mathbf{X}$ than does T_2.

DEFINITION The *relative efficiency* of T_2 with respect to T_1 is defined as the ratio

$$e(T_2, T_1) = \frac{E[(T_1 - \theta)^2]}{E[(T_2 - \theta)^2]}, \qquad (1)$$

when this ratio does not depend on θ.

EXAMPLE 8.2a Consider the problem of estimating the parameter θ in $\mathcal{U}(0, \theta)$ using a random sample of size $n = 5$. Since $\theta/2$ is both the mean and the median, it is natural to try $T_1 = 2\bar{X}$ and $T_2 = 2\tilde{X}$ (where $\tilde{X}$ is the sample median) as estimators of θ. Both are unbiased estimates of θ, so their mean squared errors are their variances. In a sample of size 5, the median is the third smallest: $\tilde{X} = X_{(3)}$. In Example 7.6c we obtained a formula for the variance of $X_{(k)}$ which (with $n = 5$ and $k = 3$) gives us var$[X_{(3)}] = 1/28$. With this we find

$$\operatorname{var} T_2 = 4\operatorname{var}[X_{(3)}] = \theta^2/7.$$

Turning to T_1, we note that the population variance is $\sigma^2 = \theta^2/12$, so
$$\operatorname{var} T_1 = 4 \operatorname{var} \bar{X} = \frac{4\sigma^2}{n} = \frac{\theta^2}{15}.$$

The relative efficiency of $2\widetilde{X}$ with respect to $2\bar{X}$ is therefore 7/15.

Another estimator, one based on the largest observation, comes to mind when we recall (Problem 7-29) that
$$E[X_{(n)}] = \frac{n}{n+1}\theta.$$
Clearly, the quantity $T_3 = \frac{n+1}{n} X_{(n)}$ is an unbiased estimator of θ. To find its variance we recall (also from Problem 7-29) that
$$\operatorname{var}[X_{(n)}] = \frac{n\theta^2}{(n+1)^2(n+2)}.$$
The m.s.e. of T_3 is thus
$$\operatorname{var} T_3 = \left(\frac{n+1}{n}\right)^2 \operatorname{var} X_{(n)} = \frac{\theta^2}{n(n+2)}.$$
In the present example, where $n = 5$, this yields $\operatorname{var} T_3 = \theta^2/35$. So T_3 is more efficient than either of the first two estimators:
$$\operatorname{eff}(T_1, T_3) = 15/35, \text{ and } \operatorname{eff}(T_2, T_3) = 7/35.$$
As the sample size increases, the relative efficiencies of T_1 and T_2 with respect to T_3 tend to 0. (See Problem 8-14.) ∎

EXAMPLE 8.2b Consider a random sample of size n from a population with finite variance σ^2. The linear combination $T = \sum a_i X_i$ is an unbiased estimate of the population mean when $\sum a_i = 1$. The most efficient of these estimators is the one that minimizes the variance:
$$\operatorname{var}\left(\sum a_i X_i\right) = \sum a_i^2 \operatorname{var} X_i = (\operatorname{var} X) \sum a_i^2.$$
To find the minimum in the class of unbiased linear combinations we substitute $a_n = 1 - a_1 - \cdots - a_{n-1}$ and minimize with respect to $a_1, \ldots, a_{n-1}$. Being bounded below, the variance has a minimum; and at the minimum, the partial derivatives with respect to the a_i must vanish:
$$\frac{\partial}{\partial a_i} \sum_1^n a_j^2 = 2(a_i - a_n) = 0, \quad i = 1, \ldots, n-1.$$
This says that the a_i are all equal at the minimum, so $a_i = 1/n$. Thus, the most efficient linear combination is the sample mean. ∎

262 Chapter 8 Estimation

In the last example we found the most efficient estimator in a restricted class of estimators. We might well ask whether there is one which is most efficient among *all* estimators. In this generality the answer is "no;" but in the rather wide (but still restricted) class of *unbiased* estimators, there is a lower bound of variances—a "best" variance, which can sometimes be achieved.

THEOREM 1 (Information inequality) *The variance of every unbiased estimator $T = t(\mathbf{X})$ of a parameter θ is bounded below by the reciprocal of the information:*

$$\operatorname{var} T \geq 1/I(\theta), \tag{2}$$

where

$$I(\theta) = \operatorname{var}\left[\frac{\partial}{\partial \theta} \log f(\mathbf{X} \mid \theta)\right]. \tag{3}$$

The inequality (2), also called the *Cramér-Rao inequality* (although apparently first given by Fréchet), is a particular case of the Schwarz inequality, applied to the variables T and V, the score function [(2) of §7.10]. To see this we first calculate the covariance of T and V, when $E(T) = \theta$, recalling that $EV = 0$:

$$\operatorname{cov}(V, T) = E(TV) = E\left\{T\frac{\partial}{\partial \theta} \log f(\mathbf{X} \mid \theta)\right\}$$

$$= \int t(\mathbf{x}) \frac{1}{f(\mathbf{x} \mid \theta)}\left(\frac{\partial}{\partial \theta} f(\mathbf{x} \mid \theta)\right) f(\mathbf{x} \mid \theta) d\mathbf{x}$$

$$= \frac{d}{d\theta} \int t(\mathbf{x}) f(\mathbf{x} \mid \theta) d\mathbf{x} = \frac{d}{d\theta} E(T) = \frac{d\theta}{d\theta} = 1.$$

Then [from (4) of §4.8, p. 117]

$$\rho_{V,T}^2 = \frac{1}{(\operatorname{var} V)(\operatorname{var} T)} \leq 1. \tag{4}$$

This establishes the theorem—if we assume conditions of regularity that permit the interchanges of integral and derivative, the existence and integrability of the various partial derivatives, and the nonvanishing of $I(\theta)$.[1]

DEFINITION The (absolute) *efficiency* of an unbiased estimator T of a parameter θ (when the Cramér-Rao inequality is valid) is

$$e(T) = \frac{1/I(\theta)}{\operatorname{var} T}, \tag{5}$$

---

[1] See E. L. Lehmann, *Theory of Point Estimation*, New York, John Wiley & Sons (1983), page 122.

if this ratio is independent of θ. When $e(T) = 1$, T is said to be *efficient*.

EXAMPLE 8.2c Consider estimation of the variance θ of a normal population with *known* mean μ. The variance is the population second central moment; its m.l.e. is the corresponding sample second moment: $\hat{\theta} = \frac{1}{n}\sum(X_i - \mu)^2$. This is sufficient and unbiased, and its variance is

$$\operatorname{var} \hat{\theta} = \frac{\operatorname{var}[(X-\mu)^2]}{n} = \frac{1}{n}\Big(E[(X-\mu)^4] - \{E[(X-\mu)^2]\}^2\Big) = \frac{2\theta^2}{n}.$$

[The fourth central moment of a normal distribution is $3\sigma^4$, according to (11) of §6.8.] In Example 7.10b we found the information in a random sample from this population to be $n/(2\theta^2)$. This is the reciprocal of $\operatorname{var} \hat{\theta}$, so the lower bound in (2) is achieved, and $\hat{\theta}$ is *efficient*. ∎

The distribution in Example 8.2c is a member of the exponential family, and one might speculate that it is the very special nature of this family that results in efficiency. Indeed, when f is in the exponential family:

$$f(x \mid \theta) = B(\theta) h(x) \exp[Q(\theta) R(x)],$$

the statistic $T = \sum R(X_i)$ is sufficient, and

$$V = \frac{\partial}{\partial \theta} \log L(\theta) = n \frac{B'(\theta)}{B(\theta)} + Q'(\theta) T.$$

Since V and T are linearly related, the information inequality (2) is actually an *equality*. In the class of estimators with the same bias, T has the smallest variance [see Problem 8-18]. If it is unbiased, it is efficient.

Conversely, suppose T is efficient: $\rho_{T,V} = \pm 1$. This means that (with probability 1) V is a linear function of T. The coefficients in this relationship may depend on θ, so we can write it as follows:

$$V = C'(\theta) T + D'(\theta).$$

When we integrate each side with respect to θ, the additive constant can involve $\mathbf{X}$, so

$$\log f(\mathbf{X} \mid \theta) = C(\theta) T + D(\theta) + K(\mathbf{X}). \tag{6}$$

Thus, $\mathbf{X}$ has a distribution in the exponential family, and T is *sufficient* for θ.

In §7.9 we mentioned the fact that (subject to certain conditions of regularity) if the minimal sufficient statistic T for a real parameter θ is one-dimensional, the population distribution is in the exponential family. But one cannot then jump to the conclusion that if T is unbiased, it is efficient. However, if an unbiased T occurs in the likelihood function as in (6)—a function of θ times T, then it is efficient.

When the efficiency $e(T_n)$, defined for each n, tends to a finite limit as n becomes infinite, this limit is called the **asymptotic efficiency**:

$$\lim_{n \to \infty} e(T_n) = \lim_{n \to \infty} \frac{1/I(\theta)}{n(\operatorname{var} T_n)}, \tag{7}$$

where $I(\theta)$ is the information in one observation. This limit will be a positive finite number only when $\operatorname{var} T_n$ behaves asymptotically like $1/n$. Asymptotic efficiency can be defined in some cases where efficiency is not defined for finite n. (It may be that T_n is biased, or does not have a variance.)

DEFINITION When $T_n \approx \mathcal{N}(\theta, c^2/n)$, its *asymptotic efficiency* is $[c^2 I(\theta)]^{-1}$. If this asymptotic efficiency is 1, we say that T_n is *asymptotically efficient*.

EXAMPLE 8.2d In Problem 8-4 we considered the following estimator of a Bernoulli parameter p: $T_n = \frac{Y+1}{n+2}$, where Y denotes the number of successes in n independent trials. The first two moments are

$$E(T_n) = \frac{np+1}{n+2}, \quad \operatorname{var} T_n = \frac{npq}{(n+2)^2}.$$

The bias is $\frac{1-2p}{n+2}$, which is not 0 but tends to 0 as n becomes infinite. Now,

$$T_n - ET_n = \frac{y+1}{n+2} - \frac{np+1}{n+2} = \frac{Y-np}{n+2}.$$

And then

$$\frac{T_n - ET_n}{\sqrt{\operatorname{var} T_n}} = \frac{Y-np}{\sqrt{npq}} \approx \mathcal{N}(0, 1).$$

So $T_n \approx \mathcal{N}(p, pq/n)$, and the "$c^2$" in the definition of asymptotic efficiency is $c^2 = pq$. But [from Problem 7-68(b)] $I(p) = 1/pq$, so $c^2 I(p) = 1$. That is, T_n is asymptotically efficient. ∎

PROBLEMS

8-13. Find the relative efficiency of $T_1 = (X_1 + 2X_2)/3$ with respect to the mean, $T_2 = (X_1 + X_2)/2$, when X_1 and X_2 are i.i.d., each with finite variance σ^2.

8-14. Referring to Example 8.2a, estimating the parameter θ in $\mathcal{U}(0, \theta)$,
(a) find the relative efficiency of $2\bar{X}$, which is unbiased, with respect to

the unbiased estimator $\frac{n+1}{n} X_{(n)}$.

(b) find the relative efficiency of $2\widetilde{X}$ with respect to $2\bar{X}$, and the limiting value as n becomes infinite. (For simplicity assume n is odd.)

8-15. Show that the sample mean is an efficient estimator of the population mean, for random samples from each of the following populations, (i) by comparing its variance with the sample information [see Problem 7-68] and (ii) by reasoning in terms of the exponential family:

(a) $N(\mu, 1)$.   (c) $\text{Exp}(1/\theta)$.   (e) $\text{Geo}(p)$.
(b) $\text{Ber}(p)$.   (d) $\text{Poi}(\lambda)$.

[Hint for (e): Find $I_n(\theta)$, where $\theta = 1/p$.]

8-16. Show that if $\xi = a\theta + b$, where $a \neq 0$, and if T is an efficient estimator of θ, then $aT + b$ is an efficient estimator of ξ. [Hint: See Problem 7-70.]

8-17. When X, Y, Z are i.i.d. $N(0, \theta)$, the distance U from the origin to the point (X, Y, Z) has the *Maxwell distribution* (Problem 6-84):

$$f_U(u \mid \theta) \propto \frac{u^2}{\theta^{3/2}} e^{-u^2/(2\theta)}, \quad u > 0.$$

Show that the statistic $T = \frac{1}{3}\overline{U^2} = \frac{1}{3n}\sum U_i^2$ is efficient in estimating θ. [Hint: Show that f is in the exponential family.]

8-18. Adapt the proof of the information inequality to derive the following is version of the inequality, in which T may be biased:

$$\text{var}\, T \geq \frac{[\text{cov}\,(V,\,T)]^2}{\text{var}\, V} = \frac{[1 + b'_T(\theta)]^2}{I(\theta)}.$$

8-19. Based on a random sample of size n from $X \sim \text{Exp}(\lambda)$, an obvious estimator of $\lambda = 1/\mu$ is $1/\bar{X} = n/U$, where $U = \sum X_i$. In Problem 8-10 we found a multiple of $1/U$ which is unbiased and calculated its variance. Find the efficiency of this estimator, given [from Example 7.10c] the information: $I_n(\lambda) = n/\lambda^2$.

8.3 Using sufficiency for variance reduction

Suppose a statistic T is sufficient for θ. The distribution in the sample space, given $T = t$, is then independent of θ, and for any estimator U, the conditional mean of U given $T = t$ depends only on t:

$$g(t) - E(U \mid T = t). \tag{1}$$

Thus $V = g(T)$ is a *statistic*, being a function of sample observations

alone. It has the same mean value as U:
$$EV = E[E(U \mid T)] = EU,$$
according to the iterated expectation formula [(6) in §4.3], so it has the same bias as U. In particular, if U is unbiased, so is V. Moreover, according to Theorem 14 of §4.5 (page 109),
$$\operatorname{var} U = \operatorname{var} V + E[\operatorname{var}(U \mid T)] \geq \operatorname{var} V. \tag{2}$$
We formalize this result as follows:

THEOREM 2 (Rao-Blackwell) *If U is an unbiased estimator of θ, and $T = t(\mathbf{X})$ is sufficient for θ, then $V = E(U \mid T)$ is an unbiased estimator of θ, and its variance is no larger than the variance of U:* $\operatorname{var} V \leq \operatorname{var} U$. *And* $\operatorname{var} V = \operatorname{var} U$ *if and only if U is essentially a function of T.*

We reason the conclusion about equality as follows: If $U = h(T)$, then the variable $U \mid t$ is a constant, and $\operatorname{var}(U \mid t) = 0$, so that $\operatorname{var} U = \operatorname{var} V$. To establish the converse, suppose $\operatorname{var} U = \operatorname{var} V$. Then $E[\operatorname{var}(U \mid T)] = 0$, and $P[\operatorname{var}(U \mid T) = 0] = 1$. [See Theorem 5 of §4.2.] But if $\operatorname{var}(U \mid t) = 0$, the conditional distribution of U given $T = t$ is concentrated at its mean value, $E(U \mid t) = g(t)$. Now, for any $\mathbf{x}$, let $t(\mathbf{x}) = t$; at this point $\mathbf{x}$, the random variable $U = u(\mathbf{X})$ has the value $u(\mathbf{x}) = g(t) = g(t(\mathbf{x}))$. So $U = g(T)$, with probability 1.

EXAMPLE 8.3a Let $\mathbf{Y}$ denote the order statistic derived from a random sample $\mathbf{X}$. The statistic X_1 (the first sample observation) is an unbiased estimator of the population mean: $E(X_1) = \mu$. According to Corollary 2 to Theorem 12, §7.8, page 232, the order statistic $\mathbf{Y}$ is sufficient, and
$$P(X_1 = y_i \mid \mathbf{Y} = \mathbf{y}) = 1/n.$$
(The first observation is just as likely to be the smallest as it is the second smallest, etc.) Hence,
$$V = E(X_1 \mid \mathbf{Y}) = \frac{1}{n}\sum Y_i = \bar{X},$$
which is unbiased. Also, $\operatorname{var} V = \sigma^2/n$,— smaller than $\operatorname{var} X_1 = \sigma^2$ when $n > 1$.

If the population is symmetric about its mean, the midrange is another unbiased estimate of μ: $U = \frac{1}{2}(Y_1 + Y_n)$. But $E(U \mid \mathbf{Y}) = U$, and "Rao-Blackwellization" does not produce a better estimator. This is because U is already a function of the sufficient statistic $\mathbf{Y}$. ∎

Suppose we find an estimator by Rao-Blackwellizing U, conditioning on a sufficient statistic T: $V = E(U \mid T)$. Then V is a function

8.3 Using sufficiency for variance reduction

of T. Is there perhaps another function of T with the same mean, and possibly a smaller variance? Rao-Blackwellizing V would of course yield V, but perhaps one could start with a different U. The answer, in part, is that in some cases, V is the *only* function of T with the same mean as U, as we've seen in several problems (e.g., 8-11 and 8-12). The concept of "completeness" of a family of probability distributions will provide a basis for a uniqueness theorem.

DEFINITION The family of distributions of T, $\{F_T(t \mid \theta)\}$, is called *complete* if 0 is the only unbiased estimate of 0, that is, if the condition

$$E_\theta[h(T)] = 0 \text{ for all } \theta$$

implies that $P[h(T)=0] = 1$.

The adjective "complete" is sometimes attached to the statistic T instead of to the family of its distributions, and it is said that T is a "complete statistic."

[This use of the term "complete" is like its use in the case of a set of basis vectors: A set of vectors $\{\mathbf{u}_n\}$ (which could be functions) is complete if the only vector with zero components in all "directions" $\mathbf{u}_n$ is the zero vector:

$$(\mathbf{V}, \mathbf{u}_n) = 0 \text{ for all } n \text{ implies } \mathbf{V} = \mathbf{0},$$

where $(\,,\,)$ denotes inner product. The expected value of $h(T)$ is like an inner product:

$$E_\theta[h(T)] = \int h(t) dF_T(t \mid \theta).$$

If this component of h is zero in each "direction" $F_T(t \mid \theta)$, then h must be 0, when $\{F_T(t \mid \theta)\}$ is complete.]

EXAMPLE 8.3b Let $Y \sim \text{Geo}(p)$: $f(n) = p(1-p)^{n-1}$, $n = 1, 2, \dots$, and suppose that $T = t(Y)$ is an unbiased estimate of 0:

$$ET = \sum_1^\infty t_n p(1-p)^{n-1} = 0.$$

We can write the sum as a power series in $q = 1 - p$:

$$ET = \sum_{n=1}^\infty t_n q^{n-1}(1-q)$$

$$-\sum_{n=1}^\infty t_n q^{n-1} \sum_{k=1}^\infty t_k q^k - \sum_{n=1}^\infty (t_n - t_{n-1}) q^{n-1} = 0,$$

where $k = n-1$ and $t_0 = 0$. Since power series expansions are unique, it follows that all coefficients $(t_n - t_{n-1})$ are 0, which implies $t_n = 0$ for all n. That is, $t(Y) = 0$, and Y is complete. ∎

The distribution of Y in Example 8.3b is in the exponential family of distributions. It can be shown that for the one-parameter exponential family, with p.f. or p.d.f. $f(x \mid \theta) = B(\theta)h(x)\exp\{Q(\theta)R(x)\}$, the family of distributions of the sum $\sum R(X_i)$ is complete.

THEOREM 3 *When the distribution family of a statistic T is complete, there can be only one estimator of the form $h(T)$ with a specified mean value. In particular, an unbiased estimator $h(T)$ would be the unique unbiased estimator based on T.*

If there were another function $k(T)$ with the same mean as that of $h(T)$, the difference $h(T) - k(T)$ would be an unbiased estimator of 0. This would imply that the functions h and k are essentially the same (agreeing everywhere except on a set with probability 0 for all θ).

EXAMPLE 8.3c For a random sample of size n from $\mathcal{N}(\mu, 1)$, the sample sum $\sum X_i$ is minimal sufficient and complete. As in Example 8.3a, let U denote the sample midrange (average of the smallest and largest observations). From completeness it follows that Rao-Blackwellizing must produce the sample mean: $E(U \mid \sum X_i) = \bar{X}$, since $\bar{X}$ is an unbiased estimator which is a function of $\sum X_i$. (This conditional mean of U can be calculated directly, but only with some difficulty.) ∎

EXAMPLE 8.3d In Problem 8-10, concerned with random samples from $\text{Exp}(\lambda)$, we found the variance of the unbiased estimator T, defined as $(n-1)/\sum X_i$, to be

$$\text{var}\, T = \frac{\lambda^2}{n-2}.$$

This is greater than the Cramér-Rao lower bound for the variance of unbiased estimators: $[nI(\lambda)]^{-1} = \lambda^2/n$. So T is not efficient:

$$\text{eff}(T) = \frac{[nI(\lambda)]^{-1}}{\text{var}\, T} = \frac{n-2}{n} < 1.$$

But $\text{Exp}(\lambda)$ is in the exponential family, so the family of distributions of T is complete; there is therefore no other function of T which is an unbiased estimator of λ. Also, T is sufficient for λ, so the variance of any unbiased estimator U is at least as large as $\text{var}\, V$, where V is $E(U \mid T)$; but $V = T$, by Theorem 3, so there is no "efficient" estimator of λ. ∎

THEOREM 4 *Suppose T is sufficient for θ, and its distribution family is complete. Then if U is any unbiased estimator of θ, the statistic $V = E(U \mid T)$ is the unique minimum variance unbiased estimator of θ.*

Consider any other unbiased estimator U_1 and let $V_1 = E(U_1 \mid T)$, which is a function of the sufficient statistic T. Then by completeness of T, the estimator V_1 is essentially the same as V, and by Theorem 2,

$$\text{var } U_1 \geq \text{var } V_1 = \text{var } V.$$

So U_1 can't have a smaller variance than V. And if it has the same variance as V, it must be a function of T—which means that it is equal to V.

That an estimator is unbiased with minimum variance does not mean that it is absolutely optimal. For one thing, there may be a biased statistic with a smaller mean squared error. Also, as in Example 8.3d above, the minimum variance unbiased estimator need not be efficient.

PROBLEMS

8-20. Consider a random sample $\mathbf{X}$ of size n from a geometric distribution: $f(x \mid p) = p(1-p)^{x-1}$, $x = 1, 2, \ldots$. Define the estimator U as the indicator function of the event $\{X_1 = 1\}$.
  (a) Show that U is an unbiased estimator of p.
  (b) Find a sufficient statistic, T.
  (c) Calculate $E(U \mid T)$.

8-21. Consider a random sample of size n from $\mathcal{N}(\mu, 1)$, and define U to be the indicator function of the event $[X_1 \leq \lambda]$. Show that U is unbiased in estimating $F_X(\lambda)$ for a given λ, and use the sufficient statistic $\bar{X}$ to obtain an estimate with smaller variance than that of U. *Given:* The conditional distribution of X_1 given $\bar{X} = t$ is $\mathcal{N}(t, 1 - 1/n)$, as will be shown in Chapter 12.

8-22. Show that these distribution families are complete:
  (a) Poi(λ).   (c) $f(y \mid \theta) = ny^{n-1}/\theta^n$, $0 < y < \theta$.
  (b) Bin(p).
[Hint for (c): If $EY = 0$ for all θ, its derivative vanishes for all θ.]

8-23. For the case of random samples from $\mathcal{U}(0, \theta)$, show that

$$E(2\bar{X} \mid X_{(n)}) = \frac{n+1}{n} X_{(n)}.$$

[Hint: $Y = X_{(n)}$ has the p.d.f. given in (c) of the preceding problem.]

8-24. Suppose we have a single observation X from Poi(λ) for estimating $\theta = e^{-2\lambda}$ (the probability of no "arrivals" in two time units). We saw in

Example 8.1d that the only unbiased estimator of θ is T, defined as 1 or -1, according as X is even or odd.
(a) Calculate the information in X and the variance of the estimator T defined above, and show that the efficiency of T tends to 1 as $\lambda \to 0$.
(c) Suppose instead of X, we are given Y = number of arrivals in four time units; find an unbiased estimate of $\theta = e^{-2\lambda}$, and its efficiency.

8-25. The mean of a random sample from $\mathcal{U}(0, \theta)$ is an unbiased estimate of $\theta/2$, so $U = 2\bar{X}$ is an unbiased estimate of θ. Use the sufficient statistic $X_{(n)}$ to improve on U by Rao-Blackwellization. [Hint: Write the mean as

$$\bar{X} = \frac{1}{n}[X_{(n)} + (n-1)\bar{X}']$$

where $\bar{X}'$ is the mean of $X_{(1)}, \ldots, X_{(n-1)}$, and use the fact that

$$E(\bar{X}' \mid X_{(n)}) = X_{(n)}/2.$$

(This is intuitively reasonable, but you might try to show it.)]

8.4 Consistency

The essence of the notion of consistency of an estimator is that when the estimator is applied to the whole population as the sample, it produces the true value of the parameter being estimated. This notion can be formalized in more than one way, although it is usually the case that the estimator in question is defined for samples of any size.

EXAMPLE 8.4a The sample mean is defined for a sample of any size: $\bar{X} = \Sigma X_i/n$. If a population is finite, and the population members are equally likely (as they are when sampling is random), the formula for sample mean can be applied to the population: $\bar{X}_N = \Sigma X_i/N$, where N is the population size. But this formula is exactly the formula we use to find the population mean μ. We say the estimator $\bar{X}_n$ is "consistent." ∎

For infinite populations we need to examine the behavior of an estimator T_n, defined for each n, as the sample size grows without limit.

DEFINITION An estimator T_n (defined for every n) is *consistent* as an estimator of θ if it tends to θ in probability, that is, if for any $\epsilon > 0$,

$$\lim_{n \to \infty} P(|T_n - \theta| > \epsilon) = 0. \tag{1}$$

8.4 Consistency

EXAMPLE 8.4b Suppose $X \sim \mathcal{N}(\mu, 1)$, and consider a random sample of size n. We know that $\bar{X}_n \sim \mathcal{N}(\mu, 1/n)$, so

$$P(|T_n - \theta| > \epsilon) = 1 - P(\mu - \epsilon < T_n < \mu + \epsilon)$$

$$= 1 - [\Phi(\sqrt{n}\,\epsilon) - \Phi(-\sqrt{n}\,\epsilon)],$$

where Φ denotes the standard normal c.d.f. As n becomes infinite, this tends to zero. We have thus shown that $\bar{X}_n$ is consistent as an estimate of the mean of a normal population. (We have not sacrificed generality in assuming that $\sigma = 1$.) ∎

It is not always as easy, as in Example 8.4b, to show that (1) holds in order to establish consistency. It is often easier to show convergence in a different sense:

DEFINITION: When an estimator T_n converges to θ in quadratic mean:

$$\lim_{n \to \infty} E[(T_n - \theta)^2] = 0, \qquad (2)$$

we say that it is *consistent in quadratic mean* as an estimator of θ.

THEOREM 5 *If T_n is consistent in quadratic mean, it is consistent. Equivalently, a sufficient condition for consistency is that the bias and the variance of T_n both tend to 0 as n becomes infinite:*

$$E(T_n) \to \theta \quad \text{and} \quad \text{var}\, T_n \to 0. \qquad (3)$$

This theorem follows from the fact [Theorem 1 of §5.1] that convergence in quadratic mean implies convergence in probability, because condition (3) follows from the fact that

$$E[(T - \theta)^2] = \text{var}\, T + (ET - \theta)^2,$$

[from equation (1) in §8.1]. However, because convergence in quadratic mean requires the existence of second moments, defining consistency in terms of convergence in probability is more generally applicable.

EXAMPLE 8.4c According to Example 8.1b, the variance of a random sample of size n from $\mathcal{N}(\mu, \sigma^2)$ has mean squared error equal to $2\sigma^4/(n-1)$. This tends to 0 as n becomes infinite, so the variance is consistent in mean square. If the population is not normal but has fourth moments, then according to Theorem 7 of Chapter 7 (page 210), var $S^2 = O(1/n)$—that is, that the m.s.e. tends to zero at the same rate as $1/n$. So the sample variance is a (mean-square) consistent estimator of the population variance if $E(X^4) < \infty$. [Actually, it is consistent in the sense of convergence in probability if only $E(X^2) < \infty$, as we'll explain in the next section.] ■

It was stated as Theorem 8 of Chapter 7 (given without proof) that sample moments converge in probability to corresponding population moments when the latter are finite. In the language of this section, sample moments are consistent estimators of the corresponding population moments. The next example is that of a population that doesn't have a mean value, and for which the sample mean does not converge to any constant.

EXAMPLE 8.4d Consider the family of Cauchy distributions with a location parameter, defined by the p.d.f.

$$f(x \mid \theta) = \frac{1/\pi}{1 + (x - \theta)^2}.$$

The distribution is symmetric about θ, but the mean does not exist because of the heavy tails. The characteristic function (from Problem 4-76) is

$$\phi_X(t) = E(e^{itX}) = \exp[-|t| + it\theta].$$

For random samples, the characteristic function of the sample mean is

$$\phi_{\bar{X}}(t) = [\phi_X(t/n)]^n = [\exp(-|t/n| + it\theta/n)]^n = \phi_X(t).$$

So the distribution of the mean is the same as that of the population, and the mean does not converge to a constant (in any sense). ■

8.5 Method of moments

So far, we have studied the performance of estimators that have suggested themselves on intuitive grounds. If intuition does not suggest an estimator (and even when it does), it helps to have some sort of method of "deriving" estimators. One such method is the **method of moments**.

To estimate a single function of population moments, the method of moments is to use that same function of the corresponding sample moments. In particular, a population mean is estimated as the sample mean, and a population variance, as the sample second moment we have denoted by V: $\frac{1}{n}\sum(X_i - \bar{X})^2$. To estimate k parameters, the method is first to express those parameters in terms of the first k population moments and then to replace those population moments with corresponding sample moments.

EXAMPLE 8.5a The population variance is $\sigma^2 = E(X^2) - (EX)^2$. This same function of the corresponding sample moments is $\overline{X^2} - (\bar{X})^2$, which is $(n-1)S^2/n = V$ (see Example 8.1b). So V is the estimator of σ^2 produced by the method of moments. ∎

EXAMPLE 8.5b To estimate the parameters of $\text{Gam}(\alpha, \lambda)$, based on a random sample, we first express the population mean and variance in terms of the parameters:

$$\mu = \alpha/\lambda, \quad \text{var } X = E(X^2) - [EX]^2 = \alpha/\lambda^2.$$

Solving, we obtain formulas for the parameters in terms of the first two population moments:

$$\lambda = \frac{EX}{E(X^2) - [EX]^2}, \quad \alpha = \lambda\mu = \frac{[EX]^2}{E(X^2) - [EX]^2}.$$

To get the method of moments estimators we replace EX by $\bar{X}$ and $E(X^2)$ by $\overline{X^2} = \sum X_i^2/n$:

$$\tilde{\lambda} = \frac{n\bar{X}}{(n-1)S^2}, \quad \tilde{\alpha} = \frac{n\bar{X}^2}{(n-1)S^2}. \quad \blacksquare$$

THEOREM 6 *Estimators obtained by the method of moments are consistent.*

As stated in §8.4, sample moments of the form $\overline{X^k}$ are consistent as estimators of the corresponding population moments, μ_k. Theorem 6 follows from Theorem 3 of Chapter 5, which states roughly that a continuous function of variables which converge in probability converges in probability to that same function of the limits.

PROBLEMS

8-26. Show that for random samples from $\text{Ber}(p)$, the sample proportion is mean-square consistent in estimating the population proportion p.

8-27. Show consistency in mean square for the method of moments estimator of $\sigma^2 = \text{var}\, X$, assuming $E(X^4) < \infty$.

8-28. Find the method of moments estimator of λ, for a random sample of size n from $\text{Exp}(\lambda)$ and show directly that it is consistent in quadratic mean. (See Problem 8-10.)

8-29. For random samples of size n from a population with p.d.f. $\theta x\, e^{-\theta x^2/2}$, $x > 0$, find the method of moments estimator of θ. [See Problem 6-84.]

8-30. Find the method of moments estimator of θ based on a random sample of size n from a population with p.d.f. $\theta x^{\theta-1}$, $0 < x < 1$.

8-31. For estimating the parameter θ of the uniform distribution on $(0, \theta)$ based on a random sample of size n,
   (a) find the method of moments estimator.
   (b) show that the estimator in (a) is consistent.

8-32. For random samples from $\mathfrak{U}(0, \theta)$, show that the largest observation $X_{(n)}$ is consistent. [Hint: In Problem 7-26 we found the mean and variance of the largest observation in a random sample $\mathbf{U}$ from $\mathfrak{U}(0, 1)$:

8-33. Let Y denote the number of successes in n independent trials of $\text{Ber}(p)$, as in Problem 8-4. Define $T_2 = (Y+1)/(n+2)$. Show that this estimator of p is consistent.

8-34. Let Y_n denote the midrange of a random sample $\mathbf{U}$ of observations on $\mathfrak{U}(\theta - \frac{1}{2}, \theta + \frac{1}{2})$:
$$Y_n = \tfrac{1}{2}[U_{(1)} + U_{(n)}].$$
   (a) Show that $U_{(n)}$ is a consistent estimate of $\theta + \frac{1}{2}$, and $U_{(1)}$ is a consistent estimate of $\theta - \frac{1}{2}$. [Hint: $U - \theta + \frac{1}{2} = X \sim \mathfrak{U}(0, 1)$.]
   (b) Show that Y_n is unbiased as an estimator of θ.
   (c) Show that Y_n is consistent in estimating θ. [Hint: You could show $Y_n \xrightarrow{P} \theta$ using the result in Problem 5-5; otherwise, you could show mean square consistency using the joint p.d.f. of the smallest and largest observations from Problem 7-33 (page 219) to find $\text{var}\, Y_n$.]

8.6 Maximum likelihood estimators

According to the maximum likelihood "principle" enunciated in §7.7, one should estimate a parameter to have the value $\hat\theta$, a value of θ that maximizes the likelihood function, $L(\theta)$. So when the problem of

8.6 Maximum likelihood estimators

inference is one of estimation, we call $\widehat{\theta}$ a *maximum likelihood estimator* (m.l.e.). This is not necessarily unique.

EXAMPLE 8.6a Consider a random sample of size n from a uniform distribution on the interval $(\theta - \frac{1}{2}, \theta + \frac{1}{2})$. The likelihood function is

$$L(\theta) = \begin{cases} 1, & X_{(n)} - \frac{1}{2} < \theta < X_{(1)} + \frac{1}{2}, \\ 0, & \text{otherwise}. \end{cases}$$

Each θ between $X_{(n)} - \frac{1}{2}$ and $X_{(1)} + \frac{1}{2}$ gives $L(\theta) = 1$. Any such number therefore maximizes the likelihood and is an m.l.e. ∎

We saw in §7.7 that $\widehat{\theta}$ can be a point in the parameter space at which the derivative of the likelihood function is not zero or does not exist. In regular cases, however, $\widehat{\theta}$ satisfies the *likelihood equation*:

$$\frac{\partial}{\partial \theta} \log L(\theta) = 0. \tag{1}$$

In the case of a vector parameter, $\boldsymbol{\theta} = (\theta_1, \ldots, \theta_k)$, the necessary condition for a maximum is that all the first order partial derivatives vanish:

$$\frac{\partial}{\partial \theta_i} \log L(\theta_1, \ldots, \theta_k) = 0, \quad i = 1, \ldots, k. \tag{2}$$

When a solution of this system of equations, $(\widehat{\theta}_1, \ldots, \widehat{\theta}_k)$, is a maximum point, we say that its components are joint m.l.e.'s of the k parameters.

Satisfying the condition (1) or (2) does not guarantee that a solution is an m.l.e., and if there is any doubt on this point, it should be investigated. A common situation is that the likelihood function L, a product of probabilities or densities, is bounded above and continuous, and that the likelihood equation has only one solution, which must then maximize L.

EXAMPLE 8.6b To find the joint m.l.e.'s of the parameters of $\mathcal{N}(\mu, \theta)$, for a random sample of size n, we take the log of the likelihood function:

$$\log L(\mu, \theta) = -\frac{n}{2} \log \theta - \frac{1}{2\theta} \sum (X_i - \mu)^2.$$

Setting the partial derivatives with respect to μ and θ equal to zero, we obtain the following simultaneous system of equations that must be satisfied by the joint m.l.e.'s.

$$\begin{cases} \frac{1}{\theta}\sum(X_i - \mu) = 0 \\ -\frac{n}{2\theta} + \frac{1}{2\theta^2}\sum(X_i - \mu)^2 = 0. \end{cases}$$

To solve this system of equations we observe that the first involves only μ, and solve it to obtain $\mu = \bar{X}$. We then substitute this value for μ in the second equation. The latter is then satisfied by $\theta = V$, the sample second moment about the sample mean. Thus, for a maximum, the unknowns μ and θ must have the values determined in this way. It can be checked that the second derivative condition for a maximum of a function of two variables is satisfied at $(\bar{X}, V)$, so there is at least a local maximum at that point, and this is the only critical point on the interior of the parameter space. It is not so clear that there is not a larger maximum value on the boundary. However, V is the smallest second sample moment, so $L(\bar{X}, \theta) \geq L(\mu, \theta)$. We need only verify that V maximizes $L(\bar{X}, \theta)$. But

$$L(\bar{X}, \theta) = \theta^{-n/2} e^{-nV/(2\theta)},$$

which has a maximum value at $\theta = V$. Thus, $\hat{\mu} = \bar{X}$ and $\hat{\theta} = V$ are the joint m.l.e.'s of μ and σ^2, respectively. ∎

EXAMPLE 8.6c Estimating the category probabilities of a discrete variable with a finite number of possible values (categories) is an important problem of inference. Consider an experiment of chance that can result in any of k possible ways: $A_1, A_2, ..., A_k$, and let $p_i = P(A_i)$, where $\sum p_i = 1$. The corresponding frequencies Y_i in a random sample of size n have a multinomial distribution, so the likelihood function is

$$L(p_1, \ldots, p_k) = p_1^{Y_1} \cdots p_k^{Y_k}, \quad 0 \leq p_i \leq 1.$$

Taking logarithms we have

$$\mathcal{L} = \log L(p_1, \ldots, p_k) = \sum_{i=1}^{k} Y_i \log p_i.$$

At this point we need to realize that there are only $k - 1$ free parameters, since the p_i sum to 1. Substituting for p_k in terms of $p_1, \ldots, p_{k-1}$, we have

$$\mathcal{L} = \sum_{i=1}^{k-1} Y_i \log p_i + Y_k \log(1 - p_i - \cdots - p_{k-1}).$$

Assume first that all frequencies Y_i are positive. Differentiation with respect to each of $p_1, \ldots, p_{k-1}$ and setting the derivatives equal to zero yields

$$\frac{Y_i}{p_i} = \frac{Y_k}{p_k}, \quad i = 1, \ldots, k-1.$$

Thus, at the maximum point, the p's are proportional to the Y's; since they sum to 1, the solutions of the likelihood equations are the relative frequencies $\hat{p}_i = Y_i/n$. The likelihood function is clearly bounded above by 1. If its maximum does not occur on the boundary of its support, it is at $(\hat{p}_1, \ldots, \hat{p}_k)$, and these are the joint m.l.e.'s of the probabilities p_i.

If any $Y_i = n$, the likelihood function is p_i^n, with a maximum when $p_i = 1$ (and the other p's equal 0). If some Y_i is 0, the corresponding p_i factor in the likelihood is missing. For instance, when $k = 3$ and if $Y_1 = 0$, $Y_2 = 4$, $Y_3 = 5$, the likelihood function is

$$L(p_1, p_2, p_3) = p_2^4 p_3^5.$$

To maximize this subject to the constraints on the p's, let $p_1 = r^2$, $p_2 = s^2$, and $p_3 = 1 - s^2 - r^2$. (This takes care of the constraints $p_i \geq 0$.) Then

$$L = s^8(1 - s^2 - r^2)^5,$$

and an (interior) maximum point is obtained by setting derivatives of the log-likelihood (with respect to r and s) equal to 0. Doing this we find: $r = 0$, $s^2 = \hat{p}_2 = 4/9$, and $\hat{p}_3 = 5/9$. So the m.l.e.'s are again the corresponding relative frequencies. ∎

The method of maximum likelihood is not guaranteed to produce a good estimator, but it often does. We note the following important properties:

PROPERTIES OF M.L.E.'s:

1. If T is sufficient for θ, a unique solution of the likelihood equation is a function of T only.
2. If there is an efficient estimator of θ, the method of maximum likelihood will produce it.
3. Under certain conditions of regularity, a maximum likelihood estimator is asymptotically efficient, having an asymptotically normal distribution with variance $1/I(\theta)$.

To show Property 1, suppose $T = t(\mathbf{X})$ is sufficient:

$$L(\theta) = f(\mathbf{X} \mid \theta) = g[t(\mathbf{X}), \theta] h(\mathbf{X}).$$

Taking logs on both sides we have

$$\log g[t(\mathbf{X}), \theta] + \log h(\mathbf{X}).$$

This has a maximum with respect to θ wherever the first term has a maximum, and the value of θ that maximizes the first term can depend on $\mathbf{X}$ only through the value of $t(\mathbf{X})$.

Property 2 is seen as follows: If T is efficient, the score is a linear function of T:

$$V = \frac{\partial}{\partial \theta} \log f(\mathbf{X} \mid \theta) = C(\theta)T + D(\theta).$$

This vanishes at $\theta = \widehat{\theta}$, the m.l.e.:

$$C(\widehat{\theta})\,T + D(\widehat{\theta}) = 0.$$

But the mean score is 0 for *all* θ:

$$EV = C(\theta)\,\theta + D(\theta) = 0,$$

since an efficient estimator is unbiased. So in particular, for $\theta = \widehat{\theta}$,

$$C(\widehat{\theta})\,\widehat{\theta} + D(\widehat{\theta}) = 0,$$

from which it follows that

$$T = -\frac{D(\widehat{\theta})}{C(\widehat{\theta})} = \widehat{\theta}.$$

Proof of Property 3 is beyond our scope.[2]

EXAMPLE 8.6d Example 8.6a dealt with a uniform distribution on an interval of length 1 whose midpoint is the parameter θ. We found that there can be many m.l.e.'s. In particular, the midrange

$$M = \tfrac{1}{2}[X_{(1)} + X_{(n)}]$$

is a m.l.e. It is unbiased and has variance $[2(n+2)(n+1)]^{-1}$. So it is clearly consistent. Its asymptotic variance is of *smaller* order than the $1/n$ called for in the definition of asymptotic efficiency—smaller than the order implied in Property 3 for m.l.e.'s in regular cases. ∎

---

[2]See E. L. Lehmann, *Theory of Point Estimation*, John Wiley & Sons, New York (1983), page 417, or H. Cramer, *Mathematical Methods of Statistics*, Princeton U. Press, Princeton, N. J. (1946), page 500.

8.7 Bayes estimators

In the Bayesian approach to inference we treat unknown parameters as random variables—as we treat any quantities about which we are uncertain. The current distribution of a parameter θ, whether a prior or a posterior, can be used to suggest a value in the parameter space as an estimate of θ. Intuitively, it would seem that a value of θ somewhere near the middle of its distribution should be a good estimate. The mean and median suggest themselves, or possibly the mode (in the case of typical, unimodal distributions). These are *Bayes estimators*.

In §8.1 we defined the "error" in estimating θ to have the value a to be $a - \theta$. The absolute error will be a minimum, on average, if we choose a to be the *median* of our current distribution for θ. The squared error will be a minimum, on average, if we choose a to be the *mean* of that distribution. The particular measure of error we adopt is a kind of penalty for using the corresponding Bayes estimator. (See also Chapter 15.) We'll refer to $(a - \theta)^2$ as *squared error loss* or *quadratic loss*, and to $|a - \theta|$ as *absolute error loss*.

The mode of the distribution of θ is in a sense the "most probable" value of θ. If the prior p.d.f. or p.f. is constant, the mode is simply the value of θ that maximizes the likelihood function—the maximum likelihood estimator.

Bayes estimators, using the data only through the likelihood function, are procedures that follow the likelihood principle, as enunciated in Chapter 7.

EXAMPLE 8.7a In Example 7.11b we obtained the posterior distribution for a Bernoulli parameter, given Y successes in n trials and assuming a Beta(r, s) prior distribution for p:

$$h(p \mid Y) \propto p^{Y + r - 1}(1 - p)^{n - Y + s - 1}, \quad 0 < p < 1.$$

The mode of this posterior is the exponent on p divided by the sum of the exponents:

$$\text{Mode} = \frac{Y + r - 1}{n + s + r - 2}.$$

For a uniform prior ($r = s = 1$), this is the m.l.e., $\widehat{p} = Y/n$.

With absolute error loss, the Bayes estimator of p is the posterior median—a number x such that

$$\frac{1}{B(r + Y, n - Y + s)} \int_0^x p^{Y + r - 1}(1 - p)^{n - Y + s - 1} dp = \frac{1}{2}.$$

Finding the median x usually requires a numerical approximation.

With quadratic loss, the Bayes estimator is the posterior mean:
$$E(p \mid Y) = \frac{r+Y}{r+s+n}.$$

For a uniform prior ($r = s = 1$), this conditional mean reduces to
$$E(p \mid Y) = \frac{Y+1}{n+2} = \left(\frac{n}{n+2}\right)\hat{p} + \left(1 - \frac{n}{n+2}\right)\cdot\frac{1}{2},$$

a weighted average of the sample proportion of successes and the prior mean. As n becomes infinite, the weight on $\hat{p}$ tends to 1—the sample information takes over.

To take another particular case, suppose $r = 3, s = 6, n = 10,$ and $Y = 4$. The posterior mode is $6/17 \doteq .353$, and the posterior mean is $7/19 \doteq .368$. The median is an x such that

$$\frac{1}{B(7, 12)} \int_0^x p^6(1-p)^{11} dp = .5.$$

Using PC statistics software we found $x \doteq .364$. ∎

PROBLEMS

8-35. Find the m.l.e. of λ, the rate parameter of a Poisson process, based on $(X_1, ..., X_n)$, the numbers of "events" in n successive intervals of length t_0.

8-36. Find the m.l.e. of λ, the rate parameter of a Poisson process, based on n observations of T, the time to the next "event."

8-37. Given: n independent pairs (X_i, Y_i), each with joint p.d.f.
$$f_{X, Y}(x, y \mid \theta) = e^{-\theta x - y/\theta} \text{ for } x > 0, y > 0,$$
where $\theta > 0$.
(a) Find the minimal sufficient statistic.
(b) Find the m.l.e. of θ. (Is it sufficient?)

8-38. Given a random sample of size n from $N(0, \sigma^2)$, find the maximum likelihood estimate of σ^2.

8-39. Referring to Example 8.6d,
(a) Show that $X_{(1)} + \frac{1}{2}$ is a consistent estimator of θ.
(b) Define the statistic $T_n = X_{(1)} + \frac{1}{2} - \frac{1}{n+1}$ and calculate its efficiency and asymptotic efficiency of T_n with respect to the midrange M, in estimating the parameter θ..

8-40. Consider independent observations $Y_1, ..., Y_n$, where each Y_i is $N(\alpha + \beta x_i, 1)$, for given *constants* $x_1, ..., x_n$. Find the joint m.l.e.'s of the parameters (α, β).

8-41. Given a random sample of size n from $\text{Exp}(\lambda)$, find the Bayes estimate of λ, given the prior $\lambda \sim \text{Exp}(\beta)$ [as in Example 7-11c], assuming quadratic loss.

8-42. Suppose we observe a Bernoulli process with parameter p and find that it takes 15 trials to get the 4th success. If our prior for p is Beta(4, 2) and we assume quadratic loss, find the Bayes estimate of p?

8-43. Let $\bar{X}$ denote the mean of a random sample of size $n = 5$ from $N(\mu, 1)$. Given that $\mu \sim N(5, 1)$ and $\bar{X} = 4$, find the Bayes estimate assuming absolute error loss.

8-44. Suppose a lot of 10 items has M defectives, and a simple random sample of size four includes exactly one defective. Find the Bayes estimate of M, when $M \sim \text{Bin}(10, 1/2)$, , assuming quadratic loss.

8.8 Interval estimates

Announcing a single value as one's estimate of a parameter θ calls for some auxiliary indication of how much that value can be trusted—a measure of its reliability. The notion of "standard error" introduced in §8.1 is a commonly used measure, and sometimes estimates are given as a point estimate (single value) plus or minus a standard error. The standard error is not a maximum error, of course, so one could not expect the true value of the parameter to be within one standard error of the point estimate in every instance. Sometimes it will, and sometimes it won't.

EXAMPLE 8.8a Consider estimating a population mean using the sample mean. One might report the result as $\bar{X} \pm$ s.e. We can't know, given a particular sample and its mean, whether μ is within the interval with endpoints $\bar{X} \pm$ s.e. But we can calculate the probability that we'll get a sample for which this is the case.

When n is large, the sample mean is approximately normal, and the Z-score is approximately standard normal (according to Corollary 3 of Theorem 8 in §7.4, page 211):

$$Z = \frac{\bar{X} - \mu}{S/\sqrt{n}} \approx N(0, 1). \tag{1}$$

The denominator is the standard error, and (from Table I)

$$P(\bar{X} - \text{s.e.} < \mu < \bar{X} + \text{s.e.}) \doteq .683.$$

Similarly,

$$P(\bar{X} - 2\,\text{s.e.} < \mu < \bar{X} + 2\,\text{s.e.}) \doteq .954. \quad \blacksquare$$

The intervals we obtained in the above example, defined by the limits $\bar{X} \pm \text{s.e.}$ and $\bar{X} \pm 2\,\text{s.e.}$, are referred to, respectively, as 68.3% and 95.4% (large sample) **confidence intervals** for μ. The percentages are **confidence levels**. The interpretation is as follows: In stating that the random interval $(\bar{X} - 2\,\text{s.e.}, \bar{X} + 2\,\text{s.e.})$ is an "interval estimate" for μ, we are using a technique that has a 95.4 percent chance of success.

The quantity Z in (1) has (approximately) a distribution which is free of any population parameters. A function of data and parameters whose distribution does not depend on any parameters is called a **pivotal quantity** (or for short, a "pivotal"). We can use pivotals to construct confidence intervals, and we'll do this in the next section for the parameters of a normal population. The following is another situation in which we can find a pivotal quantity, illustrating the technique for constructing confidence limits from a pivotal.

EXAMPLE 8.8b Let $X \sim \text{Exp}(1/\theta)$. Suppose we want 90 percent confidence limits for θ based on a random sample of size n. The sample sum has a gamma distribution: $\sum X_i \sim \text{Gam}(n, 1/\theta)$, so (according to the result in Problem 6-81) the variable $Y = 2\sum X_i/\theta$ has a chi-square distribution with $2n$ degrees of freedom. Since the distribution of Y does not involve θ, Y is pivotal. To construct confidence limits, we use the chi-square table (Table Vb) to find numbers A and B such that

$$P(A < 2\sum X_i/\theta < B) = .90,$$

and solve the inequalities for θ. For instance, suppose $n = 10$. The fifth and ninety-fifth percentiles of $\text{chi}^2(20)$ are $A = 10.9$, $B = 31.4$. Thus,

$$P(10.9 < 2\sum \frac{X_i}{\theta} < 31.4) = .90,$$

or (upon solving for θ)

$$P\left(\frac{2\sum X_i}{10.9} > \theta > \frac{2\sum X_i}{31.4}\right) = .90.$$

The extremes in the extended inequality are 95 percent confidence limits for the parameter θ. $\blacksquare$

The method illustrated in this last example depends on one's finding a pivotal quantity. However, when the point estimator T has a continuous distribution with c.d.f. F_T, the quantity $F_T(T \mid \theta)$ is pivotal, since it is distributed as $\mathcal{U}(0, 1)$, independent of θ.

EXAMPLE 8.8c Suppose we have a random sample of size $n = 10$ from $\mathcal{U}(0, \theta)$ in which the largest observation is $X_{(n)} = 2.80$ and want a 95 percent confidence interval for the parameter θ. Since

$$F_{X_{(n)}}(t \mid \theta) = [F_X(t)]^n = \left(\frac{t}{\theta}\right)^n,$$

it follows that $(T/\theta)^n \sim \mathcal{U}(0, 1)$. So

$$.95 = P\left(.025 < \left(\frac{T}{\theta}\right)^n < .975\right) = P\left(\frac{T}{.9975} < \theta < \frac{T}{.6915}\right).$$

When $T = 2.80$, the desired interval is $(2.81, 4.05)$. ■

Clearly, there is arbitrariness in constructing a confidence interval—in the choice of the point estimator T, in the setting of the confidence level η, and in the allocation of the probability $1 - \eta$ to the tails of the distribution of T. In Examples 8.8b and 8.8c we put half in each tail, for no obvious reason. Putting all of the probability η in one tail would produce a one-sided interval, or a *confidence bound* for the parameter. (This will be illustrated in Example 8.9a.) The question naturally arises, is there any criterion by which some of this arbitrariness can be resolved? Various notions of "good" confidence intervals have been considered: A confidence interval that has a minimal probability of covering wrong parameter values is termed "uniformly most accurate." Another criterion of optimality is that of minimum length of the interval.[3]

When one constructs a confidence interval and finds, say, that the interval $(13, 17)$ is a 95 percent confidence interval for θ, it is tempting to say that $P(13 < \theta < 17) = .95$. But there is no theoretical basis for such a statement in the framework in which the interval was obtained. However, when θ is viewed as a random variable (which it is *not*, in the confidence interval setting), it does make sense to calculate such things as $P(13 < \theta < 17)$, in whatever current distribution θ may have. This probability may or may not be equal to the confidence level (but usually not).

EXAMPLE 8.8d Suppose we observe 15 successes in 25 independent trials of $Ber(p)$, and that our prior for p is Beta(10, 10). The posterior is

---

[3]Lehmann, E. L., *Testing Statistical Hypotheses*. 2nd Ed., 1986, Belmont, CA: Wadsworth & Brooks/Cole, pages 90, 330.

$$h(p \mid \widehat{p}) \propto p^{15}(1-p)^{10} \cdot p^9 (1-p)^9 = p^{24}(1-p)^{19}.$$

This is Beta(25, 20), and with statistical software now available one can easily determine the probability of any interval of values of p. For instance, in the beta posterior we just obtained,

$$P(.410 < p < .696) \doteq .950.$$

So (.410, .696) is a 95 percent "probability interval" for p. On the other hand, a 95 percent confidence interval for p is (.407, .766) [see Problem 8-48], and in the above posterior distribution for p this interval has approximate probability .976. ■

PROBLEMS

8-45. Measurement of a certain dimension on each of n parts yields a sample with mean $\bar{X} = 2.30$ and standard deviation $S = .100$. Construct confidence intervals for the population mean dimension for these combinations of sample size and confidence level η:
(a) $n = 50$, $\eta = .95$.   (c) $n = 400$, $\eta = .95$.
(b) $n = 50$, $\eta = .99$.   (d) $n = 400$, $\eta = .99$.

8-46. It is known from long experience that the reliability of a certain chemical measurement is indicated by a standard deviation $\sigma = .005$ gm/ml. Find a sample size n such that a 99 percent confidence interval for μ has a width of .001 gm/ml.

8-47. A TV program has a rating of 23, which means that 23 percent of the viewers in a sample were watching it. Assume a random sample of 1000 to find approximate 95 percent confidence limits for p, the proportion of watchers in the population sampled. [If $X \sim \text{Ber}(p)$, $\mu_X = p$.]

8-48. Find 95 percent confidence limits for p, given that we have observed 15 successes in 25 independent Bernoulli trials. [Use the fact that, even for as few as 25 observations, $\bar{X} \approx \mathcal{N}(p, pq/n)$ so that the corresponding Z-score is an approximate pivotal.]

8-49. Construct confidence limits for the mean of a Poisson population with rate parameter λ, based on a random sample size n, in terms of the sample mean $\bar{X}$ and the confidence level η. (Assume that n is large enough that a Z-score provides an approximate pivotal.)

8-50. Show that for a single observation from the Cauchy population with p.d.f. $f(x \mid \theta) \propto (1 + [x - \theta]^2)^{-1}$ the quantity $X - \theta$ is pivotal, and use this fact to give 90 percent confidence limits for θ.

8-51. Given the following observations from Exp(λ), obtained as a ran dom sample:

1.098, .137, 7.209, 4.242, 1.163, 6.350, .082, .527, 5.154, 1.416,

(a) find 90 percent confidence limits for λ.
(b) find the probability of the confidence interval in (a) if $\lambda \sim \text{Exp}(1)$.

8.9 Estimating normal population parameters

The joint maximum likelihood estimators for the mean and variance of a normal population (from Example 8.6b) are $\hat{\mu} = \bar{X}$ and $\hat{\sigma}^2 = V = \frac{1}{n}\sum(X_i - \bar{X})^2$. The standard error of the mean and confidence intervals for the mean as given in §8.1 and §8.8, respectively, apply to the case of normal populations, when n is large. In Example 8.9a, following the method of §8.8 we'll find a (one-sided) confidence interval for σ^2. To do this we use the fact that the quantity $(n-1)S^2/\sigma^2$ is pivotal (from §7.5, Theorem 11, page 213), having a chi$^2(n-1)$ distribution—a distribution that is independent of any population parameters.

Although it is questionable whether one really would want to estimate a variance, the standard deviation may well be of interest, being expressed in the same units as the measurements themselves. An obvious estimator of σ is the square root of the sample variance. However, even though S^2 is unbiased as an estimator of σ^2, S is biased as an estimate of σ (Problem 8-3).

The distribution of S can be found from the distribution of $(n-1)S^2/\sigma^2$, which is chi$^2(n-1)$. In particular,

$$E(S) = E\left[\left(\frac{(n-1)S^2}{\sigma^2}\right)^{1/2}\right] \cdot \frac{\sigma}{(n-1)^{1/2}}$$

$$= \frac{\sigma}{(n-1)^{1/2} 2^{(n-1)/2} \Gamma(\frac{n-1}{2})} \int_0^\infty x^{1/2 + (n-3)/2} e^{-x/2}\, dx$$

$$= \frac{\Gamma(\frac{n}{2})}{\Gamma(\frac{n-1}{2})} \cdot \left(\frac{2}{n-1}\right)^{1/2} \sigma = \alpha_n \sigma.$$

The estimator $T = S/\alpha_n$ is thus an unbiased estimator of σ, and

$$\text{var } T = \frac{1}{\alpha_n^2}(\text{var } S) = \left(\frac{1}{\alpha_n^2} - 1\right)\sigma^2. \tag{1}$$

But $\alpha_n S$ is the multiple of S with the smallest m.s.e. (Problem 8-57).

EXAMPLE 8.9a To check the variability in explosion times of a type of detonator, these measurements were obtained (in milliseconds short of 2.7 seconds):

$$11, 23, 25, 9, 2, 6, -2, 2, -6, 8, 9, 19, 0, 2.$$

The sample variance is $S^2 = 85.9$. Suppose we want an upper 90 percent confidence bound for σ, assuming normally distributed observations. The quantity $13S^2/\sigma^2$ is pivotal—it has a chi-square distribution, independent of any population parameters. From the chi-square table (Table Va) we find $\chi^2_{.10}(13) = 7.04$, so

$$.90 = P\left(\frac{13S^2}{\sigma^2} > 7.04\right) = P\left(\sigma^2 < \frac{13S^2}{7.04}\right).$$

The bound for σ^2 is $13S^2/7.04 = 158.6$.

For a point estimate of σ we could use simply the sample standard deviation, $S = 9.269$. When $n = 14$, $\alpha_n = .98097$, so the unbiased estimate S/α_n is 9.449, and the multiple of S with minimum m.s.e. is $\alpha_n S = 9.093$. (Using the unbiased estimator based on the sample range, from Example 8.1c, yields $R/a_n \doteq 9.12$.) To obtain a 90% confidence bound for the population standard deviation, we simply take the square root of the bound for σ^2: $\sigma < \sqrt{158.6} \doteq 12.6$. ∎

A **confidence region** for the parameter pair (μ, σ^2) can be constructed in terms of the sample mean and variance—a random region that, with specified probability, will cover the actual (μ, σ^2). For this we use the facts that $\bar{X} \sim N(\mu, \sigma^2/n)$ and $(n-1)S^2/\sigma^2 \sim \chi^2(n-1)$ along with the independence of mean and variance to write

$$P\left\{\left(\frac{\bar{X}-\mu}{\sigma/\sqrt{n}}\right)^2 < z^2_{1-\delta}, \ \chi^2_\epsilon < \frac{(n-1)S^2}{\sigma^2} < \chi^2_{1-\epsilon}\right\}$$

$$= P\left\{\left(\frac{\bar{X}-\mu}{\sigma/\sqrt{n}}\right)^2 < z^2_{1-\delta}\right\} \cdot P\left\{\chi^2_\epsilon < \frac{(n-1)S^2}{\sigma^2} < \chi^2_{1-\epsilon}\right\}$$

$$= (1-2\delta)(1-2\epsilon).$$

EXAMPLE 8.9b Suppose we want a 90 percent confidence region for the pair (μ, θ) in Example 8.9a, where $n = 14$ and $\bar{X} = 7.714$. With

$$\delta = \epsilon = .025, \ (1-2\delta)(1-2\epsilon) \doteq .90,$$

the inequalities defining the confidence region are

8.9 Estimating normal population parameters

$$\frac{(\mu - 7.714)^2}{\theta/14} < 1.96^2, \text{ and } 5.01 < \frac{13 \times 85.9}{\theta} < 24.7.$$

These define the region between the lines $\theta = 45.2$ and $\theta = 222.9$ and inside the parabola $(\mu - 7.714)^2 < .2744\,\theta$, shown in Figure 8-2. ∎

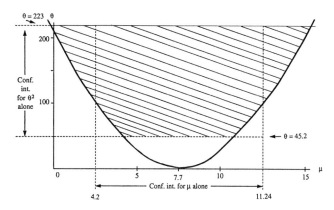

Figure 8-2

In the construction of small sample confidence limits, the mean is somewhat more difficult to deal with than the variance, since the distribution of the sample mean depends not only on μ but also on σ^2. In Example 8.8a we used the ratio $\sqrt{n}\,(\bar{X} - \mu)/S$ as an approximate pivotal for large n. However, when n is not large, the distribution of this ratio is not well approximated by $\mathcal{N}(0, 1)$. Yet, it *is* pivotal, as we now show by deriving its distribution. We define the following notation:

$$T = \sqrt{n}\,\frac{\bar{X} - \mu}{S} = \sqrt{n-1}\,\frac{Z}{U}, \text{ where } Z = \frac{\bar{X} - \mu}{\sigma/\sqrt{n}}, \quad U = \sqrt{\frac{(n-1)S^2}{\sigma^2}}.$$

Since Z and U are independent, their joint p.d.f. is the product of their marginal p.d.f.'s, a standard normal p.d.f. for Z and a "chi" p.d.f. for U—the p.d.f. of the square root of a chi-square variable with $n - 1$ d.f. (See Problem 8-59.) The c.d.f. of $V = Z/U$ is then

$$F_V(v) = P(Z/U < v) = K \int_0^\infty \int_{-\infty}^{vu} e^{-z^2/2} u^{n-2} e^{-u^2/2} dz\, du,$$

where the constant K involves only the sample size n. Differentiating with respect to v produces the p.d.f. of V:

$$f_V(v) = K \int_0^\infty e^{-(uv)^2/2} u^{n-2} e^{-u^2/2} u \, du$$

$$= K \int_0^\infty e^{-(1+v^2)u^2/2} u^{n-2} u \, du.$$

With the substitution $w = (1+v^2)\,u^2/2$, $dw = (1+v^2)\,u\,du$, we obtain

$$f_V(v) = K' \int_0^\infty e^{-w} \left(\frac{w}{1+v^2}\right)^{\frac{n-2}{2}} \frac{dw}{1+v^2} = \frac{K''}{(1+v^2)^{n/2}}.$$

[Without the factor $(1+v^2)^{-n/2}$, the w-integral is a constant: $\Gamma(n/2)$.]
Thus,

$$f_T(t) = f_V\left(\frac{t}{\sqrt{n-1}}\right)\frac{1}{\sqrt{n-1}} \propto \frac{1}{\left(1+\frac{t^2}{n-1}\right)^{n/2}}. \tag{2}$$

This is the p.d.f. of "Student's t-distribution with $n-1$ degrees of freedom,"[4] and we write $T \sim t(n-1)$. Observe that it is symmetric about $t = 0$. [The constant of proportionality in (2), chosen so that the area under the p.d.f. is equal to 1, will not be useful to us.]

Probabilities for T are areas under the t-density, and these have been calculated numerically and tabled. (See Table III.) As n becomes infinite, the t-density (2) converges to the standard normal density:

$$\left(1+\frac{x^2}{n}\right)^{-(n+1)/2} = \left\{\left(1+\frac{x^2}{n}\right)^{n/x^2}\right\}^{-x^2/2}\left(1+\frac{x^2}{n}\right)^{-1/2} \to \exp\left(-\frac{1}{2}x^2\right).$$

Thus, the t-percentiles tend towards the corresponding percentiles of the standard normal distribution, and this tendency is evident in Table III. For this reason it suffices for t-tables to have entries for degrees of freedom up to about 30 or 40.

The fact that the distribution of T depends only on n means that it is pivotal, and we can use this fact to construct confidence intervals for μ. For a two-sided interval with confidence coefficient $1-\eta$, we find the points $\pm t^*$ in the $t(n-1)$ distribution such that there is $\eta/2$ in each of the two tails: $t^* = t_{1-\eta/2}(n-1)$. Then

$$P\left(-t^* < \sqrt{n}\,\frac{\bar{X}-\mu}{S} < t^*\right) = 1-\eta.$$

Solving for μ, we obtain

$$P\left(\bar{X} - t^*\frac{S}{\sqrt{n}} < \mu < \bar{X} + t^*\frac{S}{\sqrt{n}}\right). \tag{3}$$

---

[4]W. S. Gosset, a statistician with the Guiness Brewery in Dublin, obtained this distribution and published it in 1901 under the pseudonym "Student."

8.9 Estimating normal population parameters

The extremes in the extended inequality are $1 - \eta$ confidence limits for μ.

EXAMPLE 8.9c The mean of the 14 observations in Example 8.9a is $\bar{X} = 7.714$, and the standard deviation is $S = 9.269$. For 90 percent confidence limits we need the ninety-fifth percentile of $t(13)$: $t^* = 1.77$, from Table III. The desired limits are then $7.714 \pm 1.77 \times 9.269/\sqrt{14}$, and the interval is $3.33 < \mu < 12.1$. ∎

In principle, we can use the *Bayesian* approach to estimate μ and σ^2. When both parameters are unknown, we'd need a bivariate prior for (μ, σ^2). When the variance is known, the conjugate family of priors for μ is normal (Problem 7-75). Given a random sample of size n, the likelihood function is

$$L(\mu) = \exp\left[-\frac{n}{2\sigma^2}(\mu - \bar{X})^2\right],$$

and if we assume $\mu \sim \mathcal{N}(\nu, \tau^2)$, the posterior is normal with mean

$$E(\mu \mid \mathbf{X}) = \frac{\tau^2 \bar{X} + \nu(\sigma^2/n)}{\tau^2 + (\sigma^2/n)}. \quad (4)$$

This is the Bayes estimate of μ with either quadratic or absolute error loss.

The posterior variance of μ is a measure of reliability of the estimate:

$$\text{var}(\mu \mid \mathbf{X}) = \frac{\sigma^2 \tau^2}{n\tau^2 + \sigma^2}. \quad (5)$$

The reciprocal variance, called the *precision*, is

$$\pi_{po} = \frac{n}{\sigma^2} + \frac{1}{\tau^2} = \pi_s + \pi_{pr}, \quad (6)$$

the sum of the precision provided by the sample, $\pi_s = n/\sigma^2$, and the precision of the prior, $\pi_{pr} = 1/\tau^2$.

The posterior mean is a weighted average of the sample mean and the prior mean in which the weights are the precision of the prior and the precision of the data. When n is large, or when the prior variance τ is large, the weight is more on the sample mean; when the prior variance τ is small, or when the sample size is small, the weight is more on the prior mean.

[The distribution of the sample variance does not involve μ, so for a given sample, a likelihood function (of σ^2) is defined by the p.d.f. of S^2. This involves σ^2 in an awkward way; a conjugate family is not obvious.[5]]

PROBLEMS

8-52. Ten children were treated with a therapy involving the drug ethosuximide, in an attempt to increase IQ scores. Increases for the ten children were

$$16, 7, -5, 24, 19, 11, 6, 2, 10, 30.$$

Suppose that the population of such increases in score is $N(\mu, \sigma^2)$.
(a) Find 95 percent confidence limits for μ.
(b) Find a 95 percent upper confidence bound for μ.
(c) Given that $\mu \sim N(5, 16)$, find the probability of the confidence interval found in (a).
(d) Given $\mu \sim N(5, 16)$, find the probability that μ does not exceed the confidence bound found in (b).

8-53. In Example 8.1c it was shown that for random samples from a normal population, the quantity $W = R/\sigma$ is pivotal, where R is the sample range. For $n = 10$ and $R = 6.5$, use the percentiles of W in Table XIII to construct 90 percent confidence limits for σ.

8-54. Outside diameters of 21 supposedly identical parts drawn at random from a production line were measured and found to have mean 9.318 and s.d. .0676 (in hundredths of an inch greater than 1 inch).
(a) Find 95 percent confidence limits for μ.
(b) Find 95 percent confidence limits for σ.
(c) Find a 90 percent confidence region for (μ, σ^2).

8-55. Suppose our prior for μ in $N(\mu, 1)$ is $N(10, .25)$, and that a sample of size n has mean $\bar{X} = 10.2$. Assuming quadratic loss, find the Bayes estimate and the posterior precision, when
    (a) $n = 4$.   (b) $n = 100$.

8-56. For the two cases in the preceding problem, find 95 percent confidence intervals for μ and their posterior probabilities.

8-57. Find the multiple of the sample standard deviation S with the smallest mean squared error in estimating σ in $N(\mu, \sigma^2)$.

8-58. For estimating the standard deviation of a normal population based on the sample range, and assuming random samples, find the multiple of R with the smallest mean squared error.

8-59. Find the p.d.f. of $\sqrt{X}$, where $X \sim \text{chi}^2(\nu)$.

---

[5]Ferguson, T. S., *Mathematical Statistics*, New York: Academic Press, Inc., 1967, page 163, or M. H DeGroot, *Optimal Statistical Decisions*, New York: McGraw-Hill Book Co, 1970.

9
Testing Hypotheses

A **hypothesis** is a statement or claim about some unknown aspect of the state of nature. Scientific investigators, industrial quality control engineers, market researchers, governmental decision makers, among others, will often have hypotheses about the particular facets of nature of immediate concern to them. They gather data and look to the data for evidence that will either support or cast doubt on their hypothesis.

A **test** of a hypothesis is a procedure, based on sample information, that culminates in an inferential statement about the hypothesis and possibly, in some situations, in a decision as to what action to take.

If one has perfect knowledge of a population—of the population distribution, the correct inference or action is clear. Yet, a sample provides only an approximate knowledge of the population, and this may or may not be close enough to the truth to lead one to a correct inference. It must be remembered that statistical procedures, generally, can lead to wrong conclusions by "the luck of the draw."

This text focuses on the *theory* of statistical inference, and indeed, this chapter will include some theorems. But to what extent the mathematical theory relates to or supports statistical practice is not at all clear, and is still subject to considerable controversy. For that matter, not all statisticians agree that testing a hypothesis is the way to approach a real problem.[1] Some champion estimation and others, prediction, as the proper mode of inference.

In this chapter we take up the two "classical" paradigms for the testing of hypotheses—testing as evaluation of the evidence provided by a set of data, for or against a hypothesis (§9.3), and testing as the use of data in an informed choice of action in a two-action decision problem (§9.4). The former is nearly universally used in scientific investigations, and the latter is appropriate when making decisions in business, industry,

---

[1] An eminent biostatistician, the late Dr. Joseph Berkson of the Mayo Clinic, once said that he had never tested a hypothesis!

9.1 Some generalities

A **statistical hypothesis** is a statement about the probability model for an experiment of chance. It is **simple** if it completely specifies the model, otherwise, **composite**. A composite hypothesis comprises the various simple hypotheses satisfying the conditions which define it.

EXAMPLE 9.1a The statement $X \sim \mathcal{N}(0, 1)$ is a simple hypothesis about the distribution of the random variable X, being a complete specification of it. The statement $X \sim \mathcal{N}(\mu, 1)$ is a composite hypothesis, since the value of μ is not specified; it is a composite of all the normal models with variance 1. The hypothesis that a random variable Y has mean 0 and variance 1, on the other hand, is composite because these two moments do not alone characterize a distribution. ∎

A natural question to ask is this: If the hypothesis to be tested is not true, is there some other explanation of the sample result? Standard terminology refers to the hypothesis being tested as the **null hypothesis**, and to the set of other possible models as the **alternative hypothesis**. Common notations for these are, respectively, H_0 and either H_A or H_1. Each of these can be either simple or composite, but frequently the null hypothesis is simple and the alternative, composite. In *parametric* inference, simple hypotheses are identified by the one or more real parameters that index a family of distributions.

Specifying the hypothesis to test is not always as simple as it sounds. In some cases the hypothesis is clear, notably in such settings as genetics, where assuming certain probabilistic rules for combination leads to the specification of a model.

EXAMPLE 9.1b The biologist Gregor Mendel's theory of inheritance led him to the claim that in crossings of round-yellow pea plants with wrinkled-green pea plants, the progeny would turn out to be round-yellow, round-green, wrinkled-yellow, and wrinkled-green in the proportions 9:3:3:1. In terms of probability, the model he claimed is this:

Type	RY	RG	WY	WG
Probability	9/16	3/16	3/16	1/16

This model is H_0, and the alternative H_A is simply that H_0 is false. ∎

In some settings a null hypothesis is implied in a performance specification, or in a claim about some product. Whether an action is called for depends on the setting. In checks of a production process, concluding that the process is out of "specs" would call for a shutdown and a search for an assignable cause, whereas testing a claim of tar content of cigarettes might only result in publication of the results.

EXAMPLE 9.1c Suppose a machined steel bearing is to be 10 cm in diameter. In the manufacture of parts such as this, there are inevitable variations in dimension that make a random variable the appropriate model for the dimension. The specification would have to be that the *mean* diameter is 10 cm. (One might also specify the variability, but ignore this for the present.) Both the supplier and the consumer of this bearing would want to know if the specification is being met—would want a test of the hypothesis $\mu = 10$. This is a null hypothesis, and the relevant alternative is $\mu \neq 10$, since bearings that are either too small or too large would give trouble. In this situation one would need to take some action on the basis of the outcome. ∎

A common statistical problem is that of determining whether or not a treatment produces an effect. The result of a treatment may be measured on some performance scale, or as a categorical result (e.g., success-failure). Whether the treatment is applied to animals, humans, or other pieces of experimental material, it is usually the case that the result varies from individual to individual. The overall effect of the treatment is therefore expressed in terms of some parameter of the population of individual results, usually an *average* (in the case of a numerical scale) or a *probability* of success.

EXAMPLE 9.1d Does a certain treatment for hypertension actually lower blood pressure? The change in blood pressure after treatment is a quantity that will vary from patient to patient. If the treatment is completely ineffective—the same as no treatment—the average change over all patients would be $\mu = 0$. This is a null hypothesis. (The term "null" perhaps derives from such applications, corresponding to *no effect*.) What is the alternative? Conceivably, the treatment could either raise or lower the blood pressure. But surely the investigator would not be interested in a treatment that raises it and would take $\mu < 0$ as the alternative of special interest. Although this phrases the problem neatly in mathematically terms, it does not take into account that a treatment which lowers the average by only a tiny amount is no better than one for which $\mu = 0$. It would be hard to draw a fine line, saying that μ on one side of that line is effective, and on the other side, ineffective. ■

In situations like those of the preceding examples, where hypotheses are phrased in terms of a population parameter θ, it is typical that H_0 is some set of θ-values, and H_A is the complementary set. The dividing line is sharp, and this sharpness is an idealization that may not do justice to the real situation. We'll see later that a fundamental weakness of traditional hypothesis-testing as a mode of inference in scientific investigations is the need for a precise specification of the null hypothesis. We'll also see that this precision is not a factor in applications to decision-making, since the exact specification of H_0 and H_A is really irrelevant in those situations.

Testing a hypothesis involves comparing sample results with some aspect of a model as defined by the hypothesis being tested. The comparison is based on some measure of discrepancy or inconsistency between data and model, on what we'll call the *test statistic*, T. Two obvious questions: (1) How does one choose an appropriate and efficient measure? (2) What conclusion is drawn for the value of T observed in a particular instance?

With respect to (1), we'll look at two possibilities. Sometimes one's intuition immediately suggests a measure to use. For instance, when H_0 specifies a population parameter, intuition would suggest that the difference between that value and a sample estimate of the parameter is indicative. Thus, if H_0 specifies $\mu = \mu_0$, then the discrepancy between $\bar{X}$ and μ_0 is an obvious candidate. And if H_0 specifies a population proportion $p = p_0$, then the difference between that value and the sample proportion $\hat{p}$ seems relevant. But there are situations in which a deeper intuition is required to invent a measure of discrepancy. A second, somewhat more systematic way of "deriving" a measure of inconsistency between sample and population is possible when H_0 and H_A are defined as sets in the parameter space of a parametric family of distributions. That way is to base the measure on a comparison of likelihoods, which we'll do in §9.2. (This approach, too, is intuitive—and doesn't always work out as one might expect.)

With respect to question (2), we'll study two kinds of inference. In scientific investigations, the test statistic T (or some function of it) is taken as a measure of evidence against H_0, and the results of the experiment are published as tentative findings—possibly to be revised after further experimentation. This type of inference will be studied in §9.3. In settings where decisions must be made and action taken, the test statistic can be used as an aid in choosing between two possible actions, and we'll introduce this kind of inference in §9.4. We'll see that both of these inferences from the data involve subjective elements—in quantifying and interpreting the "strength" of evidence, and in balancing the two types of error one can make in choosing between two actions.

The classical, commonly used approaches to testing, which we present in §9.3 and §9.4, have their problems. In the first type of application, the measure of evidence is not easy to interpret and subject to misinterpretation. Also, one may question whether it really addresses the scientific question of interest. In the second type of application, where a decision is required, the traditional approach offers no objective way of balancing the two types of error. On the other hand, a Bayesian statistician starts with a prior distribution for the state of nature, and of course this much is subjective. But from there on, calculating a posterior distribution is an objective process.

One catch in using the value of a test statistic T in inferences of any type is that T is a random variable. Whereas one sample may lead to one conclusion, another sample from the same population can suggest a different conclusion. So we need to know how T varies—need to know its sampling distribution, at least under H_0.

The distribution of a test statistic T depends on the population distribution, because T is a function of the sample observations. In many situations, the null hypothesis, even though it may not be simple, will define the distribution of the test statistic, and we call this its *null distribution*. Indeed, the null hypothesis is usually stated with just enough specificity that T will have a well-defined null distribution. For instance, with a family of models indexed by a parameter θ, one might really want to take as H_0 an *interval* of θ-values such as (9, 11); but the distribution of T would ordinarily depend on which value of θ in that interval is assumed. Taking $\theta = 10$ as H_0 may be sufficiently specific to define the null distribution of T.

EXAMPLE 9.1e Suppose we are to test $\mu = 1$ against some alternative specified in terms of μ, using the information in a large sample. Means of large samples tend to be close to the corresponding population means, so comparing these two means is a natural thing to do. A test of $\mu = 1$ might be based on how close the sample mean is to that value of μ, that is, on the difference $\bar{X} - 1$.

For large samples, $\bar{X} \approx N(\mu, \sigma^2/n)$, where σ is the population standard deviation, and the approximate *null* distribution of $\bar{X}$ is $N(1, \sigma^2/n)$. However, this distribution depends on the population variance, which is more often than not unknown. So whether an observed value of $\bar{X}$ is considered to be far from $\mu = 1$ depends on something unknown. This difficulty can be handled when n is large by resorting to sample information about population variability and

296   Chapter 9   Testing Hypotheses

measuring the discrepancy $\bar{X} - 1$ against the standard error of $\bar{X}$ as

$$Z = \frac{\bar{X} - 1}{S/\sqrt{n}}, \tag{1}$$

called a **Z-score** or *standard score*. According to Corollary 3 to Theorem 8 of Chapter 7 (page 211), $Z \approx \mathcal{N}(0, 1)$ under H_0. ∎

9.2 The likelihood ratio

The likelihood function is a sufficient summary of sample information, and the likelihood principle says that inferences should be based on likelihood functions. When there are competing models, as possible explanations of a set of data, the likelihood approach is to compare their likelihoods.

When H_0 and H_1 are both *simple* hypotheses, we compare their likelihoods by means of the **likelihood ratio statistic**:

$$\Lambda^* = \frac{L(H_0)}{L(H_1)}. \tag{1}$$

Indeed, we have seen earlier that Λ^* is minimal sufficient in this situation (Problem 7-65). Intuition suggests that we should prefer the explanation offered by H_1 when Λ^* is small—when H_1 is much more "likely" than H_0. In the next and later sections we'll study tests based on Λ^* as a measure of discrepancy between H_0 and H_1.

It is often the case that the statistic Λ^* is awkward to use because of the difficulty of finding its sampling distribution. However, it may be that Λ^* can be expressed as a function of a statistic that is more convenient and perhaps more intuitive or familiar.

When the null and alternative hypotheses are defined by two specific values of a parameter θ indexing a family included in the exponential family:

$$f(x \mid \theta) = B(\theta) h(x) \exp[Q(\theta) R(x)], \tag{2}$$

the joint p.d.f. of the observations in a random sample is

$$\prod f(x_i \mid \theta) = B^n(\theta) \exp[Q(\theta) \sum R(x_i)] \prod h(x_i),$$

and the likelihood ratio statistic is

$$\Lambda^* = \frac{B^n(\theta_0) \exp\left[Q(\theta_0) \sum R(X_i)\right]}{B^n(\theta_1) \exp\left[Q(\theta_1) \sum R(X_i)\right]} \propto \exp[\{Q(\theta_0) - Q(\theta_1)\} \sum R(X_i)].$$

9.2 The likelihood ratio

In many subfamilies the function Q is monotonic. In particular, then, if $Q(\theta_0) > Q(\theta_1)$, the ratio Λ is an increasing function of $T = \sum R(X_i)$. This statistic is also minimal sufficient, and in many cases it is a more familiar statistic and easier to work with than Λ.

EXAMPLE 9.2a Consider a random sample $\mathbf{X}$ of size n from $\mathcal{N}(\mu, 1)$, for use in testing μ vs. μ'. The likelihood function is given by (6) in Example 7.7d (page 223), from which we find the likelihood ratio statistic (1) to be

$$\Lambda^* = \frac{L(\mu)}{L(\mu')} = \frac{\exp\{-\frac{n}{2}(\mu - \bar{X})^2\}}{\exp\{-\frac{n}{2}(\mu' - \bar{X})^2\}} = \exp\left\{-\frac{n}{2}(\mu^2 - \mu'^2) + n(\mu - \mu')\bar{X}\right\}.$$

When $\mu < \mu'$, this is clearly monotone decreasing in $\bar{X}$: The statistic $\bar{X}$ is large when Λ^* is small and small when Λ^* is large. [$\mathcal{N}(\mu, 1)$ is a member of the exponential family (2), with $Q(\mu) = -\mu$ and $R(x) = x$.] As a test statistic, $\bar{X}$ is more intuitive and more mathematically tractable than Λ^*. ∎

When H_0 and H_A are *not* both simple, it is natural to turn to the maximum likelihood principle, and compare the largest likelihood in H_0 with the largest likelihood in H_A. Thus, we might take the ratio of these largest likelihoods as our test statistic:

$$\Lambda^* = \frac{\sup_{H_0} L(\theta)}{\sup_{H_A} L(\theta)}. \tag{3}$$

[Of course, when H_0 and H_A are both simple, Λ^* is the ratio of their likelihoods, as defined in (1).]

In many situations it is convenient to use a modification of Λ^* that compares the largest likelihood for $\theta \in H_0$ with the largest likelihood over all θ, that is, for $\theta \in H_0 \cup H_A$. We define the (generalized) **likelihood ratio statistic**:

$$\Lambda = \frac{\sup_{H_0} L(\theta)}{\sup_{H_0 \cup H_A} L(\theta)}. \tag{4}$$

Clearly, if $\Lambda^* < 1$, then $\Lambda = \Lambda^*$. But if $\Lambda^* \geq 1$, then $\Lambda = 1$.

When just H_0 is simple: $\theta = \theta_0$, the likelihood ratio statistic is

$$\Lambda = \frac{L(\theta_0)}{L(\hat{\theta})}, \tag{5}$$

where $\hat{\theta}$ is the maximum likelihood estimate of θ.

As in the case of (1), the distribution of the generalized statistic (3) is often quite complicated and so very awkward to use. But sometimes, like Λ^* in (1), the generalized version (4) of Λ can be expressed as a function of some more convenient sample statistic, one whose distribution we know or can find.

EXAMPLE 9.2b Again consider a random sample of size n from $\mathcal{N}(\mu, 1)$, this time for the purpose of testing the null hypothesis $\mu = \mu_0$ against the two-sided alternative $\mu \neq \mu_0$. As in Example 9.2a, the likelihood function is

$$L(\mu) = \exp\left\{-\frac{n}{2}(\mu - \bar{X})^2\right\}.$$

Its "sup" over H_0 is $L(\mu_0)$. The union $H_0 \cup H_A$ is the real line, and the maximum of $L(\mu)$ occurs at $\hat{\mu} = \bar{X}$, where $L(\bar{X}) = 1$. So (5) becomes

$$\Lambda = \exp\left\{-\frac{n}{2}(\mu_0 - \bar{X})^2\right\}. \tag{6}$$

This is a decreasing function of the squared distance from $\bar{X}$ to μ_0: $(\mu_0 - \bar{X})^2$. Thus, Λ is large when $\bar{X}$ is close to μ_0 and small when $\bar{X}$ is far from μ_0. ∎

In the two preceding examples we see that the intuitive notion of likelihood leads us to essentially the same statistic as does our intuition based on the consistency of sample moments. So these examples do not make clear what the likelihood approach accomplishes. In later chapters we'll encounter examples in which a simpler approach is not evident, and we need to work with Λ itself, or some simple function of it. When Λ is not readily expressible as a function of some familiar sample statistic, it is useful to know that in certain standard situations and for large samples, the null distribution of the quantity $-2 \log \Lambda$ is well approximated by a familiar distribution. The following is given without proof[2]:

THEOREM 1 *When $H_0 \cup H_A$ depends on r free real parameters, and H_0 assigns values to k of those parameters, then $-2 \log \Lambda \approx \mathrm{chi}^2(k)$.*

Here, k can also be thought of as the dimension of $H_0 \cup H_A$ minus the dimension of H_0: $r - (r - k)$.

EXAMPLE 9.2c In testing $\mu = \mu_0$ against $\mu \neq \mu_0$ in $\mathcal{N}(\mu, 1)$ with a

---

[2] See M. Kendall and A. Stuart, *The Advanced Theory of Statistics, Vol. II: Inference and Relationship*, 4th Ed., New York: Macmillan Pub. Co.

9.2 The likelihood ratio 299

random sample of size n, as in the preceding example, we have

$$-2\log\Lambda = n(\mu_0 - \bar{X})^2 = \left(\frac{\bar{X} - \mu_0}{1/\sqrt{n}}\right)^2.$$

As the square of a standard normal variable, this is *exactly* chi$^2(1)$. (Theorem 1 above says only that $-2\log\Lambda$ is asymptotically chi-square. Since H_A is one-dimensional, and H_0 assigns a value to the one free parameter, the degrees-of-freedom parameter is $k = 1$, the same as the d.f. of the exact distribution.) ∎

PROBLEMS

9-1. Determine whether the given hypothesis is simple or composite, in each of the following cases:
(a) A pair of dice is "straight" (six equally likely sides).
(b) The dice in (a) are crooked (not straight).
(c) $E(X) = 0$.   (d) $X \sim \text{Exp}(3)$.   (e) $E(X) = E(Y)$.

9-2. For a test of $p = .5$ against $p = .7$, find the likelihood ratio statistic as a function of Y, the number of successes in n independent trials of Ber(p).

9-3. Consider testing H_0: $\theta = 1$ vs. H_1: $\theta = 2$ with a random sample of size n from $\mathcal{N}(0, \theta)$.
(a) Find the likelihood ratio and express it as a monotone function of a sample moment.
(b) Find the distribution of the moment in (a) under H_0.

9-4. Let Z denote a discrete random variable having one or the other of the two distributions shown in this table of probabilities:

	z_1	z_2	z_3	z_4	z_5
H_0	.2	.3	.1	.3	.1
H_1	.3	.1	.3	.2	.1

Calculate Λ^* for each z_i and list the z_i's in order of increasing Λ.

9-5. Find the likelihood ratio statistic for testing H_0: $\theta = \theta_0$ vs. H_A: $\theta \neq \theta_0$ using a random sample of size n from $\mathcal{N}(0, \theta)$. Plot $\Lambda^{2/n}$ as a function of $\overline{X^2}/\theta_0$ and so conclude what kinds of values of $\overline{X^2}$ would point to H_A.

9-6. Find the likelihood ratio statistic for testing H_0: $\lambda = \lambda_0$ vs. H_A: $\lambda \neq \lambda_0$ using a random sample of size n from Exp(λ). Plot $\Lambda^{1/n}$ as a function of $\lambda_0 \bar{X}$ and so conclude what ranges of values of $\bar{X}$ would point to H_A.

9-7. Consider testing $p = \frac{1}{2}$ vs. $p \neq \frac{1}{2}$, based on a random sample from Ber(p).
(a) Calculate Λ, the likelihood ratio (3), for $Y = 0, 1, 2, 3$, where Y is the number of successes in three independent trials.
(b) Find the null distribution of Λ in (a).
(c) Show that for general n, $\log \Lambda$ increases when $\hat{p} < \frac{1}{2}$ and decreases when $\hat{p} > \frac{1}{2}$, so that Λ is small when $\hat{p}$ is close to 0 or close to 1.

9-8. For testing equality of two population means: $\mu_1 = \mu_2$, based on independent random samples of sizes n_1 and n_2, with means $\bar{X}_1$ and $\bar{X}_2$ and variances S_1^2 and S_2^2, construct a standard score as the ratio of the expected mean difference divided by its standard error (see Problem 8-9), and give its null distribution. (Assume large samples.)

9-9. A study of pain and activity levels for good and poor sleepers yielded the following statistics: Among 28 good sleepers, the hours of activity had mean $\bar{X} = 10.7$ and s.d. $S_X = 4.8$. For 70 poor sleepers the statistics were $\bar{Y} = 8.6$ and $S_Y = 4.8$. Determine an approximate Z-score for the mean difference that one could use in testing the hypothesis that $\mu_X = \mu_Y$.

9-10. Polls of 1015 adults in 1987 and 1000 adults in 1986 found 55 and 54 percent, respectively, who thought that homosexual relations between consenting adults should not be legal. Calculate a Z-score appropriate for use in testing equalities of the population proportions in 1986 and 1987.

9.3 Tests: assessing evidence

The basic idea underlying a statistical test is that a random sample from a population should look something like the population. Of course, when the sample size is less than the population size, samples will usually *not* look exactly like the population—even with random sampling. It then becomes a matter of deciding how much discrepancy between sample and population casts serious doubt on the hypothesis. In the typical situation, the test statistic T that measures this discrepancy has a null distribution that is "nice"—unimodal, tailing off in one or both directions away from the mode. In assessing the credibility of H_0, the approach of *significance testing* is to look at the null distribution of T and draw conclusions on the basis of how far the observed T is from what is typical when H_0 is true. For instance, if most of the null distribution of T lies between 10 and 12, and we happen to observe $T = 15$, this is apt to be deemed *significant*—signifying that perhaps the population distribution violates H_0. The other possible explanation of $T = 15$ is that this is merely the result of sampling variability—perhaps unusual or surprising under H_0, but possible.

If values of T far from the middle of the distribution in one or the other direction (or both) are more easily accounted for by H_A, these are thought of as "extreme" and taken as evidence against H_0. Naturally, the farther out in the tail of its null distribution, the stronger the evidence against H_0 is provided by an observed value of T. On the other hand, values of T more or less near the middle of its null distribution would not argue against H_0.

When using the generalized likelihood ratio statistic Λ [(4) of § 9.2] as a test statistic, *small* values of Λ are considered "extreme," since small values suggest that the explanations of the data in H_A are to be preferred over those in H_0.

EXAMPLE 9.3a Consider testing $\mu = \mu_0$ against $\mu \neq \mu_0$ using the statistic $\bar{X}$. The null distribution of the statistic (1) in Example 9.1e:

$$Z = \frac{\bar{X} - \mu_0}{S/\sqrt{n}}$$

is approximately $\mathcal{N}(0, 1)$ when n is large. This distribution is unimodal, centered at 0. A value of Z far from 0 would be more easily accounted for by a value of μ other than μ_0, so we consider large values of $|Z|$ as extreme.

If we can assume a particular parametric family of distributions for X, we can use the likelihood approach. In particular, if $X \sim \mathcal{N}(\mu, 1)$, the likelihood ratio statistic, determined in Example 9.2c, is

$$\Lambda = \exp\left\{-\frac{n}{2}(\mu_0 - \bar{X})^2\right\}.$$

This is small (extreme for Λ) when $|\bar{X} - \mu_0|$ is large, and again, from the point of view of likelihoods, we see that values of $\bar{X}$ far from μ_0 are extreme. ■

The scale of values of a test statistic T is usually peculiar to the particular problem at hand, so the distance from the middle of a distribution is not very satisfactory as a universal measure of discrepancy between sample and population. It is true that many of the commonly used statistics have distributions that are close to normal, and a Z-score (like that in Example 9.3a) *is* a rather standard and well understood measure. However, it is customary and generally useful to indicate the amount of discrepancy between the observed and expected values of T by giving the area under the null p.d.f. of T over the set of values that are at least as extreme as the observed T.

DEFINITION The **P-value** corresponding to an observed $T = t_0$ is the total probability at and beyond t_0, in the direction of more extreme values.

This definition is generally accepted when T is an estimate of θ that is used for testing $\theta = \theta_0$ against an alternative hypothesis of the form $\theta > \theta_0$ (or $\theta < \theta_0$). How to define "P-value" when the alternative is two-sided: $\theta \neq \theta_0$, is quite controversial.

EXAMPLE 9.3b Consider a large sample obtained to test H_0: $\mu = 10$ against the alternative H_A: $\mu > 10$. Using $\bar{X}$ as the relevant sample statistic, we observe that, as pointed out in Example 9.1e,

$$Z = \frac{\bar{X} - 10}{S/\sqrt{n}} \approx \mathcal{N}(0, 1).$$

Suppose then we obtain a sample of size $n = 100$ and find $\bar{X} = 10.8$, $S = 2.50$. The Z-score is $.8/.25 = 3.2$. That is, the sample mean is 3.2 standard errors to the right of the population mean under H_0, and the P-value is $1 - \Phi(3.2)$ or about .0007. There are certainly values of μ in H_A corresponding to models with which the sample result is much less discrepant. Figure 9-1 shows the graph of $P(\mu)$, the P-value corresponding to $\bar{X} = 10.8$ when testing the hypothesis that the population mean is μ.

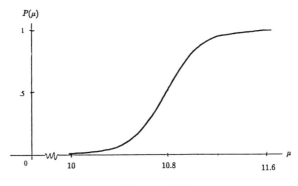

Figure 9-1

If the problem had been posed as testing $\mu = 10$ against the "two-sided" alternative $\mu \neq 10$, the values of $\bar{X}$ that would be considered extreme are those which are either much smaller or much larger than 10. Some statisticians feel strongly that in such a case the P-value should be the area above the region $|Z| > 3.2$, or twice the area that we calculated before: $P = 2 \times (.0007) = .0014$. ∎

R. A. Fisher says,[3] "The actual value of p . . . indicates the strength of the evidence against the null hypothesis." This seems intuitively correct within the context of a given experiment: A Z-score of 2 with $P = .0228$ is surely stronger evidence against H_0 than a Z-score of 1 with $P = .1587$. But it is not so clear in comparing or combining experimental results. The next example shows that P-values can be equal when the inferences are clearly not the same.

EXAMPLE 9.3c Suppose we test $\mu = 5$ vs. $\mu > 5$ with the mean of a random sample of size n from $\mathcal{N}(\mu, 1)$. Then $Z = \sqrt{n}(\bar{X} - 5) \sim \mathcal{N}(0, 1)$ when H_0 is true. If $n = 4$ and we find that $\bar{X} = 6$, then $Z = 2$, and the P-value is $.0228$. If $n = 400$ and we find that $\bar{X} = 5.1$, then again $Z = 2$ and $P = .0228$. Although the P-values are the same, there is surely different information in results of the two experiments. ∎

In reports of scientific investigations there is a commonly used convention of language that takes certain arbitrarily chosen values of P as meaningful. In this convention, a result for which $P < .05$ is called *statistically significant* and, in the published report, to mark it with an asterisk. When $P < .01$, the result is called *highly statistically significant* and marked with two asterisks. Adopting .05 as a boundary between what is called significant and what is not significant is said to be testing "at the .05 level of significance."

Clearly, it is less informative to say that a result is statistically significant than it is to give the P-value. So why this terminology? There is a natural desire on the part of scientists to have an objective procedure—one that would say: this much discrepancy (or more) between the data and the null hypothesis means that the null hypothesis is not true, and that what we have observed is not explainable as mere sampling variability. But .05 is quite *arbitrary*, and interpreting a P-value is a subjective matter. Indeed, as Fisher himself writes,[4] "... in fact

---

[3]R. A. Fisher, *Statistical Methods for Research Workers*, 13th Ed. (1958), Hafner Publishing Co., New York, page 80.

[4]R. A. Fisher, *Statistical Methods and Scientific Inference*, 3rd Ed. (1973), Hafner Press, New York, p. 45.

no scientific worker has a fixed level of significance at which, from year to year and in all circumstances, he rejects hypotheses; he rather gives his mind to each particular case in the light of his evidence and his ideas."

The language of "significance" is often misinterpreted. A *statistically* significant result can have little *practical* significance: With enough data, a test result will be found to be statistically significant when the true mean, for instance, is 5.001 instead of the null hypothesis value of 5.000, when in fact a difference of .001 may be of no practical importance Also, with a small sample size, a practically important difference can fail to be detected as a statistically significant result, simply because the sample information is too meager.

EXAMPLE 9.3d A 1965 case in Alabama (*Swain* v. *Alabama*) dealt with alleged discrimination against blacks in jury-selections. Among 1,050 called for duty were only 177 blacks, whereas the census showed about 25 percent blacks. In a test of H_0: $p = .25$, the observed ratio $177/1050 \doteq .17$ implies a *P*-value of about 6×10^{-10}, which would seem convincing evidence against H_0. But it was not deemed by the court to be *prima facia* evidence of discrimination. A law professor writes,[5] "That trivial differences can appear statistically significant only underscores the admonition that the *p*-value should not be considered in a vacuum. The courts are not likely to lose sight of the question of practical significance and to shut their eyes to the possibility that the degree of discrimination is itself *de minimis*." ∎

Besides the unaccounted for element of subjectivity that is involved, and the frequent misunderstanding of what "statistically significant" really means, there are other problems with assessing the evidence provided by an experimental result in terms of *P*-values. One is the continuing controversy over whether or not to double a *P*-value when the alternative is two-sided. Also, the same *P*-value may have different implications in different situations, as we saw in Example 9.3c. And different *P*-values can be obtained when the likelihood functions are the same; this violates the likelihood principle.

EXAMPLE 9.3e Suppose we want to test $p = .5$ in a Bernoulli model and consider (a) sampling to a preset number of trials, and (b) sampling to a preset number of successes. If an experiment results in 3 successes in 10 trials, the *P*-value under (a) is .172, the probability of $Y \leq 3$ when $Y \sim \text{Bin}(10, 0.5)$. But under (b), the *P*-value is .090, the probability that

---

[5]Cited by D. Kaye in "Statistical evidence of discrimination," *J. Amer. Stat. Assn.* 77 (1982), 773-783.

it takes at least 10 trials to obtain 3 successes when $p = .5$. So the P-value depends on how one decides to stop sampling, even though the likelihoods are the same and the end result is the same—three successes in ten trials. ∎

Research papers in scientific journals are often riddled with P-values. If one calculates 20 P-values, and the null hypotheses involved are all true, one of the 20 sample results (on average) will be declared statistically significant. How to combine P-values properly to summarize the evidence in several experiments is problematical. Various ways have been proposed and studied but are not completely satisfactory.

If one adopts .05 as a significance level, this is the probability that H_0 will be "rejected" when it is true. If two (independent) tests are carried out, the overall probability of at least one false rejection, when both null hypotheses are true is $1 - (.95)^2 = .0975$. If, instead, each test were judged at the .0253 level, the overall probability of at least one false rejection would be about .05. (But why should one make a judgment on one experiment dependent on whether or not another is being judged simultaneously?)

A common misinterpretation of a P-value is that it is the probability of the null hypothesis. Of course, this cannot be correct: In the present framework we do not attach probabilities to hypotheses. Rather, P is the probability that, if H_0 is true, a value of the test statistic would be obtained which is at least as extreme as the one actually observed. The tendency to misinterpret P suggests that what investigators really want is the probability of the null hypothesis. To find this is to require that hypotheses—models—are assigned probabilities, and for the Bayesian this not a problem (see §9.5)

PROBLEMS

9-11. Suppose X has the triangular distribution with p.d.f.
$$f(x \mid \theta) = 1 - |x - \theta|, \quad |x - \theta| < 1.$$
Given the observation $X = .7$, find the P-value for testing $\theta = 0$ against the alternative $\theta \neq 0$.

9-12. To test a claim of ESP, a psychologist selects a card at random, one of four, each with a different symbol on it. The subject, without seeing the card, states which symbol has been selected. This is repeated 12 times, and the subject correctly identifies 6 of the selections. Set up appropriate hypotheses and determine the P-value. (Would this convince *you*, one way or the other?)

9-13. Consider again the situation and data of Problem 9-10.
(a) Find the P-value, assuming independent random samples..
(b) Is the increase in percentage from 54 to 55 "statistically significant" at the five percent level?
(c) At what "level" is the increase just on the border of significance?

9-14. Blood platelet counts of a sample of 153 cancer patients averaged 359.9, and the sample s.d. was $S = 169.9$ (in $1000/\text{mm}^3$). The mean for healthy males is assumed to be 235. Find the P-value for a test of the hypothesis that the mean count for cancer patients is $\mu = 235$ vs. the hypothesis that $\mu > 235$.

9-15. An investigation studied thousands of births and found 145 cases in which conception was determined to have occurred two days after ovulation. Among these 145 there were 95 male births, whereas among the rest, the proportion of males was .52. Assess the evidence (using a P-value) against the hypothesis that the timing of conception has nothing to do with the sex of the baby, assuming that the sex of a baby is a Bernoulli trial, and that the 145 trials are independent.

9-16. Consider a Poisson process with rate parameter λ/min, and H_0: $\lambda = 1$ vs. H_1: $\lambda < 1$. Find the P-value for each of the following experimental results:
(a) We observe the process for six minutes and record exactly three arrivals.
(b) We observe the process until the third arrival and note that it takes six minutes. [Hint: Problem 6-56 may be helpful.]

9-17. In a three-year study of the effectiveness of sodium monofluorophosphate (MFP) as a decay-fighting agent in toothpaste,[6] the following summary statistics were reported:

Dentifrice	No. of children	Av. cavities/child	S.D.
MFP	208	19.98	10.61
Control	201	22.39	11.96

Find the P-value in testing the hypothesis that there is no difference between MFP and the control toothpaste, on average. (See Problem 9-8.)

9-18. At one point in the historical development of probability theory it was contended by some that when two coins are tossed, the probabilities of 0, 1, and 2 heads are equally likely. Suppose two coins are tossed 100 times and 25 trials result in no heads, 25 result in 2 heads, and 50 in

[6] Frankl, S. N., and J. E. Alman, *J. Oral Therapeut. Pharmacol.* 4 (1968), 443-449.

exactly 1 heads.
   (a) Give the likelihood function in terms of $p_i = P(i \text{ heads })$.
   (b) Calculate the likelihood ratio for H_0: $p_0 = p_1 = p_2 = 1/3$ and find the corresponding P-value using Theorem 1.

9.4 Tests: decision rules

A statistical test can be used as a guide for deciding between two actions A and B when there is incomplete knowledge of some relevant population. Suppose action A is the better action when H_0 is true, and action B is the better action when H_A is true (and H_0, false). We say that in taking action A we are *accepting* H_0, and in taking action B we are *rejecting* H_0. Thus, we have specific meanings of "accepting" and "rejecting" H_0 in such situations.

As in the preceding section, we assume here that as the basis of a test, we have a sample $\mathbf{X} = (X_1, X_2, ..., X_n)$, the result of carrying out the experiment about which the various admissible hypotheses make a statement.

DEFINITION In a two-action decision problem based on sample information, a **test** of H_0 vs. H_A is a *rule* that assigns one of the actions A (accept H_0) and B (reject H_0) to each sample point $\mathbf{x}$. The **critical region** of the test is the set of $\mathbf{x}$-values which call for rejecting H_0 according to the rule.

When the sample $\mathbf{X}$ is reduced to a statistic $T = t(\mathbf{X})$, a test is often expressed in terms of values of T, and the set of T-values calling for rejection of H_0 is then the critical region of the test. (Of course, a set C in the space of T-values defines a set of $\mathbf{x}$-values, the pre-image of C, and one could call either one the critical region. The context will usually make the usage clear.)

The tests or rules we'll be studying are based on some statistic T that measures the discrepancy between the sample and the population defined by H_0, as developed in §9.1 and §9.2. Intuition suggests that a critical region should be a set of those T-values that present the strongest evidence against the null hypothesis. In particular, when working with the likelihood ratio statistic, a critical region would be of the form $\Lambda < K$. (Here and in what follows, we'll often use K as a generic constant.)

EXAMPLE 9.4a In testing H_0: $X \sim \text{Exp}(1)$ against H_1: $X \sim \text{Exp}(2)$

with a random sample of size n, the likelihood ratio is

$$\Lambda = \frac{e^{-\Sigma X_i}}{2^n e^{-2\Sigma X_i}} = 2^{-n} e^{n\bar{X}}.$$

This is an increasing function of $\bar{X}$, so a critical region $\Lambda < K$ is equivalent to one of the form $\bar{X} < K'$. This is precisely the type of critical region for $\bar{X}$ that intuition would suggest, since the population mean is $1/\lambda$—a small value of $\bar{X}$ would point to the larger value of λ. ∎

Carrying out a statistical test can result in a wrong decision, since the information in a sample is generally imperfect. A sample can lead one astray in two distinct ways:

DEFINITION Rejecting H_0 when it is true is a *type I error*; accepting H_0 when it is false is a *type II* error.

It is natural to ask how likely it is that using a given test will result in an error (of either type). Unless the state of nature is regarded as a random variable, we can only answer conditionally, *given* a particular state of nature.

EXAMPLE 9.4b Let p denote the proportion of manufactured parts of a certain type which are defective. Suppose it has been decided that the manufacturing process is "in control" if $p \leq .05$. Let H_0 denote the hypothesis $p \leq .05$, and H_A, the hypothesis $p > .05$. The correct action is to allow the process to continue if H_0 is true, but to shut down and look for an assignable cause if H_A is true.

Suppose we take a random sample of 15 parts. Let Y denote the number of defective parts among the 15. One rule might be to shut down (reject H_0) if $Y > 1$. Following this rule would result in a type I error if more than 1 part is defective when in fact $p \leq .05$. The probability of this error, *given* H_0, depends on p. For instance,

$$P(\text{rej. } H_0 \mid p = .05) = P(Y > 1 \mid p = .05) \doteq .171.$$

Similarly, the rule leads to a type II error if $p > .05$ and it turns out that $Y \leq 1$. The probability of this also depends on p; it is .167 when $p = .2$ and .035 when $p = .3$, and so on. (We'll examine this dependence in greater detail in §9.10.) ∎

It is common practice to design a test so as to control the type I error, labeling hypotheses so that the type I error is the error of greater

9.4 Tests: decision rules

concern. For a particular critical region C, we define its **size** as

$$\sup_{\theta \in H_0} P(C \mid \theta) = \alpha. \tag{1}$$

This is also called the **significance level** of the test. For instance, in the preceding example, $\alpha = .17$ for the given test. To control the type I error to be at most of size α^*, we'd choose C so that $\alpha \leq \alpha^*$.

The mathematical theory of testing is somewhat neater when we introduce a new type of strategy—a **randomized test**. Suppose, instead of assigning either action A or action B to each possible outcome $\mathbf{x}$, we assign each outcome a *probability*, and take action A or B according as a (biased) coin with that probability of "heads" lands heads or tails.

DEFINITION A *randomized test* of H_0 is a function ϕ on the sample space such that $0 \leq \phi(\mathbf{x}) \leq 1$. To carry out the test, obtain a sample $\mathbf{X} = \mathbf{x}$, and then sample $Y \sim \text{Ber}(\phi(\mathbf{x}))$, rejecting or accepting H_0 according as $Y = 1$ or $Y = 0$.

The test defined by a critical region C is a special case of a randomized test in which the auxiliary Bernoulli randomization is singular: The "ϕ", in this case, is just the indicator function of C, so the "coin" we toss has two heads ($\phi = 1$ for $\mathbf{x} \in C$) or two tails ($\phi = 0$ for $\mathbf{x} \in C^c$).

EXAMPLE 9.4c For testing $p \leq .05$ on the basis of Y, the number of successes in 15 independent trials (as in Example 9.4b), consider this rule: Reject H_0 if $Y > 2$, accept H_0 if $Y < 2$, and reject H_0 with probability γ if $Y = 2$. For this randomized test,

$$\phi(Y) = \begin{cases} 1 \text{ when } Y > 2, \\ 0 \text{ when } Y < 2, \\ \gamma \text{ when } Y = 2. \end{cases}$$

When $p = .05$, the probability of incorrectly rejecting H_0 is

$$P(\text{rej}) = E\{P(\text{rej} \mid Y)\} = 1 \cdot P(Y > 2) + \gamma \cdot P(Y = 2) + 0 \cdot P(Y < 2).$$

$$= .036 + .135\gamma, \ 0 \leq \gamma \leq 1.$$

This is a number between .036 ($\gamma = 0$) and .171 ($\gamma = 1$), and in particular, we can make $\alpha = .05$ by taking $\gamma = .104$. ∎

A test for $\theta = \theta_0$ is implicit in the construction of a confidence interval, namely: Reject θ_0 if this value is not in the confidence interval. The type I error size of the test, the probability that the confidence interval does not include θ_0 when in fact θ_0 is the true value of θ, is simply 1 minus the confidence level. For example, a 95 percent confidence interval used in this way implies a test at the five percent significance level.

There is another side to the coin: Given a family of tests, one can construct a confidence interval, as follows. Suppose $C(\theta_0)$ is a critical region of size α for testing $\theta = \theta_0$, defined for each θ_0, so that

$$P[\mathbf{X} \in C(\theta_0) \mid \theta_0] = \alpha.$$

(It would be necessary, of course, that θ uniquely define the distribution in the sample space.) For each sample point $\mathbf{x}$, define a set

$$I(\mathbf{x}) = \{\theta : \mathbf{x} \notin C(\theta)\}. \tag{2}$$

The events $[I(\mathbf{X}) \ni \theta]$ and $[\mathbf{X} \notin C(\theta)]$ are equivalent, so

$$P[I(\mathbf{X}) \ni \theta \mid \theta] = 1 - P[\mathbf{X} \in C(\theta) \mid \theta] = 1 - \alpha.$$

Thus, $I(\mathbf{X})$ is a confidence set for θ with confidence coefficient $1 - \alpha$. If the critical regions C are two-sided, then typically the confidence region will be a finite interval. A one-sided C, on the other hand, will usually result in a one-sided confidence interval—a confidence bound for θ (upper or lower, depending on which end of the interval is finite).

EXAMPLE 9.4d To test $\mu = \mu_0$ in $\mathcal{N}(\mu, 1)$ against $\mu > \mu_0$, one might use a one-sided critical region: $C(\mu_0) = \{\mathbf{x} \mid \bar{x} > K\}$, where K is determined so that

$$P(C \mid \mu_0) = P(\bar{X} > K) = 1 - \Phi\left(\frac{K - \mu_0}{1/\sqrt{n}}\right) = \alpha.$$

Thus,

$$K = \mu_0 + \frac{z_{1-\alpha}}{\sqrt{n}},$$

where z_γ is the 100γ percentile of $\mathcal{N}(0, 1)$. Now define the set (2):

$$I(\mathbf{X}) = \{\mu : \bar{X} < K\} = \{\mu \mid \mu > \bar{X} - z_{1-\alpha}/\sqrt{n}\}.$$

This is a one-sided confidence interval, and $\bar{X} - z_{1-\alpha}/\sqrt{n}$ is a lower confidence bound for μ, with confidence level $1 - \alpha$. ∎

A significance tester who adopts a particular level such as .05 or .01

as critical—as a criterion for "statistical significance"— can be thought of as making a kind of decision by saying that if $P < .05$, he or she "rejects" H_0, and otherwise "accepts" H_0—or "does not reject" H_0. (What is meant by these phrases and the semantic difference between "accepting" and "not rejecting" are not completely clear.) The particular critical level used (.05, .01, or whatever) is called the *significance level* of the test, and the P-value is called the *observed significance level*—the level at which the observed sample result is just on the boundary of significance.

PROBLEMS

9-19. A laboratory, checking its equipment for measuring white blood cell count at the beginning of a day's runs, follows the rule of proceeding with the day's work if a single measurement of a standard supply labeled "3800" is between 3000 and 4600, otherwise stopping to look for an assignable cause. Assuming that the measured count is normal with s.d. $\sigma = 400$, find the α of their rule. (State regulations would call for an acceptance region from 2600 to 5000 for the single measurement; is the lab justified in using the smaller interval?)

9-20. Let M denote the number of defective items in a carton of five, and consider a sample of two, drawn without replacement, to test the null hypothesis $M \leq 1$ against the alternative $M > 1$.
(a) List all possible tests, sensible or not.
(b) For the test that rejects the null hypothesis if and only if any defectives are found in the sample, find the probability of a type I error for each M in H_0 and the probability of a type II error for each M in H_A.
(c) Is any of the tests in (a) better (in terms of both kinds of error) than the test given in (b)?

9-21. Consider the following rejection regions: $C_1 = \{|\bar{X} - 10| > .5\}$, $C_2 = \{|\bar{X} - 10| > .8\}$. Which one has the larger α?

9-22. Consider testing $\lambda_0 = .1$ vs. $\lambda_1 = .3$, based on the sum of the observations in a random sample of size $n = 10$ from $\text{Poi}(\lambda)$, and on the intuitive notion that a large sum suggests a large value of λ.
(a) Find a randomized test of level $\alpha = .10$.
(b) Find the probability that the test in (a) accepts $\lambda = .1$ when in fact $\lambda = .3$.

9-23. An observation Z takes one of four values according to one of the three distributions shown in the following table of probabilities:

	θ_0	θ_1	θ_2
z_1	.2	.5	.3
z_2	.3	.1	0
z_3	.1	.2	.4
z_4	.4	.2	.3

Consider testing H_0: $\theta = \theta_0$ against H_1: $\theta = \theta_1$ or θ_2.
(a) For each value of Z find the value of Λ^* [as given by (3), §9.2].
(b) Determine all critical regions for Z defined by rules of the form $\Lambda^* < K$.
(c) Find α for each test in (b).
(d) Compare the two critical regions $\{z_2\}$ and $\{z_1, z_3\}$ in terms of type I and type II errors when $\theta = \theta_2$. Which is the better test?

9-24. A test based on a discrete statistic T calls for rejecting H_0 when $T < 2$, accepting H_0 when $T > 2$, and tossing a (fair) coin when $T = 2$.
(a) Express this rule as a function $\phi(t) = P(\text{rej. } H_0 \mid T = t)$.
(b) If H_0 defines a specific distribution for T, express α in terms of f, the p.f. of that null distribution.

9-25. Find the critical region for $-2 \log \Lambda$ in Problem 9-18 to achieve $\alpha = .05$. Is H_0 rejected at this level, given the data of that problem?

9-26. In Example 8.8b we found a 90 percent confidence interval for θ:
$$2 \sum X_i / 10.9 > \theta > 2 \sum X_i / 31.4,$$
based on a random sample of size ten from $\text{Exp}(1/\theta)$.
(a) Using this interval, give a critical region for testing $\theta = 1$ in terms of $\sum X_i$, and find the "size of the type I error" (size of the critical region).
(b) Use the same technique to find a two-sided critical region with $\alpha = .05$ for testing $\theta = 2$.

9-27. Obtain an approximate upper confidence bound for μ, at the 90 percent level, based on a random sample of size $n = 400$ with $\bar{X} = 12.0$, $S = 2.40$, and $n = 400$.

9-28. Consider a random sample of size $n = 25$ from $\text{Poi}(\lambda)$, and assume that $n\lambda$ is large enough that the mean $\bar{X}$ is approximately normal.
(a) Find K such that the critical region $[\bar{X} > K]$ has (approximate) size .05 when $\lambda = \lambda_0$. (K will depend on λ_0.)
(b) Use the result in (a) to construct an approximate 95 percent lower confidence bound for λ.

9-29. In Problem 9-6 we found the likelihood ratio for testing $\lambda = \lambda_0$ against $\lambda \neq \lambda_0$ using a random sample from $\text{Exp}(\lambda)$, and expressed as a function of the sample mean. Make a rough plot of the relationship $\Lambda = g(\bar{X})$ and determine the nature of the critical region for $\bar{X}$ corresponding to $\Lambda < K$.

9-30. A single observation is obtained from a Cauchy population with location parameter θ:
$$f(x \mid \theta) = \frac{1/\pi}{1 + (x - \theta)^2},$$
for testing $\theta = 0$ against $\theta = 2$, and let $\Lambda^* = f(x \mid 0)/f(x \mid 2)$.
(a) Determine the critical region for X according to the maximum likelihood principle.
(b) Express the critical region $\Lambda^* < 5$ in terms of X, and calculate α.
(c) Find a test of the form $X > K$ with the same α as the test in (b) and compare their probabilities of erroneously accepting $\theta = 0$.
(d) Express the critical region $\Lambda^* < 17/37$ in terms of X.

9-31. For one observation X from the population of the preceding problem, find the critical region for X determined by $\Lambda < K$, where Λ is the (generalized) likelihood ratio for $\theta = 0$ against $\theta \neq 0$.

9.5 Bayesian testing

When hypotheses are phrased in terms of population parameters, the Bayesian statistician represents uncertainty about a parameter in terms of probabilities and can calculate, for any given prior, the probability of the null hypothesis. However, in testing a hypothesis such as $p = \frac{1}{2}$ or $\mu = 10$, even the Bayesian has a problem. We usually think of p and μ as measured on continuous scales and model our uncertainty with continuous distributions. And if μ is continuous, then whatever the data may have to tell us, $P(\mu = 10) = 0$. This is awkward, but the problem here is not with the Bayesian framework so much as with the choice of H_0 as a single point hypothesis. Usually, this choice is simply a mathematical idealization to make the analysis of classical procedures easier, rather than an accurate formulation of a researcher's or a decision maker's problem.

In a more realistic approach, H_0 would be an interval, which can have positive probability in a continuous distribution. If the alternative to $\theta = \theta_0$ is one-sided, say $\theta > \theta_0$, then it may make sense to change H_0 to $\theta \leq \theta_0$, even in the approach of significance testing, since a sample that is strong evidence against $\theta = \theta_0$ is even stronger evidence against a θ which is less than θ_0.

EXAMPLE 9.5a Consider testing $p \leq .5$ against $p > .5$ in Ber(p), based on Y, the number of successes in 15 independent trials. Suppose the prior for p is Beta(9, 5), with p.d.f. $g(p) \propto p^8(1-p)^4$. In this prior, before collecting any data, the probability of the null hypothesis is

$$P_g(p \leq .5) = \frac{1}{B(9,5)} \int_0^{.5} p^8(1-p)^4 dp \doteq .133.$$

[We evaluated the integral using statistical software, although one could multiply out $p^8(1-p)^4$ and evaluate the integral of each term by hand.]

Now, suppose we find $Y = 12$. With the observation of 12 successes in 15 trials, the likelihood function is $p^{12}(1-p)^3$, so the posterior p.d.f. for p is

$$h(p \mid Y = 12) \propto L(p)g(p) \propto p^{20}(1-p)^7.$$

Thus, given $Y = 12$, $p \sim$ Beta(21, 8), and

$$P_h(H_0 \mid Y = 12) = P_h(p \leq .5) = .0063.$$

Similarly, with a uniform prior, $P(H_0 \mid Y = 12) = .011$; and if the prior is Beta(4, 8), then $P(H_0 \mid Y = 12) = .163$.

According to the significance tester, with 12 successes in 15 trials, the P-value is $P(Y \geq 12 \mid p = .5) = .0175$. This measure of disagreement of sample result and model is not intended to be interpreted as the probability of the null hypothesis—note the role of the p in each probability statement. On the other hand, there may be some prior for which $P(H_0) = .0175$. ∎

To handle a decision problem, a Bayesian needs a loss function. (Indeed, who does not? A rational comparison of choices is impossible without knowledge of the costs involved.) Suppose that when the state of nature is θ, taking action a results in a loss (or negative gain) in the amount $\ell(\theta, a)$. If the distribution of Θ is defined by a p.f. or p.d.f. g, the expected loss is $E_g[\ell(\theta, a)] = B(a)$, termed the *Bayes loss*. This defines an ordering of actions and we can then choose the action with the *smallest* Bayes loss. To take data into account, we simply use the posterior distribution of Θ given the data, rather than the prior, when calculating Bayes losses. The action with the minimum Bayes loss is then the Bayes action

EXAMPLE 9.5b In the setting of the preceding example, suppose we assume the following losses:

$$\ell(p, \text{rej}) = \begin{cases} .9 - 2p, \ p < .45, \\ 0, \ p > .45, \end{cases}$$

and

$$\ell(p, \text{acc}) = \begin{cases} 0, \ p < .55, \\ p - .55, \ p > .55. \end{cases}$$

According to this loss function, shown in Figure 9-2, no loss is incurred for either action when p is between .45 and .55. Also, when $p > .55$, accepting H_0 is incorrect, and the loss in accepting H_0 is proportional to how much larger than .55 it is. When $p < .45$, rejecting H_0 is incorrect, and the loss in rejecting H_0 is proportional to how much smaller than .45 it is.

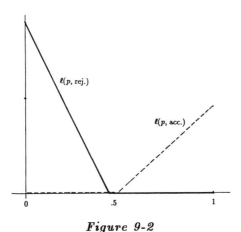

Figure 9-2

Without the data, the Bayes losses are as follows, where E_g denotes prior expectation:

$$B(\text{rej}) = E_g[\ell(p, \text{rej})] = \int_0^{.45} (.9 - 2p) \frac{p^8 (1-p)^4}{B(9, 5)} dp \doteq .016,$$

$$B(\text{acc}) = E_g[\ell(p, \text{acc})] = \int_{.55}^1 (p - .55) \frac{p^8 (1-p)^4}{B(9, 5)} dp \doteq .47.$$

Thus, with the given prior, the Bayesian would reject H_0, the action that incurs the smaller expected loss. Using the data of the preceding example (12 successes in 15 tries), similar calculations result in a posterior in which $B(\text{rej}) = .0008$ and $B(\text{acc}) = .47$. ∎

PROBLEMS

9-32. In the setting of Example 9.5a, with the null hypothesis H_0: $p \leq \frac{1}{2}$ and the alternative H_A: $p > \frac{1}{2}$, we observe eight successes in ten independent trials.
 (a) Find the prior probability of H_0, given the prior $g(p) \propto p^7(1-p)^7$.
 (b) Give the likelihood function.
 (c) Find the posterior probability of H_0 given the prior in (a).
 (d) Find the P-value (for the given data) in testing $p = \frac{1}{2}$ vs. $p > \frac{1}{2}$.
 (e) Find the posterior probability of H_0 given $g(p) \propto p^3(1-p)^9$.

9-33. Consider the null hypothesis H_0: $.48 < p < .52$, where p is the parameter of a Bernoulli distribution. Given a uniform prior on $(0, 1)$, find the probability of the null hypothesis
 (a) in the prior distribution.
 (b) after we observe four successes in five independent trials.

9-34. Suppose a minicourse is directed at an aptitude test whose average score without the special course has been found to be 500. To evaluate its effectiveness we test H_0: $\mu \leq 500$ vs. H_A: $\mu > 500$, where μ is the mean aptitude score of those who have taken the special course. With this prior for μ: $\mu \sim \mathcal{N}(520, 400)$, and assuming that $\sigma = 80$,
 (a) find the probability of the null hypothesis in the prior distribution.
 (b) find the probability of H_0 after a random sample of 50 students yields $\bar{X} = 513$. [See (4) and (5) in §8.9.]
 (c) find a P-value for testing $\mu = 500$, given the data in (b).
 (d) find the probability that the mean increase is at least 10, given the data in (b).

9-35. A uniform prior for μ in the preceding problem is improper, but using a formal calculation of posterior as prior times likelihood you'll find a proper distribution as posterior. With this, find the probability of H_0 given the data in that problem.

9-36. With a random sample of size $n = 10$ from Poi(λ) for testing H_0: $\lambda \leq 5$ against H_A: $\lambda > 5$, we find $\sum X_i = 40$. If $\lambda \sim$ Gam(10,2) prior to obtaining the data, find $P(H_0|\text{data})$. [Hint: See Problem 6-56 and Example 6.5c.]

9-37. In five trials, tossing a coin until heads appears, we observe 2, 0, 1, 1, 0, as the numbers of tails before the first appearance of heads. Find $P(p \leq \frac{1}{2})$,
 (a) assuming the prior distribution of p is $\mathcal{U}(0, 1)$.
 (b) assuming the prior distribution of p is Beta(2, 5).

9.6 Normal populations—the variance

Intuition suggests tests for normal parameters, and the likelihood ratio technique leads to essentially the same tests; it is these that are in common use and will be studied in this and the next section, as well as in the later chapter on linear models.

Because the distributions of usual estimators of σ^2 in a normal population do not involve μ, the theory for tests of hypotheses about σ^2 is simpler than those for hypotheses about μ. Yet, testing $\sigma^2 = v_0$ against $\sigma^2 = v_1$ is *not* a problem of testing one simple hypothesis versus another, because specifying the variance does not define a normal population. The likelihood ratio is

$$\Lambda^* = \frac{L(\bar{X}, v_0)}{L(\bar{X}, v_1)} = \left(\frac{v_1}{v_0}\right)^{n/2} \exp\left\{-\frac{1}{2}(n-1)S^2\left(\frac{1}{v_0} - \frac{1}{v_1}\right)\right\}.$$

When $v_1 > v_0$, the inequality $\Lambda^* < K$ is equivalent to $S^2 > K'$. Thus, since $(n-1)S^2/\sigma^2 \sim \text{chi}^2(n-1)$, P-values for given data and critical values of S^2 for a given α can be found in the chi-square table.

A test can also be based on the sample range, using the distribution of R/σ discussed in Problem 7-32. Intuition suggests that the critical region $R > K$ is suitable for testing $\sigma = \sigma_0$ against $\sigma > \sigma_0$. For a given critical value K, the type I error size is

$$\alpha = P(R > K \mid \sigma = \sigma_0) = 1 - F_W(K/\sigma_0),$$

where $W = R/\sigma$, the c.d.f. for which is given in Table XIII of the Appendix. This would also be the P-value if one were to observe $R = K$, since it is *large* values of R that tend to point to the alternative $\sigma > \sigma_0$.

Situations calling for gathering evidence against a hypothesis of the form $\sigma = \sigma_0$ are rare. However, there may be a need to test such hypotheses, as a guide in making decisions in the manufacture of products whose variability is to be controlled.

EXAMPLE 9.6a An "R-chart" is often used in conjunction with an "$\bar{X}$-chart" in controlling a continuing production process. After each of a sequence of samples, say of size $n = 5$, the mean and range of the latest sample are plotted on the charts. On each chart there is an upper control limit (UCL) and a lower control limit (LCL). When a point falls outside these limits, corrective action is called for. In the case of the range R, one might take UCL $= (a_n + 2b_n)\sigma_0$ and LCL $= (a_n - 2b_n)\sigma_0$, where σ_0 is the specified value of the standard deviation and a_n and b_n are as defined in Example 8.1c and given in Table XIII. (These are termed "2-sigma" limits, being two standard deviations of R on either side of its mean.) If it can be assumed that the population is $\mathcal{N}(\mu, \sigma^2)$, the probability of finding R between these control limits (when $n = 5$) is

$$P(.598 < R/\sigma_0 < 4.054) \doteq .955,$$

determined with the aid of Table XIII. So plotting each point on the chart amounts to a test of $\sigma = \sigma_0$ with $\alpha \doteq .045$. ∎

Consider next the comparison of two variances based on independent random samples. Rather than look at the difference of variances, we take their *ratio* to be the relevant combination, since the variance is a scale parameter. Accordingly, we phrase the null hypothesis $\sigma_1^2 = \sigma_2^2$ in the form $\sigma_1^2/\sigma_2^2 = 1$. The obvious sample statistic to use in a test is the corresponding ratio of the sample variances, S_1^2 and S_2^2, and these (except for constant factors) are independent chi-square variables. So we first need to examine the distribution of such ratios.

Let U and V denote independent chi-square variables with k and m degrees of freedom, respectively. Their joint p.d.f. is the product of their marginal p.d.f.'s:

$$f_{U,V}(u, v) = \frac{2^{-(k+m)/2}}{\Gamma(k/2)\Gamma(m/2)} u^{k/2-1} v^{m/2-1} e^{-(u+v)/2}, \quad u > 0, \; v > 0.$$

Using this we can find the distribution of the ratio U/V. The c.d.f. is

$$F_W(w) = P(W < w) = P\left(\frac{U}{V} < w\right)$$

$$= \iint_{u/v < w, \; u > 0} f_{U,V}(u, v) \, du \, dv$$

$$= \frac{2^{-(k+m)/2}}{\Gamma(k/2)\,\Gamma(m/2)} \int_0^\infty e^{-v/2} v^{m/2-1} \left\{ \int_0^{vw} e^{-u/2} u^{k/2-1} \, du \right\} dv.$$

To obtain the p.d.f. we differentiate with respect to w. The outer integral has constant limits, so we move the derivative inside. The inner integral involves w only in the upper limit, so the derivative is the integrand times the derivative of the upper limit with respect to w:

$$f_W(w) = \frac{2^{-(k+m)/2} w^{k/2-1}}{\Gamma(k/2)\Gamma(m/2)} \int_0^\infty e^{-v(1+w)/2} v^{(k+m)/2-1} \, dv.$$

The integral is a multiple of a gamma function, namely,

$$\frac{2^{(k+m)/2}}{(1+w)^{(k+m)/2}} \Gamma\left(\frac{k+m}{2}\right),$$

so the p.d.f. of W reduces to

$$f_W(w) = \frac{w^{k/2-1}}{B\left(\frac{k}{2}, \frac{m}{2}\right)(1+w)^{(k+m)/2}}, \quad w > 0. \tag{1}$$

This defines a distribution that is related to the beta distribution family, one of the type obtained in Problem 6-65 (page 178). Its mean is $\frac{k}{m-2}$.

For the ratio of sample variances we need the ratio of independent chi-square variables each divided by its number of degrees of freedom. Thus, let

$$F = \frac{U/k}{V/m} = \frac{m}{k} W. \tag{2}$$

The p.d.f. of F derives from that of W (see Theorem 7, page 64):

$$f_F(x) = \frac{k}{m} f_W\left(\frac{k}{m} x\right) \propto \frac{x^{k/2-1}}{\left(1+\frac{k}{m} x\right)^{(k+m)/2}}, \quad x > 0. \tag{3}$$

This p.d.f. defines an **F-distribution** with k numerator degrees of freedom and m denominator degrees of freedom, and we may write $F \sim F(k, m)$. The expected value of F exists for $m > 2$:

$$E(F) = E\left(\frac{m}{k} W\right) = \frac{m}{k} E(W) = \frac{m}{m-2}.$$

Some percentiles of the F-distribution are given Table IV of the Appendix. It is only necessary to give the lower or the higher percentiles, since the former imply the latter. This follows from the fact that if $X \sim F(k, m)$, then $1/X \sim F(m, k)$. Thus, for instance,

$$.05 = P(X < F_{.05}) = 1 - P(X > F_{.05}) = 1 - P\left(\frac{1}{X} < \frac{1}{F_{.05}}\right),$$

so the fifth percentile of X is the reciprocal of the ninety-fifth percentile of an F-distribution with degrees of freedom reversed.

Returning now to the problem of testing $\sigma_1 = \sigma_2 = \sigma$, we observe that under this null hypothesis, the distribution of

$$F = \frac{S_1^2}{S_2^2} = \frac{\dfrac{(n_1-1)S_1^2}{\sigma^2} \div (n_1-1)}{\dfrac{(n_2-1)S_2^2}{\sigma^2} \div (n_2-1)} \tag{4}$$

EXAMPLE 9.6b Determinations of percent nickel in solution by an alcohol method (Al) and by an aqueous (Aq) method were made with results as follows:

Al: 4.28, 4.32, 4.32, 4.29, 4.31, 4.35, 4.32, 4.33, 4.28, 4.27, 4.38, 4.28
Aq: 4.27, 4.32, 4.29, 4.30, 4.31, 4.30, 4.30, 4.32, 4.28, 4.32.

The variances, respectively, are 1.081×10^{-3} and 2.989×10^{-4}, and the test ratio is

$$F = \frac{S_{Al}^2}{S_{Aq}^2} = \frac{10.81}{2.989} \doteq 3.6.$$

Since $F_{.95}(11, 9) \doteq 3.1$, the result is significant at $\alpha = .05$ in a one-tail test. Of course, had we taken the variance ratio the other way, the end result would be the same: $1/F = 2.989/10.81 \doteq 1/3.6$, and the five percent critical value of $F(9, 11)$ is $1/3.1$. ∎

A likelihood ratio test of $\sigma_1 = \sigma_2 = \sigma$ is equivalent to a two-sided test based on the ratio of sample variances, as we show next. The m.l.e.'s under H_0 are the sample means together with

$$\hat{\sigma}^2 = \frac{(n_1-1)S_1^2 + (n_2-1)S_2^2}{n_1 + n_2}.$$

Under $H_0 \cup H_A$, where H_A is the two-sided alternative $\sigma_1 \neq \sigma_2$, the m.l.e.'s are the sample means and sample variances. Then (see Problem 9-45)

$$\Lambda^2 \propto \frac{Z^{n_1}}{(1+Z)^{n_1 + n_2}}, \quad \text{where } Z = \frac{(n_1-1)S_1^2}{(n_2-1)S_2^2}. \tag{5}$$

A critical region of the form $\Lambda < K$ defines a two-sided critical region for Z, or equivalently for F defined by (4), calling for rejection of H_0 if $F < K'$ or $F > K''$. This is evident in the sketch in Figure 9-3, showing the typical kind of relation between Λ and F. The significance level of the region shown is

$$\alpha = P(F < K' \mid H_0) + P(F > K'' \mid H_0).$$

It is not easy to determine K' and K'' to achieve a specified α, so in practice one aribtrarily chooses these critical constants so that each term is $\alpha/2$.

As a final note: It has been found in empirical studies that the assumption of population normality, which has based our derivation of the distribution of F, is fairly critical. The above test is *not* robust in the sense that when the populations are not normal, the F-distribution may not give even an approximate relation between error sizes and critical values. (Some statisticians go so far as to say one should not use the test!)

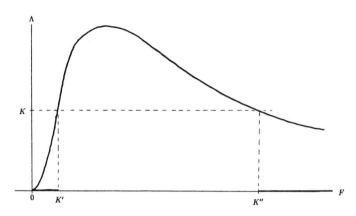

Figure 9-3

PROBLEMS

9-38. An experiment conducted to check the variability in explosion times of detonators intended to explode simultaneously resulted in the following times to detonation in microseconds:

$$2.689,\ 2.677,\ 2.675,\ 2.691,\ 2.698,\ 2.694,\ 2.702,$$
$$2.698,\ 2.706,\ 2.692,\ 2.691,\ 2.681,\ 2.700,\ 2.698.$$

The specification is that $\sigma \leq .007$; is it met? Test H_0: $\sigma = .007$ against the alternative $\sigma > .007$, assuming a normal population and a random sample,
(a) using the sample standard deviation.
(b) using the sample range.
(Give your conclusions as P-values.)

9-39. Consider random samples of size $n = 10$ from $N(\mu, \sigma^2)$ for testing H_0: $\sigma = 3$.
(a) Determine K_1 and K_2 so that each of the critical regions $[S > K_1]$ and $[R > K_2]$ has $\alpha = .1$.

(b) Find the probability, for each of the tests in (a), that H_0 is accepted when $\sigma = 5$.

9-40. Consider random samples from $N(\mu, \sigma^2)$.
(a) Show that for large n, $S^2 \approx N(\sigma^2, 2\sigma^4/n)$.
(b) Construct a large-sample, level-α test for $\sigma^2 = \sigma_0^2$ vs. $\sigma^2 > \sigma_0^2$.
(c) Find the P-value when $n = 80$, $\sigma_0^2 = 4$, and we observe $S^2 = 5$.

9-41. Construct a large sample test for equal population variances based on the difference between sample variances and apply it in the following situation: A study of intellectually gifted junior high school students reported statistics on math scores for boys and for girls:

	Number	Mean	S.D.
Boys	2046	436	87
Girls	1628	404	77

9-42. Amounts of water filtered with a particular filtration device using two methods of operation were recorded and these summary statistics calculated:

	Number	Mean	S.D.
Method 1	20	202.0	74.35
Method 2	29	278.3	79.03

Test the hypothesis that the variances are the same under the two methods using the ratio of the sample variances as a test statistic.

9-43. Calculate the variance of the F-distribution. [Cf. (17) in §6.7.]

9-44. Show that if $T \sim t(\nu)$, then $T^2 \sim F(1, \nu)$. [See (2), §8.9.]

9-45. Obtain the expression (5) for Λ^2.

9-46. Use the transformation formula [Theorem 8, §3.5, page 65] to find the p.d.f. of $1/F(n_1, n_2)$, thus demonstrating directly that this is the p.d.f. of $F(n_2, n_1)$.

9.7 Normal populations—the mean

For testing H_0: $\mu = \mu_0$, intuition suggests that a test statistic should be based on the deviation $\bar{X} - \mu_0$. Large positive values of $\bar{X} - \mu_0$ would point to $\mu > \mu_0$ and large negative values, to $\mu < \mu_0$. The likelihood ratio approach yields such a test statistic: The numerator of Λ is the maximum likelihood with respect to σ^2, holding μ fixed at μ_0:

9.7 Normal populations—the mean

$$\sup_{\sigma^2} L(\mu_0, \sigma^2) = L\left(\mu_0, \frac{1}{n}\sum(X_i - \mu_0)^2\right).$$

For the alternative $\mu \neq \mu_0$, we need the maximum of the likelihood function over both parameters for the denominator of Λ:

$$\sup_{(\mu, \sigma^2)} L(\mu, \sigma^2) = L(\bar{X}, V),$$

where V is the second central sample moment (the biased version of sample variance). The likelihood ratio is thus

$$\Lambda = \left\{1 + \frac{(\bar{X} - \mu_0)^2}{V}\right\}^{-n/2}.$$

Small values of Λ provide evidence against $\mu = \mu_0$, as then do *large* values of

$$T^2 = \frac{(\bar{X} - \mu_0)^2}{S^2/n},$$

where S^2 is the (unbiased) sample variance: $S^2 = \frac{n}{n-1}V$.

To use T^2 as a test statistic we need to know its distribution. To derive this we first observe that it can be expressed in terms of the ratio of two chi-square variables:

$$T^2 = \frac{\frac{(\bar{X} - \mu_0)^2}{\sigma^2/n}}{\frac{(n-1)S^2}{\sigma^2} \div (n-1)}. \tag{1}$$

Under the hypothesis $\mu = \mu_0$ the numerator is the square of a standard normal variable, and hence is chi$^2(1)$. The denominator is a chi$^2(n-1)$ variable divided by its number of degrees of freedom (whether or not $\mu = \mu_0$). These chi-square variables are independent [by the Corollary to Theorem 10 of Chapter 7, page 212], so T^2 is of the form (2) in the preceding section. Thus, $T^2 \sim F(1, n-1)$. Moreover, the symmetrically distributed square root has a t-distribution:

$$T = \frac{\bar{X} - \mu_0}{S/\sqrt{n}} \sim t(n-1), \tag{2}$$

as we saw in §8.9. (See also Problem 9-44.) This statistic has the form of a "Z-score," except that its null distribution is approximately standard normal only for large samples. Recall (from §8.9) that for large n, the t-distribution is closely approximated by a normal distribution.

In summary, the distribution of T given by (2) is exactly $t(n-1)$

under H_0 (and the general assumptions of this section—normal population, random sampling); and it is nearly normal for very large samples T—even when the population is not normal. Moreover, it has been found in empirical studies that when n is not large, $t(n-1)$ is a fairly good approximation to the actual distribution of T when the population is *not* normal, provided the nonnormality is not too extreme.

In a **t-test** of $\mu = \mu_0$ against $\mu \neq \mu_0$, large values of T^2 are taken as extreme, and P-values are found as right-tail areas of $F(1, n-1)$. Working instead with T, one would take either large negative or large positive values of T as extreme, and perhaps double the single tail area beyond an observed T to define a P-value. And in responding to a specified α, the t-critical region would be defined with $\alpha/2$ in each tail of the null distribution of T.

On the other hand, suppose our alternative hypothesis is one-sided, say, $\mu > \mu_0$, a t-test would regard only large positive values of T as extreme and define P-value as one-sided. And an α, specified for a one-sided test against $\mu > \mu_0$, would be put into the right-hand tail of the null distribution of T. (One-sided tests are not implied by our derivation of the likelihood ratio test for a two-sided alternative.)

Incidentally, the two-sided t-test we've described is equivalent to a test which rejects μ_0 if it is not included in a two-sided confidence interval at the confidence level $1 - \alpha$.

EXAMPLE 9.7a In a drug study, a clinician found that an intravenous injection to ten heart patients reduced the number of ventricular premature beats by the following amounts: 0, 7, –2, 14, 15, 14, 6, 16, 19, 26. For a test of $\mu = 0$ against $\mu > 0$, we calculate (2), assuming that the population is close enough to normal that under H_0, $T \approx t(9)$:

$$T = \frac{11.5 - 0}{8.67/\sqrt{10}} \doteq 4.2.$$

The P-value is less than .002 (d.f. $= 9$), fairly strong evidence against $\mu = 0$ (assuming random sampling). ∎

In Example 7.6d (page 218) we pointed out that for large random samples from $\mathcal{N}(\mu, \sigma^2)$, the *median* of a random sample is approximately normal: $\widetilde{X} \approx \mathcal{N}(\mu, \pi\sigma^2/(2n))$. In testing $\mu = \mu_0$, we may then use the Z-score

$$Z = \frac{\widetilde{X} - \mu_0}{\sqrt{\pi}S/\sqrt{2n}}$$

in conjunction with the standard normal table. The denominator is larger than the denominator of the Z-score based on $\bar{X}$, so a deviation from μ_0 equal to the deviation of $\bar{X}$ from μ_0 would yield a smaller Z-score. Nevertheless, basing the test on the median might be preferred if there is a possibility of "outliers" in the data—observations that are corrupted by recording or other errors, and not really from the true population distribution. (The median does not depend on the values of the observations in the tails of the sample distribution.)

We turn now to the two-sample problem of testing the equality of means of two normal populations, using independent random samples. We use the notation of the preceding section, and denote the two sample means by $\bar{X}_1$ and $\bar{X}_2$. The simplest case is that in which the two population variances are equal: $\sigma_1^2 = \sigma_2^2 = \sigma^2$. Under H_0: $\mu_1 = \mu_2 = \mu$, the observations in the two samples constitute a random sample of size $n_1 + n_2$ from $N(\mu, \sigma^2)$, and the m.l.e.'s are the mean and variance (the biased version, denoted by V in §7.2) of the combined sample:

$$\hat{\mu} = \frac{n_1 \bar{X}_1 + n_2 \bar{X}_2}{n_1 + n_2}, \quad \hat{\sigma}_0^2 = \frac{n_1 V_1 + n_1(\bar{X}_1 - \hat{\mu})^2 + n_2 V_2 + n_2(\bar{X}_2 - \hat{\mu})^2}{n_1 + n_2}.$$

The maximum (sup) of the likelihood function under H_0, $L(\mu, \mu; \sigma^2)$, is

$$L(\hat{\mu}, \hat{\mu}; \hat{\sigma}_0^2) = (\hat{\sigma}_0^2)^{-(n_1+n_2)/2} e^{-(n_1+n_2)/2} \tag{3}$$

And under $H_0 \cup H_A$, the maximum of the likelihood function, $L(\mu_1, \mu_2; \sigma^2)$ is

$$L(\bar{X}_1, \bar{X}_2; \hat{\sigma}^2) = (\hat{\sigma}^2)^{-(n_1+n_2)/2} e^{-(n_1+n_2)/2}, \tag{4}$$

where

$$\hat{\sigma}^2 = \frac{n_1 V_1 + n_2 V_2}{n_1 + n_2}. \tag{5}$$

Dividing (3) by (4) we find $\Lambda = \hat{\sigma}_0^2 / \hat{\sigma}^2$, and from this,

$$\Lambda^{-2/(n_1+n_2)} = 1 + \frac{n_1(\bar{X}_1 - \hat{\mu})^2 + n_2(\bar{X}_2 - \hat{\mu})^2}{n_1 V_1 + n_2 V_2}. \tag{6}$$

This is large and Λ is small when the statistic

$$T^2 = \frac{n_1(\bar{X}_1 - \hat{\mu})^2 + n_2(\bar{X}_2 - \hat{\mu})^2}{(n_1 V_1 + n_2 V_2)/(n_1 + n_2 - 2)} \tag{7}$$

is large. With some simple algebra we rewrite the numerator:

$$n_1(\bar{X}_1 - \hat{\mu})^2 + n_2(\bar{X}_2 - \hat{\mu})^2 = \frac{(\bar{X}_1 - \bar{X}_2)^2}{\frac{1}{n_1} + \frac{1}{n_2}}, \tag{8}$$

and thus find that

where

$$T^2 = \frac{(\bar{X}_1 - \bar{X}_2)^2}{S_p^2\left(\frac{1}{n_1} + \frac{1}{n_2}\right)}, \tag{9}$$

$$S_p^2 = \frac{n_1 V_1 + n_2 V_2}{n_1 + n_2 - 2} = \frac{(n_1 - 1)S_1^2 + (n_2 - 1)S_2^2}{n_1 + n_2 - 2}, \tag{10}$$

and S_i^2 is the unbiased version of the variance of sample i. The statistic S_p^2 is called the **pooled variance**, an unbiased estimate of σ^2.

To interpret the value of the two-sample T^2 defined by (9), in a particular case, we need to know at least its null distribution. For this we observe that $T^2 = Z^2/U$, where

$$Z = \frac{\bar{X}_1 - \bar{X}_2}{\sigma\sqrt{\frac{1}{n_1} + \frac{1}{n_2}}}, \quad U = \frac{\frac{(n_1 - 1)S_1^2}{\sigma^2} + \frac{(n_2 - 1)S_2^2}{\sigma^2}}{(n_1 + n_2 - 2)}. \tag{11}$$

In (11), $Z \sim \mathcal{N}(0, 1)$ when $\mu_1 = \mu$ and $\sigma_1^2 = \sigma_2^2 = \sigma^2$. The numerator of U is the sum of two independent chi-square variables with $n_1 - 1$ and $n_2 - 1$ degrees of freedom, respectively, so $U \sim \text{chi}^2(n_1 + n_2 - 2)$. Moreover, Z^2 and U are independent, since they are, respectively, functions of means and of variances. (We know these to be independent in the case of normal populations, again from the corollary to Theorem 10, §7.5.) So, under the null hypothesis,

$$T^2 = \frac{Z^2}{U} \sim F(1, n_1 + n_2 - 2),$$

and we use the F-table in interpreting the value of T^2 in a particular case—or in setting the critical value of T^2 to achieve a specified α.

We have arrived at T^2 as appropriate for the two-sided alternative $\mu_1 \neq \mu_2$, but (as in the one-sample case) the square root

$$T = \frac{\bar{X}_1 - \bar{X}_2}{S_p\sqrt{1/n_1 + 1/n_2}}$$

is intuitively suitable for a one-sided alternative. Under the null hypothesis, $T \sim t(n_1 + n_2 - 2)$ [again see §8.9], and for H_A: $\mu_1 < \mu_2$, values of T in the left tail of this null distribution argue against H_0. Although we derived a critical region for T^2 using the likelihood ratio method and our intuition that small Λ-values point to H_A, the result is a test that seems intuitively correct on the face of it. As in the one-sample case, the statistic T is a standardized measure of discrepancy between data and null hypothesis model, where the standardization is provided by the **standard error** of the difference $\bar{X}_1 - \bar{X}_2$ (as in Problem 8-9):

$$\text{s.e.}(\bar{X}_1 - \bar{X}_2) = \sqrt{\frac{S_p^2}{n_1} + \frac{S_p^2}{n_2}}. \tag{12}$$

Indeed, with this standard error, one can construct confidence limits for the difference in population means using the $t(n_1 + n_2 - 2)$ distribution.

EXAMPLE 9.7b The following are determinations of rate of flow of a certain gas through two different soil types:

Soil A: 21, 28, 19, 24, 22, 30
Soil B: 21, 29, 34, 32, 26, 28, 36, 26.

The sample means are $\bar{X}_1 = 24$, $\bar{X}_2 = 29$, and the standard deviations, $S_1 = 4.243$ and $S_2 = 4.870$. The pooled variance is

$$S_p^2 = \frac{5 \times 4.243^2 + 7 \times 4.870^2}{6 + 8 - 2} = 21.335 = 4.619^2,$$

so the two-sample t-statistic is

$$T = \frac{24 - 29}{4.619\sqrt{1/6 + 1/8}} \doteq -2.0.$$

The corresponding P-value is the area under the p.d.f. of $t(12)$ to the left of -2.0, or about .034—assuming that the populations are not too non-normal and have equal variances. ■

When it is not reasonable to assume equal population variances, the problem of comparing means is called the *Behrens-Fisher* problem, with a history of proposed solutions and controversy.[7] There is no universally accepted solution, but some statistical software packages incorporate one or the other of the various possibilities.

In Example 9.7a, the one-sample t-test was applied to *differences* from "before and after" data, in an experiment where each subject provided both control and treatment responses. But we could not compare the set of control measurements and the set of treatment measurements in a two-sample t-test because the measurements on a given individual are correlated: The samples are not independent. Such pairing of control and treatment measurements is often created by design as a device intended to increase sensitivity of the test by eliminating, at least partially, individual differences that contribute to variability and

---

[7]See H. Scheffé, "Practical solutions of the Behrens-Fisher problem," *J. Am. Stat. Assn.* 65 (1970), p. 1501.

tend to obscure treatment effects. Thus, one might use twins, pigs from the same litter, or even pairs of individual subjects that are matched on various characteristics that might affect the response, and randomly assign one member of each pair to the treatment and the other to be a control. In §9.11 we'll see how and under what circumstances this kind of design can be effective in increasing sensitivity in terms of "power."

PROBLEMS

9-47. An early study on the effects of smoking marijuana produced the following changes in scores in a "digit substitution" test, before and after a marijuana smoking session: 5, −17, −7, −3, −7, −9, −6, 1, −3. Assuming that the population is not too far from normal, carry out a t-test of the hypothesis that the mean decrease in score is zero.

9-48. Verify in several instances that the 95th percentiles of $F(1,k)$ are the squares of the 97.5 percentiles of $t(k)$.

9-49. Carry out the algebra
(a) verifying that the m.l.e. of σ^2 for a random sample from $N(\mu_0, \sigma^2)$ is what is asserted at the beginning of §9.7: $\frac{1}{n}\sum(X_i - \mu_0)^2$.
(b) that leads from (7) to (8) (page 325), in our development of the two-sample T-statistic for comparing normal population means.

9-50. Given the data in Example 9.7b, find 90 percent confidence limits for the mean difference in flow rates for the two types of soil.

9-51. Given the data in Problem 9-42, construct 95 percent confidence limits for the difference in population means.

9-52. In a study of the effect of tracheal muscle stimulation on the strength of the trachea, six newborn lambs were used as subjects. For each lamb, the tracheal compliance was measured before and after stimulation:

Lamb #	1	2	3	4	5	6
Before	.029	.043	.022	.012	.020	.034
After	.020	.011	.008	.005	.009	.009

Test the hypothesis of "no treatment effect." [See Example 9.7a and the discussion just ahead of this problem set.]

9-53. A 1969 study measured antero-posterior chest diameters of 16 "normal" males and 25 male patients with emphysema, with these results:

	n	$\bar{X}$	S
Normal	16	20.2	2.0
Emphysema	25	23.0	2.4

Carry out a test of the hypothesis of no difference in average chest diameters of normal males and males with emphysema.

9-54. A study of the effects of anesthesia reported the heart rates of four pigs under anesthesia: 116, 85, 118, 118. The normal mean heart rate for pigs is 114. Test the hypothesis of no (mean) effect of the anesthesia, using the four given rates, and again with the 85 deleted. Comment on the strange results.

9.8 Simple H_0 vs. simple H_A

The more specific structure of a test of hypotheses as a two-action decision problem with data lends itself to a rather detailed theoretical study, and it is this we take up in most of what follows in this chapter. We study first the case in which both H_0 and H_1 are simple:

$$H_0: f(\mathbf{x}) = f_0(\mathbf{x}) \quad \text{vs.} \quad H_1: f(\mathbf{x}) = f_1(\mathbf{x}),$$

where f_0 and f_1 are completely defined p.d.f.'s or p.f.'s.

DEFINITION For testing simple H_0 vs. simple H_A,
The *size* of the type I error is $\alpha = P(\text{reject } H_0 \mid H_0)$;
the *size* of the type II error is $\beta = P(\text{accept } H_0 \mid H_1)$.

Given a critical region C, the error sizes are found as sums or integrals. In the continuous case,

$$\alpha = \int_C f_0(\mathbf{x})\,d\mathbf{x}, \quad \beta = 1 - \int_C f_1(\mathbf{x})\,d\mathbf{x}. \tag{1}$$

(We have previously referred to α as the *size of the critical region*.) Clearly, if C is enlarged, α will increase and β decrease.

EXAMPLE 9.8a Consider the problem of testing the simple hypothesis $H_0: X \sim \mathcal{N}(0, 4)$ against the simple alternative $H_1: X \sim \mathcal{N}(1, 4)$, using a random sample of size 25. This is the setting of Example 9.2a except that now $\sigma^2 = 4$. In this case, $\Lambda^* \propto \exp(-\Sigma X_i/4)$, which is small when $\bar{X}$ is large. So in a likelihood ratio test we use critical regions of the form $[\bar{X} > K]$. Since $\bar{X} \sim \mathcal{N}(\mu, 4/25)$, the error sizes for these tests are

$$\alpha = P(\bar{X} > K \mid \mu = 0) = 1 - \Phi\left(\frac{K-0}{2/5}\right), \tag{2}$$

$$\beta = P(\bar{X} < K \mid \mu = 1) = \Phi\left(\frac{K-1}{2/5}\right). \tag{3}$$

For the particular critical region $\bar{X} > .4$, Figure 9-4 shows α and β as areas under the p.d.f. for $\bar{X}$, under H_0 and H_1, respectively. As K increases, α decreases and β increases, and as K decreases, α increases and β decreases.

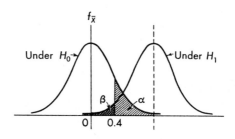

Figure 9-4

Equations (2) and (3) for α and β are parametric equations, in terms of the parameter K, for a curve in the $\alpha\beta$-plane. This is shown in Figure 9-5. In addition to the curve for the case $n = 25$, we show the corresponding curve for the same family of critical regions when $n = 4$. For the larger sample size the curve is closer to $(0, 0)$. Indeed, we ought to find that both error sizes are smaller in the larger sample if we have made intelligent use of the data. Figure 9-5 also shows the $\alpha\beta$-plot, when $n = 4$, for the family of critical regions $\bar{X} < K$, which are counterintuitive. Observe the poor performance. ∎

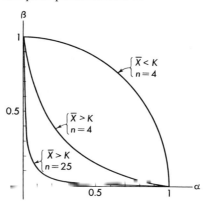

Figure 9-5

9.8 Simple H_0 vs. simple H_1

In searching for optimal tests, we look at all possible tests, including randomized tests. To compare tests we compare their error sizes. For a randomized test ϕ, the error sizes are

$$\begin{cases} \alpha = E[\phi(\mathbf{X}) \mid H_0], \\ \beta = E[1 - \phi(\mathbf{X}) \mid H_1]. \end{cases} \quad (4)$$

[The expected value of a function that is 0 or 1 is just the probability that it has the value 1; in the case of ϕ this expected value is the probability of rejection.]

Figure 9-6 shows a typical set of points representing all tests of a simple H_0 vs. a simple H_1. The type I error size of the test defined by $\phi \equiv \alpha$ is α, and the type II error size is $\beta \equiv 1 - \alpha$; so these tests, which ignore the data, are represented by points on the line $\alpha + \beta = 1$.

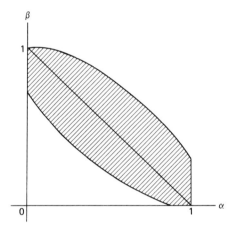

Figure 9-6

THEOREM 2 *The set of $\alpha\beta$-representations of randomized tests of H_0 is convex, is included in the closed unit square, and includes the points $(0, 1)$ and $(1, 0)$.*

[A set is *convex* if and only if contains all line segments joining two of its points. A convex set has no indentations or "dimples."]

To prove Theorem 2, let ϕ_1 and ϕ_2 denote any two tests, with corresponding error sizes (α_1, β_1) and (α_2, β_2). Define the randomized test

$$\phi(\mathbf{x}) = \gamma \phi_1(\mathbf{x}) + (1 - \gamma)\phi_2(\mathbf{x}), \quad 0 \le \gamma \le 1. \quad (5)$$

The error sizes for ϕ are

$$\begin{cases} \alpha = \dot{E}[\phi(\mathbf{X}) \mid H_0] = \gamma\alpha_1 + (1-\gamma)\alpha_2, \\ \beta = E[1 - \phi(\mathbf{X}) \mid H_1] = \gamma\beta_1 + (1-\gamma)\beta_2. \end{cases} \quad (6)$$

Equations (6) are precisely the parametric equations for the line segment joining the $\alpha\beta$-points representing ϕ_1 and ϕ_2. Every point on this line segment corresponds to a value of γ on $[0, 1]$ and defines a test, included in the set of all tests. Then, since line segments joining any two points of the set of $\alpha\beta$ representations of all tests are included in that set, it is convex, as claimed.

Points in the plane are not totally ordered, and we cannot always say that one $\alpha\beta$-pair corresponds to a "better" test than another. However, it is clear that tests that have the same α are ordered according to the value of β, and we should prefer the one with the smallest β.

DEFINITION A test ϕ for simple H_0 vs. simple H_1 is *most powerful* if and only if it has the smallest β among all tests whose α's are no larger than that of ϕ.

(We think of $1 - \beta$, the probability of rejecting H_0 when it should be rejected, as the *power* of the test for detecting H_1; and if β is smallest, then $1 - \beta$ is largest—most powerful.) It should be fairly clear, in examining pictures such as that in Figure 9-6, that the most powerful tests are represented by points on the lower left section of the boundary of the set representing all possible tests. It should be equally clear that tests on that boundary have the smallest α among tests with no larger β.

In the next section we'll see that these "best" tests are likelihood ratio tests, but that leaves the question of which one to use. One way to choose a test is to specify an α. This selects a test on the curve of best tests, controlling the type I error size and letting the type II error size do as best it can with the amount of available data. Another way of choosing a test is available to the Bayesian:

Suppose we have losses for incorrect decisions, as follows: $l(H_0, \text{rej. } H_0) = a$, and $l(H_1, \text{acc. } H_0) = b$, with no loss for a correct decision. And suppose we assume prior probabilities $\gamma_0 = P(H_0)$ and $\gamma_1 = P(H_1)$. For a test with error sizes (α, β), the expected loss is

$$E(\text{loss}) \doteq (a\gamma_0)\alpha + (b\gamma_1)\beta. \quad (7)$$

In the $\alpha\beta$-plane, this has the constant value k on a line with slope $m = -\dfrac{a\gamma_0}{b\gamma_1}$. Increasing k moves the line upward. Now, think of moving a line with slope m upward (from below the set of points representing

9.8 Simple H_0 vs. simple H_1

possible critical regions) until it first makes contact with the set. That point of contact represents the test with the smallest expected loss (7) among tests considered.

EXAMPLE 9.8b In the setting of Example 9.8a and the notation of (7) above, let $a = 20$, $b = 10$, $\gamma_0 = .2$, and $\gamma_1 = .8$. Then the slope of the relevant family of lines is $-4/8$. The point of first contact in moving a line with this slope up to the set representing tests of the form $\bar{X} > K$ is the point at which the boundary curve has the same slope (Figure 9-7):

$$-\frac{1}{2} = \frac{d\beta}{d\alpha} = \frac{d\beta}{dK} \bigg/ \frac{d\alpha}{dK} = \frac{\Phi'[2.5(K-1)]}{-\Phi'[2.5K]}.$$

The standard normal p.d.f. is $\Phi'(z) = (2\pi)^{-1/2} e^{-z^2/2}$, so

$$-\frac{1}{2} = \frac{\exp\{-[2.5(K-1)]^2/2\}}{-\exp\{-[2.5K]^2/2\}},$$

or (upon taking logs)

$$-\frac{1}{2}[2.5(K-1)]^2 = -\frac{1}{2}(2.5K)^2 - \log 2,$$

or $K \doteq .389$. With the given losses and prior probabilities, the Bayes critical region (among those considered) is $\bar{X} > .389$, with $\alpha = .165$ and $\beta = .063$. ∎

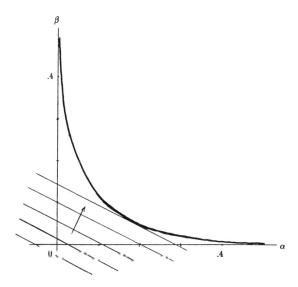

Figure 9-7

PROBLEMS

9-55. Consider testing $p = .5$ against $p = .8$, where p is the parameter of Ber(p), which we sample twice in independent trials. In Problem 9-20 you listed the eight possible tests based on a statistic Y with possible values 0, 1, 2.
(a) Find α and β for each possible test (in the present context).
(b) Plot the tests as points in the $\alpha\beta$-plane, and indicate on this plot the set of $\alpha\beta$-points for all tests, including randomized tests.

9-56. For a Poisson arrival process with rate parameter λ/min, let Y denote the number of arrivals in a ten-minute period. Find α and β for a test with critical region $Y > 12$, in testing $\lambda = 1.0$ against $\lambda = 1.5$.

9-57. Consider the critical region $[X > k]$ for a single observation, to test H_0: $X \sim \mathcal{U}(0,1)$ against H_1: $f(x) = 2x$, $0 < x < 1$.
(a) Find α and β as functions of k.
(b) Plot the $\alpha\beta$-curve defined by the parametric functions in (a).
(c) Find k so that $\alpha = 4\beta$.

9-58. For testing $\mu = 0$ against $\mu = 1$ in $\mathcal{N}(\mu, 4)$, consider critical regions of the form $\bar{X} > K$, as in Example 9.8a.
(a) Find the error sizes in terms of n and K.
(b) Determine n and K so that $\alpha = \beta = .01$.
(c) Determine n and K so that $\alpha = .01$, $\beta = .10$.

9-59. Referring to Example 9.8a, calculate the error sizes α and β for the test that rejects H_0 unless $0 < \bar{X} < 1$. [Notice where the point (α, β) falls in the plot of Figure 9-5.]

9-60. Referring to Example 9.8b, find the Bayes test if the two types of error are equally costly, and H_1 and H_0 are equally likely *a priori*.

9-61. Consider H_0: $X \sim \mathcal{N}(0, 1)$, and H_1: $X \sim \mathcal{N}(0, 4)$. Find the type I and type II error sizes for the critical region $T > 12$, where $T = \sum X^2$, the sum of squares of five independent observations on X.

9.9 The Neyman-Pearson lemma

We continue the setting of simple H_0 and simple H_1, and in the notation of the preceding section. The functions f_0 and f_1 assign numbers to each **x** in the data space. In order to exploit your intuition in constructing a good critical region, it may help to think of f_0 as a "cost" and f_1 as a "return," assigned to each sample point **x**. An optimal critical region is one that maximizes total return for a given total cost.

9.9 The Neyman-Pearson lemma

Choosing a region C so that it has size α amounts to putting points $\mathbf{x}$ into C until the total cost of C is α:

$$\alpha = \int_C f_0(\mathbf{x}) d\mathbf{x} = \text{ total cost of } C. \tag{1}$$

Then, among regions C with this same total cost α, we look for the one with the largest return $1 - \beta$ (or smallest β):

$$1 - \beta = \int_C f_1(\mathbf{x}) d\mathbf{x} = \text{ total return from } C. \tag{2}$$

It is intuitively clear that the way to choose points $\mathbf{x}$ to maximize (2) is to put first into C those points with the largest return per unit cost: $f_1(\mathbf{x})/f_0(\mathbf{x})$, or equivalently, the smallest cost per unit return: $f_0(\mathbf{x})/f_1(\mathbf{x})$. [This is the likelihood ratio encountered of §9.2.] So we line up sample points $\mathbf{x}$ according to the value of this ratio and, starting with points with the smallest ratio f_0/f_1, put into C as many points as possible without exceeding the restriction (1). Thus, the desired region C is one of the form

$$\Lambda^* = \frac{f_0(\mathbf{X})}{f_1(\mathbf{X})} < K, \tag{3}$$

where K is a constant chosen to satisfy (1). This region C should maximize (2), which is "power," subject to the size condition (1). This suggests the following theorem:

NEYMAN-PEARSON LEMMA *In testing f_0 against f_1, the critical regions*

$$C_K = \left\{ \boldsymbol{x} : \frac{f_0(\boldsymbol{x})}{f_1(\boldsymbol{x})} < K \right\}, \tag{4}$$

*where $K > 0$, are **most powerful**. That is, the power of any test $\phi(\boldsymbol{x})$ of size α_ϕ no larger than the size of C_K does not exceed the power of C_K.*

We'll prove this using the notation of randomized tests so as to be sure that the region (4) is most powerful in the class of *all* tests, randomized or not. So let ϕ_K denote the indicator function of C_K, with type I error size α_K and type II error size β_K, and let α_ϕ and β_ϕ denote the corresponding error sizes of a test ϕ for which

$$\alpha_K - \alpha_\phi = E(\phi_K - \phi \mid H_0) \geq 0 \tag{5}$$

We need to show

$$(1 - \beta_K) - (1 - \beta_\phi) \geq 0,$$

or

$$E(\phi_K - \phi \mid H_1) \geq 0 \qquad (6)$$

[using (4) in Section 9.8]. Observe first that for $\mathbf{x} \in C_K$,

$$\phi_K - \phi = 1 - \phi \geq 0, \quad \text{and} \quad f_1 > \frac{f_0}{K},$$

so that for such values of $\mathbf{x}$,

$$(\phi_K - \phi)f_1 \geq \frac{1}{K}(\phi_K - \phi)f_0. \qquad (7)$$

Likewise, for $\mathbf{x} \notin C$,

$$\phi_K - \phi = 0 - \phi \leq 0 \quad \text{and} \quad f_1 \leq \frac{f_0}{K},$$

so that (7) again holds—holds, therefore, for all $\mathbf{x}$. Integrating both sides of (7), we obtain [using (5)] the desired result:

$$E(\phi_K - \phi \mid H_1) \geq \frac{1}{K} E(\phi_K - \phi \mid H_0) \geq 0.$$

EXAMPLE 9.9a We return to the setting of Example 9.8a—now testing $X \sim \mathcal{N}(0, 4)$ against $X \sim \mathcal{N}(\mu, 4)$ where $\mu > 0$, based on a random sample of size $n = 25$. In that example we found that the likelihood ratio tests are of the form $\bar{X} > K$ (a constant). The Neyman-Pearson lemma guarantees that these tests are most powerful. This means that in Figure 9-5, the (α, β) points representing any other tests are to the upper right of the curve which is the locus of the likelihood ratio tests. ■

Consider testing $\theta = \theta_0$ vs. $\theta = \theta_1$, where $\theta_0 < \theta_1$ and θ is the parameter of a member of the exponential family [(2) in §9.2]:

$$f(x \mid \theta) = B(\theta) h(x) \exp[Q(\theta) R(x)].$$

If Q is an increasing function of θ, $Q(\theta_0) < Q(\theta_1)$, then (as we saw in §9.2) the likelihood ratio Λ is a decreasing function of $\sum R(X_i)$. So the Neyman-Pearson critical region is of the form $T = \sum R(X_i) > K$.

EXAMPLE 9.9b Suppose we want to test $\theta = .3$ vs. $\theta = .6$ in Ber(θ) using a random sample of size $n = 3$. For this member of the exponential family,

$$Q(\theta) = \log\frac{\theta}{1-\theta}, \quad R(x) = x.$$

Since Q is an increasing function of θ, the likelihood ratio Λ is a decreasing function of $T = \sum X_i$. So the Neyman-Pearson critical region is of the form $T > K$. Since $T \sim \text{Bin}(3, \theta)$, it has only four possible values, so there are just five critical regions of the form $T > K$. These most powerful regions are shown together with the corresponding error sizes in the following table:

C	α	β
$\emptyset$	0	1
$\{3\}$	.027	.784
$\{3, 2\}$	.216	.352
$\{3, 2, 1\}$	.657	.064
Ω	1	0

If ϕ_1 and ϕ_2 are indicator functions of two successive C's in this list, the randomized test defined as $\phi = \gamma\phi_1 + (1-\gamma)\phi_2$ is represented in the $\alpha\beta$-plot (Figure 9-8) as the line segment joining the points representing the two given C's. As will be shown next, these randomized tests are also most powerful. ∎

NEYMAN-PEARSON LEMMA (extended version): *In testing f_0 against f_1, the randomized test*

$$\phi^* = \begin{cases} 1, & \Lambda^* < K, \\ \gamma, & \Lambda^* = K, \\ 0, & \Lambda^* > K \end{cases}$$

where $K \geq 0$ and $0 \leq \gamma \leq 1$, is most powerful—has the greatest power in the class of tests with no larger α. To achieve a specified $\alpha > 0$,

(a) if there is a K such that $P(\Lambda^* < K) = \alpha$, use that K with $\gamma = 0$;
(b) if not, choose K so that $P(\Lambda^* < K) < \alpha \leq P(\Lambda^* \leq K)$, and let

$$\gamma = \frac{\alpha - P(\Lambda^* < K)}{P(\Lambda^* = K)}.$$

The proof of this extended version is an easy modification of the earlier proof. There, the essential inequality was

$$E(\phi^* - \phi \mid H_1) \geq \frac{1}{K} E(\phi^* - \phi \mid H_0).$$

338   Chapter 9   Testing Hypotheses

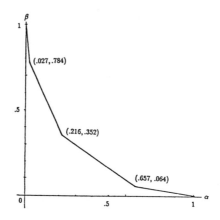

Figure 9-8 α-β plot for Example 9.9b

Again this holds for all **x**: The regions $\Lambda < K$ and $\Lambda > K$ are handled as before; for $\Lambda = K$, $(\phi^* - \phi)f_0/K = (\phi^* - \phi)f_1$. Finding α is straightforward:

$$\alpha = E(\phi^* \mid H_0) = P(\Lambda < K) + \gamma P(\Lambda = K).$$

Likelihood ratio tests of the form $\Lambda^* > K$ are clearly *least* powerful, represented by $\alpha\beta$-points on the *upper* boundary of the set of all tests. Tests of the form $\phi \equiv \gamma$ (a constant) are represented by the line segment joining (0, 1) and (1, 0).

EXAMPLE 9-9c. For the testing problem in Example 9.9b [$\theta = .3$ vs. $\theta = .6$, in Ber(θ), using three independent trials], the *least* powerful nonrandomized tests are of the form $\bar{X} < K$. There are just five of them, defined by the critical regions $\emptyset$, $\{0\}$, $\{0,1\}$, $\{0,1,2\}$, Ω. Figure 9-9 shows the (α, β)-points representing these tests and the line segments joining them, representing all least powerful randomized tests. The region between the least powerful and most powerful tests includes all other tests. ■

Neyman-Pearson tests are based on a comparison of likelihoods, and the likelihood ratio is minimal sufficient (Example 7.8f). And N-P tests include the test we'd get by choosing the hypothesis which has the larger likelihood, that is, by rejecting H_0 when $\Lambda^* < 1$. However, N-P tests with $K \neq 1$ do not follow the maximum likelihood principle, although they have the flexibility of allowing various ways of balancing the two error sizes, α and β.

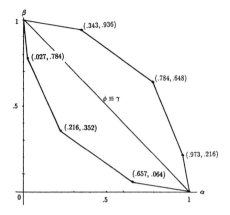

Figure 9-9 α-β plot for Example 9.9c

In this simple vs. simple setting, Bayes tests are Neyman-Pearson tests, and therefore are most powerful: Suppose (as in the preceding section) we assume 0 loss for correct decisions and the following losses for incorrect decisions: $\ell(H_0, \text{rej.}\ H_0) = a$, and $\ell(H_1, \text{acc.}\ H_0) = b$, and prior probabilities γ_0 and γ_1 for f_0 and f_1, respectively. The posterior odds are then $\gamma_0 f_0 : \gamma_1 f_1$, and the expected posterior losses are

$$B(\text{acc.}\ H_0) = \frac{b\gamma_1 f_1}{f_0 \gamma_0 + f_1 \gamma_1}, \quad B(\text{rej.}\ H_0) = \frac{a\gamma_0 f_0}{f_0 \gamma_0 + f_1 \gamma_1}.$$

The Bayes action is the one corresponding to the smaller of these, namely, to reject H_0 if and only if $a\gamma_0 f_0 < b\gamma_1 f_1$, or

$$\Lambda^* = \frac{f_0(\mathbf{x})}{f_1(\mathbf{x})} < \frac{b\gamma_1}{a\gamma_0}.$$

So a Neyman-Pearson test is Bayes for some losses and some prior. Note that if $b\gamma_1 = a\gamma_0$, the Bayes test is a maximum likelihood test.

PROBLEMS

9-62. Find the most powerful test at $\alpha = .10$ for $\lambda = 1$ against $\lambda = 2$ in Poi(λ) in terms of Y, the sum of the observations in a random sample of size $n = 3$.

9-63. Let Z denote a discrete random variable having one or the other of the two distributions shown in this table, as in Problem 9-4.

	z_1	z_2	z_3	z_4	z_5
H_0	.2	.3	.1	.3	.1
H_1	.3	.1	.3	.2	.1

(a) Find the most powerful (nonrandomized) tests. (See Problem 9-4.)
(b) Compare the most powerful test of level $\alpha = .3$ with the critical region $Z = z_4$.

9-64. As in Problem 9-55, consider testing $p = .5$ against $p = .8$, where p is the parameter of Ber(p), which we sample twice in independent trials. Determine the most powerful test with $\alpha = .2$. [In Problem 9-55 you listed the eight possible (nonrandomized) tests together with an α and β for each.]

9-65. Consider testing $\sigma = \sigma_0$ against $\sigma = \sigma_1$ using a random sample of size n from $\mathcal{N}(\mu, \sigma^2)$, where μ is known.
(a) Find the most powerful tests.
(b) Determine n and the critical boundary to achieve $\alpha = \beta = .01$ when $\sigma_0 = 2$ and $\sigma_1 = 3$.

9-66. Example 9.9b gave a randomized test as a mixture of two nonrandomized N-P tests: $\phi = \gamma \phi_1 + (1 - \gamma)\phi_2$, where $0 \leq \lambda \leq 1$ and ϕ_1 and ϕ_2 correspond to successive critical regions, as ordered by values of Λ^*. Show that ϕ is equivalent to an extended Neyman-Pearson procedure.

9-67. Find the most powerful critical regions for testing θ_0 against θ_1, where $\theta_1 > \theta_0$, based on a random sample of size n from Exp(θ). Give them in terms of the sample sum or a function thereof.

9-68. Find the most powerful critical regions for testing $f(x) = 2x$, $0 < x < 1$ against $f(x) = 3x^2$, $0 < x < 1$, with a random sample of size n.

9-69. Consider testing H_0: $X \sim \mathcal{U}(0, 1)$ against H_1: $f_X(x) = 6x(1-x)$ for $0 < x < 1$. Find α for the N-P test $\Lambda^* < K$ (in terms of K), where Λ^* is the likelihood ratio for a single observation.

9-70. For the family of Neyman-Pearson tests in a particular problem, show that the relationship between α and β is an inverse one—the larger the α the smaller the β. [Hint: Write the error sizes as integrals of the p.d.f.'s $f_i(\lambda)$ of Λ under H_i, and calculate $d\beta/d\alpha$.]

9.10 The power function

When a hypothesis H is not simple, the expression $P(C \mid H)$ is usually not defined. But $P(C \mid \theta)$ is defined for each simple hypothesis θ in H. When $\theta \in H_A$, the probability of C^c is a type II error size, and the probability of C is what we've called power. The power of C depends on which θ is assumed to hold, and as a function of θ it is termed the **power function**. The complementary function, the probability of the acceptance region, is the **operating characteristic function**.

DEFINITION The *power function* of a test with critical region C is

$$\pi(\theta) = P(C \mid \theta). \tag{1}$$

The *operating characteristic function* is

$$OC(\theta) = P(C^c \mid \theta) = 1 - \pi(\theta). \tag{2}$$

The function $\pi(\theta)$ is defined for simple hypotheses in H_0 as well as those in H_1, although the term "power" is particularly meaningful for $\theta \in H_1$. We'll usually work in terms of the power function $\pi(\theta)$, but the OC-function is traditional in some areas of application.

The power function (or its complement) tells the whole story of how a test performs as a decision rule, and two tests can be compared on the basis of their power functions, although not in all cases is one test a clear winner over the other. We might point out that a power function does not depend on how we happened to have defined the null and alternative hypotheses in terms of θ, but only on the critical region.

EXAMPLE 9.10a Consider tests based on Y, the number of "successes" in a random sample of size $n = 3$ from Ber(p). Let $C_1 = [Y \geq 2]$, and $C_2 = [Y = 3]$. The corresponding power functions are

$$\pi_1(p) = P(Y \geq 2 \mid p) = 3p^2(1-p) + p^3,$$

and

$$\pi_2(p) = P(Y = 3 \mid p) = p^3.$$

These are plotted on common axes in Figure 9-10. Both power functions have large values for large p and small values for small p. This means that either is suitable for testing H_0: $p \leq p_0$ vs. H_A: $p > p_0$, since we naturally want a low probability of rejecting H_0 when it is true and a high probability of rejecting it when it is false. The test C_2 has a smaller "α," but this low power when p is small is achieved by sacrificing power at larger values of p. ■

The ideal power function would be 1 on H_A and 0 on H_0. Those shown in Figure 9-10 have this tendency, but they are not very close to ideal because of the relatively small amount of information when $n = 3$.

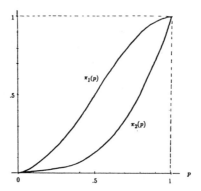

Figure 9-10 Power function for Example 9.10a

EXAMPLE 9.10b For testing $p = .5$ against $p \neq .5$ in a Bernoulli population, we try a two-sided critical region: It is only natural to prefer the explanation $p \neq .5$ if the number of successes in a sample is either unusually large or unusually small. For $n = 5$, let $C = [Y = 0, 1, 4,$ or $5]$, where, as in the preceding example, Y is the number of successes. The power function is

$$\pi_C(p) = 1 - P(Y = 2 \text{ or } 3) = 1 - 10p^2(1-p)^3 - 10p^3(1-p)^2$$

$$= 1 - 10p^2(1-p)^2 = \tfrac{3}{8} + 5(p - .5)^2 - 10(p - .5)^4.$$

This is shown in Figure 9-11. It is symmetrical about $p = .5$, where it has its minimum value of 3/8. Since H_0 is simple, the size of the type I error is uniquely defined: $\alpha = \pi(.5) = 3/8$.

Consider now a random sample of size $n = 100$, and the test defined by the critical region $C = [\,|\,Y - 50\,| > 10]$. Since $Y \approx \mathcal{N}(100p, 100pq)$,

$$\pi(p) = P(|Y - 50| > 10 \mid p) \doteq 1 - \Phi\left(\frac{60.5 - 100p}{10\sqrt{pq}}\right) + \Phi\left(\frac{39.5 - 100p}{10\sqrt{pq}}\right).$$

This is also shown in Figure 9-11. The larger sample has resulted in a type I error size closer to 0: $\alpha = .036$, and the power is greater for large and small values of p. ∎

To make comparisons with randomized tests we observe that for a randomized test ϕ the power is

$$\pi(\theta) = P(\text{rej } H_0 \mid \theta) = E[\phi(\mathbf{X}) \mid \theta]. \tag{3}$$

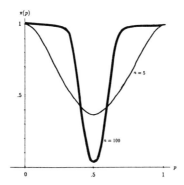

Figure 9-11 Power function for Example 9.10b

THEOREM 3 *Given any test ϕ and a sufficient statistic T, there is a test based on T (only) with the same power function as that of ϕ.*

Theorem 3 says that a test may as well be based on a sufficient statistic. To prove it, we note that if T is sufficient for θ and ϕ is any test, then the test

$$\phi^*(t) = E[\phi(\mathbf{X}) \mid T = t]$$

has the power function

$$\pi_{\phi^*}(\theta) = E\{E[\phi(\mathbf{X}) \mid T = t]\} = E[\phi(\mathbf{X})] = \pi_\phi(\theta).$$

So ϕ^*, based on the sufficient T, performs in the same way as ϕ.

In §9.4 we defined the type I error size for a composite H_0 as the largest probability of the critical region for simple hypotheses in H_0:

$$\alpha = \sup_{\theta \in H_0} P(C \mid \theta) = \sup_{\theta \in H_0} \pi(\theta). \tag{4}$$

A similar definition for β is

$$\beta = \sup_{\theta \in H_A} P(C^c \mid \theta) = \sup_{\theta \in H_A} [1 - \pi(\theta)]. \tag{5}$$

But in a typical setting H_0 and H_A have a common boundary, and $\pi(\theta)$ is monotone; then $\alpha + \beta = 1$, as shown in Figure 9-12. A finite sampling experiment could not be expected to discriminate between H_0 and H_A when θ is near the boundary point. We could then establish an indifference zone, say $I = (\theta', \theta'')$, where there is no penalty for being wrong.

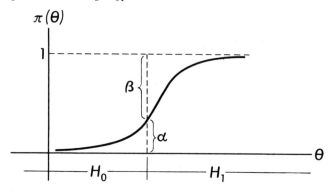

Figure 9-12

Defining α and β as maxima over $H_0 - I$ and $H_A - I$ makes it possible to achieve error sizes closer to the ideal value of 0, as in Figure 9-13. Specifying α and β as defined in this way amounts to fixing two points on the power curve. If $\pi(\theta)$ is monotone increasing, $\alpha = \pi(\theta')$ and $\beta = 1 - \pi(\theta'')$. We might then replace the given problem ($\theta \leq \theta_0$ vs. $\theta > \theta_0$) by that of testing $\theta = \theta'$ vs. $\theta = \theta''$. After finding the best test for the latter problem, we'd want to obtain its power function to see how it performs more generally. The next example illustrates these ideas.

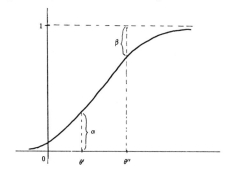

Figure 9-13

Example 9.10c Suppose $X \sim N(\mu, 9)$, and we want a sample size and critical region such that the probability of rejecting $\mu \leq 1$ when $\mu = .8$ is .05, and the probability of accepting $\mu \leq 1$ when $\mu = 1.2$ is .10. For the simple hypotheses $\mu = .8$ and $\mu = 1.2$, the most powerful test is $\bar{X} > K$. For this type of test, the given criteria are obtained by translating the given conditions on power into the following two simultaneous equations:

$$.05 = P(\bar{X} > K \mid \mu = .8) = 1 - \Phi\left(\frac{K - .8}{3/\sqrt{n}}\right),$$

and
$$.10 = P(\bar{X} < K \mid \mu = 1.2) = \Phi\left(\frac{K - 1.2}{3/\sqrt{n}}\right).$$

These simultaneous equations are easy to solve using Table I, to obtain $n = 481$ and $K = 1.025$. The corresponding power function is

$$\pi(\mu) = P(\bar{X} > 1.025 \mid \mu) = 1 - \Phi\left(\frac{1.025 - \mu}{3/\sqrt{481}}\right) = \Phi\left(\frac{\mu - 1.025}{3/\sqrt{481}}\right).$$

This is an increasing function of μ. Its value at $\mu = .8$ is $.05$ and its value at $\mu = 1.2$ is $.90$. ∎

Given a certain loss structure and prior for θ, a Bayesian can use the power function in finding an expected loss for each critical region. A best critical region can then be identified as one with minimal expected loss. Thus, if $\ell(\theta, \text{rej. } H_0) = a(\theta)$ and $\ell(\theta, \text{acc. } H_0) = b(\theta)$, the expected loss incurred in using a critical region C depends on θ:

$$R(\theta, C) = a(\theta)\pi_C(\theta) + b(\theta)[1 - \pi_C(\theta)],$$

called the **risk function**. The expected value of this function with respect to the distribution of θ is called the **Bayes loss** assigned to the region C

$$B(C) = E_g R(\theta, C) = E_g\{a(\theta)\pi_C(\theta) + b(\theta)[1 - \pi_C(\theta)]\}.$$

The Bayesian chooses a critical region to minimize the Bayes loss.

EXAMPLE 9.10d To test $p \leq \frac{1}{3}$ against $p > \frac{1}{3}$, consider a random sample of size $n = 4$ from $\text{Ber}(p)$, and the critical region $C = [Y > K]$, where Y is the number of successes in the four trials. For $K = 2$, the power function is

$$\pi(p) = P(Y > 2 \mid p) = 4p^3(1 - p) + p^4.$$

Now assume these losses:

$$\ell(p, \text{rej. } H_0) = 2 \text{ for } p < \tfrac{1}{4}, \quad \ell(p, \text{acc. } H_0) = 1 \text{ for } p > \tfrac{1}{2},$$

and 0 otherwise. The expected loss using C, for a particular p, is

$$R(p, C) = \begin{cases} 2[4p^3(1-p) + p^4], & p < 1/4, \\ 1 - [4p^3(1-p) + p^4], & p > 1/2, \\ 0, & 1/4 < p < 1/2. \end{cases}$$

Averaging with respect to the prior yields the Bayes loss. If that prior is

Beta(2, 2), the Bayes loss is

$$B(C) = \int_0^{1/4} 2[4p^3(1-p) + p^4] 6p(1-p)dp$$

$$+ \int_{1/2}^1 [1 - 4p^3(1-p) - p^4] 6p(1-p)dp \doteq .0378.$$

Similar calculations yield the following:

C	∅	Y > 3	Y > 2	Y > 1	Y > 0	Ω
B(C)	.5	.315	.0378	.0372	.145	.3125	■

9.11 Power functions of t-tests

In §9.7 we considered tests of $\mu = \mu_0$ in $\mathcal{N}(\mu, \sigma^2)$ based on the one-sample t-statistic [(2) in §9.7]:

$$T = \frac{\bar{X} - \mu_0}{S/\sqrt{n}}. \tag{1}$$

If σ were known, the critical region $Z > K$ [where Z is obtained by using the known σ in (1) in place of the unknown S] is equivalent to one of the form $\bar{X} > K'$. Finding the power function is then straightforward— in terms of Φ. But when σ is unknown, the power function for $T > K$ is

$$\pi(\mu) = P\left(\frac{\bar{X} - \mu_0}{S/\sqrt{n}} > K\right). \tag{2}$$

Under the null hypothesis, $T \sim t(n-1)$. But the distribution of T under any alternative is not so simple—the numerator of T does not have mean zero when $\mu \neq \mu_0$:

$$E(\bar{X} - \mu_0 \mid \mu) = \mu - \mu_0.$$

In this case, the numerator of T^2 as shown in (1) of §9.7 is not chi$^2(1)$.

When $\mu \neq \mu_0$, the distribution of T is one we've not yet encountered, a **noncentral** t-distribution, with $n-1$ d.f. and depending on a **noncentrality parameter**: $n(\mu - \mu_0)^2/(2\sigma^2)$. The p.d.f. has no simple closed form, but can be given as an infinite series of terms involving t-densities.[8] Tables of the distribution are available.[9]

---

[8] Kendall, M. and A. Stuart, *The Advanced Theory of Statistics*, Vol. I.

[9] Lieberman, G., and G. Resnikoff, *Tables of the Non-central t-distribution*, Stanford Calif. Stanford U. Press, 1957.

9.11 Power functions of t-tests

When $Z_1, \ldots, Z_k$ are independent, with $Z_i \sim \mathcal{N}(\mu_i, 1)$, the sum of their squares is said to have a **noncentral chi-square** distribution with k degrees of freedom. The distribution depends on the k means only through the value of the *noncentrality parameter*:

$$\lambda = \tfrac{1}{2}(\mu_1^2 + \cdots + \mu_k^2). \tag{3}$$

To see this we first write the m.g.f. of a single Z^2:

$$\psi_{Z_i^2}(t) = \frac{1}{\sqrt{2\pi}} \int_{-\infty}^{\infty} \exp\left\{tz^2 - \tfrac{1}{2}(z - \mu_i)^2\right\} dz = (1 - 2t)^{-1/2} \exp\left(\frac{\mu_i^2 t}{1 - 2t}\right).$$

The m.g.f. of $Z_1^2 + \cdots + Z_k^2$ is then the product of these for $i = 1, \ldots, k$:

$$(1 - 2t)^{-k/2} \exp\left(\frac{2\lambda t}{1 - 2t}\right),$$

which depends only on the combination λ of the various means. It reduces to a central chi-square m.g.f. when $\lambda = 0$.

A fraction in which the numerator is noncentral chi-square divided by its d.f. and the denominator is central chi-square divided by its d.f. is said to have a noncentral F distribution. A fraction in which the numerator is a normal variable with mean μ and variance 1, and the denominator is the square root of a central chi-square variable divided by its d.f., is said to have a noncentral t-distribution. (Its square is then noncentral F.) Tables of the various noncentral distributions are available.[10]

Returning to t-tests, we write the test statistic (1) in a form that displays its structure as a ratio of variables with known distributions:

$$T = \frac{\dfrac{\bar{X} - \mu_0}{\sigma/\sqrt{n}}}{\sqrt{\dfrac{(n-1)S^2}{\sigma^2} \div (n-1)}}.$$

The numerator is a normal variable with mean $\sqrt{n}(\mu - \mu_0)/\sigma$ and variance 1, so the distribution of T is noncentral t, with noncentrality parameter $\tfrac{1}{2}n(\mu - \mu_0)^2/\sigma^2$ and $n - 1$ d.f. For instance, the power function of the critical region $T > K$ is

$$P(T > K \mid \mu) = 1 - F_T(K),$$

---

[10] For instance: G. Lieberman and G. Resnikoff, *Tables of the Non-central t-distribution*, Stanford, CA: Stanford University Press, 1957; E. Fix, "Tables of Noncentral χ^2," Univ. Calif. Publ. Stat. 1, 15-19 (1949); M. Fox, "Charts of the Power of the F-test," *Ann. Math. Stat.* 28, 484-497 (1956).

obtainable from tables of the noncentral t. And 95 percent confidence limits for μ are

$$\bar{X} \pm t_{.975} \frac{S}{\sqrt{n}},$$

where $t_{.975}$ is the 97.5 percentile of the noncentral t.

In §9.7 we obtained a t-statistic for testing $\mu_1 - \mu_2 = 0$, in the case of independent random samples from normal populations with a common s.d. σ:

$$T = \frac{\bar{X}_1 - \bar{X}_2}{S_p\sqrt{1/n_1 + 1/n_2}}.$$

When $\mu_1 \neq \mu_2$, this has a noncentral t-distribution with $n_1 + n_2 - 2$ degrees of freedom and noncentrality parameter

$$\frac{\frac{1}{2}(\mu_1 - \mu_2)^2}{\sigma^2/n_1 + \sigma^2/n_2}. \tag{4}$$

On the other hand, suppose we were to test $\mu_1 = \mu_2$ using *paired* data—a sample of n differences, where the s.d. of a single difference is σ_d. The noncentrality parameter of the t-statistic is

$$\frac{\frac{1}{2}(\mu_1 - \mu_2)^2}{\sigma_d^2/n}.$$

For a two-sample t-statistic based on independent samples each of size n, the noncentrality parameter is

$$\frac{\frac{1}{2}(\mu_1 - \mu_2)^2}{2\sigma^2/n}.$$

The noncentrality parameters for the paired t and the two-sample t—and hence the power functions—will agree if $2\sigma^2/n = \sigma_d^2/n$. If pairing can be done so that $\sigma_d^2 < 2\sigma^2$, then the paired t test is more sensitive. For instance, if one could make $\sigma_d^2 = \sigma^2/2$, the power of the paired t test would be the same at $\mu_1 - \mu_2 = \delta$ as the power of the two sample t test at $\mu_1 - \mu_2 = 2\delta$.

PROBLEMS

9-71. Find the power function of the test given by the critical region $\bar{X} < 1$, where $\bar{X}$ is the mean of a random sample of size n from $\text{Exp}(\lambda)$ and n is large enough that $\bar{X}$ is approximately normal. For what kind of hypotheses H_0 and H_A would this critical region be appropriate?

9.11 Power functions of t-tests

9-72. Find and sketch the power functions of the tests given by the following critical regions based on $\hat{p}$, the sample proportion of successes in a random sample of size 400 from Ber(p). Also state the kind of H_0 and H_A for which each region would be an appropriate critical region.
(a) $\hat{p} > .52$.
(b) $|\hat{p} - 1/2| > .02$.

9-73. Consider the critical region $[X > K]$ for a sample of size $n = 1$ from $f(x \mid \theta) = \theta x^{\theta - 1}$, $0 < x < 1$ (and $\theta > 0$).
(a) Plot the power function for the case $K = \frac{1}{2}$.
(b) Find K so that the critical region $[X > K]$ has power .1 when $\theta = 1$.

9-74. Suppose we reject $\mu = 0$ if $\bar{X} > K/\sqrt{n}$, where $\bar{X}$ is the mean of a random sample of size n from a population with mean μ and standard deviation $\sigma = 1$. Assuming n large enough that $\bar{X}$ is approximately normal,
(a) find the power function of this decision rule (in terms of n and K).
(b) determine n and K so that $\pi(.1) = .95$ and $\pi(-.1) = .05$.

9-75. Consider critical regions C_1 and C_2, with power functions π_1 and π_2, respectively. Consider the test in which we decide to use C_1 or C_2 according as a fair coin falls heads or tails.
(a) Express the power function of this test in terms of π_1 and π_2.
(b) Express the test as a randomized test by defining an appropriate ϕ.

9-76. For random sampling from $N(\mu, \sigma^2)$, find the power function of critical region $S^2 > K$ [using the fact that $(n-1)S^2/\sigma^2 \sim \chi^2(n-1)$], and show that it is an increasing function of σ^2.

9-77. To test $\mu = \mu_0$ we decide to reject H_0 if a 90 percent confidence interval for μ, based on a random sample of size 25 from $N(\mu, 1)$, does not contain μ_0.
(a) Find α.
(b) Find and make a rough sketch of the power function.

9-78. Consider a discrete variable Z with p.f. $f(z \mid \theta)$, where $0 \le \theta \le 1$, defined as shown in the following table:

	z_1	z_2	z_3	z_4
$f(z \mid 0)$	1/12	1/12	1/6	2/3
$f(z \mid \theta)$	$\theta/3$	$(1-\theta)/3$	1/2	1/6

(a) For testing $\theta = 0$ vs. $\theta > 0$, find the likelihood ratio Λ at each z_i.
(b) Find the likelihood ratio critical region with $\alpha = 1/6$.
(c) Compare the power function of the test in (b) with the test that

rejects $\theta = 0$ when $Z = z_3$. Which test is preferable?

9-79. Plot power functions for the critical regions of the form $R > C$ and $S > D$, with constants C and D chosen to achieve $\alpha = .05$, for testing $\sigma = .007$ against $\sigma > .007$ with a random sample of size 9 from $N(\mu, \sigma^2)$. [S is the sample s.d. and R, the sample range.]

9-80. For a two-sided test of equality of variance of two normal populations based on the ratio of sample variances, with $\alpha = .10$ divided equally between the two tails of the null distribution, find the power function (as a function of the ratio of population variances) when $n_1 = n_2 = 9$.

9-81. Calculate the mean and variance of a noncentral chi-square variable with k d.f. and noncentrality parameter λ by exploiting its structure as a sum of squares of normal variables.

9-82. Referring to the discussion of paired versus independent random samples from normal populations with equal variances σ^2 (at the end of §9.11), find the value of the correlation coefficient ρ implied by the condition $\sigma_d^2 = \sigma^2/2$.

9-83. Verify formula (3) for the noncentrality parameter of the distribution of the two-sample t-statistic.

9.12 Uniformly most powerful tests

What we saw in Example 9.9a is not unusual: A Neyman-Pearson test for a particular θ_0 against a particular θ_1 greater than θ_0 turned out to be most powerful not only against θ_1, but also against every $\theta > \theta_0$. In such a case we say that the test is **uniformly most powerful** for $\theta = \theta_0$ vs. $\theta > \theta_0$.

DEFINITION A test of a simple hypothesis H_0 vs. a composite hypothesis H_A is *uniformly most powerful* (UMP) when among all tests with no larger α it has the greatest power against each simple hypothesis in H_A.

From the discussion in §9.9, it follows that for random samples from a population in the exponential family:

$$f(x \mid \theta) = B(\theta)h(x)\exp[Q(\theta)R(x)], \qquad (1)$$

the critical region $\sum R(X_i) > K$ is UMP for a simple H_0: $\theta = \theta_0$ against a onesided H_A, say $\theta > \theta_0$, if $Q(\theta)$ is an increasing function of θ.

When the null hypothesis is not simple, we define an α for a test with critical region C as the largest of the various type I error sizes (as in §9.4):

$$\alpha = \sup_{\theta \in H_0} P(C \mid \theta). \tag{2}$$

With this we can define a UMP test for a composite H_0:

DEFINITION A test of a composite H_0 vs. a composite H_A is *uniformly most powerful* (UMP) when, among tests with no larger α [as defined by (2)], it has the greatest power against each simple hypothesis in H_A.

EXAMPLE 9.12a Suppose (as in Example 9.10d) we test H_0: $p \leq \frac{1}{3}$ against H_A: $p > \frac{1}{3}$, with a random sample of size $n = 4$ from Ber(p). Consider again the N-P critical region $C^* = [Y > 2]$, where Y is the number of successes in the four trials. This is most powerful for $p = \frac{1}{3}$ against every $p > \frac{1}{3}$ so it is UMP over H_A. The power function for C^* (from Example 9.10d) is

$$\pi(p) = p^3(4 - 3p),$$

shown in Figure 9-14. This is an increasing function of p, so the largest value of $\pi(p)$ on H_0 is at the boundary of H_0: $\alpha^* = \pi(1/3) = 1/9$. But we know that the power of C^* on H_1 is at least as great as that of any test C whose power *at* $p = \frac{1}{3}$ is at most $1/9$. It is then automatically at least as great as that of any test whose power *throughout* H_0 is at most $1/9$, that is, any test whose α is at most $1/9$. This means that C^* is UMP for $p \leq \frac{1}{3}$ against $p > \frac{1}{3}$. ∎

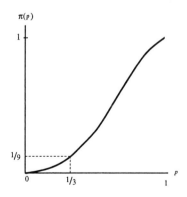

Figure 9-14

Uniformly most powerful tests do not always exist. They do, typically, in certain one-parameter problems in which the alternative is one-sided, as in the above example. It was crucial to the argument in that example that the power function is monotone, and it is useful to know that this will be the case in a large class of problems.

We saw in §9.2 that the likelihood ratio for the simple hypotheses $\theta = \theta_0$ and $\theta = \theta_1$ can often be expressed in terms of a familiar statistic T. And sometimes (as in Example 9.2b) the nature of the relationship does not depend on the particular values θ_0 and θ_1.

DEFINITION A family of distributions indexed by the real parameter θ is said to have a *monotone likelihood ratio* if there is a statistic T such that for each pair (θ, θ'), where $\theta > \theta'$, the likelihood ratio $L(\theta)/L(\theta')$ is a nondecreasing function of T.

Problem 9-87 asks you to show that if $Q(\theta)$ in (1) is monotone increasing, the one-parameter exponential family has a monotone likelihood ratio in terms of the statistic $T = \sum R(X_i)$. The next example shows that not all distribution families have a monotone likelihood ratio.

EXAMPLE 9.12b Consider a single observation from a Cauchy family indexed by a location parameter, with p.d.f.

$$f(x \mid \theta) = \frac{1/\pi}{1 + (x - \theta)^2}.$$

The likelihood ratio for testing θ against θ' is

$$\Lambda^* = \frac{1 + (X - \theta')^2}{1 + (X - \theta)^2},$$

which is not a monotone function of X. This is evident upon inspection of Figure 9-15: The numerator has the value d'^2 and the denominator, the value d^2, where $d' > d$ and $\Lambda > 1$ for the x shown in the figure. But for an x on the other side of θ', the inequality is reversed; and $\Lambda = 1$ when x is half-way between θ and θ'. As x becomes infinite in either direction, Λ tends to 1—from below as $x \to -\infty$, and from above as $x \to +\infty$. ∎

THEOREM 4 If $\{f(x \mid \theta)\}$ has a monotone likelihood ratio in terms of a statistic T, the power function of the Neyman-Pearson critical region for testing θ_0 against θ_1, with $\theta_0 < \theta_1$ is an increasing function of θ.

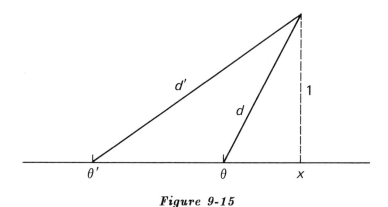

Figure 9-15

COROLLARY *If $\{f(x \mid \theta)\}$ has a monotone likelihood ratio in terms of a statistic T, the Neyman-Pearson critical region for testing θ_0 against θ_1, where $\theta_0 < \theta_1$, is UMP for $\theta \leq \theta^*$ against $\theta > \theta^*$.*

To prove the theorem we need to show that for any θ' and θ'', with $\theta' < \theta''$, the power of the Neyman-Pearson region C at θ'' is at least as great as its power at the point θ': $\pi_C(\theta'') \geq \pi_C(\theta')$. But we know that the inequality $L(\theta')/L(\theta'') < K$ is equivalent to an inequality of the form $T > K'$, which in turn is equivalent to $L(\theta_0)/L(\theta_1) < K''$ (for some constants K' and K''). This means that C is most powerful not only for θ_0 vs. θ_1, but also for θ' vs. θ''. Thus, for any test ϕ with no greater α (at θ'), the power of C at θ'' is at least as great as the power of ϕ at θ''. That this is true in particular for the test $\phi \equiv \alpha_C$ establishes the theorem:
$$\pi_C(\theta'') \geq \pi_\phi(\theta'') = E_{\theta''}(\phi) = \alpha_C = \pi_C(\theta').$$

To show the corollary we argue much as we did in Example 9.9c: The N-P region C for θ_0 vs. θ_1 is also most powerful for θ^* against any particular $\theta > \theta^*$ and is therefore UMP against the composite alternative $\theta > \theta^*$. But because the power function of C is monotone increasing, the largest "α" in $\theta \leq \theta^*$ is the power at θ^*:
$$\alpha_C = \sup_{\theta \leq \theta^*} \pi_C(\theta^*) = \pi_C(\theta^*).$$

The power of a test ϕ with $\alpha \leq \alpha_C$ does not, in particular, exceed the power of C at θ^* and is therefore uniformly smaller than the power of C throughout $\theta > \theta^*$, as asserted in the corollary.

The force of the corollary is that when the family $\{f(x \mid \theta)\}$ has a monotone likelihood ratio, we can construct a family of UMP tests of $\theta \leq \theta^*$ against $\theta > \theta^*$ as the Neyman-Pearson tests of θ_0 against θ_1 for arbitrarily chosen values θ_0 and θ_1 such that $\theta_0 < \theta_1$.

EXAMPLE 9.12c If the distribution of X is in the exponential family (1) with a monotone increasing $Q(\theta)$, then (from Problem 9-87) it has a monotone likelihood ratio in terms of $T = \sum R(X_i)$. The discussion in §9.9 showed that $T > K$ is the Neyman-Pearson region for θ_0 against any $\theta_1 > \theta_0$. According to the corollary, this test is UMP for $\theta \leq \theta^*$ against $\theta > \theta^*$.

In particular, suppose $X \sim \text{Exp}(1/\theta)$. This distribution is in the exponential family with $R(x) = x$ and $Q(\theta) = -1/\theta$, a monotone increasing function of θ. Hence, the critical region $\bar{X} > K$ is UMP for $\theta \leq \theta^*$ against $\theta > \theta^*$. ∎

When the power function of a test is monotone, and the testing problem is one-sided, the power on the null hypothesis cannot exceed the power at any point on the alternative.

DEFINITION A test whose power function is nowhere larger on H_0 than it is on H_A is said to be **unbiased**.

A UMP test is always unbiased: The randomized test $\phi \equiv \alpha$ has level α, and its power is α at each $\theta \in H_A$. The power of a UMP test must be at least α for $\theta \in H_A$, whereas it is at most α for $\theta \in H_0$.

Because UMP tests are unbiased, we can easily see that there is not always a UMP test. For the one-point null hypothesis $\theta = \theta_0$, the power function of a UMP test, which is unbiased against $\theta \neq \theta_0$, would have to have its minimum value at θ_0. But the power function of the N-P test (with the same α) for θ_0 against a θ_1 to the right of θ_0 would usually be increasing at θ_0. For θ's just to the right of θ_0, this N-P test would be more powerful than the unbiased test. The next example may help to explain this point.

EXAMPLE 9.12d Consider testing $p = .5$ against $p \neq .5$ based on Y, the number of successes in five independent trials of $\text{Ber}(p)$, with this randomized test:

$$\phi(Y) = \begin{cases} 1, & Y = 4 \text{ or } 5, \\ .6, & Y = 3, \\ 0, & Y = 0 \text{ or } 1. \end{cases}$$

The power function of the test is

$$\pi(p) = 5p^4(1-p) + p^5 + .6 \times 10p^3(1-p)^2.$$

This is shown in Figure 9-16, along with the power function of the rejection region $R = \{0, 1, 4, 5\}$. The latter test is unbiased, but it cannot be UMP. The two tests have the same α (3/8); but when $p > .5$, the one-sided ϕ is more powerful than the symmetric R, at the expense of lower power on $p < .5$. ∎

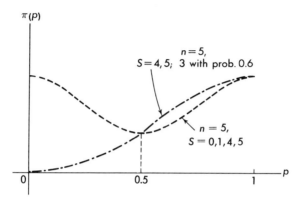

Figure 9-16

The normal family of distributions is a two-parameter family, a fact that makes it difficult to apply some of the notions we have been studying. In our treatment of UMP tests as derived in monotone likelihood ratio families using the Neyman-Pearson lemma, we considered only single-parameter families. Of course, normal distributions with known mean or known variance belong to the exponential family, and much of what we have learned does apply. Even so, in practice it is more usual that both mean and variance are unknown.

PROBLEMS

9-84. Find the UMP tests for $\mu \leq 0$ vs. $\mu > 0$, using a random sample of size n from $\mathcal{N}(\mu, 1)$.

9-85. Find the UMP tests for $\theta \leq 1$ vs. $\theta > 1$, using a random sample of size n from $f(x \mid \theta) = \theta^{-2} x e^{-x/\theta}$, $x > 0$.

9-86. Suppose X is Cauchy with location parameter θ, as in Example 9.12b.

(a) Find the power function of the critical region $[X > 1]$.

(b) Show that the power function in (a) is monotonic. (Can you thus jump to the conclusion that the test UMP for $\theta \le 0$ vs. $\theta > 0$?)

9-87. Show that if $Q(\theta)$ in the one-parameter exponential family [as defined by (1) in this section] is monotone increasing, the family has a monotone likelihood ratio in terms of the statistic $T = \sum R(X_i)$.

9-88. Show that the family of hypergeometric distributions, given the sizes of population and sample (N and n), has a monotone likelihood ratio. [Hint: Show that for any M, the ratio $f(k \mid M+1)/f(k \mid M)$ is a monotonic function of the argument k, and then use the fact that the product of increasing functions is increasing.]

9-89. An industrial sampling plan calls for accepting a lot of N articles, each classified as good or defective, if a simple random sample of size n from the lot contains no more than c defectives. Show that this test is UMP for testing $H_0\colon p = p_0$ against $p > p_0$, where p is the (unknown) lot fraction defective.

9-90. Find the power function of the test that rejects $p = \frac{1}{2}$ when more than 60 or fewer than 40 successes occur in 100 independent trials of $\text{Ber}(p)$.

9-91. Determine a test of $p = \frac{1}{2}$ based on 100 independent trials of $\text{Ber}(p)$, with the same α as in the preceding problem, but which is UMP against $p > \frac{1}{2}$.

10
Analysis of Categorical Data

In this chapter we consider experiments in which each outcome is classified into one of a finite number of categories, rather than measured on a numerical scale.[1] The simplest problems are *univariate*, involving a single scheme of classification. More generally, experimental units or population members may be classified according to each of several schemes, the interest being in relationships among the schemes.

10.1 The Bernoulli model

We have already encountered some problems of inference for categorical populations with two categories—Bernoulli populations. Estimation of p was taken up in Chapter 8, and some tests, in §9.2-9.5. The proportion of successes in a sample is a special case of a sample mean; and in this case, for testing $p = p_0$, the Z-score is

$$Z = \frac{Y/n - p_0}{\sqrt{p_0 q_0/n}} = \frac{Y - np_0}{\sqrt{np_0 q_0}}, \qquad (1)$$

where Y is the number of successes in n independent trials. According to the central limit theorem (Chapter 5), $Z \approx \mathcal{N}(0, 1)$ when $p = p_0$. We exploited this fact for testing $p = p_0$ in various problems and examples in Chapter 9.

With a one-sided alternative, say $p > p_0$, large values of Z are taken as evidence against H_0, and an approximate P-value for an observed Z is found in the standard normal table. For testing at a fixed significance level α, the critical region $Y > K'$, or equivalently, $Z > K$ is UMP against $p > p_0$, and the relationship between K and α is given by the standard normal table. If the alternative is two-sided: $p \neq p_0$, large

---

[1] For more thorough and extensive treatments, we refer to the books of Kendall and Stuart and of Bishop, Fienberg, and Holland.

values of $|Z|$ or, equivalently, large values of Z^2 are to be interpreted as evidence against H_0. For testing at a fixed significance level, the critical regions $Z^2 > K$ are unbiased and can be shown to be UMP in the class of unbiased tests.[2]

Large sample confidence intervals for the Bernoulli parameter p are essentially special cases of confidence intervals for a mean—the point estimate $\hat{p} \pm$ a multiple of the standard error. Problem 8-48 called for construction of a confidence interval when the sample size is not large enough that the standard error of $\hat{p}$ is close to its standard deviation. We now carry out the details of that construction, beginning with the approximate pivotal

$$Z = \frac{Y/n - p}{\sqrt{pq/n}}, \qquad (2)$$

whose asymptotic distribution is $\mathcal{N}(0,1)$. Let k denote the value of Z with probability $\alpha/2$ in the right tail of its distribution. Then

$$1 - \alpha \doteq P\left(\frac{|\hat{p} - p|}{\sqrt{p(1-p)/n}} < k\right).$$

We square the inequality and solve for p:

$$\frac{(\hat{p} - p)^2}{p(1-p)/n} < k^2, \quad \text{or} \quad \hat{p}^2 - 2p\hat{p} + p^2 < p\frac{k^2}{n} - p^2\frac{k^2}{n}.$$

This quadratic inequality in p is satisfied when p is between the zeros of the quadratic function

$$\left(1 + \frac{k^2}{n}\right)p^2 - 2\left(\hat{p} + \frac{k^2}{2n}\right)p + \hat{p}^2.$$

Using the quadratic formula from algebra we find the zeros to be

$$\frac{1}{1 + k^2/n}\left(\hat{p} + \frac{k^2}{2n} \pm k\sqrt{\frac{\hat{p}(1-\hat{p})}{n} + \frac{k^2}{4n^2}}\right). \qquad (3)$$

These are (approximate) confidence limits for p at the level $1 - \alpha$. If n is quite large, the terms k^2/n and $k^2/(4n^2)$ can be neglected; with these terms gone, the confidence limits reduce to

$$\hat{p} \pm k \times \text{s.e.}(\hat{p}), \qquad (4)$$

the usual large sample confidence limits for a mean. As a rule of thumb, if p is between .1 and .9 and n is 25 or so, the limits (3) can be used; if n is larger than 100 the limits (4) may be adequate.

---

[2]Lehmann, E. L., *Testing Statistical Hypotheses*, 2nd Ed. Pacific Grove, Cal.: Wadsworth & Books/Cole, 1986, page 138.

EXAMPLE 10.1a Suppose the 25 circuit boards in a random sample were tested and found to be 20 percent defective. What is the population proportion of defectives? The obvious point estimate (the m.l.e.) is .20, and the standard error is $\sqrt{.16/25} = .08$. The approximate 90 percent confidence interval given by (4) is $.07 < p < .33$. More exact limits are given by (3), where $k^2/n = .108$:

$$\frac{1}{1.108}\left(.20 + .054 \pm 1.645\sqrt{.0064 + .00108}\right) = .229 \pm .128,$$

defining the interval $.101 < p < .357$. Both intervals are rather wide, and it's not clear that being more exact serves any purpose. The problem is that in only 25 observations there is not a great deal of information about p.

When we have a prior for p, the data can be used to update a prior distribution for p, as explained in §8.7. For instance, if $p \sim \text{Beta}(2, 18)$, the posterior is $\text{Beta}(7, 38)$, and the interval $.101 < p < .357$ has posterior probability .85. ∎

The problem of comparing population proportions arises, for example, when two populations are given two different treatments and the response is either "success" or "failure." Given independent random samples from the two populations, the maximum likelihood estimate of $p_1 - p_2$ is the corresponding difference in sample proportions, $\hat{p}_1 - \hat{p}_2$, where $\hat{p}_i = f_i/n_i$ and f_i is the frequency of successes in the sample of size n_i from population i ($i = 1, 2$). The sample difference is approximately normal with mean $p_1 - p_2$ and standard deviation

$$\text{s.d.}(\hat{p}_1 - \hat{p}_2) = \sqrt{\frac{p_1(1-p_1)}{n_1} + \frac{p_2(1-p_2)}{n_2}}.$$

A standard error may be defined as the result of replacing p_i by $\hat{p}_i$:

$$\text{s.e.}(\hat{p}_1 - \hat{p}_2) = \sqrt{\frac{\hat{p}_1(1-\hat{p}_1)}{n_1} + \frac{\hat{p}_2(1-\hat{p}_2)}{n_2}}. \tag{5}$$

Approximate confidence limits for $p_1 - p_2$ are then constructed as usual:

$$\hat{p}_1 - \hat{p}_2 \pm k \times \text{s.e.}(\hat{p}_1 - \hat{p}_2), \tag{6}$$

with k found (as usual) in the standard normal table.

Problem 9-10 called for the construction of a Z-score appropriate for testing the equality of the p's in two Bernoulli populations, based on independent random samples. The construction is as follows: Denote the sample proportions by $\hat{p}_i = f_i/n_i$, ($i = 1, 2$). Under H_0: $p_1 = p_2 = p$, the variance of the difference in sample proportions is

$$\operatorname{var}(\hat{p}_1 - \hat{p}_2) = p(1-p)\left(\frac{1}{n_1} + \frac{1}{n_2}\right),$$

and the corresponding Z-score is

$$Z = \frac{\hat{p}_1 - \hat{p}_2}{\text{s.d.}(\hat{p}_1 - \hat{p}_2)} = \frac{\hat{p}_1 - \hat{p}_2}{\sqrt{p(1-p)\left(\frac{1}{n_1} + \frac{1}{n_2}\right)}}. \quad (7)$$

This is approximately $N(0, 1)$ under H_0. Because p is not assumed known, this Z is not a statistic, so we must estimate p. The m.l.e. of the common value p is

$$\hat{p} = \frac{f_1 + f_2}{n_1 + n_2}, \quad (8)$$

which is simply the success ratio in the *combined* sample. Replacing p by $\hat{p}$ in the standard deviation of $\hat{p}_1 - \hat{p}_2$, we obtain the statistic

$$Z = \frac{\hat{p}_1 - \hat{p}_2}{\sqrt{\hat{p}(1-\hat{p})\left(\frac{1}{n_1} + \frac{1}{n_2}\right)}}, \quad (9)$$

which is approximately $N(0, 1)$ under H_0. With the aid of the standard normal table we can find P-values, or critical values for fixed α tests—one- or two-sided according as the alternative is one- or two-sided.

EXAMPLE 10.1b Results of a seven-year study[3] on the effect of aspirin as a treatment for the prevention of strokes showed that among 406 men who had suffered at least one transient ischemic attack, 29 of the 200 treated with aspirin subsequently had strokes or died, whereas in the nonaspirin group of 206 men, 56 had strokes or died. The sample proportions are .145 and .272, and the question is whether the difference should be attributed to chance, or indicates a difference in population proportions—that is, a "treatment effect." The estimate of p (the common proportion under H_0) is $\hat{p} = 85/406 = .209$, and the Z-score for the difference is

$$Z = \frac{.272 - .145}{\sqrt{.209 \times .791 \times \left(\frac{1}{206} + \frac{1}{200}\right)}} = 3.15,$$

rather strong evidence against the hypothesis of no treatment effect.

To *estimate* the difference in population proportions, we take the corresponding difference in sample proportions: $.272 - .145 = .127$. The

---

[3]Described in *Newsweek*, July 24, 1978, from a study reported in the *New England Journal of Medicine*.

standard error of this estimate, given by (5), is

$$\text{s.e.}(\hat{p}_1 - \hat{p}_2) = \sqrt{\frac{.272 \times .728}{206} + \frac{.145 \times .855}{200}} = .03976.$$

Then, 95 percent confidence limits for the difference are $.127 \pm .078$, defining an interval, incidentally, that does not include 0. ∎

10.2 Pearson's chi-square for goodness of fit

Suppose that each outcome or experimental unit is classified into one of k categories: $A_1, ..., A_k$. A probability model assigns corresponding probabilities $p_j = P(A_j)$ that sum to 1. The likelihood function, given a random sample of size n, is

$$L(\mathbf{p}) = p_1^{f_1} \cdots p_k^{f_k}, \tag{1}$$

where f_j is the frequency of the category A_j among the n observations. Of course, $\sum f_j = n$. The relative frequencies f_j/n are joint m.l.e.'s of the category probabilities, and the frequency vector $(f_1, ..., f_k)$ is minimal sufficient and has a multinomial distribution. (See §6.10 and Problem 7-64.)

The *goodness of fit* problem is that of testing a set $\mathbf{p}_0$ of specific probabilities for the k categories against the alternative that at least one of these is incorrect:

$$H_0: \mathbf{p} = \mathbf{p}_0 \quad \text{vs.} \quad H_A: p_j \neq p_{j0}, \text{ for one or more } j.$$

In §10.1 we treated this problem in the special case $k = 2$, basing a large sample test on the asymptotic normality of the sample proportion of one of the two categories. We now generalize to the case of general k.

The critical region we used in testing $p = p_0$ in Ber(p) against a two-sided alternative is of the form $Z^2 > K$, where

$$Z = \frac{f/n - p_0}{\sqrt{p_0 q_0/n}},$$

and f is the frequency of "success" in a random sample of size n. In place of (p, q) we now write (p_1, p_2); in place of (p_0, q_0), we write (p_{10}, p_{20}); and in place of $(f, n - f)$ we write (f_1, f_2). With this modification of notation, Z^2 becomes

$$Z^2 = \frac{(f_1 - np_{10})^2}{np_{10}} + \frac{(f_2 - np_{20})^2}{np_{20}}, \tag{2}$$

according to Problem 10-9.

When there are k categories, it is natural simply to add a term like those in (2) for each additional category after two, to define

$$\chi^2 = \sum_{j=1}^{k} \frac{(f_j - np_{j0})^2}{np_{j0}}. \tag{3}$$

The statistic (3) was first introduced and studied by Karl Pearson in 1900. (It is now usually referred to as "Pearson's chi-square statistic.") The product np_{j0} is the *expected* frequency of category j under H_0, so the differences in the numerator of (3) measure the "fit" of the data and model. Clearly, $\chi^2 = 0$ if and only if the relative frequency of each category equals the corresponding category probability in H_0. Moreover, intuition suggests that the larger the χ^2, the stronger the evidence against H_0.

Pearson found[4] that under H_0, $\chi^2 \approx \text{chi}^2(k-1)$. When the sample size is large, this asymptotic distribution may be used to approximate the actual distribution of χ^2, which is usually rather complicated and depends on the population distribution (see Problem 10-5). When $k = 2$, the chi-square statistic is equal to Z^2 as given by (2), the square of an approximately normal score; and from this it is clear that the asymptotic distribution is $\text{chi}^2(1)$. The statistic Z^2 is suited to the two-sided alternative $p \neq p_0$, whereas Z can be used for one-sided alternatives. When $k > 2$, we lose the notion of "side" as a possibility in defining alternatives to H_0.

In applying the asymptotic distribution, a conservative rule of thumb is that np_{0j} should be at least 5. But if n is at least five times the number of categories, the chi-square distribution gives an adequate approximation even if some of the expected frequencies np_{0j} are as small as 1. When there are many categories with small expected frequencies, one might combine some of them and test the resulting cruder model. Also, to test a model in which the number of categories is not finite (as in the case of a Poisson variable), an approximate test can be carried out on the model in which some categories are combined so as to make the number of categories finite.

EXAMPLE 10.2a The six faces of an ordinary die are often taken as equally likely, when the die is rolled. In 120 rolls, then, the expected cell frequencies are $120 \cdot \frac{1}{6} = 20$. Suppose we obtain these frequencies for the six faces: 18, 23, 16, 21, 18, 24. (In the chi-square statistic it will not matter which frequency goes with which face.) The chi-square statistic

---

[4] See H. Cramér, *Mathematical Methods of Statistics*, Princeton, N. J., Princeton U. Press (1946), p. 417.

(3) is then

$$\chi^2 = \frac{(18-20)^2}{20} + \frac{(23-20)^2}{20} + \frac{(16-20)^2}{20}$$
$$+ \frac{(21-20)^2}{20} + \frac{(18-20)^2}{20} + \frac{(24-20)^2}{20} = 2.5.$$

The mean value of $\chi^2(5)$ is 5, so the value 2.5 is not even as large as one might expect under H_0. The given data offer no evidence against the assumption of a "fair" die. ∎

It is important to realize that the large sample distribution depends only on the number of categories, and not on the population distribution being tested. So, at least for large samples, the chi-square statistic provides us with a *distribution-free* test of fit of a set of data to a model.

Finding the power of a chi-square test analytically is an unmanageable task, because of the composite nature of the alternative. It can be shown that for given α, the power tends to 1 on any alternative as n tends to infinity, and the chi-square test is then said to be "consistent."[5] This consistency makes for a rather awkward situation: One is almost sure to reject H_0 if the sample size is large enough. For, it is quite unlikely that a real experiment is represented with perfect precision by the model we take as H_0, and if it is not, then a consistent test will ultimately reject H_0,—given enough data. This strange situation is the result of having to specify H_0 as a simple hypothesis. [Indeed, the same awkwardness would arise, for instance, when one uses a Z-test for $\mu = \mu_0$ in $N(\mu, 1)$, which is also a consistent test and ultimately (with enough data) will discover that μ is not *exactly* μ_0.]

The Pearson chi-square statistic can be adapted for testing a distribution *type*—a family of distributions in which the category probabilities depend on an unknown parameter $\theta = (\theta_1, ..., \theta_r)$—that is, $p_i = p_i(\theta)$. To be able to calculate the chi-square statistic we need a specific value for np_i as the expected category frequency. To get one, we estimate the unknown parameter using the data with an estimator that is consistent, asymptotically normal, and asymptotically efficient. In particular, a maximum likelihood estimator will do, and we measure lack of fit by the value of

$$\chi^2 = \sum_{i=1}^{k} \frac{[f_i - np_i(\hat{\theta})]^2}{np_i(\hat{\theta})}, \tag{4}$$

where $\hat{\theta}$ is the m l e. of θ. With this modification, it can be shown[6] that

---
[5] E. L. Lehmann, *Testing Statistical Hypotheses*, 2nd Ed. Pacific Grove, Cal.: Wadsworth & Brooks/Cole, 1986, page 142.

364   Chapter 10   Analysis of Categorical Data

if $k > r + 1$, the asymptotic distribution of χ^2 under H_0 is again of the chi-square type, but now with $k - 1 - r$ degrees of freedom, where r is the number of estimated parameters (that is, the dimension of $\boldsymbol{\theta}$).

EXAMPLE 10.2b   In 30 settings of gill nets in a lake, these counts were obtained:

Number of fish	0	1	2	3	4
Frequency	12	10	6	0	2

The question was whether a Poisson distribution describes the distribution of fish. Because a Poisson variable has infinitely many possible values, we have to combine tail values into a single class—in this case, four or more. The average number per setting is $30/30 = 1$, so we take this as an estimate of the Poisson parameter λ. The probabilities (Table II) and expected frequencies in a random sample of 30 are as follows:

Number of fish	0	1	2	3	4 or more
Probability	.368	.368	.184	.061	.019
Expected freq.	11.04	11.04	5.52	1.84	.57

Then $\chi^2 = 5.65$, with d.f. $= 5 - 1 - 1 = 3$, so $P > .10$. If we had combined the last two cells into a single cell "3 or more," then $\chi^2 = .29$ (2 d.f.). In either case we'd be testing the approximate model with finitely many categories, and (strictly speaking) not the hypothesis that the distribution is Poisson. ∎

[A fussy point: The sample average we calculated in the above example as an estimate of the Poisson parameter is not exactly the m.l.e. for the finite model, and the asymptotic distribution used is not quite correct. We should use instead the θ that maximizes the likelihood function

$$L(\theta) = \prod_{1}^{k} [p_i(\theta)]^{f_i},$$

which is usually hard to calculate. In some cases (including the Poisson), the error is not serious.[7]]

It may seem like cheating to use the data to estimate the category mean frequencies—and indeed it is, to some extent. The fit is bound to

---
[6]See H. Cramér, *Mathematical Methods of Statistics* Princeton, N. J.: Princeton U. Press, 1946, page 417.

[7]See H. Chernoff and E. Lehmann, "The use of the maximum likelihood estimates in χ^2 tests of goodness of fit," *Ann. Math. Stat.* 25, 579 (1954).

be better when we use the sample itself to define a specific set of expected frequencies to be fitted. It turns out that not all of the sample information is used up in the estimation of parameters (provided that $r < k - 1$), so there is still some left for testing fit. We pay for the "cheating" in the loss of r degrees of freedom, which shifts a critical boundary so as to require a smaller χ^2 for accepting H_0 at a given significance level. This is only fair, since we have made the data look more like the model by using the data to define the model to be tested.

The chi-square statistic is designed and especially useful for categorical variables with no special ordering of the categories. A discrete numerical variable is categorical—its possible values are the categories, so the chi-square test can indeed be used in the way we used it for testing the hypothesis that a distribution is Poisson in Example 10.2b. In the case of continuous models, the summarizing of data in a frequency distribution according to a scheme of class intervals implies a categorical approximation that can be tested using Pearson's chi-square statistic. In particular, one can test the hypothesis of normality, a hypothesis that underlies various t- and F-tests. In such cases, generally, the ordering of the categories (which is ignored by chi-square) makes other tests more suitable and usually more powerful than chi-square. Chapter 12 will take up some of these tests, and in particular, tests for normality.

EXAMPLE 10.2c The distribution of times to failure of a type of radar indicator tube is thought to be exponential. Actual failure times[8] are summarized in the following frequency distribution:

Hours	Frequency
0-100	29
100-200	22
200-300	12
300-400	10
400-600	10
600-800	9
800-up	8

The mean value is $\bar X = 300$ (given in the source), so the m.l.e. of λ is $1/300$. Using this, we find the expected frequencies of the seven class intervals to be 28.4, 20.3, 14.5, 10.5, 12.7, 6.6, 7.0—if the population distribution is Exp(1/300). The chi-square statistic is

---

[8]From D. J. Davis, "*An analysis of some failure data,*" J. Amer. Stat. Assn. 47 (1952), p. 147.

$$\chi^2 = \frac{(29 - 28.4)^2}{28.4} + \cdots + \frac{(8 - 7.0)^2}{7.0} = 2.23.$$

Since $k = 7$ and $r = 1$, d.f. $= 7 - 1 - 1 = 5$, and $P \doteq .8$. The data give no basis for doubting that the population distribution is exponential. ∎

10.3 Likelihood ratio for goodness of fit

As given by (1) in the preceding section, the likelihood function, for a random sample of size n from a categorical population, is

$$L(\mathbf{p}) = p_1^{f_1} \cdots p_k^{f_k}, \tag{1}$$

where f_j is the frequency of the category A_i among the n observations, and $\sum f_j = n$. Again, we consider testing

$$H_0: \mathbf{p} = \mathbf{p_0} \quad \text{vs.} \quad H_A: p_j \neq p_{j0}, \text{ for one or more } j,$$

this time using the likelihood ratio Λ as a test statistic. The null hypothesis is simple, so the numerator of Λ is just $L(\mathbf{p_0})$. The m.l.e. of the probability vector $\mathbf{p}$ is the vector $\hat{\mathbf{p}}$ of relative frequencies, $\hat{p}_i = f_i/n$. The likelihood ratio is then

$$\Lambda = \frac{L(\mathbf{p_0})}{L(\hat{\mathbf{p}})} = \prod_{j=1}^{k} \frac{(p_{0j})^{f_j}}{(f_j/n)^{f_j}}. \tag{2}$$

and

$$-2 \log \Lambda = -2 \sum f_j \left\{ \log p_{0j} - \log \frac{f_j}{n} \right\}. \tag{3}$$

The large sample distribution of $-2 \log \Lambda$ is approximately $\text{chi}^2(k - 1)$, where $k - 1$ is the number of (free) parameters assigned values by H_0. Small values of Λ provide evidence against H_0, and a P-value can be found as a tail area of the chi-square p.d.f.

It is not just that χ^2 and $-2 \log \Lambda$ have the same asymptotic distribution; these statistics are in fact asymptotically equal under H_0. This can be seen (in steps that could be made rigorous) as follows:

$$\log p_{0j} - \log \frac{f_j}{n} \doteq \frac{1}{f_j/n}\left(p_{0j} - \frac{f_j}{n}\right) - \frac{1}{(f_j/n)^2} \frac{(p_{0j} - f_j/n)^2}{2},$$

so that

$$-2 \log \Lambda \doteq -2 \sum n\left(p_{0j} - \frac{f_j}{n}\right) + \sum \frac{n^2}{f_j}\left(p_{0j} - \frac{f_j}{n}\right)^2,$$

or

$$-2\log \Lambda \doteq 0 + \sum \frac{(np_{0j} - f_j)^2}{np_{0j}}.$$

So the likelihood ratio test and the χ^2 test, each developed on intuitive grounds, are asymptotically equivalent. This means the chi-square test has the large-sample optimality properties of a likelihood ratio test.[9]

EXAMPLE 10.3a We return to the die-toss data of Example 10.2a frequencies (18, 23, 16, 21, 18, 24) for the six sides, and calculate:

$$-2\log \Lambda = -2n\log n - 2\sum f_j(\log p_{0j} - \log f_j)$$
$$= -2[120\log 120 + 120\log \tfrac{1}{6} - 18\log 18 - 23\log 23$$
$$\quad - 16\log 16 - 21\log 21 - 18\log 18 - 24\log 24] = 2.503.$$

This is close to 2.5, the value of χ^2 found in Example 10.2a, as is to be expected when the die is fair (and we have no evidence that it is not). ∎

PROBLEMS

10-1. According to genetic theory, if two recessive genes that give rise to phenotypes A and B, respectively, are not linked, then the proportions of phenotypes AB, AB^c, A^cB, and A^cB^c will occur in the population in the proportions 1:3:3:9. Given these data for a random sample of size 80:

Type	AB	AB^c	A^cB	A^cB^c
Frequency	8	21	12	39

Calculate Λ and χ^2 for testing the theory and find a P-value.

10-2. A research team at the Roper Center in Connecticut checked the possibility of nonresponse bias in a survey of over 10,000 professors by telephoning a sample of 500 nonresponders. (The response rate of the original survey was 53 percent.) They obtained 477 interviews and found these results for one item ("There should be a top limit on income"):

Strongly agree	86
Agree with reservation	100
Disagree with reservation	119
Strongly disagree	172

Test the hypothesis that the population of nonresponders answer with the same proportions of the four answers as population of responders.

---
[9] See S. Wilks, *Mathematical Statistics*, New York: John Wiley & Sons, Inc., 1959, page 413.

Take the latter to be the proportions in the 5300 responders: .15, .21, .23, .41.

10-3. Problem 9-18 gave these data for 100 tosses of a pair of coins:

Number of heads	0	1	2
Frequency	25	50	25

to test the hypothesis that the three outcomes (0, 1, 2) are equally likely.
(a) The likelihood ratio for that H_0 was found there to be $\Lambda = .00277$. Calculate Pearson's χ^2 (and compare with $-2 \log \Lambda$).
(b) Calculate both Λ and χ^2 for the hypothesis $P(0) = P(2) = .25$.

10-4. Suppose poll A shows 960 out of 2000 as favoring a certain proposition and poll B shows 756 out of 1500 in favor. Is there evidence that they were not sampling the same population? (Assume independent random samples.)

10-5. For a discrete population with three possible outcomes, consider testing the equal likelihood of the outcomes on the basis of six independent trials.
(a) List the possible combinations of three frequencies that total $n = 6$.
(b) Find the distribution of χ^2 under H_0.
(c) Find the distribution of Λ under H_0.
(d) Sketch the c.d.f. of chi$^2(2)$ and superpose a sketch of the c.d.f. of χ^2.
(e) Calculate the power of the test $\chi^2 > 3$ for the particular alternative with $p_2 = p_3 = .5$.

10-6. An industrial plant reported the following summary of accidents per month, over a period of 32 months:

Number of accidents	0	1	2	3
Number of months	22	6	2	2

Test the hypothesis that the number of accidents per month follows a Poisson distribution. (Combine 3, 4, ..., into a single category, "3 or more.")

10-7. In the 80 games, one season, in which a baseball player had four times at bat per game, his hits per game were as follows:

Number of hits	0	1	2	3	4
Number of games	26	35	16	2	1

Test the hypothesis that the number of hits per game is binomial.

10-8. Show that if $Y_1, ..., Y_k$ are i.i.d. Poi(np_i), their conditional joint dis-

tribution given $\sum Y_j = n$ is then multinomial with parameters $(n; p_1, ..., p_k)$.
[Note: The distribution of the category frequencies in a random sample of size n from a population with k categories is multinomial with parameters $(n; p_1, ..., p_k)$. So one would then expect that χ^2 is like the sum of squares of k variables

$$Z_i = \frac{Y_j - np_j}{\sqrt{np_j}}$$

subject to the condition $\sum Z_i = 0$. As n becomes infinite, the Z_i are approximately $\mathcal{N}(0, 1)$. We'll show in Chapter 12 that the conditional distribution of the sum of squares of independent standard normal variables, given that their sum is 0, is chi$^2(k-1)$. (This reasoning leading to the asymptotic null distribution of χ^2 was suggested by R. A. Fisher.)]

10-9. Show that in the case of a population with just two categories, Pearson's χ^2 for testing $p = p_0$ is equal to Z^2, where Z is the usual Z-statistic we would use for this purpose.

10.4 Two-way contingency tables

The outcome of an experiment on which two categorical variables A and B are defined is a pair (A_i, B_j). Data are ordinarily summarized in a two-way table—frequencies n_{ij} in a cross-classification of outcomes on each of the two variables. For instance, if $i = 1, 2, 3$ and $j = 1, 2, 3, 4$, the table is as follows:

	B_1	B_2	B_3	B_4	
A_1	n_{11}	n_{12}	n_{13}	n_{14}	n_{1+}
A_2	n_{21}	n_{22}	n_{23}	n_{24}	n_{2+}
A_3	n_{31}	n_{32}	n_{33}	n_{34}	n_{3+}
	n_{+1}	n_{+2}	n_{+3}	n_{+4}	n

(The "+" in a subscript indicates a subscript that has been "summed out.") Such a table is called a *contingency table*. For an $r \times c$ table, the notation is as follows: n_{ij} is the frequency of (A_i, B_j), $i = 1, 2, ..., r$, and $j = 1, 2, ..., c$. (That is, the contingency table has r rows and c columns.) The entries in the right and lower margin are row and column totals:

$$n_{i+} = \sum_j n_{ij}, \quad n_{+j} = \sum_i n_{ij}, \quad n = n_{++} = \sum_i \sum_j n_{ij}. \quad (1)$$

Thus, the marginal entries n_{i+} and n_{+j} are frequencies in the marginal frequency distributions of A and B, respectively.

Data in the form of a contingency table can arise in another way: Several populations or experiments with the categorical variable A defined on them are sampled a finite number of times, the frequencies of the various categories recorded—n_{ij}, as the number of times category A_i occurs in population or experiment j. With either method, the interest usually lies in whether or not there is a relationship.

EXAMPLE 10.4a The following is a cross-classification of students according to college and according to citizenship:

	Liberal Arts	Technology	Agriculture	Education	
U.S.	33	16	12	9	70
Foreign	7	14	8	1	30
	40	30	20	10	100

These counts might have been obtained according to either method described above. That is, a sample of 100 students could have been cross-classified on the two variables, college and citizenship. Or, samples of size 70 and 30 could have been drawn from lists of U. S. and foreign students, respectively. Or, samples of 40, 30, 20, and 10 could have been drawn from lists of students in Liberal Arts, Technology, Agriculture, and Education, respectively.

One way of formulating the question of relationship is to ask whether the proportions of foreign students in the (sub)populations of students in the four colleges are the same, or not. Another, are the proportions of students in the four colleges the same for foreign as for U. S. students? Still another: Are the variables "college" and "citizenship" independent? ■

When viewing the experiment as one of sampling from a bivariate categorical population, we let $p_{ij} = P(A_i, B_j)$, and

$$p_{i+} = \sum_j p_{ij}, \quad p_{+j} = \sum_i p_{ij}, \quad p_{++} = \sum_i \sum_j p_{ij} = 1. \quad (2)$$

In this bivariate model we have conditional probabilities:

$$p_{i\mid j} = P(A_i \mid B_j) = \frac{p_{ij}}{p_{+j}}. \qquad (3)$$

[The notation $p_{i\mid j}$ would be ambiguous if specific numbers were involved: $p_{2\mid 3}$ could mean either $P(A_2 \mid B_3)$ or $P(B_2 \mid A_3)$.]

When data on the bivariate pair (A, B) are obtained as a random sample and summarized in the counts n_{ij}, the joint distribution of these counts is multinomial:

$$f(n_{11}, \ldots, n_{rc}) = \frac{n!}{\prod\prod n_{ij}!} \prod\prod p_{ij}^{n_{ij}}. \qquad (4)$$

When independent samples are obtained from c (sub)populations B_j and classified according to the categories of A, the cell frequencies n_{ij} are multinomial for each j, and the joint p.f. of all cell frequencies given the sample sizes n_{+j} is the product of c multinomial p.f.'s:

$$f(n_{11}, \ldots, n_{rc} \mid n_{+1}, \ldots, n_{+c}) = \prod_j \frac{n_{+j}!}{\prod_i n_{ij}!} \prod_i (p_{i\mid j})^{n_{ij}}. \qquad (5)$$

10.5 Independence and homogeneity

Given the data in a 2-way contingency table, the null hypothesis to be tested is that there is no relationship between the variables A and B, or that the model for a subpopulations B_j is independent of j. From the former point of view, we test *independence*:

$$H_{\text{ind}}: \quad p_{ij} = p_{i+}p_{+j}, \quad \text{all } i, j. \qquad (1)$$

From the latter, we test *homogeneity*:

$$H_{\text{hom}}: \quad p_{i\mid j} \text{ is a function of } i \text{ alone.} \qquad (2)$$

These superficially different models are in fact essentially equivalent:

THEOREM 1 *Categorical variables A and B are independent if and only if the variables $A \mid B_1, \ldots, A \mid B_c$ are identically distributed.*

[By "essentially" equivalent we mean this: If only the c univariate populations are given, each with categories $A_1, \ldots, A_r$, we can construct a bivariate model for (A, B) by using *any* distribution $(p_{+1}, \ldots, p_{+c})$ to form the joint probability function:

$$p_{ij} = p_{i\mid j} p_{+j}, \qquad (3)$$

and the theorem then applies to the resulting joint distribution, assuming that the proportions p_{+j} are positive.]

The theorem is an easy consequence of the following:

LEMMA *In the matrix* (n_{ij}) *the following three conditions are equivalent:*

(a) *rows are proportional,*
(b) *columns are proportional, and*
(c): $n_{ij} = \dfrac{n_{i+}n_{+j}}{n_{++}}.$

Given (c), the n_{ij} in any row are proportional to the column totals n_{+j}—with constant of proportionality n_{i+}/n_{++}; and the n_{ij} in any column are proportional to the row totals n_{i+}, with constant of proportionality n_{+j}/n_{++}. Given (a), there are constants a_i, b_j such that

$$\begin{pmatrix} n_{1j} \\ \vdots \\ n_{rj} \end{pmatrix} = b_j \begin{pmatrix} a_1 \\ \vdots \\ a_r \end{pmatrix},$$

which says that $n_{ij} = a_i b_j$. Summing on j and again on i we find that

$$n_{i+} = a_i \sum b_j, \quad n_{+j} = b_j \sum a_i, \quad n_{++} = \sum a_i \sum b_j,$$

and these imply (c). Similarly, given (b), we again have $n_{ij} = a_i b_j$, for some constants a_i and b_j, and (c) follows.

Theorem 1 is an immediate consequence of the lemma, applied to the numbers p_{ij}, because $p_{i\,|\,j} = p_{ij}/p_{+j}$. That is, the columns of the probability array are proportional if and only if the conditional probabilities for A given j are the same for every j, and the rows are proportional if and only if the conditional probabilities for B given i are the same for every i. And either of these conditions is equivalent to the factorization of p_{ij} that defines independence.

Imposing the conditions of models (1) and (2) on the joint p.f.'s (4) and (5) of the preceding section, we obtain the next three theorems, which give needed facts about the null distribution of the test statistics that we'll need in testing the hypothesis of no relationship:

THEOREM 2 *Under the hypothesis of independence, and given a random sample from* (A, B) *of size* n,

1. *The marginal totals for A and the marginal totals for B are independent multinomial vectors, with p.f.'s*

$$f_1(n_{1+}, \ldots, n_{r+}) = \dfrac{n!}{\prod n_{i+}!} \prod (p_{i+})^{n_{i+}}, \tag{4}$$

and
$$f_2(n_{+1}, \ldots, n_{+c}) = \frac{n!}{\prod n_{+j}!} \prod (p_{+j})^{n_{+j}}. \tag{5}$$

2. The joint p.f. of the cell frequencies is
$$f_{\text{ind}}(n_{11}, \ldots, n_{rc}) = \frac{n!}{\prod \prod n_{ij}!} \prod p_{i+}^{n_{i+}} \prod p_{+j}^{n_{+j}} = f_1 f_2 f_3,$$

where f_1 and f_2 are as defined by (4) and (5), and
$$f_3 = \frac{\prod n_{i+}! \prod n_{+j}!}{n! \prod \prod n_{ij}!}. \tag{6}$$

3. The function f_3 given by (6) is the conditional p.f. of the cell frequencies given all marginal totals:
$$f_{\text{ind}}(n_{11}, \ldots, n_{rc} \mid n_{1+}, \ldots n_{+c}) = f_3.$$

THEOREM 3 Under the hypothesis of homogeneity, given independent random samples of size n_{+j} from variables B_j, $j = 1, \ldots, c$,

1. The marginal totals n_{i+} are multinomial with p.f. f_1 given by (4).

2. The joint p.f. of the cell frequencies is
$$f_{\text{hom}}(n_{11}, \ldots, n_{rc}) = \prod_j \frac{n_{+j}!}{\prod n_{ij}!} \prod p_{i+}^{n_{i+}} = f_1 f_3, \tag{7}$$

where f_3 is given by (6).

3. The conditional p.f. of the n_{ij} given all marginal totals is f_3:
$$f_{\text{hom}}(n_{11}, \ldots, n_{rc} \mid n_{1+}, \ldots n_{+c}) = f_3.$$

The p.f.'s f_1 and f_2 are in the exponential family, and this fact can be used (with extensions of results given in §8.3) to show the following[10]:

THEOREM 4 Under H_{ind} [given by (1)] the set of all marginal totals $(n_{1+}, \ldots n_{+c})$ is complete and sufficient for the set of all $r + c$ marginal probabilities $(p_{1+}, \ldots, p_{+c})$. Under H_{hom} [given by (2)], the marginal row totals $(n_{1+}, \ldots, n_{r+})$ are complete and sufficient for the corresponding marginal probabilities $(p_{1+}, \ldots, p_{r+})$.

---

[10] E. L. Lehmann, *Testing Statistical Hypotheses*, 2nd Ed. Pacific Grove, Cal.: Wadsworth & Brooks/Cole, 1986, page 142.

374   Chapter 10   Analysis of Categorical Data

THEOREM 5   *Under H_{ind} or H_{hom}, the expected cell frequencies, given all marginal totals, are*

$$E(n_{ij} \mid n_{1+}, \ldots, n_{+c}) = \frac{n_{i+} n_{+j}}{n}. \tag{8}$$

To establish Theorem 5, we observe that n_{ij} is an unbiased estimate of $n p_{i+} p_{+j}$, so (because of completeness) $E(n_{ij} \mid \text{MT})$ is the only function of the marginal totals (MT) that is an unbiased estimate of $n p_{i+} p_{+j}$. But $n_{i+} n_{+j}/n$ is also a function of the marginal totals and is an unbiased estimate of $n p_{i+} p_{+j}$ (Problem 10-17).

We'll be using a goodness of fit statistic to test H_{ind} or H_{hom}, and for either the chi-square or the likelihood ratio statistic we need m.l.e.'s.

THEOREM 6   *Under H_{ind} the m.l.e.'s of the marginal probabilities are the corresponding marginal relative frequencies:*

$$\widehat{p}_{i+} = \frac{n_{i+}}{n}, \quad \widehat{p}_{+j} = \frac{n_{+j}}{n}.$$

Under H_{hom} the m.l.e.'s of the common conditional probabilities $p_{i \mid j} = p_{i+}$ are the corresponding row-sum relative frequencies:

$$\widehat{p}_{i \mid j} = \frac{n_{i+}}{n}.$$

Under either H_{ind} or H_{hom}, the m.l.e.'s of the expected cell frequencies $e_{ij} = n p_{ij}$ (H_{ind}) or $e_{ij} = n_{+j} p_{i \mid j}$ (H_{hom}) are given by

$$\widehat{e}_{ij} = \frac{n_{i+} n_{+j}}{n}. \tag{9}$$

Proofs of the assertion in Theorem 6 are left as exercises for the reader. We use these results in forming the likelihood ratio statistic:

THEOREM 7   *The likelihood ratio statistic used for testing H_{ind} against the alternative that the marginal variables are not independent is identical with the likelihood ratio statistic for testing H_{hom} against the hypothesis that the populations defined by the B_j's are not identically distributed, namely:*

$$\Lambda = \frac{\prod \prod (n_{i+}/n)^{n_{ij}}}{\prod \prod (n_{ij}/n_{+j})^{n_{ij}}} = \prod \prod \left(\frac{n_{i+} n_{+j}/n^2}{n_{ij}/n} \right)^{n_{ij}}. \tag{10}$$

Consider first Λ_{hom}. From (5) in §10.4, the likelihood function,

10.5 Independence and homogeneity

given the count data n_{ij}, is

$$L(p_{1|1}, \ldots, p_{r|c}) = \prod \prod (p_{i|j})^{n_{ij}}. \tag{11}$$

There are $r-1$ free parameters in each of the c columns, so the number of free parameters over which to maximize is $c(r-1)$. The maximum of L over this parameter space is at $\hat{p}_{i|j} = n_{ij}/n_{+j}$. From (7), the likelihood function under H_{hom}, given the count data n_{ij}, is

$$L(p_{1+}, \ldots, p_{r+}) = \prod_i p_{i+}^{n_{i+}}. \tag{12}$$

This depends on $r-1$ free parameters, and maximizing on these we obtain the m.l.e. $\hat{p}_{i+} = n_{i+}/n$. Substituting these in the likelihood functions produces the ratio (10). Problem 10-14 calls for a derivation of the likelihood ratio for testing H_{ind}, which again leads to (10).

EXAMPLE 10.5a The contingency table for the count data in Example 10.4a are repeated here, along with the table of $\hat{e}_{ij}$'s:

33	16	12	9	70		28	21	14	7	70
7	14	8	1	30		12	9	6	3	30
40	30	20	10	100		40	30	20	10	100

Then

$$-2\log \Lambda = 2[n\log n + \sum\sum n_{ij} \log n_{ij} - \sum n_i \log n_i - \sum n_j \log n_j]$$
$$= 2[460.517 + 276.544 - 731.962] \doteq 10.2.$$

The number of degrees of freedom associated with Λ is the difference in dimension of the parameter spaces over which we maximized:

$$d.f. = c(r-1) - (r-1) = (c-1)(r-1) = 3. \blacksquare$$

The likelihood ratio is but one way of measuring the discrepancy between the data and the null hypothesis. Another is given by Pearson's chi-square statistic, in the case where the category probabilities (under either H_0) depend on unknown parameters. In testing independence: $p_{ij} = p_{i+}p_{+j}$, for all i, j, the m.l.e.'s of the expected cell frequencies are $n_{i+}n_{+j}/n$, so the chi-square statistic is

$$\chi^2 = \sum_i \sum_j \frac{(n_{ij} - n_{i+}n_{+j}/n)^2}{n_{i+}n_{+j}/n}.$$

Under the hypothesis of independence, probabilities of the rc categories (the cells of the contingency table) are defined by $(r-1)+(c-1)$ parameters, and these must be estimated. The number of degrees of freedom for the chi-square statistic is then

$$rc - 1 - (r+c-2) = (r-1)(c-1),$$

the same as the d.f. for Λ.

For testing H_{hom}, the expected cell frequencies are the same as for testing independence, so χ^2 is the same. The unrestricted model is defined by rc parameters (the cell conditional probabilities), but in each of the c columns these sum to 1. So, were there no parameters to be estimated, the chi-square statistic would have $rc - c$ d.f.; since there are $r-1$ parameters to be estimated, the number of degrees of freedom is again $(rc - c) - (r-1)$.

EXAMPLE 10.5b To find χ^2 for testing H_{ind} or H_{hom} with the data of the preceding example, we use the observed and expected frequencies given there:

$$\chi^2 = \frac{(33-28)^2}{28} + \frac{(16-21)^2}{21} + \cdots + \frac{(9-7)^2}{7} = 9.80,$$

with 3 d.f. So χ^2 is not much different from $-2\log \Lambda = 10.20$. (These statistics are asymptotically equal under H_0, but there is evidence that H_0 does not hold in this situation.) ∎

10.6 Large- and small-sample tests

Since large values of χ^2 or of $-2\log \Lambda$ are to be taken as evidence against independence (or homogeneity), the approximate P-value in a particular case is the area under the p.d.f. of chi$^2[(r-1)(c-1)]$ to the right of the observed value of the statistic. To test at fixed α the critical value of χ^2 or of $-2\log \Lambda$ is that for which the area to the right is α.

EXAMPLE 10.6a In Examples 10.5a and 10.5b we found $\chi^2 \doteq 9.8$ and $-2\log \Lambda = 10.2$. The corresponding P-values are .020 and .017, respectively. For a test at $\alpha = .05$, the critical value of either statistic is 7.81. ∎

A small-sample test, called *Fisher's exact test*, can be constructed as follows, based on f_3, the conditional p.f. of the cell frequencies given the marginal totals, as obtained in Theorems 2 and 3:

$$f_3 = \frac{\prod n_{i+}! \prod n_{+j}!}{n! \prod \prod n_{ij}!}. \tag{1}$$

10.6 Large- and small-sample tests

For a test with fixed level α, we consider all contingency tables of size $r \times c$. For each of the distinct sets of marginal totals, we choose, for our critical region, tables that are most "extreme" (as suggested by H_0 and H_A), up to the point where the total of their conditional probabilities f_3 is as close as possible to α without exceeding α. Then

$$P(\text{reject } H_0 \mid H_0) = E[P(\text{reject } H_0 \mid \text{marginal totals \& } H_0)] \leq E(\alpha) = \alpha.$$

To define "extreme," we use a measure of discrepancy between the data and H_0—either χ^2 or Λ will do. Tables with the smallest Λ-values or largest χ^2-values are used first in forming a critical set.

Although it may seem formidable to prepare a critical set of tables for *each* possible set of marginal totals, all we need is to calculate the probabilities of those tables with the same marginal totals as those of the contingency table defined by the observed data.

When $r = c = 2$, the probabilities f_3 are hypergeometric:

$$f_3 = \frac{n_{1+}!}{n_{11}!n_{12}!} \cdot \frac{n_{2+}!}{n_{21}!n_{22}!} \cdot \frac{n_{+1}!n_{+2}!}{n!} = \frac{\binom{n_{1+}}{n_{11}}\binom{n_{2+}}{n_{21}}}{\binom{n}{n_{+1}}}.$$

This defines a univariate distribution—the distribution, say, of n_{11}, since the other entries n_{ij} are determined from it by the marginal totals.

EXAMPLE 10.6b Suppose, to test $p_1 = p_2$ in two Bernoulli populations, we obtain the following contingency table:

	Success	Failure	Total
From population 1:	3	3	6
From population 2:	7	2	9
Total	10	5	15

There are six 2×2 tables with these same marginal totals, namely:

$$\begin{vmatrix} 6 & 0 \\ 4 & 5 \end{vmatrix} \quad \begin{vmatrix} 5 & 1 \\ 5 & 4 \end{vmatrix} \quad \begin{vmatrix} 4 & 2 \\ 6 & 3 \end{vmatrix} \quad \begin{vmatrix} 3 & 3 \\ 7 & 2 \end{vmatrix} \quad \begin{vmatrix} 2 & 4 \\ 8 & 1 \end{vmatrix} \quad \begin{vmatrix} 1 & 5 \\ 9 & 0 \end{vmatrix}$$

Identifying each table with the value of n_{12}, we obtain the following conditional distributions for χ^2 and Λ, given the marginal totals and H_0:

n_{12}	χ^2	$-2\log \Lambda$	Probability
0	5	6.73	.042
1	1.25	1.323	.252
2	0	0	.42
3	1.25	1.243	.24
4	5	5.178	.045
5	11.25	3.69	.002

According to $-2\log \Lambda$, we'd put points into the critical region in the order (from most to least evidence against H_0) 5, 0, 4, 1, 3, 2. If testing at the fixed level $\alpha = .10$, we'd use the critical region $\{5, 0, 4\}$ for n_{12}, which has conditional probability .089. (Adding any other values would send α above .10.) The statistic χ^2 orders the n_{12}-values in almost the same way, except that 1 and 3 are tied, and 0 and 4 are tied. For the given data, $n_{12} = 3$, for which $P \doteq .29$.

If we are merely reporting results as summarized in a P-value, we sum the probabilities for values of n_{12} that are 3 or more: $P = .58$. ∎

Fisher's exact test is an instance of a *conditional test*, in which a critical region is defined for the data *given* the value of some statistic U. Such a test may have optimality properties if the statistic U is not informative concerning the hypothesis being tested. Even in the case of a 2×2 table, the matter seems not to be resolved to everyone's satisfaction. There may be a little information in the marginal totals— but only a little, so that the conditional test is good enough for practical purposes, even if not strictly optimal.

PROBLEMS

10-10. Folk lore in the Air Force has it that fighter pilots are more likely to sire daughters than sons. A study investigating the possibility that high G-force exposure may be a factor found these data:[11]

		G-force exposure	
		High	Low
Sex of offspring	Male	295	66
	Female	287	100

Test the hypothesis of no relationship between G-force exposure and sex

---

[11] B. B. Little, et al, "Pilot and astronaut offspring: Possible G-force effects on human sex ratio," *Aviation Space & Envir. Med.* 58 (1987), 707-709.

of offspring.

10-11. An investigation of the relationship between sex and field of study used questionnaires to classify individuals as masculine (M), feminine (F), androgynous (A), or undifferentiated (U). Data were obtained as shown in the following table:[12]

	M	F	A	U
Non	8	6	2	14
Bio	11	10	2	7
Phys	4	9	8	9

Test independence of sex and field of study.

10-12. Suppose four observations on a bivariate categorical variable yield the following marginal totals:

			2
			2
1	2	1	4

(a) Find expected cell frequencies under H_{ind}, given these marginal totals.

(b) For each table of possible frequencies with the given marginal totals find the value of Λ and the probability of that table given the marginal totals. (In calculating Λ, define $x^x = 1$ when $x = 0$.)

10-13. Derive the m.l.e.'s given in Theorem 6 (§10.5).

10-14. Derive the likelihood ratio for testing H_{ind}.

10-15. For a 2×2 contingency table, determine a critical set for testing independence at $\alpha \leq .3$ given these marginal totals: $n_{1+} = 4$, $n_{+1} = 9$, $n_{2+} = 8$, $n_{+2} = 3$.

10-16. Four independent trials are carried out in each of three Bernoulli populations, with results as follows:

	Population		
	1	2	3
0	1	3	0
1	3	1	4

---

[12] R. Baker, "Masculinity, femininity, and androgyny among male and female science and non-science majors," *School Science and Math* 84 (1984). 459-467.

Carry out the conditional exact test for homogeneity—that is, for the hypothesis that the probability of success is the same in all three populations.

10-17. Show that $n_{i+}n_{+j}/n$ is an unbiased estimate of $np_{i+}p_{+j}$ under H_{hom} or H_{ind}.

10.7 Measures of association

Dividing Pearson's χ^2 by n produces a quantity that depends only on the sample proportions in the various cells:

$$\frac{\chi^2}{n} = \sum\sum \frac{\left(\frac{n_{ij}}{n} - \frac{n_{i+}}{n}\cdot\frac{n_{+j}}{n}\right)^2}{\frac{n_{i+}}{n}\cdot\frac{n_{+j}}{n}}.$$

A similar expression in terms of *population* proportions or cell probabilities defines a population parameter:

$$\theta^2 = \sum\sum \frac{(p_{ij} - p_{i+}p_{+j})^2}{p_{i+}p_{+j}}.$$

This is 0 when and only when $p_{ij} = p_{i+}p_{+j}$ for all i, j. Its value is sometimes used as the basis of a measure of departure from independence:

$$\gamma = \left[\frac{\theta^2}{\min(r-1, c-1)}\right]^{1/2}.$$

This has the property that $0 \leq \gamma \leq 1$. Moreover, $\gamma = 1$ if and only if there is exactly one nonzero element in each row of the matrix (p_{ij}) if $r \geq c$, or in each column if $r < c$.

For measuring association in a sample it would be natural to use the corresponding statistic:

$$C = \left[\frac{\chi^2}{n\min(r-1, c-1)}\right]^{1/2}.$$

When rows (and columns) of the matrix (n_{ij}) are proportional, $C = 0$. When $C = 1$, there is exactly one nonzero element in each row (if $r \geq c$). It has been argued that except in the 2×2 case, the measure C is not easy to interpret. In particular, it does not permit calculation of such things as a conditional probability of correctly classifying a population member according to variable A, given its category in variable B.

10.8 McNemar's test

In testing homogeneity, we have assumed independent random samples for each subpopulation B_j, each sample classified according to variable A. In the 2×2 case, there is another way in which data are sometimes collected, namely, using matched pairs, in which case the samples are not independent. One instance of this is that in which individuals are classified before and after a treatment—each individual is his or her own "control."

EXAMPLE 10.8a Suppose subjects are asked for an opinion (for or against) on some proposal, and are then given some background education and asked again for their opinions. Data might be summarized like this:

	Before	After	
Pro	8	13	21
Anti	12	7	19
	20	20	

Testing equality of proportions "pro" in the before and after responses using the tests of §10.5 would be inappropriate since we don't have independent samples. Instead, the 20 pairs of responses can be put into a contingency table:

		After: Pro	Anti	
Before:	Pro	7	1	8
	Anti	6	6	12
		13	7	20

(Of course, the entries in this table are not all implied by the previous table—only the marginal entries.) Now the entries are counts with a multinomial distribution, assuming that the 20 individuals constitute a random sample. Some questions one might ask are (1) whether the proportion of "pros" is changed by the education, (2) whether the proportion who change from pro to anti is the same as the proportion who change from anti to pro, and (3) whether the proportion of those who are pros before the education and change to antis is the same as the proportion of those who are antis before the education and change to pros. ■

We consider a response variable with two categories, $\{1, 2\}$, and a random sample of individuals with two responses under different circumstances or a random sample of pairs with a response from each member of the pair, modeled as a two-way table of joint probabilities, with p_{ij}-notation as for any 2×2 table. Suppose we want to test H_0: $p_{+1} = p_{1+}$, to answer question (1) in the above example. This hypothesis of exchangeability is equivalent to H_0: $p_{12} = p_{21}$ (Problem 10-22). To carry out either a chi-square test or a likelihood ratio test we need the likelihood function:

$$L(\mathbf{p}) = p_{11}^{n_{11}} p_{12}^{n_{12}} p_{21}^{n_{21}} p_{22}^{n_{22}}, \qquad (1)$$

which is maximized (subject to $\Sigma\Sigma p_{ij} = 1$) at $\hat{p}_{ij} = n_{ij}/n$. Under H_0, with $p_{12} = p_{21} = p$, the likelihood function is

$$L_0(p_{11}, p_{22}) = p_{11}^{n_{11}} p^{n_{12}+n_{21}} p_{22}^{n_{22}}, \qquad (2)$$

where $2p = 1 - p_{11} - p_{12}$. This likelihood is maximized at

$$\hat{p}_{11} = \frac{n_{11}}{n}, \quad \hat{p}_{22} = \frac{n_{22}}{n}, \quad \hat{p} = \frac{n_{12} + n_{21}}{2n}. \qquad (3)$$

(See Problem 10-23.)

Using estimates (3) to find the (estimated) expected cell frequencies, we can calculate a value of the chi-square test statistic. The null distribution of χ^2 is chi$^2(k)$, where $k = 4 - 1 - 2 = 1$. (Two parameters are estimated.)

In the likelihood ratio approach we calculate $L_0(\hat{p}_{11}, \hat{p}_{22})$, the maximum of L_0 in (2), and divide by $L(\hat{\mathbf{p}})$, the maximum of L as given by (1). Observe that once again the main diagonal elements play no role:

$$\Lambda = \frac{[\tfrac{1}{2}(n_{12} + n_{21})]^{n_{21}+n_{21}}}{n_{12}^{n_{12}} n_{21}^{n_{21}}}.$$

The number of degrees of freedom for Λ is $3 - 2 = 1$, the difference between the number of free parameters in (1) and the number in (2).

Another way of viewing the problem, when one realizes that it hinges on just the off-diagonal frequencies, is to look at the conditional probability of one of them, given their sum:

$$p^* = \frac{p_{12}}{p_{12} + p_{21}}. \qquad (4)$$

In terms of p^*, the null hypothesis is $p^* = 1/2$. A Z-score for testing this hypothesis is

or (for $H_A: p \neq p_0$)
$$Z = \frac{n_{12} - \frac{1}{2}(n_{12} + n_{21})}{\sqrt{\frac{1}{4}(n_{12} + n_{21})}},$$

$$Z^2 = \frac{(n_{12} - n_{21})^2}{n_{12} + n_{21}}. \tag{5}$$

The null distribution of Z^2 is approximately chi$^2(1)$. The test based on these facts is **McNemar's test**. It is equivalent to the chi-square test described above (see Problem 10-24).

EXAMPLE 10.8b Returning to the data in Example 10.8a, we see that when $i = j$, the estimated expected values np_{ij} are equal to the observed values, so the frequencies on the main diagonal produce 0's as the corresponding terms of χ^2. (Your intuition may have suggested that there is no information regarding H_0 in those frequencies.) The off-diagonal expected frequencies are equal:
$$n\hat{p} = \frac{1}{2}(n_{12} + n_{21}) = 3.5,$$
so that
$$\chi^2 = \frac{(3.5 - 1)^2}{3.5} + \frac{(3.5 - 6)^2}{3.5} = \frac{25}{7} \doteq 3.57.$$

The Z^2-statistic given by (5) is
$$Z^2 = \frac{(6-1)^2}{6+1} = \frac{25}{7},$$
equal, as we expect, to χ^2. The likelihood ratio is
$$\Lambda = \frac{\left(\frac{7}{40}\right)^7 \left(\frac{7}{40}\right)^{6+1} \left(\frac{6}{20}\right)^6}{\left(\frac{7}{20}\right)^7 \left(\frac{1}{20}\right)^1 \left(\frac{6}{20}\right)^6 \left(\frac{6}{20}\right)^6} = \frac{\left(\frac{7}{40}\right)^7}{\left(\frac{1}{20}\right)\left(\frac{6}{20}\right)^6} = .1379,$$

and $-2 \log \Lambda = 3.96$. The approximate P-values for Z^2 (or χ^2) and Λ are .06 and .05, respectively, from chi$^2(1)$, the asymptotic distribution. ∎

PROBLEMS

10-18. Calculate the measure C of association for each of the following:

(a) $\begin{array}{|cc|} 32 & 88 \\ 68 & 112 \\ \end{array}$

(b) $\begin{array}{|cc|} 200 & 0 \\ 0 & 600 \\ \end{array}$

10-19. Calculate the value of $-2\log \Lambda$ for testing independence, for each table in the preceding problem. (Recall that when a cell frequency is zero, the corresponding factor in the likelihood function is missing.)

10-20. For the data in Problem 10-11, calculate C and the following approximation to its standard deviation: $[n\min(r-1, c-1)]^{-1/2}$.

10-21. Each of 85 patients with Hodgkin's disease was paired with a normal sibling who did not have the disease.[13] In 26 pairs, both individuals had had tonsillectomies; in 15, only the normal individual had had a tonsillectomy; and in 7, only the one with Hodgkin's disease had had a tonsillectomy. Construct an appropriate two-way contingency table and test the hypothesis that the proportion of all those who have had tonsillectomies is the same among those with Hodgkin's disease as among those who haven't had it.

10-22. Show that for a 2×2 contingency table the hypothesis $p_{12} = p_{21}$ is equivalent to $p_{+1} = p_{1+}$.

10-23. Show that under the null hypothesis $p_{12} = p_{21}$, the maximum likelihood estimates are those given as (3) in §10.8.

10-24. Show that the chi-square statistic for testing $p_{12} = p_{21}$ in a 2×2 contingency table is equivalent to the McNemar statistic.

10-25. A multicenter trial[14] studied a treatment for blindness in premature babies. Half of 136 babies with disease in both eyes had their left eye treated and the other half, their right eye. In all but 40 babies, the outcomes in both eyes were the same. In 34 of the 40, the outcome was favorable in the treated eye and unfavorable in the untreated eye and in 6, the outcome was unfavorable in the treated eye and favorable in the untreated eye. Test the null hypothesis that the treatment is not effective.

10-26. Given the data of Example 10.8a, consider the hypothesis H_1 that Bernoulli variables A and B are i.i.d.
(a) Construct the two-way table of joint probabilities under H_1.
(b) Calculate the likelihood ratio statistic for testing H_1 against the alternative that A and B are merely exchangeable (the hypothesis H_0 in McNemar's test).

---

[13]S. Johnson and R. Johnson, "Tonsillectomy history in Hodgkin's disease," *New Engl. J. Med.* 287 (1972), 1122-1125. The article carried out the *wrong* chi-square test (as was pointed out in several letters to the editor) and found $\chi^2 = 1.53$.

[14]"Multicenter trial of cryotherapy for retinopathy of prematurity," *Archives of Ophthalmology* 106 (1988), 471-479.

(c) Calculate the likelihood ratio statistic for testing H_1 against the alternative that the cell probabilities are restricted only in that they sum to 1.

(d) Find the value of $-2\log\Lambda$ in (b) and (c), and look for a relationship among these and the value found in Example 10.8b.

10-27. Give a null hypothesis for question (3) in Example 10.8a, show that it is equivalent to a test of independence, and carry out a chi-square test, applying it to the data in that example.

10.9 Log-linear models for 2 × 2 tables

The model for a two-way contingency table can be recast in terms of the logarithms of the cell probabilities—the "natural" parameters,[15] a formulation whose extension to tables of higher dimension seems to provide useful insights and models. When independent observations (A, B) are cross-classified, the appropriate multinomial model is in the general exponential family, with log-likelihood

$$\log L = \sum\sum n_{ij}\log p_{ij} = \sum\sum n_{ij}\log e_{ij} - n\log n, \qquad (1)$$

where e_{ij} is a convenient notation for the expected cell frequency, np_{ij}.

The condition $p_{ij} = p_{i+}p_{+j}$ that defines independence of the marginal variables can be rewritten in terms of the new parameters:

$$H_{\text{ind}}: \log e_{ij} = \log e_{i+} + \log e_{+j} - \log n,$$

where $e_{i+} = np_{i+}$ and $e_{+j} = np_{+j}$. Thus, under H_{ind}, the logarithm of the expected cell frequencies is of the form

$$\log e_{ij} = \alpha_i + \beta_j + \lambda, \qquad (2)$$

an additive combination of a term depending only on i and a term depending only on j (and a constant). In such sums we can define the constant λ so that

$$\sum \alpha_i = \sum \beta_j = 0. \qquad (3)$$

Moreover, if $\log e_{ij}$ has the structure (2), then A and B are independent,

---

[15] When the term in the exponent of an exponential family density that mixes parameter and observation is of the form $\Sigma\theta_k R_k(x)$, the parameters θ are said to be *natural*.

which we can see as follows. From

$$np_{ij} = e_{ij} = e^{\alpha_i + \beta_j + \lambda},$$

it follows (upon summing) that

$$np_{i+} = e^{\alpha_i + \lambda} \sum e^{\beta_j}, \quad np_{+j} = e^{\beta_j + \lambda} \sum e^{\alpha_i}, \quad n = e^{\lambda} \sum e^{\alpha_i} \sum e^{\beta_j},$$

and therefore

$$n^2 p_{i+} p_{+j} = e^{\alpha_i + \beta_j + \lambda} \left(e^{\lambda} \sum e^{\alpha_i} \sum e^{\beta_j} \right) = n^2 p_{ij}.$$

In general, a set of constants $\{\log e_{ij}\}$ cannot be written as an additive combination of the form (2). This form, without terms that can account for "interaction," corresponds precisely to the hypothesis of independence.

10.10 Three-way classifications

Let independent observations be classified according to three variables: Denote the categories of A by A_i ($i = 1, ..., r$), of B by B_j ($j = 1, ..., s$), and of C by C_k ($k = 1, ..., t$). Let $p_{ijk} = P(A_i, B_j, C_k)$, and denote the corresponding frequency by n_{ijk}. As before, a "+" as a subscript will indicate that that subscript has been summed out. In particular,

$$p_{+++} = \sum\sum\sum p_{ijk} = 1, \quad n_{+++} = \sum\sum\sum n_{ijk} = n.$$

The contingency table for the n_{ijk} is three-dimensional, visualized as a rectangular parallelepiped divided into $r \times s \times t$ cubical cells. The third dimension, corresponding to variable C, defines what are conveniently shown as *layers* of the table, each layer being a two-dimensional table of cross-classification frequencies for (A, B). In the case of a $2 \times 2 \times 2$ table, the frequencies may be shown in layers, like this:

	B_1	B_2		B_1	B_2		B_1	B_2	
A_1	n_{111}	n_{121}	n_{1+1}	n_{112}	n_{122}	n_{1+2}	n_{11+}	n_{12+}	n_{1++}
A_2	n_{211}	n_{221}	n_{2+1}	n_{212}	n_{222}	n_{2+2}	n_{21+}	n_{22+}	n_{2++}
	n_{+11}	n_{+21}	n_{++1}	n_{+12}	n_{+22}	n_{++2}	n_{+1+}	n_{+2+}	n
	Layer C_1			Layer C_2			Marginal layer		

In the most general model the only restriction on the p_{ijk} is that they sum to 1. The likelihood function is multinomial:

$$L = \prod \prod \prod p_{ijk}^{n_{ijk}}, \qquad (1)$$

with logarithm

$$\log L = \sum \sum \sum n_{ijk} \log e_{ijk} - n \log n. \qquad (2)$$

(As before, $e_{ijk} = np_{ijk}$, the expected cell frequency.) The m.l.e.'s of the p_{ijk} are the cell relative frequencies, $\hat{p}_{ijk} = n_{ijk}/n$. In this maximization the dimension of the parameter space is $rst - 1$.

We now consider restricted models for the expected cell frequency, which have the form

$$\log e_{ijk} = \alpha_i + \beta_j + \gamma_k + \xi_{jk} + \eta_{ki} + \zeta_{ij} + \lambda, \qquad (3)$$

where

$$\sum \alpha_i = \sum \beta_j = \sum \gamma_k = 0,$$

and

$$\sum_j \xi_{jk} = \sum_k \xi_{jk} = \sum_k \eta_{ki} = \sum_i \eta_{ki} = \sum_i \zeta_{ij} = \sum_j \zeta_{ij} = 0.$$

The general model, with *only* the necessary requirement that $\sum p_{ijk} = 1$, cannot be expressed in the form (3) without the addition of a "third-order interaction" term involving all three subscripts. By successively dropping second-order interaction terms we create a hierarchy of successively simpler models. This can be done in one of six ways, one of which results in the following:

H_1: $\log e_{ijk} = \alpha_i + \beta_j + \gamma_k + \xi_{jk} + \eta_{ki} + \zeta_{ij} + \lambda$

H_2: $\log e_{ijk} = \alpha_i + \beta_j + \gamma_k + \xi_{jk} + \eta_{ki} + \lambda$

H_3: $\log e_{ijk} = \alpha_i + \beta_j + \gamma_k + \xi_{jk} + \lambda$

H_4: $\log e_{ijk} = \alpha_i + \beta_j + \gamma_k + \lambda.$

We'll refer to the most general model (with a third-order interaction term) as H_0—not to be confused with a null hypothesis we may want to test.

In searching, among the various H_i for one that in some sense best explains the data, we can test any candidate in the list against the next more complicated model—a test of the hypothesis that the additiional parameter in the latter can be dropped. Thus, starting with H_4 we can test $\zeta_{jk} = 0$ in H_3; rejecting this we can test $\eta_{ki} = 0$ in H_2, and so on. Determining the level (α) of this kind of sequential test procedure in terms of the levels of the individual tests is not easy because the test statistics in successive tests are not independent. Nevertheless, the procedure is useful in suggesting an appropriate model, provided one does not take significance levels too seriously. (A similar awkwardness will be encountered in connection with nested models in regression analyses in Chapter 14.)

A test for H_a in the context of H_b—that is, for H_a against the alternative $H_b - H_a$, where $a > b$—can be carried out using the likelihood ratio statistic:

$$\log \Lambda_{ab} = \log \left\{ \frac{\sup L(p)}{\sup L(p)} \right\} = \sum \sum \sum n_{ijk} (\log \widehat{e}_a - \log \widehat{e}_b), \quad (4)$$

where $\widehat{e}_m$ is the m.l.e. of the expected cell frequency e_{ijk} under H_m. The asymptotic distribution of $-2 \log \Lambda$ is chi-square, with d.f. equal to the dimension of H_b minus the dimension of H_a.

Model H_4 is the model for complete independence of the three variables, in which the cell probabilities factor:

$$p_{ijk} = p_{i++} p_{+j+} p_{++k}, \quad \text{all } i, j, k. \quad (5)$$

Clearly, if this condition is satisfied, $\log e_{ijk}$ is of the form H_4. But the converse is also true (Problem 10-28). The log-likelihood under H_4 is

$$\log L = \sum \sum \sum n_{ijk} \log(p_{i++} p_{+j+} p_{++k})$$

$$= \sum n_{i++} \log p_{i++} + \sum n_{+j+} \log p_{+j+} + \sum n_{++k} \log p_{++k}.$$

Maximizing over H_4 (dimension $= r + s + t - 3$) we obtain

$$\widehat{p}_{i++} = \frac{n_{i++}}{n}, \quad \widehat{p}_{+j+} = \frac{n_{+j+}}{n}, \quad \widehat{p}_{++k} = \frac{n_{++k}}{n}, \quad (6)$$

as the (sufficient) m.l.e.'s. The corresponding m.l.e.'s of the e_{ijk} are

$$\widehat{e}_{ijk} = \frac{n_{i++} n_{+j+} n_{++k}}{n^2}. \quad (7)$$

To test independence (H_4) against the general alternative that the variables are not independent (H_0), we use the statistic

$$-2 \log \Lambda_{40} = -2 \sum \sum \sum n_{ijk} (\log \widehat{e}_4 - \log \widehat{e}_0)$$

10.10 Three-way classifications

$$= -2\left\{\sum\sum\sum n_{ijk}[\log(n_{i++}n_{+j+}n_{++k}) - \log n_{ijk}] - 2n \log n\right\}. \quad (8)$$

This is asymptotically chi-square with d.f. $= (rst - 1) - (r + s + t - 3)$.

EXAMPLE 10.10a Consider the following $2 \times 2 \times 2$ table for variables (A, B, C), for 100 observations in a random sample:

6	14	20		6	4	10		12	18	**30**
4	16	20		34	16	50		38	32	**70**
10	**30**	40		**40**	**20**	60		**50**	**50**	100
	Layer C_1				Layer C_2				Marginal layer	

We have used boldface type to identify the marginal sums needed for calculating the estimates (6). These estimated expected cell frequencies (under H_4) are:

6	6		9	9
14	14		21	21

Substituting in (8) we find $-2\log\Lambda_{40} = 30.685$, a number far out in the tail of the $\text{chi}^2(4)$ distribution. So at the most used significance levels, independence would be rejected. But perhaps there is a relationship described by a model *between* H_0 and H_4. (To be continued.) ∎

Adding the second-order interaction term ξ_{jk} to the terms in the model H_0 yields the model called H_3 above. Under this hypothesis, it can be shown (Problem 10-29) that then $p_{ijk} = p_{i++}p_{+jk}$ or, equivalently,

$$P(A_i B_j C_k) = P(A_i) P(B_j C_k).$$

In this case we say that A is jointly independent of B and C. With this model, the log-likelihood is

$$\log L = \sum n_{i++}\log p_{i++} + \sum\sum n_{+jk}\log p_{+jk},$$

which has a maximum value at the m.l.e.'s:

$$\widehat{p}_{i++} = \frac{n_{i++}}{n}, \quad \widehat{p}_{+jk} = \frac{n_{+jk}}{n}. \quad (9)$$

These are sufficient for H_3 and define the expected cell frequencies under the model H_3:

$$\widehat{e}_{ijk} = \frac{n_{i++}n_{+jk}}{n}. \tag{10}$$

The parameter space of H_3 has dimension $(r-1) + (st-1) = r + st - 2$.

Adding to H_3 the second-order interaction term η_{ki} producess the model named H_2 above. This can be shown to be equivalent to the following condition on the cell probabilities:

$$p_{ijk} = \frac{p_{i+k}p_{+jk}}{p_{++k}}, \quad \text{all } i, j, k.$$

In terms of $(A, B, C,)$ this can be written as

$$P(A_i B_j \mid C_k) = P(A_i \mid C_k) P(B_j \mid C_k), \quad \text{all } i, j, k.$$

Thus, A and B are independent conditional on C—that is, or independent in the layers of C. The log-likelihood under H_2 is

$$\log L = \sum \sum n_{+jk} \log p_{+jk}$$
$$+ \sum \sum n_{i+k} \log p_{i+k} - \sum n_{++k} \log p_{++k}.$$

The m.l.e.'s of the parameters p_{i+k}, p_{+jk}, p_{++k} can be shown to combine to yield the m.l.e. of the expected cell frequency:

$$\widehat{e}_{ijk} = \frac{n_{i+k}n_{+jk}}{n_{++k}}. \tag{11}$$

The dimension of the parameter space is $(rt - t) + (st - t) + (t - 1)$.

Finally, in the model H_1 all three second-order interaction terms are present. However, this model has no simple interpretation in terms of independence. It can be shown that in the $2 \times 2 \times 2$ case the model criterion is equivalent to the constancy of the *cross-product ratio* across layers:

$$\frac{p_{111}p_{221}}{p_{211}p_{121}} = \frac{p_{112}p_{222}}{p_{212}p_{122}}.$$

The marginal totals n_{ij}, n_{ik}, n_{jk} are sufficient, but it is not possible to give direct formulas for the $\widehat{e}_{ijk}$'s in terms of them. Yet, these m.l.e.'s can be approximated by using an iterative process that will be illustrated in Example 10.10b below. The dimension of the parameter space for H_1 is $rt + st + tr - (r + s + t)$.

In (4) we have defined the likelihood ratio Λ_{ab} appropriate for testing H_a in the context of H_b. Returning to the particular hierarchy defined at the outset: H_0, H_1, H_2, H_3, H_4, we note from (4) that

10.10 Three-way classifications 391

$$-2\log\Lambda_{04} = -2[\log\Lambda_{01} + \log\Lambda_{12} + \log\Lambda_{23} + \log\Lambda_{34}].$$

The terms in this sum are log-likelihood ratios for successive comparisons of models in the hierarchy. Degrees of freedom are differences in dimension of the two parameter spaces involved in a comparison:

a	Dimension of H_a	ab	Degrees of freedom, H_a vs. H_b
0	$rst - 1$	10	$(r-1)(s-1)(t-1)$
1	$rs + st + tr - (r+s+t)$	21	$(r-1)(s-1)$
2	$rs + st - t - 1$	32	$(r-1)(t-1)$
3	$r + st - 2$	43	$(s-1)(t-1)$
4	$r + s + t - 3$		

For this hierarchy of models we can calculate the likelihood ratios for successive pairs (even though these are not independent) and take them as measures of degradation of the fit as the model is successively simplified. We give no rule (if, indeed, one can be given) for deciding how much simplification is justified. The choice of which of the six hierarchies to consider—which interactions are reasonabe to drop, and in what order— would have to be based on a knowledge of the particular the field of application.

EXAMPLE 10.10b We continue our analysis of the data in Example 10.10a, where we tested H_4 against H_0, finding $\Lambda_{40} = 30.685$. To find the other Λ_{ab}'s we need some expected cell frequencies; for H_3 and H_2 we use (10) and (11):

H_3:

3	9	**10**	12	6	**18**	15	15	**30**
7	21	**28**	28	14	**42**	35	35	**70**
10	**30**	40	**40**	**20**	60	**50**	**50**	100

H_2:

5	15	**20**	$\frac{40}{6}$	$\frac{20}{6}$	**10**			
5	15	**20**	$\frac{200}{6}$	$\frac{100}{6}$	**50**			
10	**30**	40	**40**	**20**	60			

The marginal totals that determine the m.l.e.'s for H_2 and H_3 are shown in boldface type. For H_1 we use an iterative process that will be explained below, with these results:

H_1:
$$\begin{array}{|cc|c|} \hline 5.17 & 14.82 & 19.99 \\ 4.84 & 15.17 & 20.01 \\ \hline 10.01 & 29.99 & 40 \\ \end{array} \qquad \begin{array}{|cc|c|} \hline 16.83 & 3.18 & 10.01 \\ 33.16 & 16.83 & 49.99 \\ \hline 39.99 & 20 & 60 \\ \end{array}$$

With the various estimates substituted in the appropriate likelihood function we obtain the following values of $-2\log \Lambda$:

a	b	$-2\log \Lambda_{ab}$	d.f.
4	3	17.261	1
3	2	12.654	1
2	1	.029	1
1	0	.742	1
4	0	30.685	4

Independence, and joint independence of A with B and C seem farfetched, but the model of independence in layers fits the data quite well, and it wouldn't be of much help to include all three of the interaction terms. We don't give any precise rule for model selection since, as we pointed out earlier, an overall "α" is not simply related to significance levels of the individual tests involved.

To find estimates of expected cell frequencies under H_1, we proceed as follows: Starting with 1's in place of all n_{ijk},

$$\begin{array}{|cc|c|} \hline 1 & 1 & 2 \\ 1 & 1 & 2 \\ \hline \end{array} \qquad \begin{array}{|cc|c|} \hline 1 & 1 & 2 \\ 1 & 1 & 2 \\ \hline \end{array}$$

we multiply each row by a factor that makes the row totals correct—that is, "correct" according to the marginal totals:

$$\begin{array}{|ccc|} \hline 10 & 10 & 20 \\ 10 & 10 & 20 \\ \hline 20 & 20 & \\ \end{array} \qquad \begin{array}{|ccc|} \hline 5 & 5 & 10 \\ 25 & 25 & 50 \\ \hline 30 & 30 & \\ \end{array}$$

Next, we multiply columns by a factor that will make the column totals (which are now wrong) correct—the first column by 1/2, and the second column by 1/3, in the first array, and the first column by 1/3 and the second by 2/3 in the second array:

$$\begin{array}{cc|c}5 & 15 & 20 \\ 5 & 15 & 20 \\ \hline 10 & 30 \end{array} \qquad \begin{array}{cc|c} 20/3 & 10/3 & 10 \\ 100/3 & 50/3 & 50 \\ \hline 40 & 20 \end{array} \qquad \begin{array}{cc} 35/3 & 55/3 \\ 115/3 & 95/3 \end{array}$$

Now the row and column sums are correct, but the sums across layers (far right) are not. The next adjustment, multiplying across layers by an appropriate factor, makes them correct:

$$\begin{array}{cc} 36/7 & 162/11 \\ 114/23 & 288/19 \end{array} \qquad \begin{array}{cc} 48/7 & 36/11 \\ 760/23 & 320/19 \end{array} \qquad \begin{array}{cc} 12 & 18 \\ 38 & 32 \end{array}$$

The row and column totals are now a bit off, so the whole cycle is repeated. One more repetition yields the results given earlier, in which the row and column totals are within .01 of the right values. This is usually close enough for calculating Λ. ∎

The iterative scheme illustrated in the example can be shown to give the correct m.l.e.'s, not only in the case of H_1, but (with appropriate modifications) in other cases as well. And in those instances in which the m.l.e.'s can be calculated by formulas involving the relevant marginal totals, the iterative process quickly yields the exact answers given by the formulas. For details of this admittedly sketchy account of three-way tables and a thorough study of higher-dimensional tables we refer to advanced treatises.[16]

PROBLEMS

10-28. Show that if the model for three-way tables is the one we've called H_4: $\log e_{ijk} = \alpha_i + \beta_j + \gamma_k + \lambda$, the cell probabilities satisfy the multiplicative condition for independence.

10-29. Show that if the model for three-way tables has just one second-order interaction term (H_3), then one variable of classification is jointly independent of the other two, as claimed on page 389.

10-30. Derive the m.l.e.'s of the probabilities p_{i++} and p_{+jk} under H_3.

10-31. Derive the expected cell frequency estimates under H_3 by means of an iterative scheme. (One cycle should do it.)

10-32. Verify the calculation of $-2\log\Lambda$ in at least one case, as given in

---

[16]Y. A. Bishop, S. E. Fienberg, and P. W. Holland, *Discrete Multivariate Analysis: Theory and Practice*, Cambridge, Mass: M. I. T. Press, 1975.

Example 10.10b.

10-33. Data 72 families enrolled in a California HMO included statistics on church-going habits (C), smoking status of the mother (M), and smoking status of the father (F). Define categories as follows:

M and F: $1 =$ never smoked, $2 =$ smoked but quit, $3 =$ smokes,
C: $1 =$ at least 1 parent attend regularly with children,
$2 =$ children attend regularly without parents,
$3 =$ family members attend only sporadically,
$4 =$ no one in family attends

Data on 72 families are summarized as follows, with mother's smoking status as the layering variable:

M = 1:

		C		
F:	1	2	3	4
1	11	0	5	6
2	11	1	0	3
3	4	0	1	1

M = 2:

		C		
	1	2	3	4
1	1	0	0	2
2	4	0	0	1
3	0	0	0	0

M = 3:

		C		
	1	2	3	4
1	2	0	0	4
2	2	0	3	0
3	3	1	3	3

Calculate the value of the likelihood ratio statistic for a test of
(a) complete independence (H_4 vs. H_0).
(b) joint independence of C with F and M (H_3 vs. H_0).
(c) conditional independence of F and M given C (H_2 vs. H_0).

11
Sequential Analysis

Our discussion of testing hypotheses up to this point assumed a fixed sample size n and a critical boundary, either set arbitrarily or determined by specified error sizes α and β. In situations where data are collected sequentially, it may become clear early on what the situation is, whereas the test would call for proceeding to collect all the data before coming to any conclusion. If the evidence becomes overwhelming, one way or the other, before the entire sample of size n is observed, there is nothing to prevent one from stopping the sampling and coming to a conclusion; but if this is done, the planned error sizes are not applicable.

Perhaps the first attempt at dealing with this kind of problem was the *double sampling* plan of Dodge & Romig. This plan called for a preliminary sample of size m and, on the basis of this sample, a decision either to reject H_0, to accept H_0, or to take a second sample of size $n - m$ and then make a decision. Clearly one could extend this idea to a plan with more than two stages. With such a plan, one might be able to get by with fewer observations (on average) for a given degree of reliability of the conclusion. Another advantage of using such multi-stage schemes is that early results can guide the planning of later stages.

11.1 The sequential likelihood ratio test

The sequential plan to be studied here, presented and treated extensively by Wald,[1] assumes that observations are obtained in sequence, one at a time, allowing for an examination of the results after each observation. The plan is designed to achieve specified error sizes in testing a simple null against a simple alternative hypothesis; but when these hypotheses are given as specified values of a parameter θ, the test

---

[1]Wald, Abraham, *Sequential Analysis*, New York, John Wiley & Sons, Inc., 1950.

may also be suitable for testing composite hypotheses of the form $\theta \leq \theta^*$ vs. $\theta > \theta^*$.

For testing the null hypothesis model $f_0(x)$ against the alternative $f_1(x)$, the Wald sequential likelihood ratio test (SLRT) is defined by constants A and B, where $A < B$, and is based on the likelihood ratio calculated after each observation. After n observations it is

$$\Lambda_n = \frac{f_0(X_1, \ldots, X_n)}{f_1(X_1, \ldots, X_n)}.$$

At this nth stage, the procedure is as follows: Reject H_0 if $\Lambda_n \leq A$, accept H_0 if $\Lambda_n \geq B$, but continue sampling if $A < \Lambda_n < B$. When the observations are i.i.d., the likelihood ratio is a product:

$$\Lambda_n = \prod_{i=1}^{n} \frac{f_0(X_i)}{f_1(X_i)}, \tag{1}$$

and taking logarithms converts this to a sum:

$$\log \Lambda_n = \sum_{i=1}^{n} \log \frac{f_0(X_i)}{f_1(X_i)} = Z_1 + \cdots + Z_n, \tag{2}$$

where

$$Z_i = \log \frac{f_0(X_i)}{f_1(X_i)}. \tag{3}$$

Because the X's are independent, so are the Z's.

When expressed in terms of logarithms, the inequality for continuing the sampling becomes

$$\log A < \sum_{1}^{n} Z_i < \log B. \tag{4}$$

Figure 11-1 shows a typical plot of $\log \Lambda_n$ as a function of the integer n. [The points (n, Λ_n) are connected by straight lines merely for ease in following the progress of the test.] A decision is reached when the path first crosses a boundary. The decision is then to accept H_0 if it is the upper boundary that is first crossed and to reject H_0 if it is the lower. In either case, sampling stops at this point.

As we saw in §9.2, a likelihood ratio is sometimes expressible in terms of some more familiar statistic. Thus, if the population distributions is in the exponential family, the log-likelihood ratio for H_0: $\theta = \theta_0$ vs. H_1: $\theta = \theta_1$ is

$$\log \Lambda_n = n \log \frac{B(\theta_0)}{B(\theta_1)} + [Q(\theta_0) - Q(\theta_1)] \sum_{i=1}^{n} R(X_i).$$

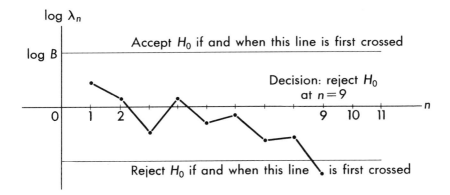

Figure 11-1

The inequality for continuing the sampling can then be expressed as

$$C + Kn < \sum_{i=1}^{n} R(X_i) < D + Kn,$$

where the constants C, D, and K depend on A, B, θ_0, and θ_1. Graphically, the critical curves for the statistic $\sum R(X)$ are two parallel lines.

EXAMPLE 11.1a Consider testing $\theta = 1$ vs. $\theta = 2$ when $X \sim \mathrm{Exp}(\theta)$. The log-likelihood ratio for a single observation is

$$Z = \log\left(\frac{e^{-X}}{2e^{-2X}}\right) = X - \log 2.$$

After n observations,

$$\log \Lambda_n = Z_1 + \cdots + Z_n = \sum_1^n X_i - (\log 2)n.$$

If $A = 1/4$ and $B = 4$, the inequality (4) for continuing sampling becomes

$$-1.386 + .693\,n < \sum_1^n X_i < 1.386 + .693\,n.$$

In Figure 11-2 we have plotted the critical lines and the sample sums for the sequence of observations (1.31, .19, .80, .10, .49, 1.2, 3.0, .33, . . .). At the seventh observation the plot of $\sum X$ crosses the upper critical line, and at this point we accept H_0. ∎

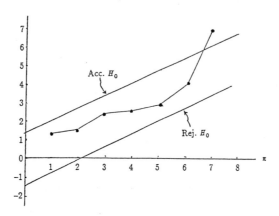

Figure 11-2

Given constants A and B, for use in (4) and the associated decision rule, a next logical question is how the decision rule or test performs. In particular, what are the error sizes, α and β? Formally, we can write

$$\alpha = P_0(\Lambda_1 \leq A) + P_0(A < \Lambda_1 < B \text{ and } \Lambda_2 \leq A) + \cdots,$$

where P_i here and in what follows denotes probability given H_i. A similar expression can be written for β. But to calculate these error sizes or to find values of A and B to satisfy a specification of α and β is a formidable task indeed. Wald devised the following scheme. Rigorous derivations are beyond our scope, since we'd need a measure on the sample space of infinite sequences. So we argue informally.

The probability of a set of sequences leading to rejection of H_0 at stage k is actually a marginal distribution in the space of the first k observations, since the decision to terminate at that stage is made on the basis of those observations. Let E_k denote the set of k-tuples $(x_1, \ldots, x_k)$ that call for rejection of H_0 at stage k. Then

$$\alpha = P_0(\text{rej. } H_0) = \sum_k P_0(E_k),$$

and

$$1 - \beta = P_1(\text{rej. } H_0) = \sum_k P_1(E_k).$$

The inequality $f_0(\mathbf{x}) \leq A f_1(\mathbf{x})$ holds at each point of E_k, so

11.1 The sequential likelihood ratio test

$$P_0(E_k) = \int_{E_k} f_0(\mathbf{x})\, d\mathbf{x} \le A \int_{E_k} f_1(\mathbf{x})\, d\mathbf{x} = A\, P_1(E_k).$$

Summing on k yields

$$\alpha = \sum_k P_0(E_k) \le A(1 - \beta),$$

or

$$\frac{\alpha}{1-\beta} \le A. \tag{5}$$

A similar argument gives this result:

$$\frac{\beta}{1-\alpha} \le \frac{1}{B}. \tag{6}$$

The relations (5) and (6) are satisfied by the α and β of the test defined by given values of A and B. In fact, they are almost equalities: When Λ_n goes outside the interval (A, B), it usually does so not by much. We show next that if we take (5) and (6) as *equalities* to define an A and B in terms of given α and β, the test so defined will have error sizes close to the specified α and β.

So, suppose sizes α and β are specified. To try to achieve these we define a sequential likelihood ratio test with

$$A = \frac{\alpha}{1-\beta}, \quad B = \frac{1-\alpha}{\beta}. \tag{7}$$

Let α' and β' denote the actual type I and II error sizes of this test. Then, according to (5) and (6),

$$\frac{\beta'}{1-\alpha'} \le \frac{1}{B} = \frac{\beta}{1-\alpha}, \quad \text{and} \quad \frac{\alpha'}{1-\beta'} \le A = \frac{\alpha}{1-\beta}.$$

Multiplying through by $1 - \alpha'$ and $1 - \beta'$, respectively, and adding the results, we obtain

$$\alpha' + \beta' \le \alpha + \beta.$$

From this it is clear that at most one of the achieved error sizes will be larger than what is specified, and indeed, for small α and β,

$$\alpha' \le \frac{\alpha'}{1-\beta'} \le \frac{\alpha}{1-\beta} \doteq \alpha(1+\beta), \quad \text{and} \quad \beta' \le \frac{\beta'}{1-\alpha'} \le \frac{\beta}{1-\alpha} \doteq \beta(1+\alpha).$$

Therefore, the one error size that may exceed the specification does not do so by more than a factor of about $1 + \alpha$ or $1 + \beta$. For instance, if $\alpha = \beta = .05$, neither α' nor β' can be larger than about .0525.

Although using the critical values A and B given by (7) results in error sizes that are at most slightly larger than the α and β specified, they may be smaller. This would be disconcerting only in that the statistician would be using more data than necessary. Wald shows[2] that the difference in sample size would be at most slight: In certain common cases it amounts to at most one or two observations in small samples and around one or two percent in larger samples.

We have glossed over a point that may have occurred to you. Suppose it turns out, for some sample sequence, that the sampling process never terminates. The probability that this happens is 0, as we demonstrate next.

Observe first that the log-likelihood Z (for a single observation) will be 0 only when f_0 and f_1 agree. If these p.d.f.'s are to define essentially different distributions, the set of points where $Z \neq 0$ would have positive probability in either distribution. So assume $P(Z = 0) < 1$, and choose $d > 0$ and $q > 0$ such that either $P(Z > d) \geq q$, or $P(Z < -d) > q$. In the former case, choose an integer r which divides n so that $rd > \log(B/A)$, and break up the sum (2) into k sections of length r:

$$\sum_1^n Z_i = (Z_1 + \cdots Z_r) + \cdots + (Z_{(k-1)r+1} + \cdots + Z_{kr}).$$

For the first group of terms,

$$P\left(\left|\sum_1^r Z_i\right| > \log\frac{B}{A}\right) \geq P\left(\sum_1^r Z_i > \log\frac{B}{A}\right) \geq P\left(\sum_1^r Z_i > rd\right)$$

$$\geq P(Z_i > d,\ i = 1,\ \ldots,\ r) \geq q^r > 0.$$

In the second case, choose r so that $-rd < -\log(B/A)$, with the same result. And of course the same result would follow for *any* of the groups of r Z's:

$$P\left(\left|Z_{(j-1)r+1} + \cdots + Z_{jr}\right| > \log\frac{B}{A}\right) \geq q^r. \tag{8}$$

(Recall that the Z's are i.i.d.)

Now let N denote the (random) sample size required for a decision. The events $[N > n]$ form a descending sequence, and

$$P(N \text{ not finite}) = P\left(\bigcap [N > n]\right) = \lim_{n \to \infty} P(N > n) = \lim_{k \to \infty} P(N > kr),$$

where r is the integer defined above. (The last equality follows from the fact that the sequence $P(N > n)$ is monotone and bounded below, so it has a limit, which is also the limit of any subsequence.) The inequality

---
[2] Wald, A., *op. cit.*, pp. 65-69.

11.1 The sequential likelihood ratio test

$N \geq kr$ implies that no decision has been reached through stage kr, so that

$$\log A < \log \Lambda_n < \log B, \quad n \leq kr.$$

This in turn means that none of the groups of Z's can exceed $\log(B/A)$:

$$|Z_{(j-1)r+1} + \cdots + Z_{jr}| < \log \frac{B}{A}, \quad j = 1, 2, \ldots, k, \qquad (9)$$

for otherwise, $\log \Lambda_n$ could not remain between $\log A$ and $\log B$ for $n \leq kr$. But from (8), the probability of (9) is not greater than $1 - q^r$. Hence,

$$P(N > kr) \leq (1 - q^r)^k. \qquad (10)$$

and so

$$P(N \text{ not finite}) \leq \lim_{k \to \infty} (1 - q^r)^k = 0,$$

as was to be shown. (Because $q > 0$, $1 - q^r < 1$.)

PROBLEMS

11-1. Consider testing $p = .5$ against $p = .2$ in a Bernoulli population with $p = P(S)$ and $1 - p = P(F)$.
(a) Express the sequential rule for $\alpha = .1$ and $\beta = .2$ in terms of Λ.
(b) Express the rule in (a) in terms of the number of successes after n trials.
(c) Suppose we observe this sequence: $S, F, F, S, F, S, F, S, F, \ldots$. What conclusion is reached, and at what stage?

11-2. Carry out a sequential test for $p = \frac{1}{2}$ against $p = \frac{2}{3}$, with error sizes $\alpha = .05$ and $\beta = .1$ using each of the following basic experiments:
(a) Toss a penny in the usual way, with $p = P(\text{heads})$.
(b) Stand a quarter on its edge, on a table, holding it with a finger of the left hand so that you can see the head is upright. Then flick the side of the quarter with the other hand so that it spins. It will slow down and finally fall flat on the table with one side showing (heads or tails). Again let $p = P(\text{heads})$.

11-3. Construct the sequential likelihood ratio test for $\mu = \mu_0$ against the alternative $\mu = \mu_1$, based on random sampling from $N(\mu, \sigma^2)$, where σ^2 is known.
(a) Express the inequality for continuing sampling in terms of the sample sum and determine the nature of the curves bounding the region where the inequality holds. Sketch these for $\mu_0 = 4$, $\mu_1 = 6$, $\alpha = \beta =$

.05, $\sigma^2 = 9$.

(b) Compare the curves in (a) with those defined for $\alpha = .1$ and $\beta = .2$ [and other parameters the same as in (a)].

(c) Compare the curves in (a) with those used for the alternative $\mu_1 = 10$ [and other parameters the same as in (a)].

11-4. Obtain the SLRT, in terms of the sample sum, for H_0: $\lambda = 1$ against H_1: $\lambda = 2$, where sampling is random from Poi(λ).

11-5. Fill in the details showing the inequality (8) when $P(Z < -d) > q$.

11-6. Write out the "similar argument" for (6).

11-7. Suppose sequential observations on a Cauchy population:

$$f(x \mid \theta) = \frac{1/\pi}{1 + (x - \theta)^2},$$

with these results: 3/2, 3, −2, 0, −1, 5, −3, 1, 1/2, −3/2, 4, ... Test the hypothesis $\theta = 0$ against $\theta = 2$, when $\alpha = \beta = .05$.

11-8. Solve equations (5) and (6) as approximate equalities for α and β.

11.2 Expected sample size

For any given SLRT, the number of observations required to reach a decision is a random variable. Call it N. The distribution of N depends on which population is being sampled, but we have seen that $P(N < \infty) = 1$. Let $p_i = P(N = i)$, where then $\sum p_i = 1$. We'll show next that the moments of N are finite.

Choose an integer r, as in the derivation of (8) of the preceding section, and group the terms that define $E(N^k)$ in bunches of r:

$$E(N^k) = [1^k p_1 + \cdots + r^k p_r]$$

$$+ [(r+1)^k p_{r+1} + \cdots + (2r)^k p_{2r}] + \cdots$$

$$\leq r^k(p_1 + \cdots + p_r) + (2r)^k(p_{r+1} + \cdots + p_{2r}) + \cdots.$$

From (10) of the preceding section we have

$$p_{ir+1} + \cdots + p_{(i+1)r} \leq \sum_{ir+1}^{\infty} p_j = P(N > ir) \leq (1 - q^r)^i.$$

Consequently,

11.2 Expected sample size 403

$$E(N^k) \le r^k + (2r)^k(1-q^r) + (3r)^k(1-q^r)^2 + \cdots.$$

The standard ratio test shows the right-hand side to be a convergent series if (as is the case) $0 < q < 1$, so $E(N^k)$ is finite.

The variables Z_i are i.i.d., and let Z denote another variable with that same distribution. The event $[N \ge i]$ is independent of any events involving Z_i, because whether $N \ge i$ or not is determined by the values of the Z_j's for $j < i$. Now define Y_i to be the indicator variable of the event $[N \ge i]$. Then Y_i and Z_i are independent, and

$$E(\log \Lambda_N) = E\left\{\sum_{i=1}^{\infty} Y_i Z_i\right\} = \sum_{i=1}^{\infty} E(Y_i Z_i) = E(Z) \sum_{i=1}^{\infty} E(Y_i), \quad (1)$$

provided that it is permissible to interchange summation and expectation. And this is the case when $E(|Z|)$ and $E(N)$ are finite. For,

$$\sum_{i=1}^{\infty} E(Y_i) = \sum_{i=1}^{\infty} P(N \ge i) = \sum_{i=1}^{\infty} \sum_{j=i}^{\infty} P(N = j)$$

$$= \sum_{i=1}^{\infty} \sum_{i=1}^{j} P(N = j) = \sum_{i=1}^{\infty} j P(N = j) = E(N), \quad (2)$$

and therefore

$$\sum_{i=1}^{\infty} |E(Y_i Z_i)| \le \sum_{i=1}^{\infty} E(Y_i) E(|Z_i|) = E(|Z|) E(N).$$

Combining (1) and (2), we find the following useful expression for the expected sample size (sometimes calles *average sample number*):

$$E(N) = \frac{E(\log \Lambda_N)}{E(Z)}. \quad (3)$$

We can approximate the numerator in (3) as follows: Λ_n is approximately a Bernoulli variable, whose value is $\log A$ if the decision is to reject H_0, and $\log B$ if the decision is to accept H_0:

$$E(\log \Lambda_N) \doteq (\log A) P(\text{rej. } H_0) + (\log B) P(\text{acc. } H_0). \quad (4)$$

Of course, (3) and (4) depend on the population distribution assumed for calculation of expected values. In particular,

$$E(N \mid H_0) \doteq \frac{1}{E(Z \mid H_0)} [\alpha \log A + (1-\alpha) \log B],$$

and

$$E(N \mid H_1) \doteq \frac{1}{E(Z \mid H_1)}[(1-\beta)\log A + \beta \log B].$$

EXAMPLE 11.2a To test $\mu = 0$ against $\mu = 1$ in $N(\mu, 1)$ with error sizes $\alpha = \beta = .01$, we use the constants $A = 1/99$ and $B = 99$. Then

$$E(\log \Lambda_N \mid H_0) \doteq (1 - 2\alpha) \log 99 = 4.503.$$

The log-likelihood ratio is

$$Z = \log \frac{f_0(X)}{f_1(X)} = \log\left\{\frac{\exp(-\frac{1}{2}X^2)}{\exp[-\frac{1}{2}(X-1)^2]}\right\} = \frac{1}{2} - X,$$

so $E(Z) = E(\frac{1}{2} - X) = \frac{1}{2} - \mu$. When $\mu = 0$,

$$E(N) = \frac{E(\log \Lambda_N \mid H_0)}{E(Z)} \doteq \frac{4.503}{1/2 - 0} \doteq 9.$$

A similar calculation yields $E(N) \doteq 9$ when $\mu = 1$.

These expected sample sizes might be compared with the sample size required in a test with fixed sample size, with the same α and β. The most powerful (Neyman-Pearson) critical region is of the form $\bar{X} > k$, and the given error sizes are achieved with a sample of size $n \doteq 22$, rather larger than the expected size with a sequential test. Of course, the sequential test *could* require more than 22 trials; it is only on average that it takes 9. ∎

11.3 Power of the SLRT

Even though we use particular values of a parameter θ to define the null and alternative hypotheses:

$$H_0: \theta = \theta_0, \quad H_1: \theta = \theta_1,$$

it may well be of interest to study the performance of the sequential test for other values of θ. This is certainly the case when the simple θ_0 and θ_1 we use are simply chosen to represent what we really want as competing hypotheses:

$$H_0: \theta \leq \theta', \quad H_1: \theta > \theta'.$$

Once we obtain the power function $\pi(\theta)$ of the SLRT, we can then find the expected sample size as given by (3) of §11.2 for other values of θ. To do this we derive $\pi(\theta)$ following an ingenious method of Wald's.

11.3 Power of the SLRT

Let θ be fixed, a value at which we want to find $\pi(\theta)$. We first determine a number $h \neq 0$ for which

$$E_\theta\left\{\left(\frac{f(X \mid \theta_0)}{f(X \mid \theta_1)}\right)^h\right\} = 1. \qquad (1)$$

[The notation E_θ means average with respect to $f(x \mid \theta)$.] The expected value on the left is surely 1 when $h = 0$, but there is usually another value of h for which it is 1. In particular, $h = -1$ works when $\theta = \theta_0$ and $h = 1$, when $\theta = \theta_1$.

EXAMPLE 11.3a Returning to the setting of Example 11.2a, testing $\mu = 0$ vs. $\mu = 1$ in $\mathcal{N}(\mu, 1)$, we write the following as the integrand in (1):

$$\left\{\frac{f(X \mid \theta_0)}{f(X \mid \theta_1)}\right\}^h = \left\{\frac{\exp(-X^2/2)}{\exp[-(X^2 - 2X + 1)/2]}\right\}^h = \left(e^{-X + 1/2}\right)^h = e^{-hX + h/2}.$$

Taking the expected value with respect to $\mathcal{N}(\mu, 1)$, we obtain

$$E_\mu\left\{e^{-hX + h/2}\right\} = e^{h/2}\psi_X(-h) = e^{h/2}e^{-\mu h + h^2/2}.$$

[Here, ψ is the m.g.f. of $\mathcal{N}(\mu, 1)$—see (12) in §6.8]. For this to be equal to 1 the exponent must be 0, and setting the exponent equal to 0 we find two solutions: $h = 0$ (as predicted), and $h = 2\mu - 1$. The latter can also be written with μ as a function of h: $\mu = \frac{1}{2}(h + 1)$. ∎

Continuing the derivation of the power function we write the expected value in (1) as an integral:

$$\int_{-\infty}^{\infty} \left(\frac{f(X \mid \theta_0)}{f(X \mid \theta_1)}\right)^h f(x \mid \theta)\, dx = 1.$$

This implies that the integrand [nonnegative, with integral 1] is a p.d.f:

$$f^*(x \mid \theta) = \left(\frac{f(X \mid \theta_0)}{f(X \mid \theta_1)}\right)^h f(x \mid \theta).$$

Consider now the auxiliary problem of testing (for the fixed θ) the simple hypotheses H^*: $f^*(x \mid \theta)$ vs. H: $f(x \mid \theta)$, and a sequential test for this problem, defined by the critical constants A^h and B^h, where A and B are the corresponding critical constants in the test we are analyzing (i.e., for f_0 vs. f_1). The inequality for continuing the sampling in the auxiliary problem is

$$A^h < \prod \frac{f^*(X_i \mid \theta)}{f(X_i \mid \theta)} = \prod \left(\frac{f(X_i \mid \theta_0)}{f(X_i \mid \theta_1)} \right)^h < B^h.$$

Upon taking the $1/h$ power, we see that this is precisely the inequality for continuing sampling in the original problem, and rejecting θ_0 is equivalent to rejecting H^*. (If h happens to be negative, we'd take the lower boundary to be B^h and the upper one to be A^h.) Thus,

$$\pi(\theta) = P_\theta(\text{rej. } H_0) = P_\theta(\text{rej. } H^*) = P_{H^*}(\text{rej. } H^*) = \alpha^*,$$

where α^* is the type I error size for the auxiliary problem. Problem 11-8 gives an approximate expression for the "α" in a SLRT in terms of the critical values A and B. From this, applied to the auxiliary problem, we have

$$\pi(\theta) = \alpha^* = \frac{B^h - 1}{B^h - A^h}. \tag{2}$$

This, together with the relation between θ and h, gives us a pair of parametric equations for the power function.

A rough sketch of the power function can sometimes be made from its plot at five points corresponding to these values of h: $-\infty, -1, 0, 1, \infty$. When $A < 1$ and $B > 1$, $\pi(-\infty) = 0$, and $\pi(\infty) = 1$. We saw earlier that θ_0 corresponds to $h = -1$ and θ_1, to $h = 1$, and of course $\pi(\theta_0) = \alpha$ and $\pi(\theta_1) = 1 - \beta$. Equation (1) does not define θ for $h = 0$, but the value of $\pi(\theta)$ when $h = 0$ can be determined by passing to the limit as h tends to 0. Similarly, the expression (2) is indeterminate when $h = 0$, but it can be evaluated by l'Hospital's rule:

$$\pi(\theta) \mid_{h \to 0} = \frac{\log B}{\log B - \log A}. \tag{3}$$

EXAMPLE 11.3b In the preceding example, for testing $N(\mu, 0)$ vs. $N(\mu, 1)$, we found that $\mu = \frac{1}{2}(h+1)$. Thus, when $\alpha = \beta$ (so that $A = 1/B$), the following points provide a rough plot:

h	$-\infty$	-1	0	1	∞
μ	$-\infty$	0	$1/2$	1	∞
$\pi(\mu)$	0	α	$1/2$	$1 - \beta$	1

Figure 11-3 shows the power function for $\alpha = \beta = .1$ and for $\alpha = \beta = .01$, which was sketched using the above points (plus a few additional points: $h = \pm 1/2, \pm 1/4$). ■

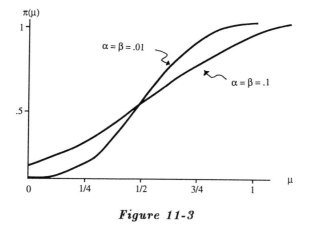

Figure 11-3

With the power function in hand, we can find the expected sample size for values of θ other than the two used to construct the test, in view of (4) of §11.2, which now becomes

$$E(\log \Lambda_N \mid \theta) \doteq (\log A)\,\pi(\theta) + (\log B)\,[1 - \pi(\theta)]. \qquad (4)$$

This is useful to know when the values θ_0 and θ_1 were chosen simply as representative values of θ in setting up a sequential test for $\theta \leq \theta^*$ against $\theta > \theta^*$. Some of the values of θ between θ_0 and θ_1 are usually in an indifference zone, where wrong decisions are not apt to be costly; yet, as we'll see in the next example, the expected sample size in this zone can be quite large. This is not unexpected, since it ought to take longer to decide which action to take when the actual θ is about as close to θ_0 as it is to θ_1. Ways of altering the procedure to deal with this annoyance will not be taken up here.

EXAMPLE 11.3c In Example 11.2a, for testing $\mu = 0$ against $\mu = 1$ in $\mathcal{N}(\mu, 1)$, we found the expected sample size under H_0 or H_1 to be about nine, compared with 22 for a sample of fixed size. We saw in Example 11.2a that $E(Z \mid \mu) = \frac{1}{2} - \mu$. With this in (3) and (4) of §11.2, we find the following when $\alpha = \beta = .01$:

μ	$-\infty$	0	1/4	1/2	3/4	1	$-\infty$
$\pi(\mu)$	0	.01	.09	.5	.91	.99	0
$E(N)$	0	9	15.1	21.1	15.1	9	0

The graph of $E(N)$ is shown in Figure 11-4 for $\alpha = \beta = .01$ and also for the case $\alpha = \beta = .10$. It is worth noting that a test of $\mu = 0$ against $\mu = 1$ with a fixed sample size requires $n = 22$ in the former case, and $n = 6$ in the latter—larger than the *expected* sample size even at $\mu = 1/2$, although of course the required sample size in a particular instance can be less or more than the expected size. ■

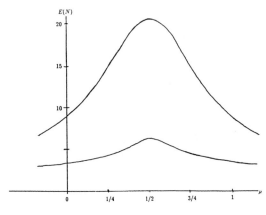

Figure 11-4

PROBLEMS

11-9. Consider a sequential likelihood ratio test of $p = p_0$ vs. $p = p_1$, where p is the probability of "success" in a Bernoulli population.
(a) Find the relation between p and h [as defined by (2)].
(b) Find $E(Z)$ [see (5) in §11.3].
(c) Plot the power function when $p_0 = .5$, $p_1 = .7$, and $\alpha = \beta = .05$.

11-10. Find the expected sample size when $p = p_0$ and when $p = p_1$ for the case in (c) of the preceding problem.

11-11. Find $E(N)$ as a function of p for the case in (c) of Problem 11-9.

11-12. For testing λ_0 against λ_1 with i.i.d. observations from Poi(λ), obtain the relation

$$\lambda = \frac{(\lambda_1 - \lambda_0)^h}{1 - (\lambda_0/\lambda_1)^h},$$

where h is defined by (3) in §11.1.

11-13. Obtain graphs of the power function and $E(N)$ for a sequential test of $\theta_0 = 2$ against $\theta_1 = 1$, based on a sequence of i.i.d. observations from Exp(θ), to achieve $\alpha = .05$ and $\beta = .10$.

12
Multivariate Distributions

Multivariate distributions were encountered in earlier chapters—basic definitions were given in Chapters 3 and 4, and the contingency tables in Chapter 10 are for several categorical variables. This chapter presents material on continuous multivariate distributions that we'll need as preparation for Chapters 13 and 14, and a brief discussion of correlation and prediction in bivariate populations. (Inference for multivariate populations is beyond our scope.)

12.1 Transformations

We first review transformations in one dimension, the understanding of which will make the case of several dimensions seem more natural. So, consider a transformation $y = g(x)$ that maps a region R of x-values into a region S of y-values. The transformation will be monotonic if the derivative is continuous and does not vanish in R:

$$\frac{dy}{dx} = g'(x) \neq 0, \quad \text{all } x \in R.$$

For each $y \in S$ there is a unique $x \in R$, called $g^{-1}(y)$, such that $g[g^{-1}(y)] = y$. The function g^{-1} is the *inverse* function, which is differentiable:

$$\frac{dg^{-1}(y)}{dy} = \frac{dx}{dy} = \frac{1}{dy/dx} = \frac{1}{g'(x)}.$$

The absolute differential coefficient $|dy/dx|$ is the change-in-length factor when going from dx to $dy = (dy/dx)dx$. Likewise, the reciprocal,

$$\frac{dx}{dy} = \frac{d}{dy} g^{-1}(y),$$

gives us the change factor in going from dy to $dx = (dx/dy)dy$. A definite integral is transformed as follows:

$$\int_R h(x)\,dx = \int_S h(g^{-1}(y))\left|\frac{dx}{dy}\right|dy. \tag{1}$$

If X is a random variable with p.d.f. f_X, the variable $Y = g(X)$ has the p.d.f.

$$f_Y(y) = f_X(g^{-1}(y))\left|\frac{dx}{dy}\right|. \tag{2}$$

In two dimensions, the transformation

$$\begin{cases} u = g(x,\,y) \\ v = h(x,\,y) \end{cases}$$

maps a region R of points in the xy-plane into a region S of points in the uv-plane. We assume now that g and h are continuously differentiable. The quantity that plays the role of the derivative is here the *Jacobian* of the transformation, defined as

$$\frac{\partial(u,\,v)}{\partial(x,\,y)} = \begin{vmatrix} \dfrac{\partial u}{\partial x} & \dfrac{\partial u}{\partial y} \\ \dfrac{\partial v}{\partial x} & \dfrac{\partial v}{\partial y} \end{vmatrix}.$$

We assume further that there is an inverse transformation:

$$\begin{cases} x = G(u,\,v) \\ y = H(u,\,v), \end{cases}$$

which takes each point $(u,\,v)$ into a unique point $(x,\,y)$ in R, such that

$$\begin{cases} g[G(u,\,v),\,H(u,\,v)] = u \\ h[G(u,\,v),\,H(u,\,v)] = v. \end{cases}$$

[These are identities in $(u,\,v)$.] If the Jacobian of the transformation does not vanish in R, the Jacobian of the inverse transformation is defined in S and is equal to the reciprocal of the Jacobian of the direct transformation:

$$\frac{\partial(x,\,y)}{\partial(u,\,v)} = \left\{\frac{\partial(u,\,v)}{\partial(x,\,y)}\right\}^{-1}.$$

It gives us the change-in-area factor for conversion of area elements. Thus, the change of variables in a double integral is accomplished as follows:

$$\int\!\!\int_R f(x,\, y)\, dx\, dy = \int\!\!\int_S f(G(u,\, v),\, H(u,\, v)) \left| \frac{\partial(x,\, y)}{\partial(u,\, v)} \right| du\, dv, \qquad (3)$$

where S is the image of R under the transformation.

If $(x,\, y)$ is a possible value of the random vector $(X,\, Y)$, the transformation being considered defines a new random vector $(U,\, V)$:

$$\begin{cases} U = g(X,\, Y) \\ V = h(X,\, Y). \end{cases}$$

The transformation induces a probability distribution in the uv-plane: If S is a set in the uv-plane and R is the set of all points in the xy-plane with images in S, then

$$P[(U,\, V) \in S] = P[(X,\, Y) \in R].$$

If the distribution of $(X,\, Y)$ is of the continuous type with joint p.d.f. $f_{X,\,Y}$, the probability of S can be expressed as an integral over R:

$$P[(U,\, V) \text{ in } S] = \int\!\!\int_R f_{X,\,Y}(x,\, y)\, dx\, dy$$

$$= \int\!\!\int_S f_{X,\,Y}(G(u,\, v),\, H(u,\, v)) \left| \frac{\partial(x,\, y)}{\partial(u,\, v)} \right| du\, dv.$$

This relation holds for each event S in the plane, so the p.d.f. of $(U,\, V)$ is the integrand of the uv-integral:

$$f_{U,\,V}(u,\, v) = f_{X,\,Y}(G(u,\, v),\, H(u,\, v)) \left| \frac{\partial(x,\, y)}{\partial(u,\, v)} \right|. \qquad (4)$$

This exactly parallels the transformation formula (2) for the univariate case.

EXAMPLE 12.1a Consider random variables (Θ, Φ) with p.d.f. $\frac{1}{2\pi} \sin \theta$ for $0 < \theta < \pi/2$ and $0 \le \phi < 2\pi$, and the transformation

$$X = \sin \Theta \cos \Phi, \qquad Y = \sin \Theta \sin \Phi.$$

The image set S in the xy-plane of the support R of $(\Theta,\, \Phi)$ is the interior of the unit circle. (See Figure 12-1.) The inverse transformation is as follows:

$$\Theta = \arcsin \sqrt{X^2 + Y^2}, \qquad \Phi = \arctan \frac{Y}{X}.$$

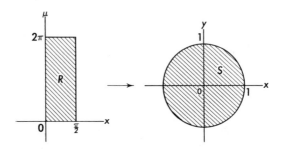

Figure 12-1

The Jacobian of the transformation from (θ, ϕ) to (x, y) is

$$\frac{\partial(x, y)}{\partial(\theta, \phi)} = \begin{vmatrix} \cos\theta\cos\phi & -\sin\theta\sin\phi \\ \cos\theta\sin\phi & \sin\theta\cos\phi \end{vmatrix} = \sin\theta\cos\theta.$$

So the p.d.f. of (X, Y) is

$$f_{X,Y}(x, y) = f_{\Theta, \Phi}(\theta, \phi)\left|\frac{\partial(\theta, \phi)}{\partial(x, y)}\right| = \frac{\sin\theta}{2\pi} \cdot \frac{1}{\sin\theta\cos\theta} = \frac{1}{2\pi\cos\theta}$$

$$= \frac{1/(2\pi)}{\sqrt{1 - (x^2 + y^2)}}, \quad x^2 + y^2 < 1. \blacksquare$$

The next example illustrates a useful technique for finding the p.d.f. of a function of random variables as the marginal p.d.f. of a joint distribution of that function and some other convenient variable(s).

EXAMPLE 12.1b Suppose X and Y are i.i.d. $\mathcal{N}(0, 1)$, and we want to find the distribution of their ratio, $U = X/Y$. We introduce the variable $V = Y$ and consider the bivariate transformation given by

$$U = \frac{X}{Y}, \quad V = Y.$$

The inverse, defined except on a set of probability 0, is

$$X = UV, \quad Y = V.$$

The Jacobian of this inverse transformation is

$$\frac{\partial(x,\,y)}{\partial(u,\,v)} = \begin{vmatrix} v & u \\ 0 & 1 \end{vmatrix} = v,$$

which is nonzero except at $v = y = 0$. The joint p.d.f. of (X, Y) is

$$f_{X,Y}(x,\,y) = \frac{1}{2\pi}\exp\left\{-\frac{1}{2}(x^2+y^2)\right\},$$

so the joint p.d.f. of (U, V) is

$$f_{U,V}(u,\,v) = f_{X,Y}(uv,\,v)\left|\frac{\partial(x,\,y)}{\partial(u,\,v)}\right| = \frac{|v|}{2\pi}\exp\left\{-\frac{1}{2}(u^2v^2+v^2)\right\}.$$

We can now find the marginal p.d.f. of U by integrating out the v:

$$f_{X/Y}(u) = \frac{1}{2\pi}\int_{-\infty}^{\infty}|v|\exp\left\{-\frac{1}{2}(u^2v^2+v^2)\right\}dv$$

$$= \frac{1}{\pi}\int_{0}^{\infty}v\exp\left\{-\frac{v^2}{2}(u^2+1)\right\}dv = \frac{1/\pi}{1+u^2}.$$

This is the p.d.f. of a Cauchy distribution with median 0, as could have been predicted in view of Theorem 9, §6.12. ∎

Linear transformations are of particular interest:

$$\begin{cases} x = au + bv \\ y = cu + dv. \end{cases}$$

The Jacobian is the determinant Δ of the coefficients:

$$\frac{\partial(x,\,y)}{\partial(u,\,v)} = \begin{vmatrix} a & b \\ c & d \end{vmatrix} = ad - bc = \Delta.$$

If $\Delta = 0$, y is a multiple of x, since then $a{:}b = c{:}d$. We rule out this degenerate case by assuming $\Delta \neq 0$. The transformation then has a unique inverse:

$$\begin{cases} u = Ax + By \\ v = Cx + Dy, \end{cases}$$

where $A = d/\Delta$, $B = -b/\Delta$, $C = -c/\Delta$, and $D = a/\Delta$. The Jacobian of the inverse transformation is

$$\frac{\partial(u, v)}{\partial(x, y)} = \begin{vmatrix} A & B \\ C & D \end{vmatrix} = AD - BC = \frac{1}{\Delta}.$$

Suppose now that we apply this linear transformation to the random vector (U, V) to obtain the random vector (X, Y):

$$\begin{cases} X = aU + bV \\ Y = cU + dV. \end{cases}$$

If (U, V) has a continuous distribution with joint p.d.f. $f_{U,V}$, then (X, Y) has the joint p.d.f.

$$f_{X,Y}(x, y) = \frac{1}{|\Delta|} f_{U,V}(Ax + By, Cx + Dy).$$

EXAMPLE 12.1c Let U and V have independent, exponential distributions with mean 1 and define

$$\begin{cases} X = U + V \\ Y = U - V. \end{cases}$$

The joint p.d.f. of (U, V) is

$$f_{U,V}(u, v) = e^{-u-v}, \quad u > 0, \ v > 0,$$

and the Jacobian of the transformation is

$$\frac{\partial(x, y)}{\partial(u, v)} = \begin{vmatrix} 1 & 1 \\ 1 & -1 \end{vmatrix} = \Delta = -2.$$

The inverse transformation, obtained by solving for U and V, is

$$\begin{cases} U = \frac{1}{2}(X + Y) \\ V = \frac{1}{2}(X - Y). \end{cases}$$

The transformation takes the first quadrant of the uv-plane into the region between $x = y$ and $x = -y$, as shown in Figure 12-2. The new joint p.d.f. is

$$f_{X,Y}(x, y) = f_{U,V}\left(\frac{1}{2}(x + y), \frac{1}{2}(x - y)\right) \cdot \frac{1}{2} = \frac{1}{2} e^{-x}, \quad -x < y < x, \ x > 0.$$

As a by-product, we can find the p.d.f. of the sum $X = U + V$ by integrating out the y. For a given x, the limits of integration for y are evident in the above inequalities defining the support of $f_{X,Y}$:

$$f_X(x) = \int_{-\infty}^{\infty} f_{X,Y}(x, y)\, dy = \frac{1}{2}\int_{-x}^{x} e^{-x}\, dy = xe^{-x}, \qquad x > 0. \quad \blacksquare$$

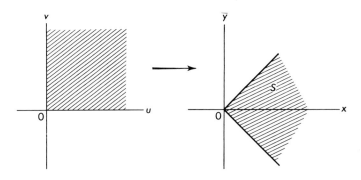

Figure 12-2

An important special case of a linear transformation is that of a *rotation* through an angle θ:

$$\begin{cases} X = (\cos\theta)\,U + (-\sin\theta)\,V \\ Y = (\sin\theta)\,U + (\cos\theta)\,V. \end{cases}$$

This transformation has the property that $x^2 + y^2 = u^2 + v^2$, and the Jacobian is $\Delta = 1$. The covariance of the new bivariate distribution is

$$\operatorname{cov}(X, Y) = \cos\theta \sin\theta\,(\sigma_U^2 - \sigma_V^2) + (\cos^2\theta - \sin^2\theta)\,\sigma_{U,V}.$$

If we choose θ so that

$$\frac{\sigma_V^2 - \sigma_U^2}{2\sigma_{U,V}} = \frac{\cos^2\theta - \sin^2\theta}{2\sin\theta\cos\theta} = \cot 2\theta,$$

then X and Y are uncorrelated. Thus we have:

THEOREM 1 *Given that (U, V) has a nonsingular bivariate distribution, there is a rotation transformation resulting in variables (X, Y) whose components are uncorrelated.*

COROLLARY *When U and V are not perfectly correlated, there is a nonsingular linear transformation such that the new variables (Z, W) are uncorrelated and have unit variances.*

The corollary follows, after obtaining (X, Y) as in the theorem, when we set $Z = X/\sigma_X$, $W = Y/\sigma_Y$, which is permissible if neither σ_X nor σ_Y is zero. But if $\rho^2 < 1$, no linear combination of U and V can be constant—have zero variance. The determinant of the transformation from (U, V) to (Z, W) is just $(\sigma_X \sigma_W)^{-1}$, so it is is nonsingular.

PROBLEMS

12-1. Express the joint p.d.f. of $X' = X + h$ and $Y' = Y + k$ in terms of the p.d.f. of (X, Y).

12-2. Find the joint p.d.f. of $2X - Y$ and $X + Y$, where X and Y are independent, standard normal variables.

12-3. Let X and Y be independent observations on $\mathcal{U}(0, 1)$, and define new variables $Z = XY$, $W = Y$. Find the p.d.f. of $Z = XY$ as a marginal p.d.f. of (Z, W).

12-4. Show that if X and Y have independent gamma distributions with $\lambda = 1$, then $X/(X + Y)$ has a beta distribution. [Hint: Let $Z = X + Y$ and $W = X/(X + Y)$.]

12-5. Suppose X and Y have independent chi-square distributions and define $U = X/Y$, $V = X + Y$.
(a) Derive the joint p.d.f. of U and V and so show their independence.
(b) Obtain the marginal p.d.f. of U.

12-6. Given that $\mu_X = \mu_Y = 1$, $\sigma_X^2 = \sigma_Y^2 = 2$, and $\sigma_{X,Y} = 1$, obtain new variables (U, V) as a linear transformation on (X, Y) so that U and V are uncorrelated, with 0 means and unit variances.

12-7. Let (X, Y) have a uniform distribution in the unit square. Find the joint distribution of the sum and the difference: $(X + Y, X - Y)$.

12-8. Suppose the distribution of (X, Y) is circularly symmetric about the origin—that is, that $f_{(X,Y)}(x, y) \propto g(x^2 + y^2)$. Let

$$Z = \sqrt{X^2 + Y^2}, \quad T = \text{Arctan}(Y/X), \quad W = \exp[-\tfrac{1}{2}(X^2 + Y^2)].$$

(a) Show that Z and T are independent.
(b) Show that T and W are independent.
(c) Show that if $g(\lambda) = e^{-\lambda}$, the variables W and T are each $\mathcal{U}(0, 1)$.

[The above can be exploited to convert a stream of independent, uniform observations into a stream of independent, standard normal observations, using the inverse transformation:

$$X = \sqrt{-2\log W}\cos(2\pi\Theta),\ Y = \sqrt{-2\log W}\sin(2\pi\Theta).$$

This is referred to as the *Box-Muller transformation*.]

12.2 The bivariate normal distribution

The joint distribution of two independent, standard normal distributions is said to be *circular normal*. The p.d.f. of this bivariate distribution is

$$f(x, y) = \frac{1}{2\pi}\exp\left(-\frac{1}{2}\{x^2 + y^2\}\right).$$

The surface is bell-shaped, and its level curves—curves of constant altitude— are circles: $x^2 + y^2 = a^2$. We'll call a distribution with this density "bivariate normal" but will extend the use of this term to include distributions obtained from the circular normal by linear transformations. The surfaces representing the p.d.f.'s will be bell-shaped in an extended sense—bells with elliptical cross-sections. Each component variable in such a distribution, being a linear combination of independent, normal variables, is univariate normal, according to Problem 6-77. It is this property that we focus on in defining a general bivariate normal distribution, an approach that permits inclusion of some singular cases as normal, albeit in a degenerate sense.

DEFINITION The random vector (X, Y) is said to have a *bivariate normal distribution* if an only if the combination $aX + bY$ is univariate normal for every choice of constants a and b.

THEOREM 2 *The marginal distributions of a bivariate normal distribution are (univariate) normal.*

THEOREM 3 *If variables U and V are obtained from the bivariate normal pair (X, Y) by a linear transformation, then (U, V) is bivariate normal.*

Theorem 2 follows from the fact that each of X and Y is a linear combination of X and Y. And the fact that the combination $cU + dV$ is a linear combination of linear combinations of X and Y establishes Theorem 1.

We derive next the moment generating function of a bivariate normal distribution. If (X, Y) is bivariate normal, then for any r and s, the variable $U = rX + sY$ has mean and variance given by

$$\begin{cases} \mu_U = r\mu_X + s\mu_Y, \\ \sigma_U^2 = r^2\sigma_X^2 + s^2\sigma_Y^2 + 2rs\,\sigma_{X,Y}. \end{cases} \quad (1)$$

Since U is normal, its m.g.f. is

$$\psi_U(t) = E(e^{t(rX+sY)}) = \exp\left\{t\mu_U + \tfrac{1}{2}t^2\sigma_U^2\right\}. \quad (2)$$

But, by definition, the joint m.g.f. of (X, Y) is

$$\psi_{(X,Y)}(r, s) = E(e^{rX+sY}) = \psi_U(1) = \exp\left\{\mu_U + \tfrac{1}{2}\sigma_U^2\right\}.$$

Substituting from (1) we get the joint m.g.f. of (X, Y):

$$\psi_{X,Y}(r, s) = \exp\left\{r\mu_X + s\mu_Y + \tfrac{1}{2}[r^2\sigma_X^2 + 2rs\sigma_{X,Y} + s^2\sigma_Y^2]\right\}. \quad (3)$$

We can use this to prove the following important result:

THEOREM 4 *If (X, Y) is bivariate normal with correlation $\rho = 0$, then X and Y are independent.*

To see this, observe that if $\rho = 0$ in (3), the m.g.f. factors into the product of a function of r alone and a function of t alone. The asserted independence then follows from Theorem 28, page 131.

THEOREM 5 *If (X, Y) is bivariate normal with $\rho^2 < 1$, the distribution is nonsingular, with p.d.f.*

$$f(x, y) = \frac{1}{2\pi\sigma_X\sigma_Y\sqrt{1-\rho^2}} \exp\left\{\frac{-1}{2(1-\rho^2)}Q(x, y)\right\}, \quad (4)$$

where

$$Q(x, y) = \left(\frac{x-\mu_X}{\sigma_X}\right)^2 - 2\rho\left(\frac{x-\mu_X}{\sigma_X}\right)\left(\frac{y-\mu_Y}{\sigma_Y}\right) + \left(\frac{y-\mu_Y}{\sigma_Y}\right)^2. \quad (5)$$

In showing this we assume first that $\mu_X = \mu_Y = 0$. According to the corollary to Theorem 1 (§12.1), there is a nonsingular linear transformation:

$$\begin{cases} U = aX + bY \\ V = cX + dY \end{cases}$$

such that U and V each have mean zero and variance 1 and are uncorrelated, which means that they are independent, standard normal variables. The Jacobian is $\Delta = ad - bc$, and the inverse transformation is

$$\begin{cases} X = \dfrac{dU - bV}{\Delta} \\ Y = \dfrac{-cU + aV}{\Delta}, \end{cases}$$

which is uniquely defined because $\Delta \neq 0$. The p.d.f. of (X, Y) is

$$f_{X,Y}(x, y) = f_{U,V}(ax + bY, cx + dy) \left| \frac{\partial(u, v)}{\partial(x, y)} \right|$$

$$= \frac{1}{2\pi} \exp\left(-\tfrac{1}{2}[(ax + by)^2 + (cx + dy)^2]\right) \cdot |\Delta| \qquad (6)$$

$$= \frac{1}{2\pi} \exp\left(-\tfrac{1}{2}[(a^2 + c^2)x^2 + (b^2 + d^2)y^2 + 2xy(ab + cd)]\right) \cdot |\Delta|.$$

Now,

$$\sigma_X^2 = \frac{d^2 + b^2}{\Delta^2}, \quad \sigma_Y^2 = \frac{a^2 + c^2}{\Delta^2}, \quad \sigma_{X,Y} = \frac{-(ab + cd)}{\Delta^2}, \qquad (7)$$

so

$$\sigma_X^2 \sigma_Y^2 = \frac{a^2 d^2 + a^2 b^2 + c^2 d^2 + c^2 b^2}{\Delta^4}, \quad \sigma_{XY}^2 = \frac{a^2 b^2 + c^2 d^2 + 2abcd}{\Delta^4},$$

and

$$\sigma_X^2 \sigma_Y^2 (1 - \rho^2) = \sigma_X^2 \sigma_Y^2 - \sigma_{XY}^2 = \frac{(ad - bc)^2}{\Delta^4} = \frac{1}{\Delta^2}. \qquad (8)$$

The Jacobian in (6) is therefore

$$\left| \frac{\partial(u, v)}{\partial(x, y)} \right| = |\Delta| = \frac{1}{\sigma_X \sigma_Y \sqrt{1 - \rho^2}}, \qquad (9)$$

Substituting from (7), (8), and (9) in (6) yields the formula of Theorem 5 when the means are both 0. The more general formula is obtained by a simple translation.

A bivariate distribution is often pictured by means of *level curves* or *contour curves*. These are the two-dimensional graphs of the relationship defined by setting $f(x, y)$ equal to a constant. (Contour maps of a lake bottom or of a mountain are commonly used instances of such graphs, in which a contour curve is drawn on a plane map for each of several selected values of a constant depth or altitude.) For a bivariate normal distribution, contour curves are obtained by setting the bivariate p.d.f. (4) [in Theorem 5] equal to a constant, or equivalently, setting

Q in (5) equal to a constant:

$$\left(\frac{x-\mu_X}{\sigma_X}\right)^2 - 2\rho\left(\frac{x-\mu_X}{\sigma_X}\right)\left(\frac{y-\mu_Y}{\sigma_Y}\right) + \left(\frac{y-\mu_Y}{\sigma_Y}\right)^2 = \text{const.} \quad (10)$$

This quadratic equation in (x, y) is an *ellipse* if the discriminant is negative:

$$\left(\frac{\rho}{\sigma_X\sigma_Y}\right)^2 - \frac{1}{\sigma_X^2}\cdot\frac{1}{\sigma_Y^2} = \frac{\rho^2 - 1}{\sigma_X^2\sigma_Y^2} < 0,$$

which *is* the case when the distribution is nonsingular. Some level curves are indicated in a typical case in Figure 12-3.

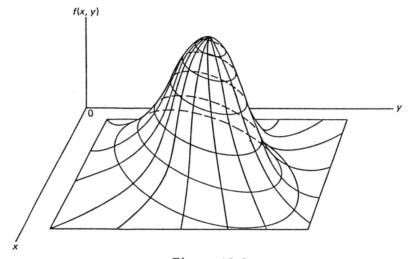

Figure 12-3

THEOREM 6 *The level curves of a bivariate normal p.d.f. for (X, Y) are ellipses centered at (μ_X, μ_Y). The axes of these ellipses are parallel to the coordinate axes if and only X and Y are independent.*

THEOREM 7 *With a suitable choice of multiplicative constant K, any quadratic function of the form*

$$f(x, y) = K\exp[-\tfrac{1}{2}(ax^2 + 2bxy + cy^2 + 2dx + 2ey)]$$

can serve as a bivariate normal p.d.f. if $a > 0$ and the discriminant of the quadratic terms is negative: $b^2 - ac < 0$.

To show Theorem 7, we first observe that linear terms in x and in y can be eliminated by a translation: $x = x' - h$, $y = y' - k$, if h and k

satisfy
$$\begin{cases} ah + bk = d \\ bh + ck = e. \end{cases}$$

(These equations have a unique solution since the determinant of coefficients is just $ac - b^2 \neq 0$.) And the translation does not change the coefficients of the second degree terms. It is then simply a matter of identifying the coefficients a, b, c with the corresponding coefficients in the p.d.f. given by (4). Doing this we find that

$$\sigma_X^2 = \frac{c}{ac - b^2}, \quad \sigma_Y^2 = \frac{a}{ac - b^2}, \quad \rho = \frac{-b}{\sqrt{ac}}. \tag{11}$$

Note that $\rho^2 < 1$ is equivalent to $b^2 - ac < 0$.

Another way to see Theorem 7 is to find a linear transformation that results in a new p.d.f. with no cross-product term (by completing squares). Then, since the transformed variables are independent and normal (and since the inverse transformation is linear), the original p.d.f. must be that of a bivariate normal distribution. The next example illustrates this.

EXAMPLE 12.2a Consider the following function as the basis of a bivariate normal p.d.f.:

$$\exp[-\tfrac{1}{2}x^2 + xy - y^2 - x + 2y] = \exp[-\tfrac{1}{2}(x^2 - 2xy + 2y^2 + 2x - 4y)].$$

Here $a = 1$, $b = -1$, $c = 2$, so $b^2 - ac = -1 < 0$. Then from (11), $\sigma_X^2 = 2$, $\sigma_Y^2 = 1$, and $\rho = 1/\sqrt{2}$. The necessary multiplicative constant is

$$\frac{1}{2\pi\sigma_X\sigma_Y\sqrt{1-\rho^2}} = \frac{1}{2\pi}.$$

Alternatively, we can rearrange terms:

$$x^2 - 2xy + 2y^2 + 2x - 4y = (x-y)^2 + y^2 + 2(x-y) - 2y$$
$$= (x - y + 1)^2 + (y - 1)^2 - 2.$$

It is now apparent that the linear transformation

$$\begin{cases} U = X - Y + 1 \\ V = Y - 1, \end{cases}$$

will result in independent standard normal variables U and V. The various parameters of (X, Y) are then easily found from the inverse transformation:

$$\begin{cases} X = U + V \\ Y = V + 1. \end{cases} \blacksquare$$

The regression function, $E(Y \mid x)$, is of importance in applications. To find it we study the conditional distributions of a bivariate normal distribution. The conditional p.d.f. of $Y \mid x_0$ is proportional to the joint density in the cross-section at x_0:

$$f(y \mid x_0) \propto f(x_0, y).$$

With x held fixed at x_0 in a bivariate normal distribution, the p.d.f. is essentially an exponential function with a quadratic in y as the exponent:

$$f(x_0, y) = K \exp[-\tfrac{1}{2}(ax_0^2 + 2bx_0 y + cy^2 + 2dx_0 + 2ey)]$$

Such a density is univariate normal. (See Figure 12-4.)

EXAMPLE 12.2b Consider again the bivariate normal p.d.f. of the preceding example:

$$f(x, y) \propto \exp[-\tfrac{1}{2}(x^2 - 2xy + 2y^2 + 2x - 4y)].$$

We now hold x fixed and complete the square in y:

$$2y^2 - 2(x+2)y + x^2 + 2x = 2\left\{y^2 - 2\left(\tfrac{x}{2}+1\right)y + \left(\tfrac{x}{2}+1\right)^2\right\} + \text{terms in } x.$$

With this we see that

$$f(y \mid x) \propto \exp\left\{-\left(y - \tfrac{x+2}{2}\right)^2\right\}.$$

This is clearly a univariate normal p.d.f. with mean

$$E(Y \mid x) = \frac{x+2}{2}.$$

We find the variance by observing that the coefficient of the quadratic term is

$$-\frac{1}{2\sigma_{Y \mid x}^2} = -1,$$

so $\text{var}(Y \mid x) = 1/2$. $\blacksquare$

By treating the general bivariate normal p.d.f. as we just did a special case in the preceding example, one obtains the following:

12.2 The bivariate normal distribution

THEOREM 8 *In a bivariate normal distribution the conditional distributions of $X \mid y$ and $Y \mid x$ are (univariate) normal. The regression functions are linear functions of the conditioning variable, and the conditional variances are constant:*

$$E(Y \mid x) = \mu_Y + \frac{\sigma_{X,Y}}{\sigma_X^2}(x - \mu_X), \quad \text{var}\,(Y \mid x) = \sigma_Y^2(1 - \rho^2).$$

(For analogous formulas for $X \mid y$, interchange the roles of x and y.)

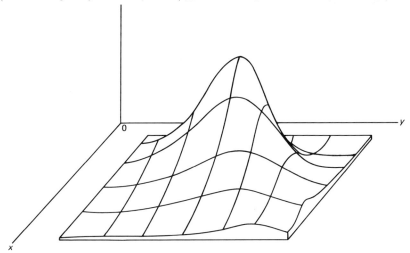

Figure 12-4

PROBLEMS

12-9. Write out the p.d.f. of a bivariate normal distribution for (X, Y) with $EX = 0$, $EY = 4$, $\text{var}\,X = 1$, $\text{var}\,Y = 9$, and $\text{cov}(X, Y) = 2$.

12-10. Suppose the random pair (X, Y) has the following joint p.d.f.:

$$f(x, y) \propto \exp(-x^2 + xy - 2y^2).$$

(a) Find the first and second moments of the distribution.
(b) Find the constant of proportionality.
(c) What are the level curves?
(d) Find the conditional distributions given x, and given y.

12-11. Given that (X, Y) has a bivariate p.d.f. proportional to

$$\exp\left\{-\tfrac{1}{2}x^2 + xy - y^2 - x + 2y\right\},$$

find the regression functions of Y on X and of X on Y.

12-12. Given that U and V are independent, standard normal variables, determine the joint distribution of (X, Y), where $X = U + 2V$ and $Y = 3U - V$.

12-13. Given that the m.g.f. of (X, Y) is

$$\psi_{X,Y}(r, s) = \exp[4r^2 - 4rs + 9s^2 - 8r + 6s],$$

find the regression function of Y on X.

12-14. Let (X, Y) be bivariate normal and define U as the deviation of X from its mean and V as the deviation of Y from the regression function of Y on X: $U = X - \mu_X$, $V = Y - E(Y \mid X)$. Show that U and V are independent.

12-15. Show that if **X** is a random sample from a normal population, the two linear combinations $U = \sum a_i X_i$ and $V = \sum b_i X_i$ are jointly bivariate normal
(a) from the basic definition.
(b) using moment generating functions.

12-16. Use the result of the preceding problem to show that if **X** is a random sample from a normal population, then $E(X_1 \mid \bar{X}) = \bar{X}$. [This is an instance of "Rao-Blackwellization" of X_1 using the sufficient statistic $\bar{X}$. See §8.3.]

12-17. Suppose X and Y are independent, standard normal variables. Show that variables (U, V) obtained by a rotation transformation are again independent, standard normal variables.

12-18. Suppose (X, Y) has a bivariate normal distribution with unit standard deviations: $\sigma_X = \sigma_Y = 1$. Show that the correlation between X^2 and Y^2 is the square of $\rho_{X,Y}$,
(a) by brute-force integration to calculate $E(X^2 Y^2)$, etc.
(b) by making a linear transformation that eliminates dependence and then calculating $E(X^2 Y^2)$ from moments of the new variables. [Hint: A rotation will do it, as will also a transformation of the form $U = aX + bY$, $V = Y$, with a and b chosen to make $\operatorname{var} U = 1$ and $\operatorname{cov}(U, V) = 0$.]

12.3 Correlation and prediction

In the study of more than one random variable, relationships among the variables are important. In this section we take up the case of two random variables, asking whether there is a relationship, and if so, how strong it is. We'll also see how to exploit a relationship in predicting one variable from a knowledge of the other.

It is also important, in practice, to know about *causal* connections between variables, and sometimes a stochastic relationship can suggest the possibility of a cause-effect connection. However, this is not a statistical matter and can only be studied within a particular field of application.

Predicting the value of a random variable has its element of risk. If we predict the value of a random variable Y to be k, we're apt to be wrong, and the *prediction error* is (by definition) $Y - k$. As in estimation, we can only assess the error in terms of some kind of average. A commonly used criterion for measuring the performance of a predictor is the **mean squared prediction error** (m.s.p.e.): $E[(Y-k)^2]$. This second moment is minimized (as we learned in §4.5) when $k = EY$, and the minimum m.s.p.e. is var Y.

Next, suppose we want to predict the value of Y given the value of a random variable X related to Y, say x_0. Clearly, the choice dictated by the criterion of minimum m.s.p.e. is the mean of the *conditional* distribution of Y, namely, $E(Y \mid x_0)$. The minimum m.s.p.e. is then the conditional variance.

EXAMPLE 12.3a In the case of a bivariate *normal* distribution, we found the conditional mean of Y given x in §12.2:

$$E(Y \mid x) = \mu_Y + \frac{\sigma_{X,Y}}{\sigma_X^2}(x - \mu_X), \qquad (1)$$

which happens to be a *linear* function of x. The m.s.p.e. for this prediction is the conditional variance (from §12.2):

$$\text{var}(Y \mid x) = \sigma_Y^2(1 - \rho^2). \qquad (2)$$

Observe that this does not depend on x, so if we think of X as random,

$$E[\text{var}(Y \mid X)] = \sigma_Y^2(1 - \rho^2).$$

[This same result follows using Theorem 14 of Chapter 4, page 109]:

$$E[\text{var}\, Y \mid X) = \sigma_Y^2 - \text{var}\{E(Y \mid X)\},$$

where

$$\text{var}\{E(Y \mid X)\} = \text{var}\left\{\frac{\sigma_{X,Y}}{\sigma_X^2}(X - \mu_X)\right\} = \sigma_Y^2 \rho^2.]$$

Thus, using the conditional mean to predict Y, given $X = x$, reduces the m.s.p.e. from σ_Y^2 (the m.s.p.e. when X is not used) by the factor $1 - \rho^2$.

For instance, suppose $\rho = .6$—not an uncommon level of linear correlation in aptitude test and performance scores. The r.m.s. prediction error can be reduced by a factor of $\sqrt{1 - .36} = .8$ by using the aptitude score as a predictor variable. ∎

Suppose, to achieve simplicity of model, we look for the best *linear* predictor—a function of the form $a + b(X - \mu_X)$. The m.s.p.e., averaged over X as well as over Y, is

$$E\{[Y - a - b(X - \mu_X)]^2\} = a^2 - 2a\mu_Y + b^2\sigma_X^2 - 2b\sigma_{XY} + E(Y^2)$$

$$= (a - \mu_Y)^2 + \sigma_X^2\left(b - \frac{\sigma_{XY}}{\sigma_X^2}\right)^2 + \sigma_Y^2 - \frac{\sigma_{XY}^2}{\sigma_X^2}.$$

Choosing $a = \mu_Y$ and $b = \sigma_{XY}/\sigma_X^2$ clearly minimizes this expression, so the minimum m.s.p.e., among linear predictors, is achieved when we use this one:

$$\mu_Y + \frac{\sigma_{X,Y}}{\sigma_X^2}(X - \mu_X), \tag{3}$$

and the minimum m.s.p.e. is $\sigma_Y^2(1 - \rho^2)$, as in the normal case.

As we saw in the above example, the linear function (3) is the best predictor (in the mean square sense) among *all* predictors when the population is normal, being the conditional mean of Y in that case. In general, if the regression function happens to be linear, then the best linear predictor is the best predictor; but if it is not linear, the amount of reduction possible using a best linear predictor depends on the magnitude of the (linear!) correlation ρ.

In this discussion we have assumed that the parameters of the bivariate model are known. And perhaps there are situations where the wealth of data and experience is practically as good as knowing the true model (for example, in educational testing). But if the true model is not known, we must resort to using sample data for basing our predictions. This process will be taken up in Chapter 14.

Since ρ is one of the parameters of a bivariate normal population, it is of interest to see how we might estimate ρ or test hypotheses about ρ. The likelihood function for a random sample $\{(X_1, Y_1), ..., (X_n, Y_n)\}$ from a bivariate population depends on the population means and variances as well as on the population correlation. It is a simple matter to write out this likelihood function but a rather tedious exercise to maximize it. We simply state that the joint m.l.e.'s of the five parameters are the corresponding moments of the sample distribution [which puts mass $1/n$ at each (X_i, Y_i)], namely, the two sample means $\bar X$ and $\bar Y$, the two sample variances V_X and V_Y (the biased versions), and the sample correlation coefficient:

$$r = \frac{\sum (X_i - \bar X)(Y_i - \bar Y)}{\left\{\sum (X_i - \bar X)^2 \sum (Y_i - \bar Y)^2\right\}^{1/2}} = \frac{\sum X_i Y_i - n\bar X \bar Y}{\sqrt{\left\{\sum X_i^2 - n\bar Y^2\right\}\left\{\sum Y_i^2 - n\bar X^2\right\}}} \quad (4)$$

[If we were to define the average product of deviations from the means with divisor $n-1$, calling it S_{XY}, then $r = S_{XY}/(S_X S_Y)$. This parallels the formula for ρ as given by (1) in §4.8.]

The statistic r tends to ρ in probability, but to use it for inference we'd need to know its distribution. For the hypothesis $\rho = 0$ (equivalent to independence of X and Y), we'll show in Chapter 14 that the null distribution of the statistic

$$T = \sqrt{n-2}\,\frac{r}{\sqrt{1-r^2}} \quad (5)$$

is $t(n-2)$. Using this we can find a P-value or find the critical value for a test at a specified level of significance.

For large samples, R. A. Fisher proposed testing values of ρ other than 0 using this score, which he showed[1] to be asymptotically normal:

$$Z = \tfrac{1}{2} \log \frac{1+r}{1-r}. \quad (6)$$

The parameters of the asymptotic distribution are

$$E(Z) \doteq \tfrac{1}{2} \log \frac{1+\rho}{1-\rho} \quad \text{and} \quad \operatorname{var} Z \doteq \frac{1}{n-3}. \quad (7)$$

Large-sample confidence intervals can also be formed using this distribution.

EXAMPLE 12.3b If a sample correlation coefficient is $r = .2$ in a random sample of size $n = 27$, could it be that $\rho = 0$? Yes—but to test $\rho = 0$ we calculate (5):

$$T = \sqrt{25}\,\frac{.2}{\sqrt{1-.04}} \doteq 1.02.$$

Looking in Table IIIa for $t(25)$, we find $P \doteq .16$ (one-sided). This is not reason to doubt $\rho = 0$. But suppose the sample size were $n = 100$. Then

---

[1]Kendall, M. G., and A. Stuart, *The Advanced Theory of Statistics*, Vol. 1, 3rd ed., 1969, New York: Hafner Press, page 391

$T \doteq 2.02$, and $P \doteq .02$.

To obtain 95 percent confidence limits for ρ, say when $n = 100$, we start with

$$P\left\{\left|\frac{Z - \mu_Z}{\sigma_Z}\right| < 1.96\right\} \doteq .95,$$

where Z, μ_Z, and σ_Z^2 are given by (6) and (7). When $r = .2$, the inequality on the left becomes

$$\left|\frac{1}{2}\log\frac{1.2}{.8} - \frac{1}{2}\log\frac{1+\rho}{1-\rho}\right| < 1.96\sqrt{\frac{1}{97}}.$$

To solve this for ρ we first rewrite it as

$$.2027 - .199 < \frac{1}{2}\log\frac{1+\rho}{1-\rho} < .2027 + .199.$$

This is equivalent to $.0037 < \rho < .38$. ∎

The best linear predictor of Y given X, from (3), is

$$Y = \mu_Y + \frac{\sigma_{X,Y}}{\sigma_X^2}(X - \mu_X), \qquad (8)$$

Since $\rho = \sigma_{X,Y}/(\sigma_X\sigma_Y)$, we can rewrite (8) in the form

$$Z_Y = \frac{Y - \mu_Y}{\sigma_Y} = \rho \cdot \frac{X - \mu_X}{\sigma_X} = \rho Z_X. \qquad (9)$$

This says that for any pair (X, Y), the Z-score for the predicted Y is ρ times the Z-score for the X, a phenomenon known as the *regression effect*.

EXAMPLE 12.3c The pioneering geneticist Sir Francis Galton, in studying the relationship between sizes of parent and progeny, looked at data on heights of fathers and heights of sons. He noticed the following: The son of a father of below average height tended to be not as far below average in height as his father, and the son of a father of above average tended to be not as far above average in height as his father. Since the correlation is not perfect, this phenomenon is to be expected, according to (9). For instance, suppose $\rho = .7$ and consider a father's whose height is two s.d.'s above the mean of fathers' heights: $Z_X = 2$. The predicted height of the son would be only $Z_Y = .7 \times 2$, or 1.4 s.d.'s above the mean of sons' heights. ∎

PROBLEMS

12-19. Let X and Y denote, respectively, the score on a first exam and the score on a second exam in a certain course. Given that $\rho = .9$, $\mu_X = \mu_Y = 65$, and $\sigma_X = \sigma_Y = 12$, predict the second exam score Y of a student whose first exam score is
(a) 45 (b) 95 (c) unknown,
and give the r.m.s. prediction error in each case.

12-20. A study interested in the question of whether weight gain over a period of time on a certain high-protein diet depends on the initial weight reported data from 15 female rats. The sample correlation coefficient between weight gain and initial weight was $r = .49$. Test the hypothesis that these variables are actually uncorrelated: $\rho = 0$. (Assume a bivariate normal distribution)

12-21. Suppose (X, Y) is uniformly distributed on the triangle bounded by $y = 0$, $y = 1 + x$, and $y = 1 - x$.
(a) Find the best predictor for Y given $X = x$.
(b) Find the best linear predictor for Y given $X = x$.

12-22. Suppose (X, Y) has the p.d.f. $f(x, y) = e^{-y}$ for $0 < x < y$.
(a) Find the regression function of Y on X.
(b) Find the best *linear* predictor of Y given $X = x$.

12-23. In a study of the performance characteristics of a popular method of blood glucose measurement, it was found that the correlation between results of that method and results of a glucose hexokinase method was $r = .988$, based on a sample (of pairs of measurements) of size 246. Find 95 percent confidence limits for ρ, the population correlation coefficient.

12-25. Show the following:
(a) If the regression function of Y on x is linear: $E(Y \mid x) = a + bx$, then $\sigma_{X,Y} = b\sigma_X^2$. [Hint: $E\{g(X, Y) \mid X = x\} = E\{g(x, Y) \mid X = x\}$.]
(b) If the regression function in (a) is linear *and* $\sigma_{Y \mid x}^2$ is constant, that constant is $\sigma_Y^2(1 - \rho^2)$, whether or not the population is normal.

12-24. Show that if $E(Y \mid x)$ is linear in x, then it must be given by (1), even if the population is not normal. [Hint: Use iterated expectations.]

12.4 Multivariate transformations

We turn now to the multivariate case and sketch some results analogous to those of the bivariate case. Functions g_i define a transformation from $\mathbf{x} = (x_1, ..., x_n)$ to $\mathbf{u} = (u_1, ..., u_n)$:

$$\begin{cases} u_1 = g_1(x_1, \ldots, x_n) \\ \vdots \qquad \vdots \\ u_n = g_n(x_1, \ldots, x_n). \end{cases} \tag{1}$$

The Jacobian of the transformation is the determinant of partial derivatives:

$$\frac{\partial(u_1, \ldots, u_n)}{\partial(x_1, \ldots, x_n)} = \det\left(\frac{\partial u_i}{\partial x_j}\right).$$

Suppose this does not vanish in a region R and there is an inverse transformation defined on S, the image of R:

$$\begin{cases} x_1 = G_1(u_1, \ldots, u_n) \\ \vdots \qquad \vdots \\ x_n = G_n(u_1, \ldots, u_n), \end{cases} \tag{2}$$

whose Jacobian is the reciprocal of that of the direct transformation. Then a change of variable from $\mathbf{x}$ to $\mathbf{u}$ in a multiple integral over R is accomplished by substituting $\mathbf{x}$ from (2), transforming the element of volume as follows:

$$dx_1 \cdots dx_n \rightarrow \left|\frac{\partial(x_1, \ldots, x_n)}{\partial(u_1, \ldots, u_n)}\right| du_1 \cdots du_n$$

and integrating over S.

Substituting the components of a random vector $\mathbf{X} = (X_1, \ldots, X_n)$ in (1) defines a new random vector with components $U_i = g_i(X_1, \ldots, X_n)$. And if S is the image in the $\mathbf{u}$-space of the set R in the $\mathbf{x}$-space, then

$$P(\mathbf{U} \in S) = P(\mathbf{X} \in R),$$

and (in the continuous case),

$$P(S) = \int_R f(\mathbf{x})\, d\mathbf{x}$$

$$= \int\int_S f(G_1(u_1, \ldots, u_n), \ldots, G_n(u_1, \ldots, u_n)) \left|\frac{\partial(x_1 \ldots, x_n)}{\partial(u_1 \ldots, u_n)}\right| du_1 \cdots du_n.$$

The integrand of this last integral in $\mathbf{u}$-space is then the p.d.f. of $\mathbf{U}$.

EXAMPLE 12.4a Let $X_1, \ldots, X_n$ denote successive interarrival times in a Poisson arrival process with rate parameter λ. These variables are i.i.d., with common p.d.f. e^{-x}, $x > 0$. The times to the n arrivals as measured from time $t = 0$ (the reference time for X_1) are sums of the

12.5 The general linear transformation

interarrival times:

$$\begin{cases} Y_1 = X_1 \\ Y_2 = X_1 + X_2 \\ \vdots \qquad \vdots \\ Y_n = X_1 + \cdots + X_n. \end{cases}$$

The inverse transformation is

$$\begin{cases} X_1 = Y_1 \\ X_2 = Y_2 - Y_1 \\ \vdots \qquad \vdots \\ X_n = Y_n - Y_{n-1}, \end{cases}$$

and the Jacobian of the transformation (either way) is 1. The support of **X** is the first orthant (all $x_i > 0$), and the support of **Y** is the set of points for which $0 < y_1 < y_2 < \cdots < y_n$. The joint p.d.f. of the Y's is

$$f_\mathbf{Y}(\mathbf{y}) = f_\mathbf{X}(y_1, y_2 - y_1, \ldots, y_n - y_{n-1})$$
$$= \exp[-y_1 - (y_2 - y_1) - \cdots - (y_n - y_{n-1})]$$
$$= e^{-y_n}, \text{ for } 0 < y_1 < y_2 < \cdots < y_n.$$

With this we can find the marginal p.d.f. of Y_n, the time to the nth arrival, by integrating out the variables $y_1, \ldots, y_{n-1}$:

$$f_{Y_n}(y_n) = e^{-y_n} \int_0^{y_n} \int_0^{y_{n-1}} \cdots \int_0^{y_2} dy_1 dy_2 \cdots dy_{n-1}$$
$$= \frac{y_n^{n-1}}{(n-1)!} e^{-y_n}, \quad y_n > 0. \blacksquare$$

Up to this point we've considered transformations from $\Re^n$ to $\Re^n$—creating exactly as many new variables as old. One may be interested in a smaller number of new variables, say $m \ (< n)$:

$$\begin{cases} u_1 = g_1(x_1, \ldots, x_n) \\ \vdots \qquad \vdots \\ u_m = g_m(x_1, \ldots, x_n). \end{cases} \qquad (3)$$

If $m > n$, then either some U's would be unnecessary or there would be inconsistent definitions of some U's. If $m < n$, one could supply $n - m$

additional, compatible relations to obtain the distribution of $(U_1, ..., U_n)$ and then find the distribution of the m U's of interest as a marginal distribution. However, the transformation (3) induces a distribution directly in the m-dimensional space, in the usual way:

$$P[(U_1, ..., U_m) \in T] = \int \cdots \int_R f(\mathbf{x})\,d\mathbf{x},$$

where R is the set of $\mathbf{x}$'s with image points in T. Taking T to be a semi-infinite "rectangle" and differentiating yields the p.d.f. of $(U_1, ..., U_m)$.

12.5 The general linear transformation

Using matrix notation will simplify the discussion. We write a random vector $\mathbf{X} = (X_1, ..., X_n)$ as a column matrix. Then, with a prime denoting matrix transposition, $\mathbf{a}'\mathbf{X} = \sum a_i X_i$, and $\mathbf{X}'\mathbf{X} = \sum X_i^2$, each of these being a scalar quantity. And $\mathbf{X}\mathbf{X}'$ is a matrix with ij-element $X_i X_j$, which we may write as $(X_i X_j)$. In matrix notation the multivariate m.g.f. is

$$\psi_\mathbf{X}(t) = E[\exp(\sum t_i X_i)] = E(e^{\mathbf{t}'\mathbf{X}}). \tag{1}$$

We define the expected value of a matrix $\mathbf{V}$ with random elements as the matrix of expected values: $E(\mathbf{V}) = (EV_{ij})$. Thus, for a multivariate vector $(X_1, ..., X_n)$, the mean vector is

$$\boldsymbol{\mu} = E(\mathbf{X}) = \begin{pmatrix} E(X_1) \\ \vdots \\ E(X_n) \end{pmatrix}.$$

The **covariance matrix** is the matrix $\mathbf{M}$ (or $\mathbf{M}_\mathbf{X}$) whose general element is

$$\sigma_{ij} = \operatorname{cov}(X_i, X_j) = E[(X_i - \mu_i)(X_j - \mu_j)].$$

Thus, we may write $\mathbf{M}$ as

$$\mathbf{M} = E[(\mathbf{X} - \boldsymbol{\mu})(\mathbf{X} - \boldsymbol{\mu})'], \tag{2}$$

The covariance matrix is *symmetric*: $\mathbf{M} = \mathbf{M}'$, since $\sigma_{ij} = \sigma_{ji}$. If the components of $\mathbf{X}$ are pairwise uncorrelated, the covariance matrix is diagonal—which is the case, in particular, if the X_i are independent.

The operation of expected value on a matrix is linear. In particular, for any constant matrices $\mathbf{A}$ and $\mathbf{B}$ (with compatible dimensions),

$$E(\mathbf{AVB}) = \mathbf{A}E(\mathbf{V})\mathbf{B}. \tag{3}$$

We are now in a position to establish an important property of covariance matrices:

THEOREM 9 *A covariance matrix* $\mathbf{M}$ *is* **nonnegative definite**:

$$\mathbf{c'Mc} = \sum_{i=1}^{n}\sum_{i=1}^{n} c_i c_j \sigma_{ij} \geq 0 \tag{4}$$

for any vector of constants $\mathbf{c'} = (c_1, ..., c_n)$.

To show (4) we simply rewrite the quadratic form using (3):

$$\mathbf{c'Mc} = \mathbf{c'}E[(\mathbf{X}-\boldsymbol{\mu})(\mathbf{X}-\boldsymbol{\mu})']\mathbf{c}$$

$$= E[\mathbf{c'}(\mathbf{X}-\boldsymbol{\mu})(\mathbf{X}-\boldsymbol{\mu})'\mathbf{c}] = E([\mathbf{c'}(\mathbf{X}-\boldsymbol{\mu})]^2) \geq 0.$$

Moreover, provided at least one c_i is not 0, the vanishing of $\mathbf{c'Mc}$ would imply that the linear combination $\mathbf{c'}(\mathbf{X}-\boldsymbol{\mu})$ is 0 with probability 1, and we can then write one $X_i - \mu_i$ as a linear combination of the others (with probability 1). That is, the distribution of $\mathbf{X}$ is then singular—all of the probability is concentrated in a subspace of lower dimension, and there is not an n-variate p.d.f. On the other hand, if $\mathbf{X}$ is not singular, then $\mathbf{c'Mc} > 0$ unless all c_i's are 0. In this case the matrix $\mathbf{M}$ is called **positive definite**. A useful criterion for a matrix to be positive definite is that the determinants of all principal minors are positive. (A principal minor is a submatrix in which certain columns together with rows with the same numbers have been deleted from the original matrix, e.g., the first and fourth rows and the first and fourth columns.)

The determinant of $\mathbf{M}$ (det $\mathbf{M}$) is positive if $\mathbf{M}$ is positive definite. If $\mathbf{M}$ is not positive definite, then $\mathbf{c'Mc} = 0$ for some $\mathbf{c} \neq \mathbf{0}$. In this case (as pointed out above) $\mathbf{c'}(\mathbf{X}-\boldsymbol{\mu}) = 0$ with probability 1, so $\mathbf{c'M} = \mathbf{0}$; and this says that the rows of $\mathbf{M}$ are linearly dependent, which implies that det $\mathbf{M} = 0$. Conversely, if det $\mathbf{M} = 0$, then $\mathbf{c'M} = \mathbf{0}$ for some $\mathbf{c} \neq \mathbf{0}$, and so $\mathbf{c'Mc} = 0$ for that $\mathbf{c}$.

If $\mathbf{M}$ has rank r, and $r < n$, then there is a subset of r of the X_i's whose r-variate distribution is nonsingular. For, there is then a (principal) $r \times r$ submatrix of $\mathbf{M}$ which is positive definite, and this is the covariance matrix of the corresponding set of r of the X_i's. Moreover, since any larger square submatrix is singular, it follows that the remaining X_i's can be expressed as linear combinations of the r that are linearly independent.

434 Chapter 12 Multivariate Distributions

A linear transformation from a random vector $\mathbf{X}$ to $\mathbf{Y}$ is given by

$$\begin{cases} Y_1 = a_{11}X_1 + \cdots + a_{1n}X_n \\ \quad\vdots \\ Y_m = a_{m1}X_1 + \cdots + a_{mn}X_n, \end{cases} \tag{5}$$

which can be written in the compact form $\mathbf{Y} = \mathbf{AX}$, where $\mathbf{A}$ is the matrix of the transformation.

THEOREM 10 *If $\mathbf{Y} = \mathbf{AX}$, the mean and covariance matrix of $\mathbf{Y}$ are*

$$\boldsymbol{\mu}_\mathbf{Y} = \mathbf{A}\boldsymbol{\mu}_\mathbf{X}, \quad \mathbf{M}_\mathbf{Y} = \mathbf{A}\mathbf{M}_\mathbf{X}\mathbf{A}'. \tag{6}$$

The m.g.f. of $\mathbf{Y}$ is obtainable from the m.g.f. of $\mathbf{X}$ as

$$\psi_\mathbf{Y}(\mathbf{t}) = \psi_\mathbf{X}(\mathbf{A}'\mathbf{t}). \tag{7}$$

To obtain the asserted formula for the mean we exploit the linearity of the expectation operation:

$$\boldsymbol{\mu}_\mathbf{Y} = E(\mathbf{Y}) = E(\mathbf{AX}) = \mathbf{A}E(\mathbf{X}) = \mathbf{A}\boldsymbol{\mu}_\mathbf{X}.$$

That linearity also is gives us the formula for the covariance matrix:

$$\mathbf{M}_\mathbf{Y} = E[(\mathbf{Y} - \boldsymbol{\mu}_\mathbf{Y})(\mathbf{Y} - \boldsymbol{\mu}_\mathbf{Y})'] = E[\mathbf{A}(\mathbf{X} - \boldsymbol{\mu})(\mathbf{X} - \boldsymbol{\mu})'\mathbf{A}']$$

$$= \mathbf{A}E[(\mathbf{X} - \boldsymbol{\mu})(\mathbf{X} - \boldsymbol{\mu})']\mathbf{A}' = \mathbf{A}\mathbf{M}_\mathbf{X}\mathbf{A}'.$$

For the m.g.f. of $\mathbf{Y}$, in terms of the m.g.f. of $\mathbf{X}$, we substitute $\mathbf{AX}$ for $\mathbf{Y}$:

$$\psi_\mathbf{Y}(\mathbf{t}) = E(e^{\mathbf{t}'\mathbf{Y}}) = E(e^{\mathbf{t}'\mathbf{AX}}) = E(e^{[\mathbf{A}'\mathbf{t}]'\mathbf{X}}) = \psi_\mathbf{X}(\mathbf{A}'\mathbf{t}).$$

When $m = n$, the matrix $\mathbf{A}$ is square, and its determinant, $\det \mathbf{A}$, is the Jacobian of the transformation (5). If this Jacobian is not 0, the inverse transformation exists: $\mathbf{X} = \mathbf{A}^{-1}\mathbf{Y}$, where $\mathbf{A}^{-1}$ is the inverse of $\mathbf{A}$, calculated as

$$\mathbf{A}^{-1} = \frac{1}{\det \mathbf{A}} \begin{pmatrix} A_{11} & \cdots & A_{n1} \\ \vdots & & \vdots \\ A_{1n} & \cdots & A_{nn} \end{pmatrix},$$

where A_{ij} is the cofactor of a_{ij} [that is, $(-1)^{i+j}$ times the determinant of order $n - 1$ obtained by striking from $\mathbf{A}$ its ith row and jth column]. The inverse matrix has the property $\mathbf{A}^{-1}\mathbf{A} = \mathbf{A}\mathbf{A}^{-1} = \mathbf{I}$, where $\mathbf{I}$ is the identity matrix of order n, with 1's on the main diagonal ($i = j$) and 0's off that diagonal.

Again, suppose **A** is square. If $\det \mathbf{A} \neq 0$ and if the distribution of **X** is nonsingular, the distribution of $\mathbf{Y} = \mathbf{AX}$ is nonsingular. If **X** has the p.d.f. $f_\mathbf{X}(\mathbf{x})$, the induced p.d.f. of **Y** is

$$f_\mathbf{Y}(\mathbf{y}) = \frac{1}{|\det \mathbf{A}|} f_\mathbf{X}(\mathbf{A}^{-1}\mathbf{y}). \tag{8}$$

[Observe the close analogy with the transformation formula in the univariate case, given by (2) of §12.1.] The assumption of nonsingularity of **A** simply means that the component variables of **Y** are really n distinct random variables—one is not just a linear combination of the others, in which case **Y** would have a singular distribution.

12.6 Multivariate normal distributions

We extend the definition of bivariate normal in the obvious, natural way, to the case of several variables:

DEFINITION: The random p-vector **X** is said to be *multivariate normal* if and only if the linear combination

$$\mathbf{a}'\mathbf{X} = a_1 X_1 + \cdots + a_p X_p$$

is univariate normal for every choice of the constants a_i.

Indeed, there *are* such distributions. In particular, a vector of *independent* normal variables is multivariate normal: Suppose **X** is such a vector, with $X_i \sim \mathcal{N}(\mu_i, \sigma_i^2)$. The m.g.f. of $\mathbf{a}'\mathbf{X}$ is

$$\psi_{\Sigma a_i X_i}(t) = \prod \psi_{X_i}(a_i t) = \prod \exp\{a_i t \mu_i + a_i^2 t^2 \sigma_i^2 / 2\}$$
$$= \exp\left\{t \sum a_i \mu_i + \left(\sum a_i^2 \sigma_i^2\right) t^2 / 2\right\},$$

which is the m.g.f. of $\mathcal{N}(\sum a_i \mu_i, \sum a_i^2 \sigma_i^2)$.

The family of multivariate normal distributions as defined above includes some singular distributions—namely, those whose covariance matrices are nonsingular. The definition has some immediate, important consequences:

THEOREM 11 *The various marginal distributions of a multivariate normal distributions are all multivariate normal.*

THEOREM 12 *If **X** is p-variate normal, then **AX** + **b** is multivariate normal for any m-vector **b** and any m × p matrix of constants **A**.*

Theorem 11 is fairly obvious—a linear combination of marginal variables in a subset of the components of **X** is also a linear combination of all the components X_i. For Theorem 12 we observe that a linear combination $\mathbf{c}'(\mathbf{AX})$ can be written $(\mathbf{A}'\mathbf{c})'\mathbf{X}$, which is a linear combination of the X_i's. Moreover, if **Y** is m-variate normal, so is **Y** + **b**, for any m-vector **b** of constants, since a linear combination of the components of **Y** + **b** is a normal variable plus a constant, which sum is normal. If the matrix **A** is m × p, the dimension m can be less than, equal to, or larger than p. But if m > p, the distribution of **Y** is singular, since then m − p components are linear combinations of the others.

As in the bivariate case, we now derive the m.g.f. of a multivariate **X** from the fact that for any constant p-vector **s**, the (scalar) variable $U = \mathbf{s}'\mathbf{X}$ is univariate normal. We'll need the mean and variance of U:

$$EU = E(\mathbf{s}'\mathbf{X}) = \mathbf{s}'\boldsymbol{\mu}, \quad \sigma_U^2 = \text{var}(\mathbf{s}'\mathbf{X}) = \mathbf{s}'\mathbf{M}\mathbf{s}. \tag{1}$$

The m.g.f. of U is then

$$\psi_U(t) = \exp\{\mu_U t + \sigma_U^2 t^2/2\} = E(e^{tU}) = E(e^{t\mathbf{s}'\mathbf{X}}).$$

By definition, the m.g.f. of **X** is

$$\psi_\mathbf{X}(\mathbf{s}) = E(e^{\mathbf{s}'\mathbf{X}}) = \psi_U(1) = \exp\{\mu_U + \sigma_U^2/2\}. \tag{2}$$

Substituting from (1) into (2) we obtain the desired formula:

THEOREM 13 *If **X** is multivariate normal, its m.g.f. is*

$$\psi_\mathbf{X}(\mathbf{s}) = \exp[\mathbf{s}'\boldsymbol{\mu} + \tfrac{1}{2}\mathbf{s}'\mathbf{M}\mathbf{s}]. \tag{3}$$

COROLLARY *If **X** is multivariate normal and its second moment matrix **M** is diagonal—all covariances equal to 0—then the components X_i are independent.*

To show the corollary we observe that when **M** is diagonal, the quadratic form $\mathbf{s}'\mathbf{M}\mathbf{s}$ is simply a sum of squares, so

$$\psi_\mathbf{X}(\mathbf{s}) = \exp\left\{\sum \mu_i s_i + \sum \sigma_i^2 s_i^2\right\} = \prod \exp\{\mu_i s_i + \sigma_i^2 s_i^2\}.$$

And this is the product of the m.g.f.'s of the variables $X_i \sim \mathcal{N}(\mu_i, \sigma_i^2)$.

THEOREM 14 *If **X** is p-variate normal, and its second moment matrix **M** is nonsingular, then **X** has a density function:*

$$f_{\mathbf{X}}(\mathbf{x}) = \frac{1}{(2\pi)^{p/2}\sqrt{\det \mathbf{M}}} \exp\left\{-\tfrac{1}{2}(\mathbf{x}-\boldsymbol{\mu})'\mathbf{M}^{-1}(\mathbf{x}-\boldsymbol{\mu})\right\}.$$

We know that if $\mathbf{M}_{\mathbf{X}}$ is nonsingular, it is positive definite. Then, according to a theorem of matrix theory, there is a nonsingular matrix $\mathbf{A}$ with the property $\mathbf{A}\mathbf{A}' = \mathbf{M}_{\mathbf{X}}$. Let $\mathbf{Y} = \mathbf{A}^{-1}(\mathbf{X}-\boldsymbol{\mu})$. This is p-variate normal with mean vector $\mathbf{0}$ and covariance matrix

$$\mathbf{M}_{\mathbf{Y}} = \mathbf{A}^{-1}\mathbf{M}_{\mathbf{X}}(\mathbf{A}^{-1})' = \mathbf{A}^{-1}\mathbf{A}\mathbf{A}'(\mathbf{A}^{-1})' = \mathbf{I}.$$

Thus,

$$f_{\mathbf{Y}}(\mathbf{y}) = \frac{1}{(2\pi)^{p/2}} e^{-\mathbf{y}'\mathbf{y}/2}.$$

From this we can get the p.d.f. of $\mathbf{X}$:

$$f_{\mathbf{X}}(\mathbf{x}) = \frac{1}{(2\pi)^{p/2}|\det \mathbf{A}|} \exp\left\{-\tfrac{1}{2}(\mathbf{x}-\boldsymbol{\mu})'(\mathbf{A}^{-1})'\mathbf{A}^{-1}(\mathbf{x}-\boldsymbol{\mu})\right\}.$$

Since $(\mathbf{A}^{-1})'\mathbf{A}^{-1} = (\mathbf{A}\mathbf{A}')^{-1} = \mathbf{M}^{-1}$, and $|\det \mathbf{A}| = \sqrt{\det \mathbf{M}}$, the theorem follows.

COROLLARY *If the covariance matrix **M** of a p-variate normal variable **X** has rank $r \leq p$, there is an $p \times r$ matrix **C** and a p-vector **d** such that $\mathbf{X} = \mathbf{C}\mathbf{Y} + \mathbf{d}$, where **Y** is r-variate normal with zero mean vector and identity covariance matrix.* [That is, the Y_i's are i.i.d. $\mathcal{N}(0, 1)$.]

Without loss of generality, suppose it is the first r rows of $\mathbf{M}$ that are linearly independent, so that the submatrix $\mathbf{M}_r$ of second moments of the first r X_i's is nonsingular. According to the proof of Theorem 12, there is a linear transformation on independent, standard normal variables $(Y_1, \ldots, Y_r)$ which yields the first r X_i's. Since the remaining X_i's can be expressed as linear combinations of the first r, it follows that $\mathbf{X}$ is a linear transformation of the first r X_i's, as claimed.

EXAMPLE 12.6a Let Y_1, Y_2, Y_3 be i.i.d. $\mathcal{N}(0, 1)$ and define (X_1, X_2) as

$$\mathbf{X} = \begin{pmatrix} 2 & -1 & 1 \\ 1 & -3 & 0 \end{pmatrix} \mathbf{Y}.$$

The covariance matrix of **X** is then

$$\mathbf{M} = \begin{pmatrix} 2 & -1 & 1 \\ 1 & -3 & 0 \end{pmatrix} \begin{pmatrix} 2 & 1 \\ -1 & -3 \\ 1 & 0 \end{pmatrix} = \begin{pmatrix} 6 & 5 \\ 5 & 10 \end{pmatrix},$$

and $\det \mathbf{M} = 35$. The inverse is

$$\mathbf{M}^{-1} = \frac{1}{35}\begin{pmatrix} 10 & -5 \\ -5 & 6 \end{pmatrix},$$

so the m.g.f. of **X** is

$$\psi_\mathbf{X}(\mathbf{t}) = \exp[\mathbf{t}'\boldsymbol{\mu} + \tfrac{1}{2}\mathbf{t}'\mathbf{M}\mathbf{t}] = \exp\left\{-\tfrac{1}{2}(6t_1^2 + 10t_1 t_2 + 10t_2^2)\right\}.$$

The p.d.f. is

$$f(\mathbf{x}) = \frac{1}{2\pi\sqrt{35}}\exp\left\{-\tfrac{1}{70}(10x_1^2 - 10x_1 x_2 + 6x_2^2)\right\}.$$

Suppose next we define **Z** as (for instance)

$$\mathbf{Z} = \begin{pmatrix} 1 & -1 \\ 2 & 1 \\ 1 & 2 \end{pmatrix}\mathbf{X}.$$

The distribution of **Z** must be singular, since there are more Z's than X's. Indeed, $Z_3 = Z_2 - Z_1$, so the distribution of **Z** is concentrated in the plane $z_1 - z_2 + z_3 = 0$. The covariance matrix of **Z** is

$$\mathbf{M} = \begin{pmatrix} 1 & -1 \\ 2 & 1 \\ 1 & 2 \end{pmatrix}\begin{pmatrix} 6 & 5 \\ 5 & 10 \end{pmatrix}\begin{pmatrix} 1 & 2 & 1 \\ -1 & 1 & 2 \end{pmatrix} = \begin{pmatrix} 6 & -3 & -9 \\ -3 & 54 & 57 \\ -9 & 57 & 66 \end{pmatrix}.$$

which is singular. ∎

The next example illustrates a way of factoring the covariance matrix of a nonsingular distribution in the form $\mathbf{M}_X = \mathbf{A}\mathbf{A}'$, as in the proof of Theorem 14, in a particular case.

EXAMPLE 12.6b Consider the following covariance matrix (which is positive definite):

$$\mathbf{M} = \begin{pmatrix} 5 & -6 \\ -6 & 9 \end{pmatrix}.$$

Adding 6/5 of the first row (or column) to the second row (column) puts 0 in the second row (column):

$$\begin{pmatrix} 1 & 0 \\ 6/5 & 1 \end{pmatrix}\begin{pmatrix} 5 & -6 \\ -6 & 9 \end{pmatrix}\begin{pmatrix} 1 & 6/5 \\ 0 & 1 \end{pmatrix} = \begin{pmatrix} 5 & 0 \\ 0 & 9/5 \end{pmatrix}.$$

Pre- and postmultiplying by the reciprocal of the square roots of the diagonal entries produces 1's; so let

$$\mathbf{C} = \begin{pmatrix} 1/\sqrt{5} & 0 \\ 0 & \sqrt{5}/3 \end{pmatrix} \begin{pmatrix} 1 & 0 \\ 6/5 & 1 \end{pmatrix} = \begin{pmatrix} 1/\sqrt{5} & 0 \\ 2/\sqrt{5} & \sqrt{5}/3 \end{pmatrix},$$

which has now been constructed so that $\mathbf{CMC'} = \mathbf{I}$. Then $\mathbf{M} = \mathbf{C}^{-1}(\mathbf{C}^{-1})'$, where $\mathbf{C}^{-1}$ is the "$\mathbf{A}$" in the proof of Theorem 14:

$$\mathbf{C}^{-1} = \begin{pmatrix} \sqrt{5} & 0 \\ -6\sqrt{5} & 3/\sqrt{5} \end{pmatrix}.$$

Then

$$\begin{cases} X_1 = \mu_1 + \sqrt{5}\, Y_1 \\ X_2 = \mu_2 - \dfrac{6}{\sqrt{5}} Y_1 + \dfrac{3}{\sqrt{5}} Y_2, \end{cases}$$

where Y_1 and Y_2 are i.i.d. $\mathcal{N}(0, 1)$. [This representation of $\mathbf{X}$ in terms of i.i.d. standard normal variables is not unique; for example, one possibility is $X_1 = \mu_1 + Y_1 + 2Y_2$, and $X_2 = \mu_2 - 3Y_2$ is another.] ∎

THEOREM 15 *The conditional distribution of any subset of components of a random vector* $\mathbf{X}$ *with a multivariate normal distribution, given values of the remaining components, is multivariate normal.*

We refer to more advanced texts[2] for proof of this theorem.

PROBLEMS

12-26. Show: If $\mathbf{X}$ has a p-variate distribution and $\mathbf{a}$ and $\mathbf{b}$ are p-vectors, then $\operatorname{cov}(\mathbf{a'X}, \mathbf{b'X}) = \mathbf{a'Mb}$.

12-27. We say that the vectors $(X_1, ..., X_k)$ and $(X_{k+1}, ..., X_p)$ are jointly normally distributed if $\mathbf{X} = (X_1, ..., X_p)$ is p-variate normal. Show that those two subsets of components of $\mathbf{X}$ are independent if all the component variables in one subset are uncorrelated with all the component variables of the other.

12-28. If $\mathbf{X}$ is a random p-vector, $\mathbf{A}$ is an $m \times p$ matrix of constants, and $\mathbf{b}$ is an m-vector of constants, find the mean vector and covariance matrix of the transformed variable $\mathbf{Y} = \mathbf{AX} + \mathbf{b}$.

---

[2] See for example C. R. Rao, *Linear Statistical Inference and Its Applications*, 2nd ed., New York: John Wiley & Sons, Inc., 1973, page 522.

12-29. Suppose the random vector $\mathbf{U}$ has the constant density $f_\mathbf{U}(\mathbf{u}) = n!$ on the region $0 < u_1 < \cdots < u_n < 1$. Define variables

$$X_1 = U_1, \ X_2 = U_2 - U_1, \ \ldots, \ X_n = U_n - U_{n-1}$$

and find the p.d.f. of $(X_1, \ldots, X_n)$.

12-30. Let $\mathbf{X}$ and $\mathbf{Y}$ be defined as in Example 12.4a and define

$$Z_1 = \frac{Y_1}{Y_2}, \ \ldots, \ Z_{n-1} = \frac{Y_{n-1}}{Y_n}, \ Z_n = Y_n.$$

Obtain the joint p.d.f. of the Z's and so deduce their independence.

12-31. Suppose $\mathbf{X}$ is multivariate normal with

$$\boldsymbol{\mu} = \begin{pmatrix} 3 \\ 3 \\ 0 \\ 0 \end{pmatrix}, \ \mathbf{M} = \begin{pmatrix} 2 & 0 & 2 & 0 \\ 0 & 1 & 1 & 0 \\ 2 & 1 & 5 & 1 \\ 0 & 0 & 1 & 1 \end{pmatrix}.$$

(a) Show that $\mathbf{M}$ is nonsingular.
(b) Calculate $\rho_{1,3}$, the correlation between X_1 and X_3.
(c) Find the mean vector and covariance matrix of $\mathbf{Y} = \mathbf{A}\mathbf{X} + \mathbf{c}$, where

$$\mathbf{A} = \begin{pmatrix} 1 & 1 & 1 & -1 \\ 1 & -1 & 1 & 1 \\ 1 & 0 & 1 & 0 \end{pmatrix}, \ \mathbf{c} = \begin{pmatrix} 2 \\ 0 \\ -1 \end{pmatrix}.$$

(d) Find the bivariate m.g.f. and bivariate p.d.f. of (Y_1, Y_2) in (c).
(e) Find the mean and variance of Y_3 directly, using $E(\cdot)$ calculations.

12-32. Show that if $\mathbf{X}$ is p-variate normal with mean vector $\mathbf{0}$ and covariance matrix $\mathbf{M}$, the quadratic form $Q = \mathbf{X}'\mathbf{M}^{-1}\mathbf{X}$ has a chi-square distribution with p degrees of freedom.

12-33. Show that if $\mathbf{X}$ is p-variate normal, the pair of linear combinations $(\mathbf{a}'\mathbf{X}, \mathbf{b}'\mathbf{X})$ has a bivariate normal distribution, and that these linear combinations are independent if and only if $\mathbf{a}'\mathbf{M}_\mathbf{X}\mathbf{b} = 0$.

12.7 A decomposition theorem

In a proof of Theorem 11 of Chapter 7 (page 213), we used the parallel axis theorem for a decomposition of a sum of squares of standard normal variables:

$$\sum \left(\frac{X_i - \mu}{\sigma}\right)^2 = \sum \left(\frac{Y_i - \bar{X}}{\sigma}\right)^2 + \left(\frac{\bar{X} - \mu}{\sigma/\sqrt{n}}\right)^2. \tag{1}$$

12.7 A decomposition theorem

This is of the form $Q = Q_1 + Q_2$, where each Q is a quadratic in standard normal variables. We showed that Q_1 and Q_2 are independent and in this way derived the distribution of the sample variance. In this section we obtain a generalization which will be needed in the "analysis of variance" to be taken up in Chapter 14.

Consider the n normally distributed random variables U_i, and a linear transformation to m variables L_j:

$$\begin{cases} L_1 = a_{11}U_1 + \cdots + a_{1n}U_n \\ \vdots \\ L_m = a_{m1}U_1 + \cdots + a_{mn}U_n. \end{cases} \qquad (2)$$

Let Q denote the sum of squares of these linear combinations:

$$Q = L_1^2 + \cdots + L_m^2. \qquad (3)$$

The *rank* of Q is defined as the largest number of L's among which there is no linear relation, or equivalently, as the rank of the matrix $\mathbf{A} = (a_{ij})$.

THEOREM 16 (Cochran's Theorem) *Let $Z_1, ..., Z_n$ be i.i.d., standard normal variables, and suppose there is the following identity:*

$$\sum_1^n Z_i^2 = Q_1 + \cdots + Q_k, \qquad (4)$$

where each Q_i is a sum of squares of linear combinations of the Z's. Let r_i denote the rank of Q_i. Then, if

$$r_1 + \cdots + r_k = n, \qquad (5)$$

it follows that the Q_i's have independent chi-square distributions, each with a number of degrees of freedom given by its rank.

We'll want to apply this theorem in situations in which $X_1, ..., X_n$ are i.i.d. with $X_i \sim \mathcal{N}(\mu, \sigma^2)$, the "$Z_i$" of the theorem is the standard score $(X_i - \mu)/\sigma$, and each Q is a sum of squares of homogeneous linear combinations of X's. Such a linear combination is also a linear combination of Z's:

$$L_j = a_{j1}X_1 + \cdots a_{jn}X_n = \sigma(a_{j1}Z_1 + \cdots + a_{jn}Z_n) + c_j.$$

This is homogeneous in the Z's if all $c_j = 0$. But $E(L_j) = c_j$, so we need only check to see that $E(L_j) = 0$. If we can see the ranks of all but one Q_i, finding the rank of the remaining Q_i can sometimes be found this way: If there are h linear relations among the m L's, the rank of the transformation from Z's to L's is at most $m - h$. Then the fact that the rank of a sum of quadratic forms is no larger than the sum of the ranks may lead to the conclusion that the rank of Q_i is exactly $m - h$. The following example illustrates this.

EXAMPLE 12.7a In the decomposition (1) which opened section, the left side is a sum of n independent, standard normal variables. The right side is the sum of two Q's:

$$Q_1 = \sum \left(\frac{X_i - \bar{X}}{\sigma}\right)^2, \quad Q_2 = \left(\frac{\bar{X} - \mu}{\sigma/\sqrt{n}}\right)^2.$$

Each of these is a sum of squares of linear combinations of the X_i's, and hence is a sum of squares of linear combinations of the standardized variables $Z_i = (X_i - \mu)/\sigma$. The mean of each combination, in both Q_{1i} and Q_2, is 0:

$$E\left(\frac{X_i - \bar{X}}{\sigma}\right) = \frac{\mu - \mu}{\sigma} = 0, \quad E\left(\frac{\bar{X} - \mu}{\sigma/\sqrt{n}}\right) = \frac{\mu - \mu}{\sigma\sqrt{n}} = 0,$$

so as linear combinations of the Z_i, they are homogeneous (no constant term). The sum of the combinations in Q_1 is zero—one linear relationship, which means that $r_1 \leq n - 1$. The single term Q_2 is clearly of rank 1. And the rank of the sum $Q_1 + Q_2$ is n, which is at most the sum of the ranks: $r_1 + r_2 = r_1 + 1 \geq n$. So r_1 is both at most and at least $n - 1$, or $r_1 = n - 1$. Thus, $r_1 + r_2 = n$, and Cochran's theorem tells us that Q_1 and Q_2 have independent chi-square distributions with $n - 1$ and 1 d.f., respectively. ∎

Now to the proof of Cochran's theorem: Define L's and Q as in (2) and (3), where the Q can (for the moment) represent any of the Q_i in the decomposition (4). Let **A** denote the matrix of a_{ij} in (2), and let r denote its rank, which is then the rank of Q. Without loss of generality, assume that a non-singular submatrix of rank r comes from the first r rows of **A**. Then the quantities $L_{r+1}, \ldots, L_m$ can be expressed as linear combinations of the first r L's in Q, which means that Q can be expressed as a quadratic in just those first r L's:

$$Q = \sum_{j=1}^{r} \sum_{i=1}^{r} b_{ij} L_i L_j = \mathbf{L'BL},$$

where **L** is the column matrix with entries $L_1, \ldots, L_r$, and **B** is the $r \times r$ symmetric matrix (b_{ij}). Since Q is clearly positive definite, the matrix **B** is nonsingular.

We next define an $r \times r$ nonsingular matrix **P** such that $\mathbf{P'BP = I}$, the r-dimensional identity matrix, and let $\mathbf{v = P^{-1}L}$ or $\mathbf{L = Pv}$. [Finding such a matrix **P** is another instance of finding a "square root" of a positive definite matrix: $\mathbf{P}'^{-1}\mathbf{I}\mathbf{P}^{-1} = (\mathbf{P}^{-1})'\mathbf{I}\mathbf{P}^{-1} = \mathbf{B}$. (See §12.6.) Then

$$Q = \mathbf{L}'\mathbf{B}\mathbf{L} = \mathbf{V}'\mathbf{P}'\mathbf{B}\mathbf{P}\mathbf{V} = \mathbf{V}'\mathbf{V} = V_1^2 + \cdots + V_r^2, \qquad (6)$$

where each V is a linear combination of L's and therefore a linear combination of Z's. So we have expressed Q as a sum of squares of exactly r linear combinations of Z's.

Now we go to the setting of Cochran's theorem, where we assume a random sample from $\mathcal{N}(\mu, \sigma^2)$ and a decomposition of a sum of squares of corresponding, independent Z-scores:

$$\sum_1^n Z_i^2 = \sum_1^n \left(\frac{X_i - \mu}{\sigma}\right)^2 = Q_1 + \cdots + Q_k, \qquad (7)$$

in which each Q_i is a sum of squares of of linear combinations of X_i's, with corresponding rank r_i. We then write each Q in the form (6), a sum of squares of linear combinations V_j of standard normal Z_i's:

$$Q_1 = V_1^2 + \cdots + V_{r_1}^2, \quad Q_2 = V_{r_1+1}^2 + \cdots + V_{r_1+r_2}^2, \cdots, \qquad (8)$$

so that, all in all, there are $r_1 + \cdots + r_k$ V's. If these ranks sum to n, then

$$\sum_1^n Z_i^2 = \mathbf{Z}'\mathbf{Z} = V_1^2 + \cdots + V_n^2 = \mathbf{V}'\mathbf{V}.$$

But each V is a linear combination of Z's: $\mathbf{V} = \mathbf{C}\mathbf{Z}$, so

$$\mathbf{Z}'\mathbf{Z} = (\mathbf{C}\mathbf{Z})'\mathbf{C}\mathbf{Z} = \mathbf{Z}'\mathbf{C}'\mathbf{C}\mathbf{Z}.$$

Clearly, $\mathbf{C}'\mathbf{C} = \mathbf{I}$. Then, since $\mathbf{C}$ is *square* (because of the rank condition in the theorem) and $\mathbf{C}^{-1} = \mathbf{C}'$, it follows that $\mathbf{C}\mathbf{C}' = \mathbf{I}$. But $\mathbf{V} = \mathbf{C}\mathbf{Z}$ is multivariate normal with mean $\mathbf{0}$ and covariance matrix $\mathbf{C}\mathbf{I}\mathbf{C}' = \mathbf{I}$, so the variables V_j are independent and standard normal. From their structure as given in (8), the Q's must therefore be independent and have chi-square distributions as asserted in the theorem.

13
Nonparametric Inference

Inference in problems in which the set of possible models is a parametric class, indexed by one or more parameters, is said to be **parametric**. In **nonparametric** inference, the basic problem may involve some distribution parameter such as the median or the mean, or some other descriptive quantity such as $P(X < Y)$, but these would not characterize the population. The distinction between parametric and nonparametric is not always clear-cut, and there is not a universally accepted definition of "nonparametric." Sometimes the contingency table analysis of Chapter 11 is included in the class of nonparametric methods. Most of the testing and estimation problems we've studied so far are parametric. In particular, estimating parameters is often parametric, and sounds like a parametric problem; but estimating a population median or the probability $P(X < Y)$ is perhaps nonparametric.

In this chapter we'll study mainly problems of testing hypotheses. We construct test statistics based on intuition, leading to tests that should work well without a requirement for a precisely specified population distribution. Of course, nonparametric methods can be used even when there *is* a parametric structure that would make a parametric test optimal. It often turns out that they work quite well by comparison with a parametric test especially designed for that situation. Moreover, they are apt to perform *better* than the parametric test when the actual population does not quite fit the parametric structure.

Although it is awkward to think of "power functions" without a parametric structure, in some situations we do examine power for certain classes of alternatives.

Many of the methods to be studied employ statistics whose distributions do not depend on the nature of the population distribution and so are said to be **distribution-free**. Such methods are thus inherently *robust*—not dependent on restrictive assumptions about the population.

Many nonparametric procedures are based on rank-order statistics, so we begin by looking more closely at the order statistic—the set of ordered observations in set of data from a continuous population.

13.1 Distribution of the order statistic

In Chapter 7 we saw that, under random sampling, the order statistic $(X_{(1)}, ..., X_{(n)})$ is sufficient for any family of continuous population distributions. Here we derive the joint p.d.f. of the $X_{(i)}$'s and the conditional distribution in the sample space given the order statistic. The latter is the basis for a number of nonparametric procedures.

THEOREM 1 *If $(X_1, ..., X_n)$ are exchangeable, with joint p.d.f. $f_\mathbf{X}(\mathbf{x})$, the joint p.d.f. of the ordered observations $Y_i = X_{(i)}$ is*

$$f_\mathbf{Y}(\mathbf{y}) = \begin{cases} n! f_\mathbf{X}(\mathbf{y}) & \text{for } \mathbf{y} \in R, \\ 0, & \text{otherwise.} \end{cases} \quad (1)$$

Let ν denote one of the $n!$ permutations of the integers 1, 2, ..., n. Each such permutation defines a region of sample points $\mathbf{x}$:

$$R_\nu = \{\mathbf{x} \mid x_{\nu_1} < x_{\nu_2} < \cdots < x_{\nu_n}\}. \quad (2)$$

Let R denote the particular R_ν where ν is the permutation (1, 2, ..., n). The $n!$ sets R_ν form a partition of the sample space, except for the boundaries, which have total probability 0 under the assumption of a continuous population. We show next that the R_ν are equally likely—each has probability $1/n!$.

For any particular permutation ν, consider the transformation $\mathbf{y} = T_\nu \mathbf{x}$, where $y_k = x_{\nu_k}$. The determinant of this linear transformation is either 1 or -1. (A suitable rearrangement of its rows or columns would result in the determinant of the identity matrix.) And T_ν transforms the corresponding R_ν into the region R where coordinates are in numerical order, according to (2). Now, because of exchangeability, the joint p.d.f. of $\mathbf{X}$ is a symmetric function of the x_i's, so it is unchanged by the permutation transformation ν:

$$f(\mathbf{x}) = f(x_1, ..., x_n) = f(y_1, ..., y_n) = f(\mathbf{y}).$$

Thus, making the change of variable $\mathbf{y} = T_\nu \mathbf{x}$, we obtain

$$P(R_\nu) = \int_{R_\nu} f(\mathbf{x})\, d\mathbf{x} = \int_R f(\mathbf{y})\, d\mathbf{y} = P(R).$$

13.1 Distribution of the order statistic

Finally, consider the order statistic:

$$\mathbf{Y} = t(X_1, ..., X_n) = (X_{(1)}, ..., X_{(n)}).$$

Since this function t maps each $\mathbf{x}$ into a point in R, where coordinates are in numerical order, the support of $\mathbf{Y}$ is R. For any set $A \in R$, the probability of A is the sum of the probabilities of the various pre-images. Let A_ν denote the pre-image of A in R_ν: $\{\mathbf{x} \mid \mathbf{x} \in R_\nu,\ t(\mathbf{x}) \in A\}$. Then

$$P(\mathbf{Y} \in A) = \sum_\nu P(\mathbf{X} \in A_\nu) = \sum_\nu \int_A f(\mathbf{y})\,d\mathbf{y} = \int_A n! f(\mathbf{y})\,d\mathbf{y}.$$

Since A is an arbitrary set in R, this demonstrates (1), completing the proof of Theorem 1. The next example may help to clarify the proof.

EXAMPLE 13.1a Consider a special case in which $n = 2$. There are just two permutations of $(1, 2)$; call them $\mathbf{0} = (1, 2)$, and $\mathbf{1} = (2, 1)$. The regions R_0 (which is R) and R_1 are the half-planes in which $x_1 < x_2$ and $x_2 < x_1$, respectively, as shown in Figure 13-1. Also shown in the figure is a set $A \in R$ and the set A_1 which is mapped into A by the order statistic.

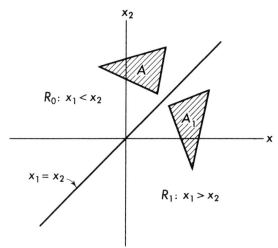

Figure 13.1

The order statistic has "values" only in R, taking each point of the plane (with $x_1 \neq x_2$) into a point in R. Thus,

$$P[(X_{(1)}, X_{(2)}) \in A] = P(\mathbf{X} \in A_0) + P(\mathbf{X} \in A_1)$$

$$= 2\,P(\mathbf{X} \in A_0) = \int_{A_0} 2\,f(x_1, x_2)\,dx_1\,dx_2.$$

The joint p.d.f. $(X_{(1)}, X_{(2)})$ is just twice the joint p.d.f. of (X_1, X_2), in the region R which supports the distribution. ∎

THEOREM 2 *If $(X_1, \ldots, X_n)$ are exchangeable, with joint p.d.f. $f_\mathbf{X}(\mathbf{x})$, the conditional distribution in the sample space given the order statistic is uniform over the $n!$ permutations of that order statistic.*

On the basis of the symmetry of the joint p.d.f., intuition clearly suggests that the various permutations of a given order statistic are equally likely to be the sample actually observed. A formal proof is a bit subtle. Consider a set B of sample points $\mathbf{x}$ and a set $A \subset R$ (in the notation of the above proof of Theorem 1). Let $\tilde{A}$ denote the union of all A_ν (the pre-images of A under the transformation defined by the order statistic). Then

$$P(\mathbf{Y} \in A,\ \mathbf{X} \in B) = P(\mathbf{X} \in \tilde{A}\,B) = \int_{\tilde{A}} I_B(\mathbf{x})\,f(\mathbf{x})\,d\mathbf{x},$$

where I_B is the indicator function of B. The set $\tilde{A}$ contains all the permutations of any one of its points, so a change of variables in the above integral by a permutation transformation T_ν would not change the region of integration—nor the value of $f(\mathbf{x})$. Thus,

$$P(\mathbf{Y} \in A,\ \mathbf{X} \in B) = \int_{\tilde{A}} I_B(T_\nu \mathbf{x})\,f(\mathbf{x})\,d\mathbf{x}.$$

This has the same value (namely, the probability on the left) for each ν, so it is equal to the average over all such permutations:

$$P(\mathbf{Y} \in A,\ \mathbf{X} \in B) = \int_{\tilde{A}} \frac{1}{n!} \sum_\nu I_B(T_\nu \mathbf{x})\,f(\mathbf{x})\,d\mathbf{x}.$$

The integrand has the same value in any A_ν as in A, so the integral over the union of the A_ν is just their number ($n!$) times the integral over A:

$$P(\mathbf{Y} \in A,\ \mathbf{X} \in B) = \int_A \left\{\frac{1}{n!} \sum_\nu I_B(T_\nu \mathbf{y})\right\} n!\,f(\mathbf{y})\,d\mathbf{y}.$$

This can be interpreted as the integral, with respect to the distribution of $\mathbf{Y}$, of the conditional probability of the event $[\mathbf{X} \in B]$ given $\mathbf{Y} = \mathbf{y}$. But A is arbitrary, and therefore

$$P(\mathbf{X} \in B \mid \mathbf{Y} = \mathbf{y}) = \frac{1}{n!} \sum_{\nu} I_B(T_\nu \mathbf{y}).$$

Since I_B is 1 on B and 0 elsewhere, this last sum simply counts the number of permutations of $\mathbf{y}$ that are in B. In particular, if B is a single point in R, the conditional probability is $1/n!$, as claimed in Theorem 2.

13.2 The transformation F(X)

Let $F(x)$ be the c.d.f. of a continuous population variable X, and let x_u denote the largest of the possible inverses $F^{-1}(u)$. Define the random variable $U = F(X)$. As we saw in §3.5, (page 66) the c.d.f. of U is

$$F_U(u) = P[F(X) \le u] = P(X \le x_u) = F(x_u) = u, \quad 0 \le u \le 1.$$

Thus, $U \sim \mathcal{U}(0, 1)$.

For a random sample $\mathbf{X} = (X_1, ..., X_n)$, let $U_i = F(X_i)$. The U's are independent, being functions of independent random variables, and each is $\mathcal{U}(0,1)$. That is, $\mathbf{U} = (U_1, ..., U_n)$ is a *random sample* from $\mathcal{U}(0,1)$. And because the transformation F is monotone non-decreasing, the ith smallest X is mapped into the ith smallest U: $U_{(i)} = F(X_{(i)})$. This means that the vector $\mathbf{V} = (V_1, ..., V_n)$, where $V_i = U_{(i)}$, is the order statistic of a random sample from $\mathcal{U}(0,1)$.

The random variable $V_k = F(X_{(k)})$ is the *area to the left* of the kth smallest observation in $\mathbf{X}$. Its distribution is that of the kth smallest observation in a random sample of size n from $\mathcal{U}(0, 1)$, whose p.d.f. was given in §7.6:

$$f_{V_k}(v) = n \binom{n-1}{k-1} v^{k-1}(1-v)^{n-k}, \quad 0 < v < 1.$$

This is the p.d.f. of Beta($k, n-k+1$), with mean value $k/(n+1)$. A useful statistic is the area between two successive, ordered observations:

$$V_k - V_{k-1} = F(X_{(k)}) - F(X_{(k-1)}),$$

with mean value

$$E(V_k - V_{k-1}) = \frac{k}{n+1} - \frac{k-1}{n+1} = \frac{1}{n+1}. \tag{1}$$

This tells us that the ordered observations tend to divide the area under the p.d.f. of X into $n+1$ equal areas.

The area between the smallest and largest observations, $V_n - V_1$, is a statistic that can be used in the construction of **tolerance intervals**. Since this difference is the *range* of a random sample from $\mathcal{U}(0, 1)$, its p.d.f. (see §7.6) is

$$f(r) = n(n-1)(1-r)r^{n-2}, \quad 0 < r < 1,$$

independent of the population distribution. The next example illustrates the notion of a tolerance interval.

EXAMPLE 13.2a Consider finding a sample size n such that at least 99 percent of a population with c.d.f. F lies between the smallest and largest observation in the sample with probability .95. Do do this we set

$$.95 = P[F(X_{(n)}) - F(X_{(1)}) > .99]$$
$$= P(V_n - V_1 > .99) = n(n-1)\int_{.99}^{1}(1-r)r^{n-2}dr$$
$$= 1 - (.99)^{n-1}(.01n + .99).$$

We can find an approximate solution of this transcendental equation for n by observing that the right side is $P(Y > 1)$, where $Y \sim \text{Bin}(n, .01)$, which we can approximate by $\text{Poi}(.01n)$. The Poisson table shows $.01n$ to be about 4.75, and a little hand calculator work yields an n of 473 as the integer closest to the solution. ∎

It will be useful in certain goodness-of-fit methods to transform the c.d.f. of a sample using the transformation $F(X)$. The sample c.d.f. is

$$F_n(x) = \frac{k}{n} \quad \text{for } X_{(k)} \leq x < X_{(k+1)} \text{ and } 1 \leq k \leq n-1,$$

with $F_n(x) = 0$ for $x < X_{(1)}$ and $F_n(x) = 1$ for $x \geq X_{(n)}$. Now define

$$F_n^*(y) = F_n(F^{-1}(y)) = \frac{k}{n}, \quad \text{for } X_{(k)} \leq F^{-1}(y) < X_{(k+1)}.$$

Taking F of each member of the last inequality we rewrite this as

$$F_n^*(y) = \frac{k}{n}, \quad \text{for } V_k \leq y < V_{k+1}, \tag{2}$$

where $\mathbf{V} = (V_1, ..., V_n)$ is the order statistic and $F_n^*(y)$, the sample d.f. of a random sample from $\mathcal{U}(0, 1)$,—whatever the distribution of X may be. We can use this result for the following theorem concerning functionals of the sample d.f. and the population c.d.f.

THEOREM 3 *For a random sample from a continuous population with an increasing c.d.f. F, these functionals are distribution-free:*

$$\int_{-\infty}^{\infty} g[F_n(x), F(x)] dF(x) \quad \text{and} \quad \sup_x g[F_n(x), F(x)]. \tag{3}$$

13.2 The transformation $F(X)$

To show the asserted independence, we make the change of variable $x = F^{-1}(y)$, which reduces the quantities (4) to

$$\int_0^1 g[F_n^*(y), y]\,dy \quad \text{and} \quad \sup_{0 < y < 1} g[F_n^*(y), y]. \tag{4}$$

Since F_n^* is based only a random sample from $\mathcal{U}(0, 1)$, the distributions of the random variables (4) are the same for any population c.d.f. F.

PROBLEMS

13-1. Consider a random sample $\mathbf{X}$ from a population with p.d.f. $f(x) = e^{-x}$, $x > 0$. Find the probability of the region $[x_2 < x_3 < x_1]$ by carrying out the integration of the joint p.d.f. over that region.

13-2. Find the joint p.d.f. of the smallest and second smallest of four independent observations from $\mathcal{U}(0, 1)$ by integrating out the unwanted variables from the joint p.d.f. of the order statistic.

13-3. Find a sample size n such that at least 90 percent of a population with c.d.f. F lies between the smallest and largest observation in the sample with probability .90.

13-4. Suppose we obtain a sample of size $n = 4$ from $\text{Exp}(1)$, with p.d.f. $f(x) = e^{-x}$, $x > 0$, and find this order statistic: $(.5, 1, 2, 4)$.

(a) For this sample evaluate $\sup_{0 < x < 1} |F_n(x) - F(x)|$.

(b) Evaluate $\int_{-\infty}^{\infty} [F_n(x) - F(x)]^2\,dF(x)$ for this sample.

[Hint for (b): Using (4) instead of (3) makes the calculation a little easier (but it's still tedious).]

13-5. Obtain the joint p.d.f. of the successive areas under the population p.d.f. between successive observations in the order statistic:

$$Y_1 = F(X_{(1)}),\ Y_2 = F(X_{(2)}) - F(X_{(1)}),\ \ldots,$$
$$Y_n = F(X_{(n)}) - F(X_{(n-1)}).$$

[Why not include $Y_{n+1} = 1 - F(X_{(n)})$?]

13-6. Let $\mathbf{X}$ be a random sample from $\mathcal{U}(0, \theta)$ and $\mathbf{Y}$, the corresponding order statistic. We know from the factorization theorem that Y_n is sufficient, so the conditional distribution of $\mathbf{X}$ given Y_n is independent of θ. To derive this distribution is beyond our scope; instead, obtain the joint p.d.f. of the first $n - 1$ Y's given Y_n, and observe that it is independent of θ.

13.3 Testing randomness

Much of the testing and estimation we've been studying has assumed a random sample—a sequence of i.i.d. variables. This assumption can be tested:

$$H_0: \quad F_X(\mathbf{x}) = F(x_1) \cdots F(x_n),$$

where $F_X(\mathbf{x})$ is the joint c.d.f. of the sample observations and $F(x)$ is an unknown population c.d.f. The most general alternative is simply that H_0 is not true, but this is too broad. Indeed, given any test with a critical region of specified size under H_0, there will be many states not in H_0 for which the power is 0. (Any multivariate distribution with zero mass on the critical region and dependent marginals is such an alternative.) So it seems best to consider more restricted classes of alternatives, consisting of possibilities that are typical ways of losing randomness.

One way of restricting the class of alternatives is to keep the assumption of independence but allow the distribution to vary from observation to observation. In particular, a *trend* alternative is one in which the joint c.d.f. of the observations is of the form

$$F_X(\mathbf{x}) = \prod_{i=1}^{n} F[x_i - g(i)],$$

for some fixed functions F and g. Under these alternatives, the distributions of successive observations are identical in shape but shifted in location. Even with such restrictions on the class of alternatives, it may not be apparent how to "derive" a suitable test from some basic principles, and most of the tests we take up were proposed on intuitive grounds. Power is usually awkward in any general sense, so our theory will consist mainly derivations of the null distributions of various test statistics.

A *run* in a sequence of symbols is a group of consecutive symbols of one kind preceded and followed by (if anything) symbols of another kind. For instance, in the following sequence of +'s and −'s, we have inserted vertical bars to separate runs of +'s and runs of −'s:

$$+++\,|\,-\,|\,++\,|\,----\,|\,++\,|\,--\,.$$

In this sequence there is a run of +'s of length 3, two runs of +'s of length 2, and runs of −'s of lengths 1, 4, and 2. We'll look first at runs of observations above and below the median observation.

Assign the letter a to an observation above the median and the letter b to an observation below the median. To simplify the discussion, we'll ignore the median itself when n is odd. Then the sequence of a's and b's will be of even length $2m$, say—m a's and m b's. Let r_a and r_b denote the number of runs of a's and b's, respectively, and let $r = r_a + r_b$, the total number of runs. An unusually large or small number of runs would suggest a lack of randomness. For instance, if there is a downward trend, the a's would tend to come at the beginning and the b's, toward the end of the sequence, and this would mean a relatively small number of runs. Or, a type of dependence might produce a systematic bouncing back and forth across the median, resulting in an unusually large number of runs.

According to Theorem 2, the $(2m)!$ arrangements of the observations are equally likely to have produced a given order statistic. Each arrangement defines a sequence of a's and b's, and the distinct arrangements of m a's and m b's are then also equally likely, since each one comes from $(m!)^2$ arrangements of the observations. So to find the probability of a given configuration of runs, we *count* arrangements of a's and b's. And to calculate $P(r_a = x, r_b = y)$, we count the arrangements with this property and divide by $\binom{2m}{m}$. There are three cases to consider: (i) $x = y + 1$, (ii) $x = y - 1$, (iii) $x = y$.

In case (i), the sequence of a's and b's begins with an a and ends with an a. To form a sequence with x runs of a's, we line up m a's and separate them into x groups by inserting slots in $x - 1$ of the $m - 1$ spaces between a's, which can be done in $\binom{m-1}{x-1}$ ways. Before putting b's into these slots, we partition them into y groups, in one of $\binom{m-1}{y-1}$ ways, putting these groups into the slots between the groups of a's. For case (ii) the argument is identical, with the roles of a and b reversed, so for these two cases,

$$P(r_a, r_b = y) = \frac{\binom{m-1}{x-1}\binom{m-1}{y-1}}{\binom{2m}{m}}, \quad x = y \pm 1.$$

In case (iii), sequences either begin with an a and end with a b, or begin with a b and end with an a. In either instance, the number of ways, calculated as for case (i), is again the product of $\binom{m-1}{x-1}$ and $\binom{m-1}{y-1}$:

$$P(r_a, r_b = y) = \frac{2\binom{m-1}{x-1}\binom{m-1}{y-1}}{\binom{2m}{m}}, \quad x = y.$$

These are conditional given the order statistic. But the result is the same for *any* order statistic, so they are also the absolute probabilities.

We next find the probability function for r, the total number of runs, from the above joint probabilities for r_a and r_b:

$$P(r=z) = \sum_{x+y=z} P(r_a = x, r_b = y). \quad (1)$$

When z is even: $z = 2k$, the sum has only one term—that in which $x = y = k$. If z is odd: $z = 2k + 1$, there are two terms in the sum—one in which $x = k$ and $y = k+1$, and one in which $x = k+1$ and $y = k$. Hence,

$$P(r=z) = \begin{cases} \dfrac{2\binom{m-1}{k-1}\binom{m-1}{k-1}}{\binom{2m}{m}}, & z = 2k \\[2ex] \dfrac{2\binom{m-1}{k}\binom{m-1}{k-1}}{\binom{2m}{m}}, & z = 2k+1 \end{cases} \quad (2)$$

for $z = 2, 3, ..., 2m$. The corresponding c.d.f. has been tabled.[1]

EXAMPLE 13.3a In samples of size $n = 8$ from a continuous population there are four observations above and four below the sample median. The number of runs can vary from 2 to 8, and the distribution of r is given by the following table of probabilities, calculated using (2):

z	2	3	4	5	6	7	8
$f(z)$	1/35	3/35	9/35	9/35	9/35	3/35	1/35

The critical region $[r < 3]$, for instance, has $\alpha = 1/35 = .0286$. ∎

It can be shown that the first two moments of r are

$$E(r) = m + 1, \quad \text{var } r = \frac{m(m-1)}{2m-1}, \quad (3)$$

and that when n is large, the statistic r is approximately normally distributed. With a continuity correction and the variance approximated by $\frac{1}{4}(2m-1)$, the percentiles of r, for large n, are approximately as follows:

$$r_p = \tfrac{1}{2}\{2m + 1 + z_p\sqrt{2m-1}\}, \quad (4)$$

where z_p is the $100p$th percentile of a standard normal distribution.

Another type of run test for randomness employs the signs of the

---

[1] F. Swed and C. Eisenhart, "Tables for testing randomness ...," *Ann. Math. Stat.* 14, 66-87 (1943).

differences of successive observations. If the population mean has, say, a rising trend, there would be a tendency for observations to increase from one observation to the next, and +'s would more often occur in groups than if there is no trend. A sample of size n defines a sequence of $n-1$ signs of differences of successive pairs of observations. Let s denote the total number of runs of +'s and −'s in such a sequence. The mean and variance of s are

$$E(s) = \frac{2n-1}{3}, \quad \text{var } s = \frac{16n-29}{90}, \qquad (5)$$

and again, for large n, the statistic s is approximately normal.[2]

Yet another test for randomness is based on the mean squared successive difference:

$$d^2 = \frac{1}{2(n-1)} \sum_{i=1}^{n-1} (X_{i+1} - X_i)^2. \qquad (6)$$

Under the assumption of randomness, the quantity $X_{i+1} - X_i$ has mean 0 and variance $2\sigma^2$, so $E(d^2) = \sigma^2$. If no trend is present, the ratio d^2/S^2 would tend to be near 1. But when there is a trend, d^2 will be about the same as without it, whereas S^2 would be larger. So small values of the ratio would favor a trend over randomness. If the sampled population is *normal*, the ratio d^2/S^2 is approximately normal[3] with mean 1 and variance $(n-2)/(n^2-1)$, when n is large.

EXAMPLE 13.3b The following are 40 successive measurements of viscosity. Is there any evidence of nonrandomness?

32.0 33.9 31.3 31.0 32.0 35.0 35.5 31.6 31.8 29.9
35.4 33.4 31.2 30.5 31.0 32.9 30.9 33.9 33.8 32.8
32.5 34.1 30.7 30.8 30.4 31.0 32.1 28.7 29.0 29.3
32.3 30.8 32.9 30.7 30.1 32.5 30.3 32.7 30.7 31.7

The median is 31.45, and there are 19 runs above and below the median. With $m = 19$, we have $E(r) = 21$, $\sigma_r = 3.12$ [from (3)]; so the Z-score is

$$Z = \frac{19.5 - 21}{3.12} \doteq .5.$$

Counting runs up and down we find $s = 27$. From (5) we obtain

---

[2] See A. M. Mood, "The distribution theory of runs," *Ann. Math. Stat.* 11, pp. 367-392 (1943).

[3] See B. I. Hart, "Significance levels for the ratio of the mean square successive difference to the variance," *Ann. Math. Stat.* 13, 445-47 (1942) and references given there.

$E(s) = 26.33$ and $\sigma_s = 2.61$, so again a Z-score would be quite small in magnitude. There is little evidence of nonrandomness in the values of the statistics r and s.

The sum of the 39 squared deviations is 159.93, so $d^2 = 2.0504$. The variance of the 40 observations is $S^2 = 2.777$, and $d^2/S^2 = .738$. The Z-score [using the given formulas for mean and variance] is $(.738 - 1)/.154 \doteq -1.7$. If the population is normal, $P \doteq .05$—some evidence against randomness. ∎

Using the test ratio d^2/S^2 as defined by (6) requires knowing (or at least assuming) the population distribution under H_0, if we are to find P-values or set a critical boundary for given α. However, a technique of randomization, suggested by R. A. Fisher, provides a test whose level does *not* depend on the population distribution. We have seen that the conditional distribution in the sample space, given the order statistic, is uniform over the $n!$ possible samples—the $n!$ permutations of the order statistic components. Suppose, for each possible order statistic, we define a critical region—a set of sample points which, if observed, call for rejecting H_0. If in each case we choose the critical set so that it has conditional probability no larger than α^*, then the size of the test based on these sets is

$$\alpha = E[P(\text{reject } H_0 \mid \text{order statistic})] \leq E\alpha^* = \alpha^*. \tag{7}$$

In carrying out such a **permutation test** test one need determine the critical set only for the order statistic actually observed.

A permutation test is a conditional test, as we used this term in Chapter 11 in describing Fisher's exact test. But it is doubtful that there is anything to be learned about randomness from the order statistic—the statistic on which we are conditioning.

An obvious first question is how to choose sample points for a critical set. We could do this in various ways, according to the various statistics one might think of that should relate to nonrandomness. In particular, we could use the value of d^2 (the mean squared successive difference) to order the $n!$ sample points, putting those with smallest d^2-values into a critical set—just enough of them so as not to exceed a specified "size" limitation (7). Another possibility for such a statistic is a *serial correlation*, of the form

$$R_h = \sum_{1}^{n-h} X_i X_{i+h}, \tag{8}$$

for some fixed h. These statistics, among others, have the disadvantage of requiring calculation of their values for each new problem, after the order statistic has been obtained. No tables can be prepared in advance. And this is where statistics based on rank-order have an advantage.

The **rank** t_i of the observation X_i is its position in the order statistic: The smallest observation has rank 1, second smallest rank 2, and so on. When the population is continuous, each sample point **X** defines a sequence of ranks $(t_1, \ldots, t_n)$—one of the $n!$ permutations of the integers $(1, 2, \ldots, n)$. Under the null hypothesis of randomness, these permutations are equally likely, independent of the population. If we then order the rank sequences according to some well chosen statistic, we can define a critical set of rank sequences which have the smallest values of that statistic, and reject independence if the rank sequence of the sample actually observed is in that critical set. But now we can get by with a single table (for a given statistic), prepared once and for all.

Two of the many rank-order statistics that have been proposed for testing randomness are these:

$$\sum_{i=1}^{n} i t_i, \text{ and } \sum_{i=1}^{n} E[Z_{(i)}] t_i,$$

where $Z_{(i)}$ denotes the ith smallest among n observations from a standard normal population.[4] Table XI gives values of these **normal scores** or **rankits**, $E[Z_{(i)}]$.

EXAMPLE 13.3c Suppose we take a random sample of size $n = 5$ from a continuous population and get this sample: $(4, 6, 12, 7, 10)$. There are 120 (5!) permutations of these numbers, each yielding the order statistic $(4, 6, 7, 10, 12)$. Some of these are given in the table below together with the corresponding rank statistic, values of $T = \sum (X_{i+1} - X_i)^2$, $\sum i t_i$, $\sum E[Z_{(i)}] t_i$, where t_i is the rank of X_i. [The values of $E[Z_{(i)}]$, from Table XI, are $-1.163, -.495, 0, .495, 1.163$.]

The entries in the table are listed in order of increasing T-values. Observe that they listing would be somewhat different if we were to use the value of $\sum i t_i$ as the basis of the ordering.

If we construct a critical region by putting into it, say, the eight rank sequences with the largest values of $\sum i t_i$ (55, 54, 53), the size of the critical region is then $\alpha = 8/120 \doteq .067$. [One rank sequence with $\sum i t_i = 53$ is not shown: (2 1 4 3 5); for this sequence (and the corresponding sample), $T = 74$.]

---

[4]See A. Stuart, "The asymptotic relative efficiencies of distribution-free tests of randomness against normal alternatives," *J. Amer. Stst. Assn.* 49, 147-157 (1954); also, I. R. Savage, "Contributions to the theory of rank order statistics ...," *Ann. Math. Stat.* 27, 968-977 (1957), and E. L. Lehmann, *Testing Statistical Hypotheses*, 2nd Ed., Pacific Grove, CA: Wadsworth & Brooks/Cole (1986), p. 349.

Chapter 13 Nonparametric Inference

In using T to order the permutations, on the other hand, we need to realize that for each of the above entries, reversing the numbers in a sequence yields the same T-value, so a one-tail critical region for T would pick up either a rising or a falling trend. For instance, the size of the critical region consisting of the permutations with the six smallest values of T would be $12/120 = .1$.

Permutation	T	Ranks	$\sum i t_i$	$\sum E[Z_{(i)}] t_i$
4 6 7 10 12	18	1 2 3 4 5	55	5.64
6 4 7 10 12	26	2 1 3 4 5	54	4.974
4 7 6 10 12	30	1 3 2 4 5	54	5.147
7 4 6 10 12	33	3 1 2 4 5	52	3.811
4 6 7 12 10	34	1 2 3 5 4	54	4.974
6 4 7 12 10	42	2 1 3 5 4	53	4.306
7 6 4 12 10	45	3 2 1 5 4	50	2.648
4 6 10 12 7	49	1 2 4 5 3	52	3.811
6 7 4 10 12	50	2 3 1 4 5	52	3.984
4 7 6 12 10	50	1 3 2 5 4	53	4.479
4 6 12 10 7	53	1 2 5 4 3	51	3.316
7 4 6 12 10	53	3 1 2 5 4	51	3.143
4 6 10 7 12	54	1 2 4 3 5	54	5.147
4 7 12 10 6	54	1 3 5 4 2	48	1.658 ■

PROBLEMS

13-7. For random samples of size $n = 6$,
(a) obtain the distribution of the total number of runs by substituting in (2), the formula for $P(r = z)$.
(b) verify the distribution of r by writing out the 20 arrangements of three a's and three b's and counting the number of runs in each.
(c) verify formulas (3) for the mean and variance in this case.

13-8. The following are successive entries in a table of "random numbers," which are used as i.i.d. observations (in simulation studies):

  15 77 01 64 69 69 58 40 81 16 60 20 00 84 22
  28 26 46 66 36 86 66 17 34 49 85 40 51 40 10

(a) Test for randomness using runs above and below the median.
(b) Test for randomness using runs up and down

13-9. Here are some (rounded) observations from a normal population:

39 42 38 53 51 30 40 28 43 46 53 55
29 24 34 53 66 43 42 38 34 57 26 33

(a) Test for randomness using runs above and below the median.
(b) Test for randomness using runs up and down.
(c) Add $i - 1$ to the ith observation and test the new sequence both ways.

13-10. Show that permuting two successive ranks in the rank sequence defined by a particular permutation of ranks changes the value of $\sum it_i$ by the amount of the difference between those two ranks. Use this fact to determine all permutations of ranks 1 to 5 for which $\sum it_i = 53$ from those with value 54 given in Example 13.3c.

13-11. Verify some of the values of $\sum E[Z_{(i)}]t_i$ shown in the table in Example 13.3c.

13-12. We found that the mean squared successive difference

$$d^2 = \frac{1}{2(n-1)} \sum_{i=1}^{n-1} (X_{i+1} - X_i)^2$$

is an unbiased estimate of σ^2. Given $\operatorname{var} d^2 = (3n-4)\sigma^4/(n-1)^2$, find the efficiency of d^2 relative to the sample variance S^2 as an estimator of σ^2 in $N(\mu, \sigma^2)$, as a function of sample size.

13-13. Show that using the statistic d^2 (in the preceding problem) is almost equivalent to using the serial correlation R_h, defined by (8), with $h = 1$.

13.4 The sign test

Consider a random sample of *pairs* of experimental units—the two sides of a leaf, the two sides of a tire, a set of twins or litter-mates, and so on—of which one (chosen at random) is treated and the other not. We assume that it can be determined which one is better or responds better than the other, not necessarily on a numerical scale. Record a + if the treatment unit is judged better, otherwise a $-$, so that the sample observations are recorded as a sequence of n signs, plus or minus. Let $p = P(+)$.

We take the null hypothesis of no treatment effect to mean that the one member of the pair is as likely to be the better as the other:

$$H_0: \; p = 1/2.$$

The alternative, in terms of p, can be one-sided ($p > 1/2$ or $p < 1/2$), or two-sided ($p \neq 1/2$). This testing problem is one that we have already studied in Chapters 9 and 10, and tests based on the number of +'s are called **sign tests**.

Suppose X and Y are the control and treatment responses, respectively, of the members of a pair measured on a continuous numerical scale, and assume that the treatment and control regimens are assigned at random within each pair. The null hypothesis of no "treatment effect" then implies

$$F_{X,Y}(x, y) = P(X < x, Y < y) = P(X < y, Y < x) = F_{Y,X}(x, y).$$

That is, X and Y are *exchangeable*, and the distribution of $X - Y$ is the same as the distribution of $Y - X$, which is then symmetrical about 0:

$$P(X - Y > 0) = P(X - Y < 0) = \tfrac{1}{2}.$$

Thus, the median of the population of differences $D = X - Y$ is zero under the null hypothesis, and we can use a sign test for the hypothesis $m_D = 0$. Clearly, we have discarded information in ignoring the *amount* by which an observation is greater or less than 0, but it may happen that the remaining information is still enough to provide strong evidence against the hypothesis of no treatment effect.

The sign test can be used as a crude test for comparing locations using paired numerical data, or as a test for a (single) population median, the usual measure of location in nonparametric inference. When X is continuous, the median is a number m such that

$$P(X > m) = P(X < m) = \tfrac{1}{2}.$$

To test H_0: $m = m_0$, using a random sample $(X_1, ..., X_n)$, replace the sample with a sequence of corresponding signs: $+$ or $-$, according as $X_i > m_0$ or $X_i < m_0$, let $p = P(X > m_0)$, and test the cruder hypothesis H'_0: $p = 1/2$.

An alternative to H_0 might be that the median is less than m_0, or that it is greater than m_0, or that $m \neq m_0$. And we can phrase these alternatives in terms of the parameter $p = P(X > m_0)$, or (for paired data) $p = P(D > d_0)$:

$$H_A: p > 1/2, \quad \text{or} \quad H_A: p < 1/2, \quad \text{or} \quad H_A: p \neq 1/2.$$

EXAMPLE 13.4h Consider the hypothesis $m = 31.0$, to be tested on the basis of these 20 observations of viscosity (taken from Example 13.3b):

35.4 33.4 31.2 30.5 31.0 32.9 30.9 33.9 33.8 32.8

32.5 34.1 30.7 30.8 30.4 31.0 32.1 28.7 29.0 29.3

The signs of the deviations from 31.0 are as follows:

+ + + − 0 + − + + + + + − − − − 0 + − − − −

Ignoring the 0's, we find 10 + 's and 8 − 's, so the one-sided P-value is

$$P(10 \text{ or more } +\text{'s}) \doteq 1 - \Phi\left(\frac{9.5 - 9}{1.5\sqrt{2}}\right) = .407.$$

On the other hand, a t-test of $\mu = 32$ yields $T = 1.67$, with $P \doteq .05$. Not surprisingly, the t-test, which is tailored for normal populations but robust enough to be applied here, has picked up a difference that the cruder sign test has not. ■

13.5 The signed-rank test

We now take up a rank test for the hypothesis $m = m_0$, where m is the median of a continuous distribution which is symmetric about its median. This can be a distribution of differences, when we deal with paired data. Unlike the sign test, the signed-rank test takes magnitudes of the observations into account, and ought therefore to be more powerful than the sign test. (But recall that the sign test can be used in situations where responses are not given on a numerical scale.) The signed-rank test is based on the following:

THEOREM 4 *If the observations in a random sample from a population symmetric about 0 are ordered by magnitude, the 2^n sequences of signs of the ordered observations are equally likely.*

Let **X** denote a random sample, and let **S** denote the vector of signs:

$$S_i = \begin{cases} +, & \text{if } X_i > 0, \\ -, & \text{if } X_i < 0. \end{cases}$$

Problem 13-20 asks for a proof of the fact that Z_i and S_i are independent for each i. The components of **S** are functions of the independent X's, so they too are i.i.d. There are 2^n possible sign sequences, and the probability of each one is $1/2^n$.

Let **S*** denote the sequence of signs of the observations *after* they are ordered by magnitude. Thus, S_i^* is the sign of that observation X which has the ith smallest magnitude. That is, **S*** is a sequence of signs obtained from **S** by a permutation transformation which depends on the vector of magnitudes, **Z**: $\mathbf{S}^* = \mathbf{P_Z S}$. A permutation transformation is always 1-1; so, given **Z**, the probability of $[\mathbf{S}^* = \mathbf{a}]$, for any pattern of signs **a**, is just the probability of its pre-image pattern:

$$P(\mathbf{S}^* = \mathbf{a} \mid \mathbf{Z}) = P[\mathbf{S} = \mathbf{P}_Z^{-1}(\mathbf{a})] = \frac{1}{2^n}.$$

Since this is the same for every vector of magnitudes **Z**, it follows that $P(\mathbf{S}^* = \mathbf{a}) = 1/2^n$. And this shows that the sign patterns $\mathbf{S}^*$ are indeed equally likely, as asserted in the theorem.

We now let $I(i)$ be 1 or 0 according as S_i^* is a + or a − and define Wilcoxon's **signed-rank statistic** as follows:

$$W_+ = \sum_{i=1}^{n} i I(i). \tag{1}$$

This is the sum of the ranks (in **Z**) of the positive observations, and is equal to

$$W_+ = \sum_{i=1}^{n} V_i, \tag{2}$$

where

$$V_i = \begin{cases} i, & \text{if } S_i^* = +, \\ 0, & \text{if } S_i^* = -. \end{cases} \tag{3}$$

The range of values of W_+ is the set of integers from 0 (when no observations are positive) to $n(n+1)/2$ (when all observations are positive).

We denote by W_- the sum of the ranks of the negative observations—the sum of the remaining ranks from among (1, 2,..., n). This sum has the same range of possible values as W_+, and

$$W_+ + W_- = 1 + 2 + \cdots + n = \frac{n(n+1)}{2}. \tag{4}$$

So if we know one of these statistics, we know the other; in any particular problem we just use one or the other—whichever is convenient.

THEOREM 5 *The distribution of the signed-rank statistic (W_- or W_+) is symmetric about $n(n+1)/4$, so that*

$$E(W_+) = E(W_-) = \frac{1}{4}n(n+1). \tag{5}$$

Under H_0, the population is symmetric about 0, so a change of + to − and − to + does not affect the distribution:

$$P(W_- = k \mid H_0) = P(W_+ = k \mid H_0).$$

From (4), then, we see that

$$P(W_+ = k \mid H_0) = P\left\{\frac{n(n+1)}{2} - W_- = k \mid H_0\right\}$$
$$= P\left\{W_- = \frac{n(n+1)}{2} - k \mid H_0\right\}.$$

This says that the distribution of W_- is symmetric about its central value, $n(n+1)/4$, and of course the same is true of W_+.

EXAMPLE 13.5a When $n = 4$, there are $2^4 = 16$ possible sign patterns. We list eight of these in the following table together with corresponding values of W_+ and W_-. The other eight can be obtained by interchanging +'s and −'s, with rank sums W_+ ranging from 5 to 0:

S*	W_+	W_-
+ + + +	10	0
− + + +	9	1
+ − + +	8	2
+ + − +	7	3
− − + +	7	3
+ + + −	6	4
− + − +	6	4
+ − − +	5	5

The distribution of W_+ follows from this array because the probability of each sequence is 1/16:

$$P(W_+ = k) = \begin{cases} 1/16, & k = 0, 1, 2, 8, 9, 10, \\ 2/16, & k = 3, 4, 5, 6, 7. \end{cases}$$

Observe that the distribution is symmetric about the value 5 (the mean) and that W_- has the same distribution. ∎

The value of W_+ will be small (and W_- large) when either (i) there are very few positive X's, or (ii) if the X's that are positive have small magnitudes. In either case, one would tend to reject H_0 in favor of the alternative. Large values of W_+ would suggest $m > m_0$, and a two-sided critical region for W_+ would be appropriate for the alternative $m \neq m_0$. Critical values for given α or P-values for a given value of W_+ can be found using Table VI when $n \leq 15$. For larger samples, we use the fact[5] that W_+ is asymptotically normal. A continuity correction helps if n is not very large. To use this approximation we need the variance of W_+:

$$\operatorname{var} W_+ = \operatorname{var} \sum_{i=1}^{n} i I_X(i) = (\operatorname{var} I_X) \sum_{i=1}^{n} i^2.$$

Since $I_X \sim \operatorname{Ber}(\frac{1}{2})$, its variance is $\frac{1}{4}$, and the sum of squares of the first n integers (a formula from algebra) is

$$1^2 + 2^2 + \cdots + n^2 = \tfrac{1}{6} n(n+1)(2n+1).$$

So the formula for the variance of W_+ (applicable also to W_-) is

$$\operatorname{var} W_+ = \tfrac{1}{24} n(n+1)(2n+1). \tag{6}$$

Although the X's in a random sample from a continuous population will involve no duplications (with probability 1), in practice there will be tied observations because of round-off. The usual procedure for breaking ties is to assign the average rank to each of the tied magnitudes. Also, the value 0 would "never" occur, ideally, but does in practice. As in the case of the sign test, we simply ignore any 0's and apply the signed-rank test to the reduced sample.

EXAMPLE 13.5b A sprinkler system is designed so that the mean activation time is 25 seconds. A series of tests yielded these times[6]:

27, 41, 22, 27, 23, 35, 30, 33, 24, 27, 28, 22, 24.

The deviations about 25 are

2, 16, −3, 2, −2, 10, 5, 8, −1, 2, 3, −3, −1.

There are five negative observations in this sample of 13, so a sign test doesn't give evidence against $\mu = 25$: $P = .29$. Magnitudes, in order, are as follows, with those of negative observations in italics:

1, *1*, 2, 2, *2*, 2, *3*, 3, *3*, 5, 8, 10, 16.

Corresponding ranks, with average rank assigned to tied magnitudes, are

---

[5] Although W_+ is a sum of independent variables, they are not identically distributed, so the central limit theorem as given in Chapter 5 does not apply. However, there is a version that is applicable. (See E. Lehman, *Nonparametrics: Statistical Methods Based on Ranks*. San Francisco: Holden-Day Inc., 1975, p.350.)

[6] From "Use of AFFF in sprinkler systems," *Fire Technology* (1976), 5.

1.5, 1.5, 4.5, 4.5, 4.5, 4.5, 8, 8, 8, 10, 11, 12, 13.

The sum of the ranks of the five negative observations is

$$W_- = 1.5 + 1.5 + 4.5 + 8 + 8 = 23.5.$$

The (one-sided) P-value, from Table VI, is about .068. The normal approximation is obtained using (5) and (6), with

$$E(W_-) = 13 \times 14/4 = 45.5, \quad \text{var}\, W_- = 13 \times 14 \times 27/24 = (14.31)^2.$$

The Z-score is then

$$Z = \frac{23.5 + .5 - 45.5}{14.31} \doteq -1.5,$$

which gives $P \doteq .067$. (So the normal approximation works quite well in this instance.)

Apparently the signed-rank statistic took into account the fact that the negative observations have small magnitudes. In a one-sample t-test with these data, $T = 1.88$ (12 d.f.), $P \doteq .042$. A quick plot of the data suggests that the population is skewed, but the sample is not large enough to rule out population normality. But whether we use T or the signed-rank statistic, H_0 includes the assumption of a symmetric distribution about the mean (or median); and in rejecting H_0 we don't know whether this happens because of asymmetry or because $\mu \neq 25$.∎

13.6 Asymptotic relative efficiency

When studying nonparametric tests, exact power calculations for finite samples are laborious and necessarily too specific. General statements are hard to make. Most tests in common use are consistent—the power approaches 1 as n becomes infinite. A large sample approach for comparing tests was introduced by Pitman,[7] who considered a sequence of alternatives approaching the null hypothesis, and compared sample sizes necessary to yield the same power. The **asymptotic relative efficiency** (ARE) of one test with respect to another test of the same size α is defined as the limiting ratio of sample sizes required in order that they have the same power on such a sequence of alternatives. This limiting ratio often turns out to be independent of α, but may depend on the particular class of alternatives considered.

Calculating the ARE of the sign test relative to the appropriate t-tests, when alternatives are normal—where the t-tests should be optimal—is fairly straightforward, and this is the only calculation we'll

---
[7] E. J. G. Pitman, unpublished Columbia Lecture Notes, 1948.

undertake. The power of the sign test is a binomial probability, and the power of the t-test is a probability in a noncentral t-distribution; but both become normal in the limit.

Let S_n denote the number of positive observations in a random sample of size n; then $S_n \sim \text{Bin}(n, p)$, where $p = P(+)$ and $ES_n = np$, $\text{var } S_n = np(1-p)$, A large-sample test of approximate size α for $p = \frac{1}{2}$ against $p > \frac{1}{2}$ employs the critical region $S_n > K$, where

$$K = \frac{n}{2} + z^* \sqrt{\frac{n}{4}},$$

with $z^* = z_{1-\alpha}$. For an alternative for which

$$p = P(+) = \frac{1}{2} + \frac{a_n}{2\sqrt{n}},$$

where $a_n \to a$, we have

$$ES_n = \frac{n}{2}\left(1 + \frac{a_n}{\sqrt{n}}\right), \quad \text{var } S_n = \frac{n}{4}\left(1 - \frac{a_n^2}{n}\right),$$

and the power of the test $S_n > K$ is

$$P(S_n > K) = 1 - \Phi\left(\frac{K - ES_n}{\sqrt{\text{var } S_n}}\right) = 1 - \Phi\left(\frac{z^*\sqrt{n/4} - a_n\sqrt{n/4}}{\sqrt{n(1 - a_n^2/n)/4}}\right)$$

$$= 1 - \Phi\left(\frac{z^* - a_n}{\sqrt{1 - a_n^2/n}}\right), \tag{1}$$

which tends to $1 - \Phi(z^* - a)$ as n becomes infinite.

Now consider the following sequence of normal alternatives that approach the null hypothesis that $X \sim N(0, \sigma^2)$:

$$H_n: X \sim N(b\sigma/\sqrt{n}, \sigma^2),$$

for some $b > 0$. For these alternatives we calculate "p":

$$p = P(X > 0) = 1 - \Phi\left(\frac{0 - b\sigma/\sqrt{n}}{\sigma}\right) = \Phi\left(\frac{b}{\sqrt{n}}\right) = \frac{1}{2} + \frac{a_n}{2\sqrt{n}},$$

where

$$a_n = 2\sqrt{n}\left\{\Phi\left(\frac{b}{\sqrt{n}}\right) - \Phi(0)\right\} \to \frac{2b}{\sqrt{2\pi}} = a.$$

Thus, as n becomes infinite, the power (1) approaches

$$1 - \Phi\left(z^* - \frac{2b}{\sqrt{2\pi}}\right).$$

Next consider a "t" critical region which has the same approximate α as the sign test above: $T_n > z^*$, where

$$T_n = \frac{\bar{X} - 0}{S/\sqrt{n}},$$

and as before, $z^* = z_{1-\alpha}$ [because T_n is asymptotically $\mathcal{N}(0, 1)$]. When $\bar{X} \sim \mathcal{N}(b'\sigma/\sqrt{n'}, \sigma^2)$, then $T_n \approx \mathcal{N}(b', 1)$, and the power function is

$$P(T_n > z^*) \doteq 1 - \Phi(z^* - b').$$

The sequence of alternatives here and the sequence of alternatives given above for the sign test will agree when $b/\sqrt{n} = b'/\sqrt{n'}$, or $n'/n = b'^2/b^2$. The asymptotic powers agree when $b' = 2b/\sqrt{2\pi}$, or $b'^2/b^2 = 2/\pi = n'/n$. This ratio of sample sizes, or about 64 percent, is the asymptotic relative efficiency of the sign test with respect to the t-test, for *normal* populations.

Pitman showed that in the case of a random sample of pairs (X, Y), the ARE of the sign test relative to the t-test against *shift* alternatives: $F_X(\lambda) = F_Y(\lambda - \theta)$, is $4\sigma^2 f^2(0)$, where $f = F'$. The ARE of the signed-rank test relative to the t-test is $3/\pi \doteq .955$ against normal shift alternatives.

13.7 Confidence intervals

Let **Y** denote the order statistic of a random sample **X** of size n from a population with median m. Then for $r < s$,

$$P(Y_r < m < Y_s) = P(r, r+1, \ldots, s-1 \ X\text{'s} < m) = \sum_{k=r}^{s-1} \binom{n}{k} \frac{1}{2^n}. \quad (1)$$

So the random interval from the rth smallest to the sth smallest X is a confidence interval for the median m, with confidence level given by the sum (1).

EXAMPLE 13.7a When $n = 20$, the probability that the interval between the fifth smallest and the sixteenth smallest observation will include the population median, from (1), is about .978. The twenty observations in Example 13.4b, arranged in order, are

28.7 29.0 29.3 30.4 30.5 30.7 30.8 30.9 31.0 31.0
31.2 32.1 32.5 32.8 32.9 33.4 33.8 33.9 34.1 35.4

So a 97.8 percent confidence interval is $30.5 < m < 33.4$. ∎

Confidence intervals for the median can also be based on the signed-rank statistic. J. Tukey has given the following graphical scheme for implementing a method of G. Noether, which is fairly easy to carry out for small sample sizes. To achieve a specified confidence level η, define an integer h such that $P(W_+ \leq h) \doteq (1-\eta)/2$. Construct an isosceles triangle whose base is the line segment from $X_{(1)}$ to $X_{(n)}$ and draw lines through the other observations (plotted on that base line) parallel to the equal sides of the triangle. Such a construction is shown for nine observations in Figure 13-2. Mark all intersections of these lines with each other and with the base line, and draw vertical lines through the $(h+1)$st intersection point from each end—that is, so that there are h intersections to the left of the left line and h to the right of the right line. The intersections of these lines with the base line define the desired confidence limits.

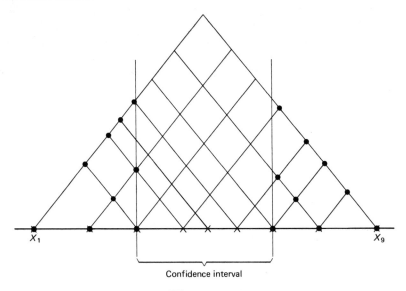

Figure 13-2

PROBLEMS

13-14. For a sign test of $p = \frac{1}{2}$ when $n = 9$, consider critical regions of the form $Y \leq c$, where Y = number of $+$'s.
  (a) Determine c so that α is close to (but not greater than) .10.
  (b) Sketch the power of the test in (a) as a function of $p = P(+)$.
  (c) Construct a randomized test with $\alpha = .10$.

13-15. Problem 9-53 gave data for six newborn lambs, in which the "after" measurements were all lower than the "before." Give the P-value, for the null hypothesis of no treatment effect, using the sign test and using the signed-rank test. (Compare with what was obtained in Problem 9-53.)

13-16. An article in *Science* (1968, page 1234) reported results of a study on the effects of smoking marijuana, including these data—changes in the number correct (in a digit substitution test) from the baseline number correct: +5, −17, −7, −3, −7, −9, −6, +1, −3. Find a (1-sided) *P*-value for testing the hypothesis of no treatment effect,
(a) using the sign test.
(b) using the signed-rank test.

13-17. The "signed-rank" of an observation is its rank in the sequence of magnitudes with the sign of the original observation attached.
(a) Find a relation between W, the sum of the signed-ranks, and W_+ as defined in §13.5.
(b) Find the mean and variance of W.
(c) Show that the Z-scores for W and for W_+ are the same.

13-18. Find the probability that the population median falls between the 40th and 60th smallest observations in a random sample of size 100 from a continuous population.

13-19. Verify the entries for $n = 6$ in Table VI.

13-20. Show that if X is continuous and symmetric about 0, the magnitude and sign of X are independent.

13-21. Twelve mineral specimens from the Black Forest were dated by a potassium-argon method, with the following results (age in millions of years):
    249 254 243 268 253 269 287 241 273 306 303 280.

(a) Find confidence limits for the median with confidence level no larger than 95 percent, based on the binomial distribution (as in Example 13.7a).
(b) Use the Tukey method to find an approximate 90 percent confidence interval for the median based on the signed-rank statistic.

13-22. For the case of a uniform population on $(\theta - \frac{1}{2}, \theta + \frac{1}{2})$, show that the ARE of the sign test of $\theta = 0$ relative to the t-test is $1/3$,
(a) using the formula $4\sigma^2 f^2(0)$.
(b) doing the calculations parallel to those given for the normal case with the sequence of alternatives $\theta_n = b/\sqrt{n}$.

13.8 Two-sample tests

We now take up several tests for the hypothesis of "no treatment effect," H_0: $F_X(\lambda) = F_Y(\lambda)$, for all λ, based on independent, random samples **X** and **Y**. The tests are directed towards shift alternatives:

$$H_A: F_X(\lambda) = F_Y(\lambda - \theta) \text{ for all } \lambda,$$

where in the one-sided case, $\theta > 0$ or $\theta < 0$, and in the two-sided case, $\theta \neq 0$. We study mainly the distributions of test statistics under H_0, with occasional mention of ARE's (asymptotic relative efficiencies). The tests are permutation tests—conditional tests given the order statistic of the combined sample $(\mathbf{X}, \mathbf{Y})$, based on Theorem 2 at the beginning of this chapter: The conditional distribution in the sample space given the order statistic is *uniform* over the possible permutations of that order statistic.

If the sample sizes are m and n, there are $(m+n)!$ permutations of the observations. If we put k of these into a critical region, the size of the test is

$$P_0(\text{reject } H_0) = E[P_0(\text{reject } H_0 \mid \text{order statistic})]$$

$$= E\left\{\frac{k}{(m+n)!}\right\} = \frac{k}{(m+n)!},$$

where the subscript 0 indicates that the probability is calculated assuming H_0. Selection of the particular permutations to use in a rejection region can be made according to whatever criterion might seem appropriate for the alternatives one has in mind.

EXAMPLE 13.8a One obvious statistic that might be used in constructing a critical region for a shift alternative is the mean difference: $\bar{X} - \bar{Y}$. Given the order statistic, we know the sum of all the observations: $\sum X_i + \sum Y_j$. So both $\sum X_i$ and $\sum Y_j$ are linear functions of $\bar{X} - \bar{Y}$ and thus would yield the same ordering of the various permutations of the combined order statistic as does $\bar{X} - \bar{Y}$. In these latter sums, the order within the sum is irrelevant, so there are $m!n!$ of the $(m+n)!$ order statistics that have the same values, say, of $\sum X_i$. We need only look at the $\binom{m+n}{n}$ distinct assignments of n of the $m+n$ observations to Y, and these are equally likely.

To illustrate, suppose there are three X's and two Y's, whose combined order statistic is $(3, 4, 6, 7, 9)$. In the 120 arrangements of these five values, there are ten distinct groups—ways we can choose three of the five observations to be X's and the others, Y's. For instance, one of these groups has $(3, 4, 6)$ as X's and $(7, 9)$ as Y's. [For each such group, the three X's appear in 3! possible ways, the two Y's, in 2! ways in the combined order statistic. Thus, there are $2! \times 3!$ or 12 order statistics with $(3,4,6)$ as X's and $(7,9)$ as Y's.] In the following table we list the ten ways of dividing the five observations into X's and Y's, along with the value of $\sum Y_j$ in each case:

X's	Y's	$\sum Y$	$6(\bar{X} - \bar{Y})$
3 4 6	7 9	16	-24
3 4 7	6 9	15	-17
3 4 9	6 7	13	-7
3 6 7	4 9	13	-7
4 6 7	3 9	12	-2
3 6 9	4 7	11	3
3 7 9	4 6	10	8
4 6 9	3 7	10	8
4 7 9	3 6	9	13
6 7 9	3 4	7	23

Each of these ten possibilities has probability 1/10, given the order statistic. Thus, for instance, if $\sum Y_j = 15$, the P-value is .2. ∎

Using $\sum Y_j$ as in the above example is awkward, since no table can be prepared in advance. On the other hand, criteria based on the relative ordering of X's and Y's in the order statistic of the combined sample are not dependent on particular data which result that ordering. In applying such criteria, we replace the sample sequence $(\mathbf{X}, \mathbf{Y})$ by a corresponding sequence of letters X and Y to indicate the samples of origin of the successive components of the order statistic. Thus, in the above example, the sequence $[(3,6,9),(4,7)]$ is replaced by the sequence (X,Y,X,Y,X). (The smallest observation is an X, the second smallest a Y, and so on.) The total number of such patterns of X's and Y's is $\binom{m+n}{n}$, and these patterns are equally likely, given the order statistic. But then they are also equally likely *unconditionally*. (Simply average over all order statistics.)

We'll take up three tests based on the patterns of X's and Y's. The first is a somewhat crude test, in a sense analogous to the sign test. For convenience, we assume that $m + n = 2k$, an even integer, so that there are k observations on either side of the median of the combined order statistic. An unusually large number of X's or large number of Y's on one side of the median would be indicative of a shift, so in a **median test** we focus on, say, the number of X's to the left of the median. The value of this determines the number of X's to the right, the number of Y's to the left, and the number of Y's to the right of the median. These four statistics are sometimes given in a contingency table with fixed marginal totals:

	Left	Right	
X's	L	$m - L$	m
Y's	$k - L$	$k - m + L$	n
	k	k	$m+n$

The L X's and $(k - L)$ Y's can appear in $\binom{k}{L}$ ways to the left of the median, and the $(m - L)$ X's and $(k - m + L)$ Y's can appear in $\binom{k}{m - L}$ ways to the right. The conditional probability of any specific pattern, given the order statistic, is the product of these numbers of ways, divided by $\binom{2k}{m}$:

$$P(L = l\ ) = \frac{\binom{k}{l}\binom{k}{m - l}}{\binom{2k}{m}}. \tag{1}$$

And (because it is independent of the condition) this is also the unconditional probability. [Notice that (1) is precisely the conditional probability of a two-way table, given all marginal totals, that we encountered in testing for independence or homogeneity. Here both marginal totals *are* fixed.]

The hypergeometric probability (1) is asymptotically normal as $m + n$ becomes infinite with $m/(m + n)$ fixed, according to a more general central limit theorem than the one we gave in Chapter 5. So the standard score

$$Z = \frac{L - km/(m + n)}{\frac{1}{2}\sqrt{mn/(m + n)}}$$

is approximately standard normal for large samples.[8]

EXAMPLE 13.8b Driving times to work for 11 trips on route A and for 5 trips on route B were reported as follows[9]:

A: 6.0 5.8 6.5 5.8 6.3 6.0 6.3 6.4 5.9 6.5 6.0
B: 7.3 7.1 6.5 10.2 6.8

---

[8] See W. Feller, *An Introduction to Probability Theory and its Applications*, Vol. 1, 3rd ed., New York: John Wiley & Sons (1968), page 194.

[9] By Gus Haggstrom, quoted in E. L. Lehmann, *Nonparametrics: Statistical Methods Based on Ranks*, San Francisco: Holden-Day, Inc., 1975, p. 83.

The AB sequence for the order statistic of the combined sample is

(A A) A (A A A) (A A) | A (B A A) B B B B,

where parentheses indicate groups of tied observations and the vertical bar, the location of the median. For the median test, the counts are as follows:

	Left	Right	
A's	8	3	11
B's	0	5	5
	8	8	16

This is the most extreme table, so its probability under H_0 is the (one-sided) P-value:

$$P = \frac{\binom{8}{8}\binom{8}{3}}{\binom{16}{11}} \doteq .013. \blacksquare$$

The next two tests we study are based on the *ranks* of the X's (or of the Y's) in the *combined* order statistic, where as before, rank 1 is assigned to the smallest, 2 to the second smallest, and so on. A statistic that makes use of these ranks ought to lead to a test that is more sensitive than the median test, which uses only the information as to whether an observation is on one side of the middle or the other. F. Wilcoxon introduced the sum of the ranks of the X's, say, as a test statistic. When the X-distribution is located to the right of the Y-distribution, the X's should occupy the higher rank positions and so have a higher average rank than if there is no difference in population distributions. We denote by R_x the sum of the ranks of the X's and by R_y, the sum of the ranks of the Y's in the order statistic of the combined sample. Clearly, if we let $N = n + m$,

$$R_X + R_Y = 1 + 2 + \cdots + N = \tfrac{1}{2}N(N+1). \qquad (2)$$

So knowing R_X is equivalent to knowing R_Y, and we use one or the other of these as our test statistic. The null distribution is implied by the fact that the $\binom{N}{m}$ distinct patterns of X's and Y's are equally likely, which we noted above.

EXAMPLE 13.8c Returning to the situation of the preceding example, we see that the ranks of the B's from right to left are 1, 2, 3, 4, 6, the three observations tied at 6.5 being assigned the average of the ranks 5, 6, 7. So the sum of the B-ranks is 16. To find the P-value, we list the most extreme sequences of B-ranks:

B-ranks	Sum
1 2 3 4 5	15
1 2 3 4 6	16
1 2 3 4 7	17
1 2 3 5 6	17
1 2 3 5 7	18
1 2 3 4 8	18
1 2 4 5 6	18

The P-value is the probability of a sum of 16 or fewer, or 2 divided by $\binom{16}{5}$. For making a list such as this, notice that (for instance) in the entry 1 2 3 4 7, increasing 4 by 1 and decreasing 7 by 1 produces the next entry, with the same sum. ∎

For samples of given size, the null distribution of a rank-sum statistic could be obtained (in principle) by the procedure carried out in part in the above example. Tables of tail probabilities have been developed using a recurrence relation, and Table VII of the Appendix gives such probabilities for sample of sizes 5 to 10. (This table assumes that, when $m \neq n$, we work with the rank-sum for observations in the smaller sample.) For samples sizes larger than 10, a normal approximation works quite well[10]; to use it we need to know the mean and variance of the distribution.

To find the mean and variance of the rank-sum statistic we first study the closely related Mann-Whitney statistic. Again working with the sequence of X's and Y's defined by the order statistic of the combined sample, we count the number of X's preceding each Y in the sequence. The sum of these counts or "X-inversions" is the Mann-Whitney statistic, U_Y. It can be written in terms of the variable

$$Z_{ij} = \begin{cases} 1, & \text{if } X_i < Y_j \\ 0, & \text{if } X_i > Y_j, \end{cases}$$

the indicator variable of the event $[X_i < Y_j]$. Adding these we obtain

---

[10] See E. Fix and J. Hodges, "Significance probabilities of the Wilcoxon test," *Ann. Math. Stat.* 26, 301-312 (1955), or J. Gibbons, *Nonparametric Statistical Inference*, New York, McGraw-Hill Book Co., 1971, p. 159. Asymptotic normality also holds under the alternative hypothesis: E. Lehmann, "Consistency and unbiasedness of certain nonparametric tests," *Ann. Math. Stat.* 22, 165-179 (1951).

$$U_Y = \sum_{i=1}^{m} \sum_{j=1}^{n} Z_{ij}. \tag{3}$$

A similar sum with 1 and 0 reversed in defining Z gives the statistic U_X as the number of "X-inversions." There are mn terms in these double sums, and adding the terms in U_Y to the terms in U_X gives a count of 1 for each of the mn terms: $U_X + U_Y = mn$. (So we need only find one or the other.)

Proposed independently, the Wilcoxon and Mann-Whitney statistics are simply related:

$$R_X = U_X + \frac{m(m+1)}{2}, \quad R_Y = U_Y + \frac{n(n+1)}{2}. \tag{4}$$

For R_Y, observe that if the rank of the smallest Y is r_1, there are $r_1 - 1$ inversions for that Y; if the rank of the next smallest Y is r_2, there are $r_2 - 2$ corresponding inversions; and so on. Adding these numbers of inversions produces the given formula. (Similar reasoning yields the formula for R_X.) Because of the relationship, of course, there is essentially only one test. In tabulating tail-probabilities we have chosen to give them for the rank-sum statistic (Table VII).

The first two moments of the U's are easy to find using (3). The indicator variable Z_{ij} is Bernoulli with $p = P(X_i < Y_j)$, and under H_0, the event $[X_i < Y_j]$ has probability $1/2$. Thus,

$$E(U_Y) = \sum_i \sum_j E Z_{ij} = \frac{mn}{2}.$$

From this and (4) we obtain the mean of the rank-sum:

$$E(R_Y) = \frac{n}{2}(m+n+1). \tag{5}$$

The formula for R_X has m and n interchanged.

For the variance we have to work a little harder, because the Z_{ij} are not independent. By Corollary 1 to Theorem 20 of Chapter 4 (page 115) we have

$$\operatorname{var} U_Y = \sum_i \sum_j \sum_h \sum_k \operatorname{cov}(Z_{ij}, Z_{hk}).$$

To find the covariance we first find the average of the products which are

$$Z_{ij} Z_{hk} = \begin{cases} 1, & \text{if } X_i < Y_j \text{ and } X_h < Y_k, \\ 0, & \text{otherwise,} \end{cases}$$

so that

$$E(Z_{ij}Z_{hk}) = P(X_i < Y_j \text{ and } X_h < Y_k) = \begin{cases} 1/2, & i = h \text{ and } j = k, \\ 1/4, & i \neq h \text{ and } j \neq k, \\ 1/3, & \text{otherwise.} \end{cases}$$

From this we find

$$\text{cov}(Z_{ij}, Z_{hk}) = \begin{cases} 0, & i \neq h \text{ and } j \neq k, \\ 1/4, & i = h \text{ and } j = k, \\ 1/12, & \text{otherwise.} \end{cases}$$

To complete the calculation of the variance it remains only to count the number of sequences in the three cases listed. There are mn terms in which $i = h$ and $j = k$, and $m^2 n$ terms in which $j = k$. Of the latter, however, mn have also $i = h$, leaving $m^2 n - mn = mn(m-1)$ in which $j = k$ and $i \neq h$. Similarly, there are $nm(n-1)$ terms with $i = h$ and $j \neq k$. Finally, then

$$\text{var } U_Y = \tfrac{1}{4} mn + \tfrac{1}{12}[mn(m-1) + mn(n-1)] = \tfrac{mn}{12}(m+n+1).$$

Since R_Y differs from U_Y by a constant, its variance is the same, as is the variance of R_X, which [from (2)] is a constant minus R_Y:

$$\text{var } R_Y = \text{var } R_X = \text{var } U_Y = \tfrac{mn}{12}(m+n+1). \tag{6}$$

From the asymptotic normality of R_Y we have (with a continuity correction),

$$P(R_y \leq k \mid H_0) \doteq \Phi\left(\frac{k + 1/2 - n(m+n+1)/2}{\sqrt{mn(m+n+1)/12}}\right). \tag{7}$$

EXAMPLE 13.8d Returning to Example 13.8b, where $R_B = 16$, we calculate the P-value using (7):

$$P = P(R_B \leq 17) \doteq \Phi\left(\frac{16.5 - 5 \times 17/2}{\sqrt{55 \times 17/12}}\right) \doteq .0016,$$

as compared with the more exact value from that example (.00046). Applying the two-sample t-test to the same data yields $T = 3.22$ (14 d.f.) with $P \doteq .003$. (Since the variability on the B-route is quite a bit larger than on the A-route, the assumption in H_0 of equal population variances may not hold. So the inequality of variances may be confounded with a shift in location to produce the large value of T. But in calculating probabilities for the Wilcoxon test we also assume identical populations, so going to a rank statistic does not get around this difficulty.) ∎

The asymptotic relative efficiency of the rank-sum test relative to the t-test for shift alternatives was shown by Pitman to be

$$\text{ARE} = 12\sigma^2 \left(\int f^2(x)\, dx \right)^2,$$

where f is the p.d.f. common to the two populations except for the shift, and σ^2 is the population variance. (The integral of f^2 would be the same for each population.) In the normal case this reduces to $3/\pi$, or about .96. Moreover, it has been shown[11] that the ARE against shift alternatives is always at least .864. And in nonnormal cases it may well be greater than 1.

Yet another test based on ranks is the Fisher-Yates test, which puts sequences of X's and Y's into a critical region according to the size of

$$c_1 = \frac{1}{n} \sum_{i=1}^{n} a_{N,\, s_i}, \tag{8}$$

where $a_{N,j} = E(Z_{(j)})$, the expected value of the jth smallest in a random sample of size $N = m + n$ from $\mathcal{N}(0, 1)$, and s_i is the rank of Y_i in the order statistic of the combined sample of X's and Y's. [The constants $a_j = E(Z_j)$ are the normal scores encountered in §13.3 and given in Table XI for $n \leq 20$.] Unusually large or unusually small values of c_1 suggest a shift alternative. We could also calculate a value of c_1 based on the ranks of the X_i, but this is always the negative of the c_1 based on the Y_i.

The normal scores test has certain optimality properties, and the ARE relative to the t-test, for shift alternatives, is always at least 1—and equal to 1 only in the case of a normal population.[12] The statistic c_1 is asymptotically normal as the sample sizes become infinite in a limiting ratio not 0 nor ∞, with mean zero and variance

$$\frac{m}{nN(N-1)} \sum_{i=1}^{N} a_{N,\, i}^2.$$

EXAMPLE 13.8e To illustrate the use of the statistic c_1, we have calculated it for some of the more extreme XY-sequences for samples of sizes $m = 6$ and $n = 4$. The following table lists these sequences, the Y-ranks, the value of nc_i, and the value of R_Y. The constants $a_{10,\, j}$ from Table XI are $-1.539, -1.001, -.656, -.376, -.123, .123, .376, .656, 1.001, 1.539$. The orderings in terms of c_1 and of R_Y are similar, but observe that c_i

---

[11] J. L. Hodges, Jr., and Lehmann, E. L., "The efficiency of some nonparametric competitors of the t-test," *Ann. Math. Stat.* 27, 324–335 (1956)

[12] See H. Chernoff and I. R. Savage, "Asymptotic normality and efficiency of certain nonparametric test statistics," *Ann. Math. Stat.* 29, 972 (1958).

discriminates among some sequences with the same rank-sum. The size of the critical region $[4c_i \geq 2.54]$ is $10/\binom{10}{4} = .0476$. For the critical region $R_Y \geq 30$ it is $12/\binom{10}{4} = .057$.

Sequence	Y-ranks	$4c_i$	R_Y
X X X X X X Y Y Y Y	7 8 9 10	3.57	34
X X X X X Y X Y Y Y	6 8 9 10	3.32	33
X X X X Y X X Y Y Y	5 8 9 10	3.07	32
X X X X X Y Y X Y Y	6 7 9 10	3.04	32
X X X Y X X X Y Y Y	4 8 9 10	2.82	31
X X X X Y X Y X Y Y	5 7 9 10	2.79	31
X X X X X Y Y Y X Y	6 7 8 10	2.69	31
X X X Y X X Y X Y Y	4 7 9 10	2.54	30
X X X X Y Y X X Y Y	5 6 9 10	2.54	30
X X Y X X X X Y Y Y	3 8 9 10	2.54	30
X X X X Y X Y Y X Y	5 7 8 10	2.45	30
X X X X X Y Y Y Y X	6 7 8 9	2.16	30 ∎

PROBLEMS

13-23. Assume the following are independent random samples:

X: (12.9, 12.4, 14.7, 13.8, 10.6, 11.8), Y: (14.3, 13.4, 14.0).

(a) Apply the rank-sum test.
(b) Apply the Fisher-Yates c_1 test.
(c) Apply the permutation test based on $\sum Y_i$.

13-24. Platelet counts (in 1000's per mm^3) for 10 normal males (N) and 14 male patients (P) with a recent thrombosis were obtained as follows:

N: 141, 185, 199, 220, 231, 237, 257, 276, 295, 319
P: 169, 188, 264, 364, 364, 368, 388, 403, 415, 426, 466, 415, 597, 681

(a) Apply the median test of the hypothesis of no treatment effect.
(b) Apply the Wilcoxon rank-sum test of the hypothesis in (a).

13-25. A study[13] on the effect of intermittent feeding on blood pressure gives blood pressures after a period of several weeks for rats on normal feeding and on intermittent feeding schedules:

Normal: 169, 155, 134, 152, 133, 108, 145

---

[13] J. Falk, M. Tang., and S. Forman, "Schedule-induced chronic hypertension," *Psychosomatic Med.* 39 (1977), 252-263.

Intermittent: 170, 168, 181, 161, 199, 207, 163

(a) Apply the median test.
(b) In giving the data we have not included a reading of 115 for a treatment rat, assuming that this outlier turned out to be a mistake. Include this value and determine a P-value using a rank-sum test.

13-26. Show that the Z-score based on the rank-sum R_X is identical with the Z-score based on the Mann-Whitney U_X.

13-27. Referring to the permutation method used in Example 13.8a, show in general that ranking sequences according to the value $\bar{X} - \bar{Y}$ is equivalent to ranking according the value of $\sum Y_i$.

13-28. Refer to the list of sequences in the table in Example 13.8b.
(a) Determine the five sequences with sum 19 (perhaps with the remark at the end of the example in mind).
(b) Calculate directly the values of the Mann-Whitney statistic U_B for the sequences with rank sums from 15 to 19 [and so verify (4)].

13-29. Verify the entries in Table VII of the Appendix for $m = 4$, $n = 6$.

13-30. Check the accuracy of the normal approximation (7) against the entries in Table VII for the case $m = 8$, $n = 10$, at $k = 57$ and $k = 49$.

13-31. Calculate the Pitman ARE of the Wilcoxon two-sample test relative to the two-sample t-test for shift alternatives,
(a) when the populations are uniform on intervals of unit length.
(b) when the populations distributions are shifts from Beta(2, 2).

13-32. Calculate the Fisher-Yates statistic c_1 for the data in Example 13.8b and give an approximate P-value using the normal approximation.

13.9 Goodness of fit

In Chapter 10 we treated the problem of testing a discrete model and pointed out that a continuous model can be tested by substituting an approximating discrete model. Here we take up methods tailored especially for the continuous case. The null hypothesis to be tested is that the population c.d.f. is a specified function $F_0(x)$.

Tests based on the sample distribution function $F_n(x)$ seem natural in view of the fact that at each point x, $F_n(x)$ converges in probability (uniformly in x) to the population c.d.f. $F(x)$ The **Kolmogorov-Smirnov** (K-S) statistic is the largest absolute difference between the sample and population d.f.'s:

$$D_n = \sup_{\text{all } x} |F_n(x) - F_0(x)|.$$

Large values of D_n indicate a large discrepancy between the data and the null hypothesis model, so these are taken as evidence against H_0.

According to Theorem 3 (§13.2), the statistic D_n is *distribution-free*, that is, independent of the c.d.f. being tested. This permits use of a single table, with only the sample size as an index. Tables have been prepared using recursion formulas for sample sizes up to $n = 30$ (see Table VIII of the Appendix). The limiting distribution of D_n is given by the the following formula for the tail probability[14]:

$$\lim_{n \to \infty} P\left(D_n > \frac{z}{\sqrt{n}}\right) = 2 \sum_{j=1}^{\infty} (-1)^{j-1} \exp(-2j^2 z^2),$$

and the approximation is provided by the first term of the sum is useful when $z > 1$:

$$P\left(D_n > \frac{z}{\sqrt{n}}\right) \doteq 2 \exp(-2z^2). \tag{1}$$

EXAMPLE 13.9a Tables of "random numbers" give what are purported to be i.i.d. observations from $\mathcal{U}(0,1)$, rounded to a certain number of decimal places. In fact, they are generated deterministically with one of various computer algorithms. The following are ten successive entries from such a table, rounded further to two decimal places:

.02 .24 .97 .06 .94 .41 .81 .28 .26 .22.

Figure 13-3 shows graphs of the sample d.f. and the uniform c.d.f. 13-3.

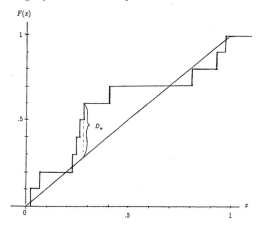

Figure 13-3

[14] See D. Darling, "The Kolmogorov-Smirnov, Cramér-von Mises tests," *Ann. Math. Stat.* 28, 823 (1957) for references.

The least upper bound of the (vertical) distances between them is evident from the graphs: $D_n = .60 - .28 = .32$. From Table VIII we see that $P \doteq .20$. [The asymptotic formula (1) with $z = \sqrt{10} \times .32 = 1.012$ yields .26.] ∎

It should be clear, from finding D_n in this example, that the maximum distance will always occur at one of the sample values, so it's not necessary to draw the picture. Simply compare the values of $F_0(X_{(i)})$ and $F_n(X_{(i)})$, as well as the values of $F_0(X_{(i)})$ and $F_n(X_{(i-1)})$, for $i = 1$ to n; the statistic D_n is the largest of the absolute differences.

For any specific alternative to H_0 the power of the test is defined, but the alternatives included in the statement that H_0 is false are too numerous for defining a "power function." One kind of result on the power of a K-S test is shown in Figure 13-4. For each value of a constant Δ it shows lower bounds on the power among the class of alternatives $F_1(x)$ for which
$$\Delta = \sup_{\text{all } x} |F_1(x) - F_0(x)|,$$
one bound for the test with $\alpha = .05$, and one for the test with $\alpha = .01$. Sampling experiments described in the paper from which the figure is taken suggest that these lower bounds are quite conservative. (It is also shown there that the K-S test is consistent and biased.)

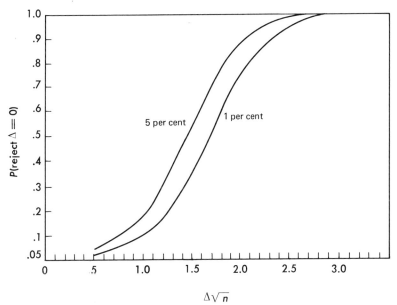

Figure 13-4

Although the K-S statistic was introduced to handle the continuous case, it can be calculated just as well when the population is discrete and numerical. It has been shown that if one uses the percentiles calculated for the continuous case to determine a rejection limit for given α, the actual α of the test does not exceed the specified value. The K-S statistic can be used for experiments in which the outcomes are numerical only by virtue of an arbitrary coding (as in the case of the roll of a die). But it must be realized that different codings will result in different values of the statistic, since the ordering of the values assigned plays a role in determining the c.d.f.

EXAMPLE 13.9b Suppose a die is rolled 120 times, to test the hypothesis that the six faces are equally likely, with the following results, where "face number" is the number of dots on the upturned face:

Face (x)	1	2	3	4	5	6
Frequency	18	23	16	21	18	24

The step heights of the sample d.f. and of $F_0(x)$, times 120, are as follows:

x	1	2	3	4	5	6
$F_n(x)$	18	41	57	78	96	120
$F_0(x)$	20	40	60	80	100	120

The largest difference is 4, so $D_n = 4/120$, or about .033, so $\sqrt{n}\, D_n = .365$.

But suppose we were to number the sides differently, say, with $1 \to 2$, $2 \to 5$, $3 \to 1$, $4 \to 4$, $5 \to 3$, $6 \to 6$. The cumulative sample frequencies are then 16, 34, 52, 73, 96, 120, and D_n is now 8/120—with the same data. ■

The use of absolute value in defining the statistic D_n makes the test "two-sided," suitable for the general alternative $F_n(x) \neq F_0(x)$. A one-sided statistic can be used in testing against the alternative $F_n(x) \geq F_0(x)$, but $F_n(x) \neq F_0(x)$ for some x. The test statistic is

$$D_n^+ = \sup_{\text{all } x}[F_n(x) - F_0(x)].$$

The null distribution of D_n^+ is known (a messy infinite series); an asymptotic formula was given by Smirnov:

$$P\left(D_n^+ > \frac{x}{\sqrt{n}}\right) \doteq \exp(-2x^2).$$

The exact distributions of the K-S statistics are known, and distribution-free, for any sample size, whereas the chi-square test is a large sample test. The K-S test is more powerful than the chi-square test for many alternatives, but it does not compare favorably with a t-test for testing a specific normal population against a shift alternative.[15] Using the K-S statistic one can construct a confidence band for the population c.d.f. (See Problem 13-36.)

Pearson's chi-square test has the important flexibility of being adaptable for testing a composite hypothesis—one consisting of a family of models indexed by one or more real parameters. The K-S test has been adapted in this way for testing the hypothesis of *normality*. The modification is the same as that used with chi-square, namely, one replaces unknown parameters in H_0 by good estimates from the sample. Lilliefors has done a Monte Carlo study[16] in which he obtains rejection limits for this important problem. Table X gives some of these results. For $n > 25$, he proposes the asymptotic value $.886/\sqrt{n}$ as the rejection limit for a test at the level $\alpha = .05$. He found, too, that this modified K-S test has greater power than the similarly modified chi-square test in all situations he considered.

Other measures of "distance" between functions lead to other statistics for tests of goodness of fit. The Cramér-von Mises statistic is

$$n\omega_n^2 = n\int_{-\infty}^{\infty} [F_n(x) - F_0(x)]^2 dF_0(x) = n\int_0^1 [F_n^*(y) - y]^2 dy,$$

where F_n^* is the transformed sample d.f. as defined by (3) in §13.2. According to Theorem 3 (in §13.2), this statistic is distribution-free, and has been adapted to testing a parametric family, with the use of a sample estimate in place of the parameter.[17]

Another measure of discrepancy between theoretical and empirical distributions is based on the fact that the area under the population

---

[15] See D. G. Chapman, "A comparative study of several one-sided goodness-of-fit tests," *Ann. Math. Stat.* 29, 655 (1959) and references contained therein.

[16] H W. Lilliefors, "On the Kolmogorov-Smirnov test for normality with mean and variance unknown," *J. Amer. Stat. Assn.* 64, 399-402 (1967).

[17] D. Darling, "The Cramér-Smirnov test in the parametric case," *Ann. Math. Stat.*, 26, 1 (1955).

p.d.f. between successive ordered observations is $1/(n+1)$ [according to (1) in §13.2]:

$$\sum_{i=1}^{n+1} \left| F_0(X_{(i)}) - F_0(X_{(i-1)}) - \frac{1}{n+1} \right|.$$

The distribution under H_0 is known and is asymptotically normal.[18]

Yet another type of measure is in common use for testing the hypothesis of population *normality*. The **Shapiro-Wilk statistic** is approximately[19]

$$W = r_{Y,a}^2,$$

where r is the correlation coefficient, **Y** is the order statistic defined by a random sample **X**, and a_i is the normal score $E[Z_{(i)}]$ used in §13.3 and listed in Table XI for $n \leq 20$. Inasmuch as the correlation coefficient is unchanged by a linear transformation on either variable, the statistic W has to do with normality, rather than with a specific normal distribution. The idea of this statistic is that if the data come from a normal population, there should be a high degree of correlation between where one expects the observations to fall and what they turn out to be. So a low correlation is taken as evidence against the hypothesis of normality. Table XII gives selected percentage points of W for selected sample sizes.

Empirical studies of power for various types of alternatives suggest that the Shapiro-Wilk test generally outperforms the K-S and the chi-square test.[20]

Some statistical software packages include the capability of producing a *rankit plot* or *normal scores plot*, showing the points (Y_i, a_i). This plot should be rather straight under the null hypothesis of normality. A study of such plots is useful as an informal check on normality, sometimes suggesting the type of nonnormality, such as heavy tails or skewness.

EXAMPLE 13.9c Tests at a coal-burning facility in Pittsburgh yielded these data on the amount of nitrogen oxide removed (in percent):

---

[18] B. Sherman, "A random variable related to the spacing of sample values," *Ann. Math. Stat.* 21, 339 (1950).

[19] In "An analysis of variance test for normality," S. Shapiro and M. Wilk proposed a test using constants for which the normal scores are approximations. They give a table of critical values for sample sizes up to 20.

[20] A. R. Dyer, "Comparisons of tests for normality with a cautionary note," *Biometrika* 61, 185 (1974). See also the preceding article on Geary's test, which uses the ratio of the mean deviation to the standard deviation as a test statistic.

91 95 90 83 91 65 55 42 55 81 89 38 20 45 58 85 78 70.

Figure 13-5 shows a rankit plot for these data. There is some evidence of skewing to the left. Using computer software we found $W = .927$. From Table XII we see that $P > .10$ (but is not much greater than .10). ∎

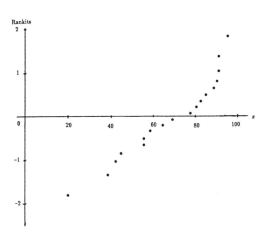

Figure 13-5

A two-sample K-S test for the hypothesis that two populations are identical: $F(x) = G(x)$, employs the test statistic

$$D = \sup_{\text{all } x} |F_m(x) - G_n(x)|,$$

where F_m is the d.f. of a random sample of size m, and G_n is the d.f. of an independent sample of size n. The test would be appropriate when the alternative to H_0 includes all ways in which the populations can differ, as opposed to situations in which, for instance, the alternative is that one population is shifted away from the other (without changing the essential shape).

Smirnov has obtained the asymptotic distribution, giving the following formula for tail probabilities:

$$P\left\{D > z\left(\frac{1}{m} + \frac{1}{n}\right)^{1/2}\right\} \doteq 2\sum_{k=1}^{\infty} (-1)^{k-1} \exp(-2k^2 z^2).$$

Again the first-term approximation is quite good and on the safe side. A test with approximate level α would thus call for rejecting H_0 when

$$D > \left\{-\frac{1}{2}\left(\frac{1}{m}+\frac{1}{n}\right)\log\frac{\alpha}{2}\right\}^{1/2}.$$

Values of this asymptotic limit for $\alpha = .05$ and $\alpha = .01$, along with small-sample rejection limits for these two levels, are given in Table IX.

PROBLEMS

13-33. Test the hypothesis that the following data were obtained as a random sample from $\mathcal{U}(-2,2)$. [Actually, they are from $\mathcal{N}(0,1)$.]

$-.676, -1.107, -1.483, .278, .493, -.442, 1.078, -.336, -.177, .057$

13-34. Suppose we toss four coins 160 times and count the number of heads showing at each toss, with these results:

x_i	0	1	2	3	4
f_i	6	38	58	47	11

Use the K-S statistic to test the hypothesis $X \sim \text{Bin}(4,\frac{1}{2})$.

13-35. Platelet counts for lung cancer patients in a VA hospital were given in Problem 13-24 as follows:

169, 188, 264, 364, 364, 368, 388, 403, 415, 415, 426, 466, 597, 681.

Test the hypothesis that these observations are from $\mathcal{N}(235, 44.6^2)$, which was assumed to describe the platelet count for the population of healthy males.

13-36. Let $\bar{X}$ be the mean of a random sample from $\mathcal{N}(\mu, 1)$. Determine the power of the critical region $|\bar{X} - \mu_0| > K$, when $n = 25$ and K is chosen so that $\alpha = .05$, at the alternative $\mu = \mu_0 + .6$. Compare this with the lower bound (from Figure 13-4) for the power of the K-S test.

13-37. Discuss the construction of a confidence band about $F_n(x)$, based on the distribution of the Kolmogorov-Smirnov statistic. This is defined by a pair of functions $A(x)$ and $B(x)$ such that

$$P[A(x) < F(x) < B(x)] = 1 - \alpha.$$

13-38. Use the data in Problem 13-35 in a Lilliefors test for normality.

13-39. Apply the Lilliefors test to the data in Example 13.9c.

13-40. A one-sided statistic for testing $F(x) = G(x)$ uses the test statistic

$$D^+ = \sup_{\text{all } x} |F_m(x) - G_n(x)|,$$

whose large-sample distribution was found by Smirnov:

$$P\{D^+ > z\left(\tfrac{1}{m}+\tfrac{1}{n}\right)^{1/2}\} \doteq e^{-2z^2}, \quad z \geq 0.$$

Show that for large samples,

$$P\left\{\frac{4(D^+)^2 mn}{m+n} > u\right\} \doteq e^{-u/2}, \quad u \geq 0.$$

13-41. Measurements of viscosity for a certain substance were made with the following results:

First day: 37.0 31.4 34.4 33.3 34.9 36.2 31.0 33.5
           33.7 33.4 34.8 30.8 32.9 34.3 33.3
Second day: 28.4 31.3 28.7 32.1 31.9 32.8 30.2 30.2 32.4 30.7

(a) Test H_0: the population has not changed from one day to the next, against the hypothesis that it has, at the level $\alpha = .01$. (Use Table IX.)
(b) Find an approximate P-value for the test result in (a).

14
Linear Models

A common statistical problem is that of comparing several normal population means when the populations are determined by either a categorical or a numerical index. An observation—a "response"—is modeled as $Y_i = \mu_i + \epsilon_i$, where $\epsilon_i \sim \mathcal{N}(0, \sigma_i^2)$ and $EY_i = \mu_i$. In the simplest situations the "error" ϵ_i has constant variance: $\sigma_i^2 = \sigma^2$, and all responses are *independent*.

We have worked with just such a model in comparing *two* means, testing equality of the means with independent random samples $(Y_{11}, ..., Y_{1,n_1})$ from Y_1, and $(Y_{21}, ..., Y_{2,n_2})$ from Y_2, assuming

$$Y_{ij} = \mu_i + \epsilon_{ij}, \quad i = 1, 2, \tag{1}$$

and $\epsilon_{ij} \sim \mathcal{N}(0, \sigma^2)$. The two populations may represent responses to two different treatments, or to a treatment and control. We'll be generalizing this problem to those in which there are more than two treatments, dealing with what are sometimes called **classification** models.

In another type of problem the index is numerical—responses Y_x depend on a controlled variable x with (in a simple instance)

$$EY_x = \alpha + \beta x. \tag{2}$$

This is a **regression** model. Although $\beta = 0$ would mean that the populations have the same mean for all values of x, it is usually the case that we don't expect equal means and are interested in just how the means vary with x.

Both of these examples are included in a model in which responses are of the form

$$Y = h(\beta) + \epsilon, \tag{3}$$

where h is a function of parameters $\beta_1, ..., \beta_p$, perhaps depending also on fixed (but given) constants $x_1, x_2, ...$ In this chapter we deal mainly with functions of the form

$$h(\beta) = \beta_1 x_1 + \cdots + \beta_p x_p, \tag{4}$$

which is *linear* in the unknown parameters β_j—a **linear model**.

In the first example described above (comparing two population means), observations are of the form

$$Y_{ij} = \mu + \tau_i + \epsilon_{ij}, \quad j = 1, \ldots, n_i, \quad i = 1, 2, \tag{5}$$

where

$$\mu = \frac{n_1 \mu_1 + n_2 \mu_2}{n_1 + n_2}, \quad \tau_i = \mu_i - \mu. \tag{6}$$

It will be convenient to write this model using matrices:

$$\begin{bmatrix} Y_{11} \\ \cdot \\ \cdot \\ \cdot \\ Y_{1n_1} \\ Y_{21} \\ \cdot \\ \cdot \\ \cdot \\ Y_{2n_2} \end{bmatrix} = \begin{bmatrix} 1 & 1 & 0 \\ 1 & 1 & 0 \\ \cdot & \cdot & \cdot \\ \cdot & \cdot & \cdot \\ 1 & 1 & 0 \\ 1 & 0 & 1 \\ 1 & 0 & 1 \\ \cdot & \cdot & \cdot \\ \cdot & \cdot & \cdot \\ 1 & 0 & 1 \end{bmatrix} \cdot \begin{bmatrix} \mu \\ \tau_1 \\ \tau_2 \end{bmatrix} + \begin{bmatrix} \epsilon_{11} \\ \cdot \\ \cdot \\ \cdot \\ \epsilon_{1n_1} \\ \epsilon_{21} \\ \cdot \\ \cdot \\ \cdot \\ \epsilon_{2n_2} \end{bmatrix}. \tag{7}$$

In the second example above, one obtains a response Y_i corresponding to selected values x_i. Assuming the model (2) we can write the data vector as

$$\begin{bmatrix} Y_1 \\ Y_2 \\ \cdot \\ \cdot \\ Y_n \end{bmatrix} = \begin{bmatrix} 1 & x_1 \\ 1 & x_2 \\ \cdot & \cdot \\ \cdot & \cdot \\ 1 & x_n \end{bmatrix} \cdot \begin{bmatrix} \alpha \\ \beta \end{bmatrix} + \begin{bmatrix} \epsilon_1 \\ \epsilon_2 \\ \cdot \\ \cdot \\ \epsilon_n \end{bmatrix}.$$

In these examples we have special cases of data **Y** generated by the linear model given by (3) and (4), which we now write in matrix notation as

$$\mathbf{Y} = \mathbf{X}\beta + \epsilon. \tag{8}$$

The matrix **X** is the *design matrix*. We'll see that the matrix **X'X** plays an important role in the theory of linear models.

The material to be presented in this chapter barely opens the door on what is an extensive area of statistics, to which complete books are devoted.[1] Sections 14.2–14.6 treat basic regression models, and Sections 14.7–14.10 introduce simple classification models. We give first a result that is generally applicable.

14.1 The likelihood function

As in the example that opened this chapter, our analyses will be mostly based on the assumption that the observations are independent and *normal*: $Y \sim \mathcal{N}(h(\boldsymbol{\beta}), \sigma^2)$, where $h(\boldsymbol{\beta}) = \beta_1 x_1 + \cdots + \beta_p x_p$. Given the observations $\mathbf{Y} = (Y_1, ..., Y_n)$, the likelihood function is

$$L(\boldsymbol{\beta}, \sigma^2) \propto (\sigma^2)^{-n/2} \exp\left(-\frac{1}{2\sigma^2}\sum [Y_i - h_i(\boldsymbol{\beta})]^2\right). \tag{1}$$

(We write "h_i" because the dependence h on the parameters can involve the design constants $\boldsymbol{x}$, which can vary with the observation.) As a first step in maximizing the likelihood we differentiate $\log L$ with respect to σ^2:

$$\frac{\partial}{\partial \sigma^2} \log L = -\frac{n}{2\sigma^2} + \frac{1}{2\sigma^4}\sum_{i=1}^{n} [Y_i - h_i(\boldsymbol{\beta})]^2,$$

which vanishes when σ^2 has the value Q/n, where

$$Q = \sum_{i=1}^{n} [Y_i - h_i(\boldsymbol{\beta})]^2. \tag{2}$$

Setting the derivatives of $\log L$ with respect to the parameters β_j equal to zero leads to equations that do not involve σ^2:

$$\frac{\partial}{\partial \beta_j}\log L = -\frac{1}{2\sigma^2}\frac{\partial Q}{\partial \beta_j} = \frac{1}{\sigma^2}\sum_{i=1}^{n}[Y_i - h_i(\boldsymbol{\beta})]\frac{\partial h_i(\boldsymbol{\beta})}{\partial \beta_j} = 0, \quad j = 1, ..., p. \tag{3}$$

These are precisely the equations we'd get in *minimizing* the sum of squared deviations of the Y_i from the h_i. Doing so is an instance of Gauss' **method of least squares**, a method of curve-fitting that has been in common use since the early 1800's.

Equations (3) are referred to as the **normal equations** of the least squares process. They are necessary conditions for a minimum; but since Q is bounded below ($Q \geq 0$), it has a greatest lower bound which, in the usual case, is a minimum. The partial derivatives vanish at any regular minimum point, and if the solution of (3) is unique, it is the minimum point.

The solution $\hat{\boldsymbol{\beta}}$ of (3) is a component of the joint m.l.e. $(\hat{\boldsymbol{\beta}}, \hat{\sigma}^2)$, and we find $\hat{\sigma}^2$ as Q/n upon substituting $\hat{\boldsymbol{\beta}}$ for $\boldsymbol{\beta}$ in (2):

---

[1] H. Scheffé, *The Analysis of Variance*, New York: John Wiley & Sons (1959); S. R. Searle, *Linear Models*, New York: John Wiley & Sons (1971).

$$\widehat{\sigma}^2 = \frac{1}{n}\sum [Y_i - h_i(\widehat{\beta})]^2. \tag{4}$$

Substituting $\widehat{\beta}$ and $\widehat{\sigma}^2$ for β and σ^2 in (1) yields the following result, which we'll need several times in what follows:

THEOREM 1 *The maximum of the likelihood function (1) is*
$$L(\widehat{\beta}, \widehat{\sigma}^2) = (\widehat{\sigma}^2)^{-n/2} e^{-n/2}.$$

There are other general results that can be derived once and for all, but these will be easier to motivate and understand after we have taken up an important special case, that of a particularly simple regression model. In regression models, the design constants x_i are numerical, usually values of one or more continuous variables, the predictor variables. The "h" in the above discussion is then a function of parameters β_i and the predictor variables x_i; as a function of the x_i it is termed the **regression function**. The function g will usually involve unknown parameters, and in a "linear model" it is a linear function of those parameters. We consider first regression problems with a single predictor variable, and denote the regression function by $g(x)$.

PROBLEMS

14-1. Find the function $\alpha + \beta x$ (representing a straight line) that best fits these points in the sense of least squares: (0, 2), (1, 1), (4, 3), (5, 2).

14-2. Use the method of least squares to obtain estimators of the parameters of each of these assumed regression functions:
  (a) $g(x) = \gamma x^2$, given data (x_i, Y_i), $i = 1, ..., n$.
  (b) $g(x, z) = \alpha x + \beta z$, given data (x_i, z_i, Y_i), $i = 1, ..., n$.

14-3. Express each of the following in matrix notation $\mathbf{Y} = \mathbf{X}\beta + \epsilon$, identifying the matrices $\mathbf{X}$ and β in each case:
  (a) The model (a) in the preceding problem.
  (b) The model (b) in the preceding problem.
  (c) The regression model with $g(x) = \beta_0 + \beta_1 x + \beta_2 x^2$.

14-4. Write the normal equations for the problem of fitting a quadratic regression function $\beta_0 + \beta_1 x + \beta_2 x^2$ to points (x_i, Y_i) by least squares, and solve them given these four data points:
$$(-1, 1), (0, 2), (1, 1), (2, -2).$$

14-5. Find the minimal sufficient statistic for independent observations Y_i at x_i, when $Y_i = \alpha + \beta x_i + \epsilon_i$, and $\epsilon_i \sim N(0, \sigma^2)$.

14.2 Simple linear regression—estimation

We take up now a regression model in which the regression function g is a function of a single predictor variable x. For the present we think of x as a controlled variable. To learn about the parameters in g, we obtain responses Y_i at selected points x_i, resulting in the n pairs (x_i, Y_i). Estimating the parameter values amounts to "fitting" a function g to these n points in the xy-plane, and this we do using method of least squares. We are assuming that, for each x, the response Y_x is random: There is a distribution of possible responses, whose mean $E(Y_x)$ is the regression function, $g(x)$. Deviations from this function are the ϵ's—random "errors," or **residuals**:

$$\epsilon_i = Y_i - g(x_i). \tag{1}$$

To measure the fit of a proposed regression curve $g(x)$ to a set of data points we use the mean squared residual, or equivalently, the sum of the squared residuals:

$$Q(g) = \sum_{i=1}^{n} [Y_i - g(x_i)]^2. \tag{2}$$

From the family of proposed functions g we choose the one with parameter values that minimize $Q(g)$.

In a *simple linear regression* we assume the regression function to be linear in x (as well in the parameters): $g(x) = \alpha' + \beta x$, where β is the slope of the line and α', the y-intercept. For convenience in studying distributions we parametrize the linear regression function g in the slightly different form

$$g(x) = \alpha + \beta(x - \bar{x}),$$

where β is the slope, as before, and the constant α is the value of g at the mean of the x_i's at which responses are observed.

To do the curve fitting, given the data points (x_i, Y_i), we try the line $y = a + b(x - \bar{x})$. The sum of squared residuals about this line is

$$Q(a, b) = \sum [Y_i - a - b(x_i - \bar{x})]^2. \tag{3}$$

Here (and throughout the rest of this chapter) it will be convenient to employ the following commonly used notation:

$$SS_{uv} = \sum (u_i - \bar{u})(v_i - \bar{v}) = \sum u_i(v_i - \bar{v}) = \sum u_i v_i - n\bar{u}\bar{v}. \tag{4}$$

First adding and subtracting $\bar{Y}$ inside the brackets in (3), we obtain

$$Q(a, b) = \sum [(Y_i - \bar{Y}) + (\bar{Y} - a) - b(x_i - \bar{x})]^2.$$

We then square the bracketed quantity as a trinomial:

$$Q(a, b) = \mathrm{SS}_{YY} + n(a - \bar{Y})^2 + b^2 \mathrm{SS}_{xx} - 2b\, \mathrm{SS}_{xY},$$

and complete the square in b:

$$Q(a,b) = \mathrm{SS}_{YY}(1 - r^2) + n(a - \bar{Y})^2 + \mathrm{SS}_{xx}\left(b - \frac{\mathrm{SS}_{xY}}{\mathrm{SS}_{xx}}\right)^2, \qquad (5)$$

where the correlation coefficient r is given by (4) in §12.3:

$$r = \frac{\mathrm{SS}_{xY}}{\sqrt{\mathrm{SS}_{xx}\mathrm{SS}_{YY}}}. \qquad (6)$$

The residual sum of squares Q is clearly minimized with the choices

$$a = \bar{Y}, \quad b = \frac{\mathrm{SS}_{xY}}{\mathrm{SS}_{xx}}, \qquad (7)$$

and the minimum value, the minimum sum of squared residuals, is

$$Q_{\min} = \mathrm{SS}_{YY}(1 - r^2). \qquad (8)$$

(Incidentally, we see again that $r^2 \leq 1$, since $Q_{\min} \geq 0$.) The expression (5) for Q is of course incorrect if $\mathrm{SS}_{xx} = 0$, but this can only occur when all responses are measured at a single x-value. To do so would be pointless (if we want to find out how the average response *varies* with x), so we assume that $\mathrm{SS}_{xx} \neq 0$.

Equation (8) says that if there is a perfect linear correlation, then the minimum of the average squared residual is 0. This would mean that each residual is 0, and the data points must lie on a line. As one might expect, the line produced by the least squares process in that case *is* the line on which the data points lie.

EXAMPLE 14.2a In an experiment designed to study the effect of dissolved sulfur on the surface tension of liquid copper, the decrease in surface tension was assumed to be a linear function of the logarithm of the percentage of sulfur. Decreases in tension were measured for several choices of sulfur percentage, yielding the data given in the following table:

14.2 Simple linear regression—estimation

$x = \log \%$ sulfur	Y = decrease in tension (deg/cm)
-3.38	308
-2.38	426
-1.20	590
$-.92$	624
$-.49$	649
$-.19$	727

Calculating the slope and intercept using (7) we obtain the following least squares line, or empirical regression line: $y = 736 + 127.6\,x$. The correlation coefficient (6) is $r = .996$, and the mean squared residual about the empirical line is 240.47. The data and the fitted line are shown in Figure 14-1. Observe the extremely close fit, reflecting the fact that the coefficient of linear correlation is nearly 1. ∎

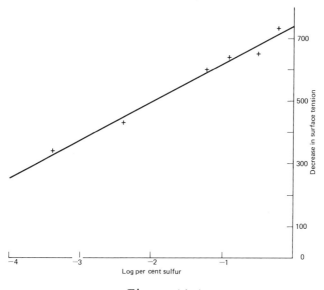

Figure 14-1

Although we derived the coefficients in the equation for the least squares regression line algebraically, we could have used the method of calculus, as we did in §14.1. In this special case of a linear regression function, the normal equations are

$$\begin{cases} \sum [Y_i - a - b(x_i - \bar{x})] = 0 \\ \sum (x_i - \bar{x})[Y_i - a - b(x_i - \bar{x})] = 0, \end{cases} \quad (9)$$

After some simplification, we have these (conditional) equations:

$$\begin{cases} \bar{Y} - a = 0 \\ SS_{xY} = b SS_{xx}. \end{cases}$$

THEOREM 2: *Given data* (x_i, Y_i), *the equation of the least squares line is* $y = \tilde{a} + \tilde{b}(x - \bar{x})$, *where*

$$\tilde{\alpha} = \bar{Y}, \quad \tilde{\beta} = \frac{SS_{xY}}{SS_{xx}}. \tag{10}$$

The corresponding sum of squared residuals about this line is

$$SSResid = \sum [Y_i - \tilde{\alpha} - \tilde{b}(x_i - \bar{x})]^2 = SS_{YY}(1 - r^2).$$

COROLLARY *With the parametrization* $a' + bx$, *the least squares coefficients are*

$$\tilde{\alpha}' = \bar{Y} - \tilde{\beta}\bar{x}, \quad \tilde{\beta} = \frac{SS_{xY}}{SS_{xx}}. \tag{11}$$

and the sum of squared residuals is the same as SSResid *above*:

$$\sum [Y_i - \tilde{a}' - \tilde{b} x_i]^2 = \sum [Y_i - \tilde{\alpha} - \tilde{b}(x_i - \bar{x})]^2.$$

Depending as they do on the data **Y**, the least squares estimators $\tilde{\alpha}$ and $\tilde{\beta}$ are random variables. They are *linear* functions of the Y_i's, and this linearity makes it easy to find the mean values. Thus,

$$E(\tilde{\alpha}) = \frac{1}{n}\sum E(Y_i) = \frac{1}{n}\sum [\alpha + \beta(x_i - \bar{x})] = \alpha,$$

$$E(\tilde{\beta}) = \frac{1}{SS_{xx}}\sum (x_i - \bar{x})E(Y_i) = \frac{1}{SS_{xx}}\sum (x_i - \bar{x})[\alpha + \beta(x_i - \bar{x})] = \beta.$$

So the least squares estimates of the regression function coefficients are *unbiased*. The linearity also makes calculation of the variances relatively easy:

$$\operatorname{var} \tilde{\alpha} = \frac{1}{n^2}\sum \operatorname{var} Y_i = \frac{\sigma^2}{n},$$

and

$$\operatorname{var} \tilde{\beta} = \frac{1}{(SS_{xx})^2}\sum (x_i - \bar{x})^2 \operatorname{var} Y_i = \frac{\sigma^2}{SS_{xx}}$$

Finally, $\tilde{\alpha}$ and $\tilde{\beta}$ are uncorrelated:

$$\operatorname{cov}\left(\bar{Y}, \frac{SS_{xY}}{SS_{xx}}\right) = \frac{1}{n(SS_{xx})} \operatorname{cov}\left(\sum_i Y_i, \sum_j (x_j - \bar{x})Y_j\right)$$

$$= \frac{1}{n(SS_{xx})} \sum_j (x_j - \bar{x})\sigma^2 = 0.$$

(This is a result of the particular parametrization chosen. See Problem 14-11.) In summary, we have the following:

THEOREM 3 *Given independent observations Y_i at x_i, $i = 1, ..., n$, where $E(Y_i) = \alpha + \beta(x_i - \bar{x})$ and $\operatorname{var} Y_i = \sigma^2$, the least squares estimates $(\tilde{\alpha}, \tilde{\beta})$ of (α, β) are unbiased and uncorrelated, with variances*

$$\operatorname{var} \tilde{\alpha} = \frac{\sigma^2}{n}, \quad \operatorname{var} \tilde{\beta} = \frac{\sigma^2}{SS_{xx}}. \tag{12}$$

COROLLARY *With the parametrization $\alpha' + \beta x$ the least squares estimates $(\tilde{\alpha}', \tilde{\beta})$ are unbiased, with variances*

$$\operatorname{var} \tilde{\alpha}' = \sigma^2 \left(\frac{1}{n} + \frac{\bar{x}^2}{SS_{xx}}\right), \quad \operatorname{var} \tilde{\beta} = \frac{\sigma^2}{SS_{xx}}.$$

and covariance

$$\operatorname{cov}(\tilde{\alpha}', \tilde{\beta}) = \frac{\bar{x}}{SS_{xx}}.$$

We notice that for an estimate of the slope to be as accurate as possible in the presence of random error, the denominator SS_{xx} should be as large as possible. There are usually practical limits on the useful range of x-values. The way to maximize SS_{xx} is to put half the x's at the right end of this range and the other half at the left end. (There has been and continues to be much research in the matter of "optimal design" for a general regression problem.)

Suppose we now add the assumption that the observations **Y** are *normally* distributed. The least squares estimates given by (10), but now called $\hat{\alpha}$ and $\hat{\beta}$, together with

$$\hat{\sigma}^2 = \frac{1}{n} \sum [Y_i - \hat{\alpha} - \hat{\beta}(x_i - \bar{x})]^2 \tag{13}$$

[from (4) of §14 1], are the joint maximum likelihood estimates of $(\alpha, \beta, \sigma^2)$. Moreover, since they are linear functions of the jointly normal Y's, the estimators $\hat{\alpha}$ and $\hat{\beta}$ have a *bivariate normal* distribution and (being uncorrelated) are *independent*. In summary:

THEOREM 4 *Given independent observations Y_i at x_i, $i = 1, ..., n$, where*
$$Y_i \sim \mathcal{N}(\alpha + \beta[x_i - \bar{x}], \sigma^2),$$

the least squares estimates $(\tilde{\alpha}, \tilde{\beta})$ are the m.l.e.'s $\hat{\alpha}$ and $\hat{\beta}$, together with $\hat{\sigma}^2$, given by (13). Further, $\hat{\alpha}$ and $\hat{\beta}$ are independent, and normal:

$$\hat{\alpha} \sim \mathcal{N}\left(\alpha, \frac{\sigma^2}{n}\right), \quad \hat{\beta} \sim \mathcal{N}\left(\beta, \frac{\sigma^2}{SS_{xx}}\right).$$

COROLLARY *With the parametrization $\alpha' + \beta x$, the m.l.e.'s (joint with $\hat{\sigma}^2$) are the least squares estimates:*

$$\hat{\alpha}' = \tilde{\alpha}', \quad \hat{\beta} = \tilde{\beta}.$$

Moreover, $\hat{\alpha}'$ and $\hat{\beta}$ have a bivariate normal distribution, in which the marginals are

$$\hat{\alpha} \sim \mathcal{N}\left(\alpha, \sigma^2\left\{\frac{1}{n} + \frac{\bar{x}^2}{SS_{xx}}\right\}\right), \quad \hat{\beta} \sim \mathcal{N}\left(\beta, \frac{\sigma^2}{SS_{xx}}\right).$$

The residuals about the actual regression function $y = \alpha + \beta(x_i - \bar{x})$ constitute a sample from the error population. So the m.l.e. $\hat{\sigma}^2$, which is the mean squared residual about the empirical function $\hat{\alpha} + \hat{\beta}(x - \bar{x})$, is an intuitively obvious choice of estimator for σ^2. The numerator of $\hat{\sigma}^2$ is the sum of squared residuals, also called the *error sum of squares* (SSE) or *residual sum of squares* (SSResid). Its distribution is implied in the next theorem.

THEOREM 5 *If $Y_i \sim \mathcal{N}(g(\boldsymbol{\beta}, \boldsymbol{x}_i), \sigma^2)$, where*

$$g(\boldsymbol{\beta}, \boldsymbol{x}) = \beta_0 + \beta_1 x_1 + \cdots + \beta_k x_k,$$

then $n\hat{\sigma}^2/\sigma^2 \sim chi^2(n - k - 1)$, and $\hat{\sigma}^2$ is independent of $\hat{\boldsymbol{\beta}}$.

We'll prove Theorem 5 in the stated generality later; for the present we'll give the proof for the simple linear regression function $\alpha + \beta(x - \bar{x})$, appealing to Cochran's theorem (Theorem 14, page 441). We start with the identity

$$Y_i - \alpha - \beta(x_i - \bar{x}) = [Y_i - \hat{\alpha} - \hat{\beta}(x_i - \bar{x})] + (\hat{\alpha} - \alpha) + (\hat{\beta} - \beta)(x_i - \bar{x}).$$

Squaring, summing on i, and dividing by σ^2, we obtain

$$\sum \left\{\frac{Y_i - \alpha - \beta(x_i - \bar{x})}{\sigma}\right\}^2 = \frac{n\hat{\sigma}^2}{\sigma^2} + \frac{(\hat{\alpha} - \alpha)^2}{\sigma^2/n} + \frac{(\hat{\beta} - \beta)^2}{\sigma^2/SS_{xx}}. \quad (14)$$

The cross-products have disappeared by virtue of the normal equations (9). The term on the left is a sum of squares of standard normal variables and so is chi$^2(n)$. The terms on the right are, respectively, a sum of squares of n linear combinations of the $\mathcal{N}(0,1)$ variables on the left, a square of *one* such linear combination, and the square of another linear combination.

Let r_i denote the rank of the ith term on the right. Then $r_2 = r_3 = 1$. And since the rank of a sum is less than or equal to the sum of the ranks: $n \leq r_1 + 1 + 1$. But we also know that r_1 is at most $n - 2$, since the squared quantities satisfy two linear constraints, namely, the normal equations (9). So $n - 2 \leq r_1 \leq n - 2$, which means that $r_1 = n - 2$, and the ranks of the terms on the right add up to the rank of the term on the left. Cochran's theorem then tells us that the terms on the right are independent chi-square variables with degrees of freedom equal to the corresponding ranks, which establishes Theorem 5 for this case.

The error variance σ^2 is usually estimated in practice as an average squared residual in which the sum of squared residuals is divided by its degrees of freedom, $n - 2$:

$$S^2 = \frac{1}{n-2} \sum [Y_i - \hat{\alpha} - \hat{\beta}(x_i - \bar{x})]^2 = \frac{\text{SSE}}{n-2}. \qquad (15)$$

This happens to be an unbiased estimator of σ^2 (Problem 14-10).

With the estimate (13) of σ^2 and formulas (12) for the variances of $\hat{\alpha}$ and $\hat{\beta}$, we can define **standard errors** of the estimators $\hat{\alpha}$ and $\hat{\beta}$:

$$\text{s.e.}(\hat{\alpha}) = \frac{\hat{\sigma}}{\sqrt{n}}, \quad \text{s.e.}(\hat{\beta}) = \frac{\hat{\sigma}}{\sqrt{\text{SS}_{xx}}}. \qquad (16)$$

Large sample confidence limits can be constructed in the usual way, as the estimate plus or minus a constant (depending on the confidence level) times the standard error. The m.l.e. $\hat{\sigma}$ is usually replaced (for s.e.'s) by

$$S = \sqrt{\frac{n}{n-2}} \hat{\sigma}. \qquad (17)$$

[The reason often given for this is that S^2 is an unbiased estimator of σ^2; but of course S is *not* an unbiased estimator of σ! (See Problem 8-3.)]

When n is not large, replacing σ by S is not so successful, but it turns out that following approximately standardized score is pivotal:

$$T = \frac{\hat{\beta} - \beta}{S/\sqrt{\text{SS}_{xx}}} = \frac{\dfrac{\hat{\beta} - \beta}{\sigma/\sqrt{\text{SS}_{xx}}}}{\sqrt{\dfrac{n\hat{\sigma}^2/\sigma^2}{n-2}}}. \qquad (18)$$

The numerator of the last fraction is $\mathcal{N}(0, 1)$, and the square of the denominator is $\text{chi}^2(n-2)$, divided by its number of degrees of freedom. Since numerator and denominator in (18) are independent, we know from §8.9 that $T \sim t(n-2)$, and this is a distribution that does not involve α, β, or σ^2. For a confidence level $1-\gamma$, we obtain confidence limits for β as usual:

$$\widehat{\beta} \pm (t_{1-\gamma/2}) \frac{S}{\sqrt{\text{SS}_{xx}}}.$$

EXAMPLE 14.2b Continuing Example 14.2a, we find the sums of squares: $\text{SS}_{YY} = 122050$, $\text{SS}_{xx} = 7.43913$, and from these, $S = 15.507$. For 90 percent confidence limits for the slope parameter β we use the multiplier $t_{.95} = 2.13$ (4 d.f.) to obtain the confidence limits:

$$127.6 \pm 2.13 \times 15.507/2.7275, \text{ or } 127.6 \pm 12.1. \blacksquare$$

PROBLEMS

14-6. Determine the distribution of the *fitted value* at x:

$$\widehat{Y} = \widehat{\alpha} + \widehat{\beta}(x - \bar{x}).$$

14-7. The following data[2] give temperatures corresponding to chirp frequencies of the striped ground cricket:

x (chirps/sec)	20	16	20	18	17	16	15	17	15	16
y (°F)	89	72	93	84	81	75	70	82	69	83

Find the least squares line and give 90 percent confidence limits for the slope.

14-8. Assuming a linear regression of amount of dissolved chemical (Y) on temperature (x), obtain the empirical regression line as fitted given these data:

Temperature (°C)	0	10	20	30	40	60
Amount dissolved (gm)	8	12	21	31	39	58

14-9. Derive the normal equations (9) for fitting the linear function $a + b(x - \bar{x})$, by differentiating the sum of squared residuals.

14-10. Show that S^2 [(15) in §14.2] is an unbiased estimator of σ^2.

14-11. Show the Corollary to Theorem 2.

---

[2] G. W. Pierce, *The Songs of Insects*, Cambridge, MA: Harvard U. Press (1949), 12.

14-12. Show the Corollaries to Theorems 3 and 4.

14-13. Show that the correlation coefficient of the observed values Y_i and the fitted values $\hat{Y}_i = \hat{\alpha} + \hat{\beta}(x_i - \bar{x})$ is $|r|$, where r is the correlation coefficient of the data pairs (x_i, Y_i). [Note: What's wanted is the "sample" r for the pairs $(Y_1, \hat{Y}_1), ..., (Y_n, \hat{Y}_n)$.]

14-14. Given data (x_i, Y_i), with not all $x_i = 0$, consider the regression model $Y_i = \beta x_i + \epsilon$, where $\epsilon \sim N(0, \sigma^2)$.
(a) Find the m.l.e.'s of the parameters, $\hat{\beta}$ and $\hat{\sigma}^2$.
(b) Obtain the distribution of $\hat{\beta}$.
(c) Obtain the distribution of $\hat{\sigma}^2$ and show that $\hat{\sigma}^2$ is independent of $\hat{\beta}$.

14-15. The m.l.e.'s $\hat{\alpha}$ and $\hat{\beta}$ are sometimes said to be consistent. Is this always the case? Suppose observations are taken at the points $x_1 = 0$, and $x_2 = x_3 = \cdots = 1$. As n becomes infinite, does $\hat{\beta}$ converge to β?

14.3 Simple linear regression—testing and prediction

Consider next the problem of testing the hypothesis $H_0: \beta = \beta^*$ against the alternative $H_A: \beta \neq \beta^*$, in the context of the model of §14.2, in which $Y_i = \alpha + \beta(x_i - \bar{x}) + \epsilon_i$, where $\epsilon_i \sim N(0, \sigma^2)$. The likelihood ratio is

$$\Lambda = \frac{\sup_{(\alpha, \sigma^2)} L(\alpha, \beta^*; \sigma^2)}{\sup_{(\alpha, \beta, \sigma^2)} L(\alpha, \beta; \sigma^2)} \tag{1}$$

So we need m.l.e.'s under H_0 and under $H_0 \cup H_A$. From §14.2 we know that the latter are the $\hat{\alpha}$, $\hat{\beta}$, and $\hat{\sigma}^2$ of that section. In particular,

$$\hat{\sigma}^2 = \frac{1}{n} \sum [Y_i - \hat{\alpha} - \hat{\beta}(x_i - \bar{x})]^2.$$

Under H_0, where we assume $\beta = \beta^*$, the m.l.e. of α is again $\hat{\alpha} = \bar{Y}$. The m.l.e. of σ^2 is

$$\hat{\sigma}_0^2 = \frac{1}{n} \sum [Y_i - \hat{\alpha} - \beta^*(x_i - \bar{x})]^2. \tag{2}$$

Appealing to Theorem 1 (at the end of §14.1), we find that (1) is evaluated as

$$\Lambda = \left(\frac{\hat{\sigma}_0^2}{\hat{\sigma}^2}\right)^{-n/2}. \tag{3}$$

To express $\Lambda < K$ in terms of familiar statistics, we first add and subtract $\widehat{\beta}(x_i - \bar{x})$ inside the brackets in (2):

$$Y_i - \widehat{\alpha} - \beta^*(x_i - \bar{x}) = [Y_i - \widehat{\alpha} - \widehat{\beta}(x_i - \bar{x})] + (\widehat{\beta} - \beta^*)(x_i - \bar{x}).$$

Upon squaring both sides and summing on i, we obtain

$$n\widehat{\sigma}_0^2 = n\widehat{\sigma}^2 + (\widehat{\beta} - \beta^*)^2 \, \text{SS}_{xx}, \qquad (4)$$

where the cross-product term has disappeared by virtue of the normal equations [(9) of §14.2]. Dividing both sides by $n\widehat{\sigma}^2$ and substituting in (3), we get

$$-\frac{2}{n} \log \Lambda = \log \left\{ 1 + \frac{(\widehat{\beta} - \beta^*)^2}{n\widehat{\sigma}^2/\text{SS}_{xx}} \right\}.$$

This will be large (and Λ small) when the second term inside the braces is large, or when the following test statistic is large in magnitude:

$$T = \frac{\widehat{\beta} - \beta^*}{S/\sqrt{\text{SS}_{xx}}}, \qquad (5)$$

where S is the estimate of σ defined in §14.2: $S = \sqrt{\frac{n}{n-2}}\,\widehat{\sigma}$. We saw at the end of the preceding section that if $\beta = \beta^*$, then $T \sim t(n-2)$. So P-values, or critical regions for type I error sizes, can be found in the t-table. For the two-sided alternative $\beta \neq \beta^*$, we could use T^2, whose whose distribution is $F(1, n-2)$.

The decomposition (4) is an instance of an **analysis of variance**, often referred to with the acronym ANOVA. Statistical software will produce the decomposition, in particular, for testing $\beta = 0$, with the terms in (4) listed in terms of "SS" notation.

The left-hand side of (4) (SS_{YY}) measures variation about the mean line, $y = \bar{Y}$, and is called the **total sum of squares** (SSTotal). This is the numerator in the formula for the variance of the Y_i's, and has $n-1$ degrees of freedom. The first term on the right is a contribution to that total variation attributed to the error term ϵ and called the **error (or residual) sum of squares** (SSE), with $n-2$ d.f. This sum of squares is $(1 - r^2)$ times SS_{YY}. The second term on the right is a contribution to total variation about $\bar{Y}$ called the **regression sum of squares** (SSReg), attributed to a nonzero slope β. This single term, which is r^2 times SS_{YY}, has d.f. $= 1$. Dividing each SS by its degrees of freedom yields "mean squares," MSReg and MSE. And the above T^2 is then MSReg/MSE. The quantity r^2, the fraction of SS_{YY} attributed to regression, is called the **coefficient of determination**. The analysis is exhibited in an **ANOVA table**:

14.3 Simple linear regression—testing hypotheses and prediction

Source	d.f.	Sum of squares	Mean square	F
Regression	1	SSReg	MSReg	MSReg/MSE
Error	$n-2$	SSE	MSE	
Total	$n-1$	SSTotal	--	

EXAMPLE 14.3a In an investigation of the damage to residential structures from blasting, data were obtained on frequency (cps) and particle displacement (in) for 13 blasts. Some statistics on log-frequency (x) and displacement (Y) are as follows:

$$\bar{Y} = .17485, \hat{\beta} = -.1038, r^2 = .65160, SS_{xx} = 9.07494, \bar{X} = 2.0021.$$

From these we can write the regression function: $y = .383 - .1038 \log x$, and find the ANOVA table entries:

Source	d.f.	SS	MS	F
Regression	1	.09778	.09778	20.6
Error	11	.05228	.004753	--
Total	12	.15006	--	--

The large F ($T \doteq 4.5$, 11 d.f., $P < .001$) is clear evidence against $\beta = 0$, but it is doubtful that anyone would expect β to be 0. The relationship accounts for about two-thirds of the total variation; the remaining one-third is attributed to "error" or unaccounted sources of variation about the regression function. ∎

Up to this point we have considered x as a controlled variable. Suppose now that the data pairs are (X_i, Y_i), independent observations on the bivariate normal vector (X, Y). The regression function $E(Y \mid X = x)$, is linear in x, and the various results we have obtained for the case of a controlled variable x are valid for studying this regression, *conditional* on given values of X. Thus, the distribution of T as defined by (5) is $t(n-2)$, given the values of X. But this t-distribution does not involve those x-values, and it follows that T has this same distribution *unconditionally*. Problem 14-18 asks you to show that for testing $\beta_0 = 0$, this T is identical to the T used in §12.3 for testing $\rho = 0$:

$$T = \sqrt{n-2} \frac{r}{\sqrt{1-r^2}}, \tag{6}$$

We have thus established what was asserted there, namely, that the null distribution of the T defined by (6) is $t(n-2)$.

In the bivariate setting, a regression function is useful as the basis of prediction. Given $X = x_0$, predicting the value of Y to be $E(Y \mid x_0)$ has the minimum mean squared prediction error, so it is natural to predict, given the data (X_i, Y_i), that Y will have the *fitted value*:

$$\widehat{Y}_0 = \widehat{\alpha} + \widehat{\beta}(x_0 - \bar{x}). \tag{7}$$

This fitted value, the estimated mean response at x_0, is a random variable with mean $\alpha + \beta(x_0 - \bar{x})$ and variance

$$\operatorname{var} \widehat{Y}_0 = \operatorname{var} \widehat{\alpha} + (x_0 - \bar{x})^2 \operatorname{var} \widehat{\beta} = \sigma^2 \left(\frac{1}{n} + \frac{(x_0 - \bar{x})^2}{SS_{xx}} \right). \tag{8}$$

(Here we have used the fact that $\widehat{\alpha}$ and $\widehat{\beta}$ are independent, and independent of the new observation Y_0 at x_0.) Since this involves the unknown population parameter σ^2, we can only approximate it by estimating σ^2. With S^2 in place of σ^2, taking the square root yields the **standard error of the fitted value** as an estimate of $\alpha + \beta(x_0 - \bar{x})$, the mean at x_0:

$$\operatorname{s.e.}(\widehat{Y}_0) = S\sqrt{\frac{1}{n} + \frac{(x_0 - \bar{x})^2}{SS_{xx}}}. \tag{9}$$

Consider now predicting the response to be $\widehat{Y}_0$ when $x = x_0$. Since $\widehat{\alpha}$ and $\widehat{\beta}$ are unbiased, the average prediction error, conditional on $x_1, \ldots, x_n$, and x_0, is

$$E(\widehat{Y}_0 - Y_0) = E[\widehat{\alpha} + \widehat{\beta}(x_0 - \bar{x})] - \alpha - \beta(x_0 - \bar{x}) = 0.$$

The mean squared prediction error is therefore the variance:

$$\operatorname{m.s.p.e.} = \operatorname{var}(\widehat{Y}_0 - Y_0) = \operatorname{var} \widehat{Y}_0 + \operatorname{var} Y_0 = \sigma^2 \left(1 + \frac{1}{n} + \frac{(x_0 - \bar{x})^2}{SS_{xx}} \right). \tag{10}$$

This differs from (8) because the prediction is in error only partly because of the error estimating the mean response, which we use for the prediction as the center of the error distribution at x_0. The response Y_0 is a sample from that error distribution; and even if we knew α and β, there would be an error of prediction with variance σ^2. To this we add the variance (8) to get the overall means squared prediction error.

Again using S as the estimate of σ [from (15), §14.2] in (10), we have the following formula for **standard error of prediction**:

$$\text{s.e.p.}(\widehat{Y}_0) = S\sqrt{1 + \frac{1}{n} + \frac{(x_0 - \bar{x})^2}{SS_{xx}}}. \tag{11}$$

Observe that this tends to be large when x_0 is far from $\bar{x}$.

EXAMPLE 14.3b In Problem 14-7 we found the regression line for temperature on chirp frequencies to be $y = \frac{32}{3}(32 + 12.2\,x)$. The correlation coefficient is $r = .9173$, convincing evidence of a relationship ($T \doteq 6.5$, 8 d.f.). The total sum of squares is $\text{SSTotal} = SS_{YY} = 589.6$, and multiplying this by $r^2 = .8415$ we get the regression sum of squares. The ANOVA table is as follows, in which $F = T^2$, with (1, 8) degrees of freedom:

Source	d.f.	SS	MS	F
Regression	1	496.15	496.15	42.48
Error	8	93.45	11.68	--
Total	9	589.6	--	--

Thus, about 84 percent of the total sum of squares (496.15) is attributed to regression and about 16 percent (93.45) to random error.

Using the empirical regression line with a chirp frequency of $x_0 = 19$/sec, we'd predict a temperature of $\widehat{Y}_0 = 87.9$. For the standard error of prediction we need an estimate of σ^2; this is the entry MSE in the table, or S^2 from (15) in §14.2: Substituting in (11) we get

$$\text{s.e.}(\widehat{Y}_0) = \sqrt{11.68\left\{1 + \frac{1}{10} + \frac{(19-17)^2}{30}\right\}} \doteq 3.8.$$

We might compare this with the standard error of $\bar{Y}$, the predictor we'd use if we did not exploit the regression line in our prediction; its standard error is $\text{s.e.}(\bar{Y}) = \sqrt{589.6/9} \doteq 8.1$. ∎

14.4 Testing linearity

In the analyses of the preceding sections, it was not required that the values x_i at which responses were measured be all distinct—as long as they were not all the same. When there is at least one value of x at which more than one response is measured, we can devise a test for *linearity*, that is, for the hypothesis H_0: $g(x) = \alpha + \beta x$, against the alternative that $g(x)$ is not a linear function of x.

It is convenient to modify the notation. Let Y_{ij} denote the jth of n_i observations taken at x_i, and let $\bar{Y}_i$ be their mean. Let $n = \sum n_i$, and define $\gamma_i = g(x_i)$. The likelihood function, assuming normal errors: $\epsilon \sim N(0, \sigma^2)$, is

$$L(\gamma_1, \ldots, \gamma_k; \sigma^2) = (\sigma^2)^{-n/2} \exp\left\{\frac{-1}{2\sigma^2} \sum_{i=1}^{k} \sum_{j=1}^{n_i} (Y_{ij} - \gamma_i)^2\right\}. \quad (1)$$

This is minimized (Problem 14-23) when the parameters have the values

$$\hat{\gamma}_i = \bar{Y}_i = \frac{1}{n_i} \sum_{j=1}^{n_i} Y_{ij}, \quad \hat{\sigma}^2 = \frac{1}{n} \sum_{i=1}^{k} \sum_{j=1}^{n_i} (Y_{ij} - \bar{Y}_i)^2. \quad (2)$$

Under H_0, the maximization of the likelihood is just as before (§14.2), but in the present notation, we have these formulas:

$$\hat{\alpha} = \bar{Y} = \frac{1}{n} \sum_{i=1}^{k} \sum_{j=1}^{n_i} Y_{ij} = \frac{1}{n} \sum_{i=1}^{k} n_i \bar{Y}_i, \quad \hat{\beta} = \frac{SS_{xY}}{SS_{xx}}, \quad (3)$$

where

$$\bar{x} = \frac{1}{n} \sum_{i=1}^{k} n_i x_i, \quad SS_{xx} = \sum_{i=1}^{k} n_i (x_i - \bar{x})^2, \quad SS_{xY} = \sum_{i=1}^{k} n_i (x_i - \bar{x}) \bar{Y}_i. \quad (4)$$

And with $\hat{Y}_i = \hat{\alpha} + \hat{\beta}(x_i - \bar{x})$,

$$\hat{\sigma}_0^2 = \frac{1}{n} \sum_{i=1}^{k} \sum_{j=1}^{n_i} (Y_{ij} - \hat{Y}_i)^2 = \hat{\sigma}^2 + \frac{1}{n} \sum_{i=1}^{k} n_i (\bar{Y}_i - \hat{Y}_i)^2. \quad (5)$$

(See Problem 14-23.)

We are now in position to find the likelihood ratio for testing the hypothesis of linearity, or more precisely, for testing

$$H_0: \gamma_i = \alpha + \beta(x_i - \bar{x})$$

against the alternative

H_A: The points (x_i, Y_i) are not collinear.

Thus,

$$\Lambda^{-2/n} = \frac{\hat{\sigma}_0^2}{\hat{\sigma}^2} = 1 + \frac{\sum n_i (\bar{Y}_i - \hat{Y}_i)^2}{\sum \sum (Y_{ij} - \bar{Y}_i)^2}. \quad (6)$$

The distribution of $n\hat{\sigma}_0^2/\sigma^2$ is chi$^2(n-2)$, and the terms on the right of (5) are quadratic expressions with ranks r_1 and r_2, where $r_1 \le n - k$ and $r_2 \le k - 2$. (See Problem 14-23.) Thus,

$$r_1 + r_2 \leq n - k + k - 2 \leq r_1 + r_2,$$

and Cochran's theorem tells us that the numerator and denominator of the last fraction in (6) are independent and proportional to chi-square variables with $k - 2$ and $n - k$ degrees of freedom, respectively. So the test ratio

$$F = \frac{\sum n_i(\bar{Y}_i - \hat{Y}_i)^2/(k-2)}{\sum\sum (Y_{ij} - \bar{Y}_i)^2/(n-k)}$$

has the F-distribution with $(k-2, n-k)$ degrees of freedom. Large values of F are taken as evidence against linearity.

The decomposition (5) is another instance of an analysis of variance. An ANOVA table would give sums of squares: *nonlinearity* with $k - 2$ d.f., *error* with $n - k$ d.f., and *total* (variation of the responses about the empirical linear regression function) with $n - 2$ d.f.

PROBLEMS

14-16. Obtain the ANOVA table entries in Example 14.3a from the statistics given in the example.

14-17. For the situation of Example 14.3a, obtain 90 percent prediction limits for the displacement when the blast frequency is 7 cps.

14-18. Show that for testing $\beta = 0$, the statistic T given by (5) in §12.3 for testing the hypothesis $\rho = 0$ with bivariate data (x_i, y_i) is identical to the statistic T defined by (6) above for testing $\beta = 0$.

14-19. Use (9) and (11) in §14.3 to construct
(a) confidence limits for the mean response at $x = x_0$, $\alpha + \beta(x_0 - \bar{x})$.
(b) prediction limits for the predicted response at $x = x_0$.

14-20. For the situation of Example 14.2a, consider the point $x_0 = -1$.
(a) Find 95 percent confidence limits for *mean* decrease in tension at x_0.
(b) Find 95 percent *prediction* limits for the decrease in tension at x_0.

14-21. Construct a t-statistic for testing $H_0: \alpha = \alpha_0$ vs. $H_1: \alpha \neq \alpha_0$ in the context of §14.3.

14-22. In Problem 14-14 you studied the regression model $Y_i = \beta x_i + \epsilon$, where $\epsilon \sim N(0, \sigma^2)$, given data (x_i, Y_i) with not all $x_i = 0$. Construct the likelihood ratio tests for $\beta = \beta_0$ against $\beta \neq \beta_0$.

14-23. Referring to the development in §14.4,
(a) obtain the maximum likelihood estimates (2).
(b) establish the decomposition (5) and show, as claimed in the paragraph following (5), that $r_1 \leq n - k$ and $r_2 \leq k - 2$.

14-24. Carry out the test for linearity of the regression function for these artificial data. Give an ANOVA table, the value of F, and a conclusion.

x_i	Y_{ij}
1	3, 4, 5
2	1, 3
3	3, 4, 5

14-25. Referring to (2) in §14.4, show directly that $n\hat{\sigma}^2/\sigma^2$ is chi$^2(n-k)$, from what you know about each of the k terms in the sum on i.

14-26. Suppose, when there are several observations at each x_i, someone suggests fitting a line to the pairs $(x_i, \bar{Y}_i)$. Would this scheme be equivalent to fitting to the points (x_i, Y_{ij})? If not, why not? Would it ever work?

14.5 Multiple regression

We now turn to a more general regression function in which the mean response depends on p (numerical) predictor variables $(x_1, ..., x_p)$:

$$E(Y \mid \beta) = \beta_1 x_1 + \beta_2 x_2 + \cdots + \beta_p x_p. \qquad (1)$$

(This expression is linear in the p parameters β_i.) The predictor variables x_i need not be functionally independent. Indeed, this model includes cases in which the x_i's are powers of a single variable x, such as the cubic function

$$\beta_0 + \beta_1 x + \beta_2 x^2 + \beta_3 x^3$$

(where $x_1 = 1$, $x_2 = x$, $x_3 = x^2$, $x_4 = x^3$), or transcendental functions such as $\log x$, $\sin x$, etc.

A sample consists of observed responses Y_i at $(x_{1i}, ..., x_{pi})$. The model for $\mathbf{Y} = (Y_1, ..., Y_n)$, in terms of the matrix notation introduced at the beginning of the chapter, is

$$\mathbf{Y} = \mathbf{X}\boldsymbol{\beta} + \boldsymbol{\epsilon}. \qquad (2)$$

(The dimensions of these matrices are $n \times 1$, $n \times p$, $p \times 1$, and $n \times 1$, respectively.) The distribution of $\mathbf{Y}$ is defined by the distribution assumed for the random component, the "errors" ϵ_{ij}.

We'll find least squares estimates of the β's by completing the square in the expression for the sum of squared residuals:

$$Q(\beta) = (\mathbf{Y} - \mathbf{X}\beta)'(\mathbf{Y} - \mathbf{X}\beta) = \sum (Y_i - [\beta_1 x_{1i} + \cdots + \beta_p x_{pi}])^2. \quad (3)$$

Expanding the matrix product and substituting $\mathbf{S} = \mathbf{X}'\mathbf{X} = \mathbf{S}'$ we get

$$Q(\beta) = \mathbf{Y}'\mathbf{Y} + \beta'\mathbf{S}\beta - 2\beta'\mathbf{X}'\mathbf{Y},$$

and to "complete the square" we add and subtract an appropriate term:

$$Q(\beta) = \mathbf{Y}'\mathbf{I}\mathbf{Y} + [\beta'\mathbf{S}\beta - 2\beta'\mathbf{S}\mathbf{S}^{-1}\mathbf{X}'\mathbf{Y} + \mathbf{Y}'\mathbf{X}\mathbf{S}^{-1}\mathbf{X}'\mathbf{Y}] - \mathbf{Y}'\mathbf{X}\mathbf{S}^{-1}\mathbf{X}'\mathbf{Y},$$

or

$$Q(\beta) = \mathbf{Y}'\mathbf{A}\mathbf{Y} + (\beta - \mathbf{S}^{-1}\mathbf{X}'\mathbf{Y})'\mathbf{S}(\beta - \mathbf{S}^{-1}\mathbf{X}'\mathbf{Y}), \quad (4)$$

where $\mathbf{A} = \mathbf{I} - \mathbf{X}\mathbf{S}^{-1}\mathbf{X}'$. This assumes that $\mathbf{S}$ has an inverse—that it is of full rank, and such is the usual case in regression problems; we assume that it is the case in what follows. It is clear from (4) that Q is minimized when the parameter vector has the value

$$\tilde{\beta} = \mathbf{S}^{-1}\mathbf{X}'\mathbf{Y}. \quad (5)$$

This is the least squares estimator of the vector β, and with this substituted into Q we obtain the minimum sum of squared residuals:

$$Q(\tilde{\beta}) = \inf_\beta (\mathbf{Y} - \mathbf{X}\beta)'(\mathbf{Y} - \mathbf{X}\beta) = (\mathbf{Y} - \mathbf{X}\tilde{\beta})'(\mathbf{Y} - \mathbf{X}\tilde{\beta}) = \mathbf{Y}'\mathbf{A}\mathbf{Y}.$$

Thus,

$$Q(\beta) = Q(\tilde{\beta}) + (\tilde{\beta} - \beta)\mathbf{S}(\tilde{\beta} - \beta). \quad (6)$$

Since $\tilde{\beta}$ is a linear transformation of $\mathbf{Y}$, we can easily get its first two moments from those of $\mathbf{Y}$. The mean vector is

$$E(\tilde{\beta}) = \mathbf{S}^{-1}\mathbf{X}'E(\mathbf{Y}) = \mathbf{S}^{-1}\mathbf{X}'\mathbf{X}\beta = \beta. \quad (7)$$

With the assumption of i.i.d. errors ϵ_i with variance σ^2, $\mathbf{M}_\mathbf{Y} = \sigma^2 \mathbf{I}$, and

$$\mathbf{M}_{\tilde{\beta}} = \mathbf{S}^{-1}\mathbf{X}'(\sigma^2\mathbf{I})\mathbf{X}\mathbf{S}^{-1} = \sigma^2 \mathbf{S}^{-1}, \quad (8)$$

from (5) of §12.5. The first two moments of the parameter estimators do not alone define the distributions of those estimators without further specification of the population distribution. Before adding the assumption of population normality, we digress to obtain the following result, which does not depend on that assumption.

THEOREM 6 (Gauss-Markov): *The least squares estimator $\tilde{\beta}$ has minimum variance in the class of unbiased, linear estimators, in the sense that the variances of its components are simultaneously smallest.*

To prove this, consider an arbitrary, unbiased linear estimator $\beta^* = C\mathbf{Y}$, and define the matrix $\mathbf{B}$ by

$$\beta^* - \tilde{\beta} = (\mathbf{C} - \mathbf{S}^{-1}\mathbf{X}')\mathbf{Y} = \mathbf{BY}.$$

We require that β^* be unbiased; and since $\tilde{\beta}$ is unbiased, it follows that $\mathbf{BX} = 0$. Now, $\mathbf{C} = \mathbf{B} + \mathbf{S}^{-1}\mathbf{X}'$, so the second moment matrix of β^* is

$$\mathbf{M}_{\beta^*} = \sigma^2(\mathbf{B} + \mathbf{S}^{-1}\mathbf{X}')(\mathbf{B} + \mathbf{S}^{-1}\mathbf{X}')' = \sigma^2(\mathbf{BB}' + \mathbf{S}^{-1}).$$

The variances of the components of β^* are the diagonal elements of $\mathbf{BB}' + \mathbf{S}^{-1}$. And since $\mathbf{S}^{-1}$ is a constant matrix, the variances are minimized when the diagonal elements of $\mathbf{BB}'$:

$$(\mathbf{BB}')_{ii} = b_{i1}^2 + b_{i2}^2 + \cdots + b_{in}^2, \quad i = 1, ..., n,$$

are minimized. This is accomplished by choosing $\mathbf{B} = 0$ (which satisfies the condition $\mathbf{BX} = 0$). So $\beta^* = \tilde{\beta}$, as was to be shown.

Returning now to the question of the distribution of the parameter estimators, we assume that the errors are i.i.d. and normal with variances that do not depend on $\mathbf{x}$:

$$\epsilon_{ij} \sim \mathcal{N}(0, \sigma^2). \tag{9}$$

This in turn means that the observed responses Y_{ij} are independent and normally distributed. We saw in §14.1 that $\hat{\beta}$, the m.l.e. of β, is just $\tilde{\beta}$, the least squares estimator (5). The m.l.e. of σ^2, from (4) of §14.1, is

$$\hat{\sigma}^2 = \tfrac{1}{n}(\mathbf{Y} - \mathbf{X}\hat{\beta})'(\mathbf{Y} - \mathbf{X}\hat{\beta}) = \tfrac{1}{n}Q(\hat{\beta}). \tag{10}$$

Since $\hat{\beta}$ is a linear transformation of $\mathbf{Y}$, it is *multivariate normal*, with mean β and covariance matrix [from (8)] $\sigma^2 \mathbf{S}^{-1}$.

As for the distribution of $\hat{\sigma}^2$, we observe first that the quadratic form (10) has rank at most $n - p$, since the n components of $\mathbf{Y} - \mathbf{X}\hat{\beta}$ satisfy p linear relations:

$$\mathbf{X}'(\mathbf{Y} - \mathbf{X}\hat{\beta}) = \mathbf{X}'\mathbf{Y} - \mathbf{X}'\mathbf{X}\mathbf{S}^{-1}\mathbf{X}'\mathbf{Y} = 0.$$

We next return to the decomposition (6), another instance of "analysis of variance":

$$Q(\boldsymbol{\beta}) = (\mathbf{Y} - \mathbf{X}\boldsymbol{\beta})'(\mathbf{Y} - \mathbf{X}\boldsymbol{\beta}) = n\widehat{\sigma}^2 + (\widehat{\boldsymbol{\beta}} - \boldsymbol{\beta})'\mathbf{S}(\widehat{\boldsymbol{\beta}} - \boldsymbol{\beta}). \quad (11)$$

[See Problem 14-29(b).] When divided by σ^2, the left-hand side of (11) is chi$^2(n)$, and its rank n does not exceed the sum of the ranks of the quadratic expressions on the right. The rank of the second term is at most p, so the ranks on the right add up to the rank on the left. This (by Cochran's theorem) suffices to show that the terms on the right, divided by σ^2, have independent chi-square distributions with d.f.'s equal to the ranks. In particular, we see that $n\widehat{\sigma}^2/\sigma^2 \sim$ chi$^2(n-p)$, which completes the proof of Theorem 5 (page 498).

For a test of the hypothesis H_a with a β's, obtained from H_b with b β's by setting $b-a$ of the parameters β in H_b equal to zero, the likelihood ratio (for the alternative that not all of those β's are 0) is

$$\Lambda = \frac{\sup L(\beta_1, \ldots, \beta_a; \sigma^2)}{\sup L(\beta_1, \ldots, \beta_b; \sigma^2)} = \left(\frac{\widehat{\sigma}_a^2}{\widehat{\sigma}_b^2}\right)^{-n/2}, \quad (12)$$

where

$$n\widehat{\sigma}_a^2 = \sum [Y_i - \widehat{g}_a(\boldsymbol{x}_i)]^2 = Q_a$$

$$n\widehat{\sigma}_b^2 = \sum [Y_i - \widehat{g}_b(\boldsymbol{x}_i)]^2 = Q_b,$$

and $\widehat{g}$ indicates the corresponding regression function with the β's replaced by $\widehat{\beta}$'s, their least squares estimates. From Theorem 4 we know that under H_b, $n\widehat{\sigma}_b^2/\sigma^2 \sim$ chi$^2(n-b)$; and since this does not depend on the values of any β's, the same distribution is correct under H_a as well. From the same theorem, we know that under H_a, $n\widehat{\sigma}_a^2/\sigma^2 \sim$ chi$^2(n-a)$. Then, with

$$Q_a = Q_b + (Q_a - Q_b), \quad (13)$$

the variable on the left is a quadratic form in standard normal variables whose rank is $n-a$. The rank of the quadratic form Q_b is $n-b$, so the rank of $Q_a - Q_b$ is at least $b-a$. We'll forego the study of this difference form and simply state:

ThEOREM 7 *When Model a is the special case of Model b in which $b-a$ β's are set equal to 0,*

(i) *under Model a, $(Q_a - Q_b)/\sigma^2 \sim$ chi$^2(b-a)$,*
(ii) *under Model b (and also under a), $Q_b/\sigma^2 \sim$ chi$^2(n-b)$,*
(iii) *under Model b (as well as a), Q_b and $(Q_a - Q_b)$ are independent.*

To test H_0: $g = g_a$ against H_A: $g = g_c$, using the likelihood ratio (12), we observe that Λ is small when the ratio

$$F = \frac{(Q_a - Q_b)/(b - a)}{Q_b/(n - b)} \tag{14}$$

is large. The null distribution of this test ratio is $F(b - a, n - b)$.

We said at the beginning of this section that the algebra involved does not rule out a case in which some predictors are functions of others. Clearly, if a predictor is a *linear* function of other predictor variables, there is no point to including that predictor. And indeed, the computer will rebel, because the β's could not be identified—there is not a unique solution. It can happen, moreover, that a predictor may be nearly linearly related to others (perhaps without the investigator's knowledge), and if the "nearly" is too close, the computer will rebel. For example, if there are two predictors, (x, z), and if $z \doteq 2x + 1$, then the points (x, z) at which responses are measured will be nearly collinear, and one could not hope for an accurate determination of the tilt of the plane representing the regression plane $y = \alpha + \beta x + \gamma z$. (A table with three nearly collinear legs is unstable indeed.)

It is evident from (5) that finding the least squares regression function involves inverting the matrix $\mathbf{S} = \mathbf{X}'\mathbf{X}$, which is of dimension $p \times p$. Doing so is very tedious by hand, but trivial for computers equipped with good statistical software. Typically, an investigator will include, as possible predictors, all potentially contributing variables for which data are available. The computer can then carry out a regression analysis for each of possible subsets of predictor variables and select the five or ten best, say, which are optimal according to various criteria, and which are thus apt to give the best predictions.

14.6 Nested models

An approach to finding a good regression model, from those involving a given set of possible predictors, is to use a step-wise procedure, either forwards or backwards—either adding predictors one at a time to the model or, from the complete set of possibilities, deleting predictors one at a time—until some sort of optimum subset is settled on. The order in which to add or delete, it is hoped, would be suggested by one's intimate knowledge of the field of application. Such procedures imply a *hierarchy* of models, in which successively simpler models are obtained from the most complicated one by successive restrictions on parameters. For instance, a hierarchy of polynomial models might be defined as follows:

$$H_1: g(x) = \beta_0,$$
$$H_2: g(x) = \beta_0 + \beta_1 x,$$
$$H_3: g(x) = \beta_0 + \beta_1 x + \beta_2 x^2,$$

and so on. Another type of possibility, with two predictor variables:

$$H_1: g(x_1, x_2) = \beta_0,$$
$$H_2: g(x_1, x_2) = \beta_0 + \beta_1 x_1,$$
$$H_3: g(x_1, x_2) = \beta_0 + \beta_1 x_1 + \beta_2 x_2,$$
$$H_4: g(x_1, x_2) = \beta_0 + \beta_1 x_1 + \beta_2 x_2 + \beta_3 x_1 x_2.$$

In H_4, the inclusion of the product $x_1 x_2$ would be intended to account for possible "interaction" between the predictors. (Although there are but two actual predictor variables, the algebra and calculation of least squares estimates treats the product $x_1 x_2$ as a third predictor.)

Suppose we assume some such hierarchy, obtained by successively setting parameters β_j in $g(\mathbf{x}) = \beta_1 x_1 + \cdots + \beta_p x_p$ equal to zero. (If the first term in g is to be a constant, define $x_1 = 1$.) A step-wise ANOVA for the hierarchy $H_1 \subset H_2 \subset \cdots \subset H_k$, where each H is obtained from the next one by setting β's equal to zero, is represented in the following table, in which m_j is the number of β's in H_j.

Source	Sum of squares	d.f.
Fitting H_2 after H_1	$Q_1 - Q_2$	$m_2 - m_1$
Fitting H_3 after H_2	$Q_2 - Q_3$	$m_3 - m_2$
$\vdots$		
Fitting H_k after H_{k-1}	$Q_{k-1} - Q_k$	$m_k - m_{k-1}$
Error	Q_k	$n - m_k$
Total (fitting H_1)	Q_1	$n - m_1$

In many applications there is a constant term ($x_1 = 1$), and the notation is modified so that the regression function is

$$E(Y \mid \boldsymbol{\beta}) = \beta_0 + \beta_1 x_1 + \beta_2 x_2 + \cdots + \beta_k x_k, \tag{1}$$

where then the number of parameters is $p = k + 1$, and k is the number of predictors. Most statistical computer software packages include a step-wise ANOVA corresponding to the adding of a predictor variable one by one, which (in part) will look something like the table above. The "H_1" in that table would be the model $Y = \beta_0$, and the total sum of squares (SSTotal) is SS$_{yy}$ or $Q_1 = \sum (Y_i - \bar{Y})^2$, with d.f. = 1. The sum of squares for fitting H_a after H_{a-1} has 1 degree of freedom, and the error sum of squares (SSE) has $n - p$ degrees of freedom. The computer output will usually give what is called a global or *overall F*; this is the F-ratio (14) of §14.5 for testing the hypothesis H_0: $\beta_1 = \cdots = \beta_k = 0$ in the context of the full model (1). The ANOVA for H_0 against the alternative that not all β's are 0 is briefly as follows:

Source	d.f.	Sum of squares
Regression	$p - 1$	$Q_1 - Q_k$ (SSReg)
Error	$n - p$	Q_k (SSE)
Total	$n - 1$	Q_1 (SSTotal)

The test ratio is the ratio of mean squares:

$$F = \frac{\text{SSReg}/(p-1)}{\text{SSE}/(n-p)},$$

whose null distribution is $F(p-1, n-p)$. The decomposition (11) of §14.5 is

$$\text{SSTotal} = \text{SSReg} + \text{SSE}, \tag{2}$$

interpreted as an amount of variability about the mean attributed to the predictor variables (or to "regression") plus an amount attributed to error. The fraction attributed to regression is called the *coefficient of determination*:

$$R^2 = \frac{\text{SSReg}}{\text{SSTotal}}. \tag{3}$$

(This can be shown to be the square of R, the *multiple correlation coefficient*, which is the maximum correlation between Y and linear functions of the x's.) Clearly,

$$F = \frac{R^2}{1 - R^2} \cdot \frac{n - p}{p - 1}. \tag{4}$$

Computer output in a multiple regression problem usually includes a value of a t-statistic for each predictor—for a test of the hypothesis that the corresponding β is 0, assuming that all other predictors are kept in the model. [This t is the square root of a corresponding $F(1, n - p)$.] In an analysis of variance, the output may give individual and/or cumulative sums of squares, for adding predictors one at a time in a specified order, and an R^2 for each model with predictors $x_1, ..., x_j$, based on the cumulative regression sum of squares up to that point.

14.6 Nested models

Adding a predictor will usually increase R^2: Since SSTotal remains the same, and some of SSE is taken out to add to SSReg for the new predictor, R^2 cannot decrease. But the cost of adding the predictor is an increase in model complexity and the loss of a degree of freedom for SSE, and predictions may be less reliable. Rather than decide the worth of an additional predictor on the basis of R^2, statisticians will use other quantities to assess the increase of the reliability of predictions. One such is the "Adjusted R^2," defined as

$$\text{Adjusted } R^2 = 1 - \frac{(n-1)\text{SSE}}{(n-p-1)\text{SSTotal}}, \quad (5)$$

where p is the number of β's in the model.

EXAMPLE 14.6a Data giving Y, the cumulative heat of hardening for cement after 180 days, and corresponding values of four predictor variables, percentages of four chemical compounds measured in batches of cement, were analyzed using a computer software package. The results were printed out as follows:

Predictor	Coefficient	t	P
Constant	62.405	.89	.40
x_1	1.5511	2.08	.07
x_2	.05101	.70	.50
x_3	.10191	.14	.90
x_4	−.14406	−.20	.84

Overall F: 111.5 ($P = .000$)

MSE: 5.983 (d.f. = 8)

R^2: .9824, Adjusted R^2: .9736

The empirical regression function is then

$$Y = 62.4 + 1.55x_1 + .510x_2 + .109x_3 - .144x_4.$$

In view of the t-values, one would probably keep x_1, but could possibly do as well without one of the other predictors. (But these t-values should not be used to conclude that one could do as well without *any* of x_2, x_3, x_4, since the test for β_2, for instance, assumes that x_3 and x_4 are still in the model.)

The ANOVA table was also printed out:

Source	d.f.	SS	MS
Regression	4	2667.9	667.0
Error	8	47.864	5.983
Total	12	2715.8	

Observe that the value of R^2 given above is the ratio SSReg/SSTotal, that $1 - R^2$ is SSE/SSTotal, and that the overall F is MSReg/MSE.

The complete ANOVA table, also given as part of the computer output, may look something like this, in part:

Step-wise Analysis of variance of Heat:

Source	Individual SS	Cumulative SS	Cum d.f.	Cum MS	Adj. R^2
x_1	1450.1	1450.1	1	1450.1	.4916
x_2	1207.8	2657.9	2	1328.9	.9744
x_3	9.7939	2667.7	3	889.22	.9764
x_4	.24698	2667.9	4	666.97	.9736
Residual	47.864	2715.8	12	226.31	

Apparently x_3 and x_4 contribute little, once x_1 and x_2 are in the model. But the step-wise ANOVA is for adding predictors *in a particular order*; some other combination of predictors may turn out to be better than the pair (x_1, x_2). ∎

PROBLEMS

14-27. Data giving Y, water absorption of wheat flour, and corresponding values of these two predictor variables: x_1 = flour protein percentage, and x_2 = starch damage, were analyzed using one computer software package with results printed out as follows:

Predictor	Coefficient	t	P
Constant	11.36	2.93	.0074
x_1	2.2265	5.94	.0000
$x_1 x_2$	-.04303	-2.43	.0231
x_2	.80962	4.13	.0004

Overall F: 272.76 $(P = .000)$

MSE: 1.0011 (d.f. $= 24$)

R^2: .9715, Adjusted R^2: .9679

(a) Predict the water absorption when $x_1 = 11$ and $x_2 = 23$.

(b) From what is given above, find the number of "cases" (data triples), the number of degrees of freedom for regression, and the regression sum of squares.

(c) Continuing, construct a brief ANOVA table and a step-wise ANOVA table, like those in Example 14.6a, when variables are entered into the model in the order x_2, x_1, $x_1 x_2$.

14-28. Find the normal equations for fitting the plane $y = \alpha + \beta x + \gamma z$ to four data points (x, z, Y): $(3, -1, 0)$, $(2, 1, 2)$, $(0, 5, 1)$, $(1, 3, 4)$. (Do you see any problem with this?)

14-29. Fill in some details in §14.5 and 14.6:
  (a) Explain why $\beta'\mathbf{X}'\mathbf{Y}$ is equal to its transpose, $\mathbf{Y}'\mathbf{X}\beta$.
  (b) Verify the "completing of the square," to obtain (4) in §14.5.
  (c) Show that $\mathbf{A}$ in (c) is symmetric and idempotent: $\mathbf{A}^2 = \mathbf{A}$.
  (d) Verify the relation between R^2 and F given by (4) in §14.6.

14-30. Use the method of calculus for minimizing a function of several variables, to obtain simultaneous equations which are the necessary condition for the sum of squared residuals (3) of §14.5 to have a minimum. Express them in matrix terms, showing then that they are satisfied (when $\mathbf{S}$ is nonsingular) by the expression for $\tilde{\boldsymbol{\beta}}$ given as (5) in §14.5.

14-31. Prove this version of the Gauss-Markov theorem for the simple linear regression $\alpha + \beta(x - \bar{x})$: The least squares estimator $\tilde{\alpha} + \tilde{\beta}(x - \bar{x})$ of the regression function is the best linear, unbiased estimator in the sense that it has minimum variance.

14-32. The data on which Example 14.3a is based are as follows:

Freq.	Displacement
2.5	.390, .250, .200
3.0	.360
3.5	.250
9.3	.077
9.9	.140
11.0	.180, .093, .080
16.0	.052
25.0	.051, .150

Plot the data, and investigate the possibility that a quadratic regression with displacement vs. frequency might produce a better fit than the straight line regression with the transformed data (i.e., with $x = \log \text{frequency}$).

14.7 A single classification model

In regression problems, populations of responses are indexed by one or more real parameters—the predictor variables, and the interest lies in the functional dependence of the population means on those parameters. We turn now to some problems in which the indexing variables—the "independent" variables—are categorical. We consider first a single variable of classification whose "values," which we refer to as *levels*, define k populations of responses. These levels typically correspond to different "treatments," and the responses are the experimental results of applying the treatments. For instance, the yield of wheat depends on the type of seed; student achievement in a course depends on the instructor; the strength of a spot weld depends on the machine used; and so on.

Variability in responses is the result of factors that affect them but are either not identifiable or simply not taken into account. A deviation of the observed response from the "true" response is thought of as experimental error. In a statistical analysis such errors are modeled as random variables. The statistical problem is that of determining whether the treatments are actually different in their effects on responses (on average), in the presence of the randomness or variability that would mask such differences.

In the simplest case, the contributions of experimental error to the responses are random variables with mean zero and *constant* variance σ^2. The notation will be as follows:

Population variable	Sample size	Sample observations	Sample means	Population means	variances
Y_1	n_1	$Y_{11}, ..., Y_{1n_1}$	$\bar{Y}_1$	μ_1	σ^2
.	.	.	.	.	.
.	.	.	.	.	.
.	.	.	.	.	.
Y_k	n_k	$Y_{k1}, ..., Y_{kn_k}$	$\bar{Y}_k$	μ_k	σ^2

The *grand mean* is the weighted average of the sample means:

$$\bar{Y} = \frac{1}{n} \sum_{i=1}^{k} \sum_{j=1}^{n_i} Y_{ij} = \frac{1}{n} \sum_{i=1}^{k} n_i \bar{Y}_i, \tag{1}$$

where $n = \sum n_i$, the size of the combined sample. A corresponding population parameter [in analogy with (6) of §14.0] is an overall population mean:

$$\mu = \frac{1}{n} \sum n_i \mu_i. \tag{2}$$

To identify the *effect* of treatment i, we define the parameter

$$\tau_i = \mu_i - \mu.$$

(This is really a differential effect, indicating a difference between the overall mean response and the mean response under treatment i.) In terms of these parameters, the jth response under treatment i is

$$Y_{ij} = \mu + \tau_i + \epsilon_{ij}, \tag{3}$$

where ϵ_{ij} is again a "random error," including the effects of the various unaccounted for factors that determine a response.

The tests to be studied are developed under the assumption that the experimental error is normal with mean 0 and constant variance:

$$\epsilon_{ij} \sim N(0, \sigma^2).$$

The likelihood function is then

$$L(\mu_1, \ldots, \mu_k; \sigma^2) = \frac{1}{(\sigma^2)^{n/2}} \exp\left\{-\frac{1}{2\sigma^2} \sum_{i=1}^{k} \sum_{j=1}^{n_i} (Y_{ij} - \mu_i)^2\right\}, \quad (4)$$

whose maximum is achieved at

$$\hat{\mu}_i = \bar{Y}_i, \quad \hat{\sigma}^2 = \frac{1}{n} \sum_{i=1}^{k} \sum_{j=1}^{n_i} (Y_{ij} - \bar{Y}_i)^2. \quad (5)$$

Since $Y_{ij} - \bar{Y}_i$ is then an estimate of $Y_{ij} - \mu_i = \epsilon_{ij}$, the "error" component of Y_{ij}, we call the sum of squared deviations in (5) the **error sum of squares**:

$$\text{SSE} = n\hat{\sigma}^2 = \sum_{i=1}^{k} \sum_{j=1}^{n_i} (Y_{ij} - \bar{Y}_i)^2. \quad (6)$$

The maximum value of the likelihood (4) is

$$L(\hat{\mu}_1, \ldots, \hat{\mu}_k; \hat{\sigma}^2) = (\hat{\sigma}^2)^{-n/2} e^{-n/2}.$$

(This is a special case of Theorem 1, page 492.) We note that the maximum likelihood estimate of μ is

$$\hat{\mu} = \frac{1}{n} \sum n_i \hat{\mu}_i = \frac{1}{n} \sum n_i \bar{Y}_i = \bar{Y},$$

and thus the m.l.e. of the treatment effect τ_i is

$$\hat{\tau}_i = \hat{\mu}_i - \hat{\mu} = \bar{Y}_i - \bar{Y}. \quad (7)$$

The null hypothesis to be tested is that there is *no* treatment effect—that the response populations are identical on average:

$$H_0: \mu_1 = \mu_2 = \cdots = \mu_k = \mu. \quad (8)$$

In terms of the parameters τ_i, this is equivalent to

$$H_0: \tau_1 = \cdots = \tau_k = 0.$$

It would be natural to base a test on the size of the estimates of the τ_i, that is, on the deviations of the sample means from the overall mean, $\bar{Y}_i - \bar{Y}$, and these will all be small when the following weighted sum of squares, the **treatment sum of squares**, is small:

$$\text{SSTr} = \sum_{i=1}^{k} n_i(\bar{Y}_i - \bar{Y})^2 = \sum_{i=1}^{k} n_i \hat{\tau}_i^2. \qquad (9)$$

We now derive a likelihood ratio test, which will be based on the size of SSTr as compared with a pooled variance estimate of the error variance.

Under H_0, the likelihood is a function of only μ and σ^2:

$$L_0(\mu, \sigma^2) = \frac{1}{(\sigma^2)^{n/2}} \exp\left\{-\frac{1}{2\sigma^2} \sum_{i=1}^{k} \sum_{j=1}^{n_i} (Y_{ij} - \mu)^2\right\},$$

whose maximum is achieved at $\hat{\mu} = \bar{Y}$ and $\hat{\sigma}_0^2 = \frac{1}{n}\text{SSTotal}$, where

$$\text{SSTotal} = \sum_{i=1}^{k} \sum_{j=1}^{n_i} (Y_{ij} - \bar{Y})^2. \qquad (10)$$

This sum of squared deviations of all the observations about their grand mean $\bar{Y}$ is referred to as the **total sum of squares**. The maximum likelihood under H_0 is

$$L_0(\hat{\mu}, \hat{\sigma}_0^2) = (\hat{\sigma}_0^2)^{-n/2} e^{-n/2},$$

and the likelihood ratio for H_0 against the alternative that the means μ_i are not all equal is

$$\Lambda = \left(\frac{\hat{\sigma}_0^2}{\hat{\sigma}^2}\right)^{-n/2}. \qquad (11)$$

An analysis of variance is obtained by adding and subtracting $\bar{Y}_i$ within the parenthesis of the expression (10) for $\hat{\sigma}_0^2$, and then squaring as a binomial:

$$\text{SSTotal} = n\hat{\sigma}_0^2 = \sum_i \sum_j [(Y_{ij} - \bar{Y}_i) + (\bar{Y}_i - \bar{Y})]^2$$

$$= n\hat{\sigma}^2 + \sum_i \sum_j (\bar{Y}_i - \bar{Y})^2 + 2\sum_i (\bar{Y}_i - \bar{Y})\sum_j (Y_{ij} - \bar{Y}_i)$$

$$= n\hat{\sigma}^2 + \sum_i n_i(\bar{Y}_i - \bar{Y})^2 + 0 = n\hat{\sigma}^2 + \sum_i \hat{\tau}_i^2 = \text{SSE} + \text{SSTr}.$$

Thus, the total sum of squares is decomposed into a sum of squares attributed to random error and a sum of squares attributed to treatment differences. The former is based on "within samples" variability and the latter, on "between samples" variability.

The quantities squared in SSTr and SSE are linear combinations of normal variables. In SSTr there are at most $k-1$ linearly independent such combinations because $\sum n_i(\bar{Y}_i - \bar{Y}) = 0$, and in SSE there are at most $n-k$ linearly independent combinations because of the k relations $\sum(Y_{ij} - \bar{Y}_i) = 0$. We know that under H_0, $n\hat{\sigma}_0^2/\sigma^2 \sim \text{chi}^2(n-1)$, and from Cochran's theorem we have the following useful result:

THEOREM 8 Under the hypothesis of no treatment effect ($\tau_i = 0$),
$$SSE/\sigma^2 \sim \text{chi}^2(n-k) \text{ and } SSTr/\sigma^2 \sim \text{chi}^2(k-1)$$
Moreover, these variables are independent.

[We could have deduced the distribution of SSE/σ^2 without Cochran's theorem—see Problem 14-38.]

Under H_0, the averages obtained by dividing by SS's by corresponding degrees of freedom:
$$\text{MSE} = \frac{\text{SSE}}{n-k} \text{ and } \text{MSTr} = \frac{\text{SStr}}{k-1}$$
are unbiased estimates of σ^2. Indeed, the MSE is simply a generalization to the case of k populations the notion of pooled variance which was introduced in defining the two-sample t-test. [We might have denoted it by S_p^2: see (10) in §9.7.] Although the treatment mean square (MSTr) is a natural estimator for σ^2 when H_0 holds, it tends to be larger when H_0 is false (Problem 14-37). Thus, intuition suggests that the ratio of MSTr to MSE would be a good test statistic.

From the point of view of likelihood ratios, and since
$$\Lambda^{-2/n} = 1 + \frac{\text{SSTr}}{\text{SSE}},$$
a critical region of the form $\Lambda < K'$ is equivalent to $F > K$, where
$$F = \frac{\text{MSTr}}{\text{MSE}} = \frac{\sum n_i(\bar{Y}_i - \bar{Y})^2/(k-1)}{\sum\sum(Y_{ij} - \bar{Y}_i)^2/(n-k)}. \tag{12}$$
Large values of F would be considered evidence against H_0. The null distribution of the test statistic F is $F(k-1, n-k)$, so critical values and P-values are found with the aid of the F-table. When $k=2$ this F-test is essentially a two-sided t-test, where t has $n-1$ d.f. and $t^2 = F$.

EXAMPLE 14.7a The following data are yields (minus a constant we subtracted for convenience of calculation) of a certain process as determined by each of four analysts. The question is, do the analysts determinations differ on average?

Analyst	Yield	n_i	$\bar{Y}_i$	$\sum(Y_{ij}-\bar{Y}_i)^2$	$n_i(\bar{Y}_i-\bar{Y})^2$
1	8, 5, −1, 6, 5, 3	6	4.33	47.33	2.67
2	7, 12, 5, 3, 10	5	7.4	53.2	28.8
3	4, −2, 1	3	1	18	48
4	1, 6, 10, 7	4	6	42	4
Totals		90	18	160.53	83.47

The decomposition of variability about the overall mean $\bar{Y}$ is usually shown in an ANOVA table:

Source	d.f.	SS	MS	F
Analysts	3	83.47	27.82	2.43
Error	14	160.53	11.47	---
Total	17	244	---	

The P-value is about .11, not much evidence of a difference. ∎

The hypothesis of no treatment effect may not be the only hypothesis of interest. Thus, if that hypothesis is rejected (or even if it is not tested), one might want to make some other comparisons involving more than one linear combination of population means. For instance, is the average of the means of populations 1 and 2 different from the average of the means of populations 3, 4, and 5? A linear combination of means such as $\frac{1}{2}(\mu_1+\mu_2)-\frac{1}{3}(\mu_3+\mu_4+\mu_5)$, in which the coefficients sum to 0, is a *contrast*. We'll present here the theory behind a method due to Scheffé for a simultaneous test of more than one such contrast. There are various other methods, to be found in texts in applied statistics, but there is no general agreement as to which is best. The presentation will be in terms of confidence interval construction. (Recall that confidence intervals define tests, and conversely.)

The expression for SSTotal as a sum of SSTr and SSE is part of a more general decomposition:

$$\sum\sum(Y_{ij}-\mu_i)^2$$
$$=\sum\sum(Y_{ij}-\bar{Y}_i)^2+\sum n_i(\bar{Y}_i-\bar{Y}-\tau_i)^2+n(\bar{Y}-\mu)^2.$$

The terms on the right, divided by σ^2, are seen to be independent chi-

14.7 A single classification model

square variables with $n-k$, $k-1$, and 1 degree of freedom, respectively, by an application of Cochran's theorem. Thus, the distribution of the variable

$$\frac{\sum n_i(\bar{Y}_i - \bar{Y} - \tau_i)^2}{(k-1)S_p^2}$$

where S_p^2 is the pooled variance [MSE], is $F(k-1, n-k)$. For a given confidence level γ, let f_γ denote the 100γ-percentile of this F-distribution:

$$P\left\{\sum n_i(\bar{Y}_i - \bar{Y} - \tau_i)^2 \le (k-1)f_\gamma S^2\right\} = \gamma. \tag{13}$$

For any k-vector $\mathbf{a} = (a_1, \ldots, a_k)$, it follows from Schwarz' inequality that

$$\left\{\sum a_i(\bar{Y}_i - \bar{Y} - \tau_i)\right\}^2 \le \sum \frac{a_i^2}{n_i} \sum n_i(\bar{Y}_i - \bar{Y} - \tau_i)^2. \tag{14}$$

Since equality holds for some $\mathbf{a}$, the right-hand side is equal to the least upper bound (over $\mathbf{a}$) of the left-hand side. Substituting from (14) [with the "sup" appended to the l.h.s.] into (13), and taking square roots of both sides of the inequality, we obtain

$$P\left\{\sup_{\mathbf{a}} \left|\sum a_i(\bar{Y}_i - \bar{Y} - \tau_i)\right| \le S\sqrt{b}\right\} = \gamma, \tag{15}$$

where

$$b = (k-1)f_\gamma \sum \frac{a_i^2}{n_i}.$$

Let A^* denote the set of vectors $\mathbf{a}$ such that $\sum a_i = 0$, so that the linear combination $\sum a_i \mu_i$ is a contrast. For any such $\mathbf{a}$,

$$\sum a_i(\bar{Y}_i - \bar{Y} - \tau_i) = \sum a_i \bar{Y}_i - \sum a_i \mu_i.$$

If, instead of over all $\mathbf{a}$, the supremum in (15) is taken over a finite set $A \subset A^*$, it is no larger, and the probability of the inequality is at least γ:

$$P\left\{\sup_A \left|\sum a_i \bar{Y}_i - \sum a_i \mu_i\right| \le S\sqrt{b}\right\} \ge \gamma.$$

The "sup" is now a simple maximum, and the inequality holds for the maximum of the absolute differences if and only if it holds for every one of them. Thus,

$$P\left(\sum a_i \bar{Y}_i - S\sqrt{b} \le \sum a_i \mu_i \le \sum a_i \bar{Y}_i + S\sqrt{b}, \text{ for all } \mathbf{a} \in A\right) \ge \gamma.$$

The inequalities within the parentheses define confidence intervals, one for each contrast in A, with the property that they simultaneously cover the corresponding contrast with probability at least γ.

EXAMPLE 14.7b Suppose we want simultaneous confidence intervals for $\mu_4 - \mu_3$, $\mu_2 - \mu_1$, and $\frac{1}{3}(\mu_1 + \mu_2 + \mu_4) - \mu_3$ at the 95 percent level. From $F(3, 14)$ we find $f_{.95} = 3.34$. For the first contrast, $b = 5.845$, for the second, 3.674, and for the last, 4.026. And then, with $S = \sqrt{11.74}$, we obtain these confidence limits (again in the same order): 5 ± 8.19, 5.93 ± 6.49, and 4.91 ± 6.796. ∎

Some situations that lend themselves to a single-classification analysis are those in which n subjects (people, animals, plots of ground, or other pieces of experimental materials) are to be used in testing k treatments. To minimize the introduction of any systematic effects it is recommended that the assignment of subject to treatments be done at random. This is called a **completely randomized design**. The randomization is considered necessary for any meaningful analysis. It also permits the construction of a permutation test (like that which we gave for the two-sample problem), a test that does not require the assumption of normal errors. Such tests are very laborious to carry out, but happily they are fairly well approximated by the normal model F-tests.[3]

14.8 A random effects model

In the model for the single classification problem, the component of response τ_i associated with treatment i was considered to be a constant—a *fixed effect*. If the treatment is, say, a machine operator, conclusions drawn from an experiment would apply only to the particular operators used in the experiment. But it may be desirable to be able to make a more sweeping inference—about the population of possible operators. In a *random effects* model one thinks of operators as drawn at random from a population of operators; the responses are then structured like this:

$$Y_{ij} = \mu + Z_i + \epsilon_{ij},$$

where the *operator effect* Z_i is now a random variable.

Another application of such a model might be in the experimental determination of a property of a product that comes in batches or sheets, with randomness from batch to batch, represented by Z_i, and random

---

[3]H. Scheffé, *The Analysis of Variance*, New York: John Wiley & Sons (1959), Chapter 9.

14.8 A random effects model

ness in the experimental determination of the property of interest for a given batch—caused either by variations from point to point in the batch or by variations in the measurement process.

In a simple random effects model, we assume that the operator or batch component Z_i is $N(0, \omega^2)$; that $\epsilon_{ij} \sim N(0, \sigma^2)$; and that all of the Z_i's and ϵ_{ij}'s are independent. Then, for instance,

$$\text{cov}(Y_{11}, Y_{12}) = \text{var } Z_i + \text{cov}(\epsilon_{11}, \epsilon_{12}) = \omega^2,$$

so the observations are not independent unless $\omega^2 = 0$. On the other hand, the null hypothesis would be that $\omega^2 = 0$. When this is the case, there is no operator effect, and the model is exactly the same as the model H_0 in the case of fixed effects. The null distribution of the F-ratio used in the fixed effects case [(12) of §14.7] is here the same as in that case, namely, $F(k-1, n-k)$.

The deviations in the denominator of the test ratio F are

$$Y_{ij} - \bar{Y}_i = \epsilon_{ij} - \bar{\epsilon}_i,$$

whose squares are used in defining the the sample variances. That is,

$$n_i S_i^2 = \sum (Y_{ij} - \bar{Y}_i)^2 = \sum (\epsilon_{ij} - \bar{\epsilon}_i)^2,$$

which when divided by σ^2 is $\text{chi}^2(n_i - 1)$, for each i, and these quantities are independent. But then their sum,

$$\text{SSE} = \sum \sum (Y_{ij} - \bar{Y}_i)^2 = \sum \sum (\epsilon_{ij} - \bar{\epsilon}_i)^2,$$

when divided by σ^2, is a sum of independent chi-square variables—which is chi-square with d.f. equal to $\sum (n_i - 1) = n - k$. So the denominator of F has the same distribution as before, whether or not $\omega^2 = 0$. However, the numerator of F has mean value

$$E(\text{MSTr}) = \sigma^2 + \frac{n\omega^2}{k-1}\left\{1 - \sum \left(\frac{n_i}{n}\right)^2\right\}, \tag{1}$$

which tends to overestimate σ^2 when $\omega^2 \neq 0$. So, again, large values of F are considered evidence against a treatment effect, and the same table as before is used in determining critical regions or P-values. But the interpretation is now in terms of the population of operators.

PROBLEMS

14-33. Given samples (4, 4, 4) from population A, (6, 6, 6, 6, 6) from B, and (7, 7) from population C, find the within samples estimate of error variance. What does this do to an F-test?

14-34. A tensile test measures the quality of a spot-weld of an aluminum clad material. To determine whether there is a "machine effect," five welds from each of three machines are tested, with these results:

Machine A: 3.2, 4.1, 3.5, 3.0, 3.1
Machine B: 4.9, 4.5, 4.5, 4.0, 4.2
Machine C: 3.0, 2.9, 3.7, 3.5, 4.2

(a) Test the hypothesis of no machine effect.
(b) Construct a 90 percent confidence interval for $\mu_B - \mu_C$.
(c) Construct simultaneous confidence intervals for $\mu_B - \mu_C$, $\mu_A - \mu_C$, and $\mu_B - \mu_A$ with confidence level .95,

14-35. Obtain the following formulas, useful in hand calculations:

(a) SSTotal $= \sum \sum Y_{ij}^2 - \frac{1}{n}(\sum \sum Y_{ij})^2$.
(b) SSTr $= \sum n_i \bar{Y}_i^2 - \frac{1}{n}(\sum \sum Y_{ij})^2$.

14-36. Show that the F-test in §14.7 reduces to the two-sided t-test of §9.7 when there are just two independent samples ($k = 2$).

14-37. With SSTr as defined by (9) in §14.7, show that $E(\text{SSTr})$ is larger when the τ_i are not all zero than when they are all zero.

14-38. Derive the distribution of SSE [given by (6) in §14.7] directly, that is, without appealing to Cochran's theorem.

14-39. The distribution of the test statistic F depends on the equality of the population variances. Construct a likelihood ratio test for this assumption.

14-40. Verify the expression (1) in §14.8 for the expected value of the numerator of the F-statistic under the random effects model.

14-41. Consider this procedure for m simultaneous confidence intervals at the level $1 - \alpha$: Let t^* denote the value of t in a t-distribution with $n - k$ d.f. with tail area $\alpha/(2m)$ beyond t^*, and let $E_{i,j}$ denote the event that the confidence interval with limits

$$\bar{X}_i - \bar{X}_j \pm t^* S_p \sqrt{\frac{1}{n_i} + \frac{1}{n_j}}.$$

Use the Bonferroni inequality [Problem 2-23, page 30] to show that the probability of simultaneous coverage of the corresponding mean differences is at least $1 - \alpha$.

14-42. For the data in Problem 14-34, carry out part (c) using the method outlined in Problem 14-41.

14.9 Two-way classifications

In studying the effect of a treatment applied at several levels, one usually is aware of other factors that may be affecting the response. If these are not taken into account, as in the case of the completely randomized design (§14.7), their effects are contributions to random error, and may obscure a treatment effect. When an experiment is designed to take such factors into account, their contribution can be extracted from the random error, and the reduced random error will make the treatment effect more noticeable. The pairing of subjects for comparing a treatment and control (§9.7) is such a design, intended to eliminate the variability among subjects that is of no particular interest.

Extending the device of pairing for the case of more than two treatment categories, one may group subjects matched *blocks*, with treatment levels assigned at random within each block. This is called a **randomized block design**.

EXAMPLE 14.9a Suppose one wants to study the differences among four seed types as they affect the yield of grain. With the realization that soil composition varies, an available piece of land may be divided into three blocks, each thought to be more homogeneous than the whole. Seeds of the four types are then planted in four subplots of each block, seed types being assigned at random to subplots within a block. The data (yield per unit area) may look like this:

		Seed type:			
		A	B	C	D
	1	22	25	20	17
Block	2	29	26	23	22
	3	42	30	32	24

Although seed type may be of primary interest, a look at the effect of the blocking factor might verify that the blocking was useful. ■

If matching or blocking is done according to the categories of a single variable, that variable is the *blocking variable*, which may be of interest in itself. When it is, there may be no reason to call one variable the treatment and the other the blocking variable. Both are factors affecting the response. The purpose in analyzing a two-way layout of responses is to sort out the factor contributions to response.

When two factors are to be taken into account (whether or not one factor happens to be a blocking variable), we may think of a "treatment" as a combination of a level of one factor with a level of the other, defining a cell in the two-way array of responses. If factor A has r levels and factor B, c levels, there are rc cells. We model the response in cell ij as a random variable Y_{ij}, and assume that responses are independent with

$$Y_{ij} = \mu_{ij} + \epsilon_{ij}, \tag{1}$$

where $\epsilon_{ij} \sim \mathcal{N}(0, \sigma^2)$. The mean μ_{ij} can always be written (as was the case with the logs of expected frequencies in §10.9) in the form

$$\mu_{ij} = \mu + \theta_i + \phi_j + \xi_{ij}, \tag{2}$$

where

$$\sum \theta_i = \sum \phi_j = \sum_i \xi_{ij} = \sum_j \xi_{ij} = 0, \tag{3}$$

simply by defining

$$\mu = \frac{1}{rc} \sum\sum \mu_{ij}, \quad \mu_{i\cdot} = \frac{1}{c} \sum_j \mu_{ij}, \quad \mu_{\cdot j} = \frac{1}{r} \sum_i \mu_{ij}, \tag{4}$$

and

$$\theta_i = \mu_{i\cdot} - \mu, \quad \phi_{\cdot j} = \mu_{\cdot j} - \mu, \quad \xi_{ij} = \mu_{ij} - \mu_{i\cdot} - \mu_{\cdot j} - \mu. \tag{5}$$

(The dot in a double subscript indicates that there has been an averaging over the subscript in that position.) The term ξ_{ij} is termed an **interaction**, a contribution to response peculiar to that particular combination of factor levels.

EXAMPLE 14.9b If there were no error, responses would be equal to the mean responses. Suppose, for $r = 2$ and $s = 3$, the means are as follows:

11	10	15
5	8	11

Formulas (4) and (5) give $\mu = 10$, $\theta = (2, -2)$, $\phi = (-2, -1, 3)$, and ξ_{ij}:

1	-1	0
-1	1	0

[Verify that then (2) and (3) are satisfied.] Sometimes the effect of "interaction" is noticed in a plot like that of Figure 14-2(a), where the connecting paths for the two factor-A levels are not parallel. This is in contrast for the same type of plot of the means wih the interactions deleted, namely, of:

$$\begin{array}{|ccc|} 10 & 11 & 15 \\ 6 & 7 & 11 \end{array}$$

shown in Figure 14-2(b). Of course, when random errors ϵ_{ij} are added in, the parallelism of paths in the latter case is only approximate. ∎

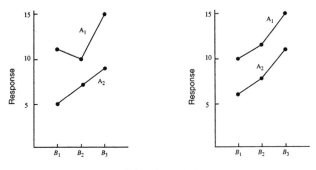

Figure 14-2

In the case of a randomized block design there would ordinarily be just one observation per cell. When studying effects of two factors, it may be possible to make several observations with each treatment—that is, per cell.

We'll consider first the case of one observation per cell and then show how cell replicates can be used to separate out an interaction as a component of the total variation. As in the case of a one-way layout, such problems could be treated using a permutation test, or in the setting of a normal, random effects model. However, we shall take up only the normal, fixed effects model.

When there is only one observation per cell, the model as defined by (1) and (2) is overly specific for the amount of data—one need but choose $\mu_{ij} = Y_{ij}$ to have a perfect fit. This model is almost surely not correct. But we might hope to fit a model with additive factor effects:

$$\mu_{ij} = \mu + \theta_i + \phi_j. \tag{6}$$

(This amounts to putting any interaction present into the "error" term.) With one observation per cell, the number of observations is $n = rc$, and the likelihood function, given (6) and the assumed independence, normality, and homoscedasticity of errors, is

$$L(\mu, \boldsymbol{\theta}, \boldsymbol{\phi}; \sigma^2) = (\sigma^2)^{-n/2} \exp\left\{-\frac{1}{2\sigma^2}\sum\sum(Y_{ij} - \mu - \theta_i - \phi_j)^2\right\}. \quad (7)$$

It is easy to show that the m.l.e.'s of the components of μ_{ij} are

$$\widehat{\mu} = \bar{Y}, \quad \widehat{\theta}_i = \bar{Y}_{i\cdot} - \bar{Y}, \quad \widehat{\phi}_j = \bar{Y}_{\cdot j} - \bar{Y},$$

where dots in a subscript indicate averaging as before. And then

$$\widehat{\sigma}^2 = \frac{1}{n}\sum\sum(Y_{ij} - \bar{Y}_{i\cdot} - \bar{Y}_{\cdot j} + \bar{Y})^2 = \frac{1}{n}\text{SSE}.$$

The term SSE is appropriate because the quantities squared are essentially estimates of the errors, ϵ_{ij}.

To find the distribution of $\widehat{\sigma}^2$ we start with the identity

$$Y_{ij} - \mu - \theta_i - \phi_j = (Y_{ij} - \bar{Y}_{i\cdot} - \bar{Y}_{\cdot j} + \bar{Y}) + (\bar{Y} - \mu)$$
$$+ (\bar{Y}_{i\cdot} - \bar{Y} - \theta_i) + (\bar{Y}_{\cdot j} - \bar{Y} - \phi_j),$$

square both sides, sum over i and j, and divide by σ^2. After doing this, the left-hand side is chi$^2(rc)$. After finding the ranks of the quadratic expressions on the right and applying Cochran's theorem, we see (among other things) that

$$\frac{\text{SSE}}{\sigma^2} \sim \text{chi}^2[(r-1)(c-1)]. \quad (8)$$

Applying Theorem 1 to the present situation (or simply substituting m.l.e.'s for parameters), we see that the maximum of the likelihood is

$$L(\widehat{\mu}, \widehat{\boldsymbol{\theta}}, \widehat{\boldsymbol{\phi}}; \widehat{\sigma}^2) = (\widehat{\sigma}^2)^{-n/2} e^{-n/2}. \quad (9)$$

One may want to test the hypothesis no factor A "effect":

$$H_0: \theta_i = 0 \text{ for } i = 1, ..., r.$$

Under this hypothesis the m.l.e.'s of the remaining parameters are $\widehat{\mu} = \bar{Y}$, $\widehat{\phi}_j = \bar{Y}_{\cdot j} - \bar{Y}$ (as under H_0), and $\widehat{\sigma}_0^2$, where

$$n\widehat{\sigma}_0^2 = \sum\sum(Y_{ij} - \bar{Y}_{\cdot j})^2 = n\widehat{\sigma}^2 + \sum_i c(\bar{Y}_{i\cdot} - \bar{Y})^2.$$
$$= \text{SSE} + \text{SSA}.$$

Under H_0, according to the above application of Cochran's theorem, the sums of squares SSE and SSA are independent, and the distribution of

14.9 Two way classifications

SSA/σ^2 is chi$^2(r-1)$. The likelihood ratio for testing H_0 against the alternative that not all $\theta_i = 0$ is $\Lambda = (\hat{\sigma}_0^2/\hat{\sigma}^2)^{-n/2}$, so that

$$\Lambda^{-2/n} = \frac{\hat{\sigma}_0^2}{\hat{\sigma}^2} = 1 + \frac{\text{SSA}}{\text{SSE}}.$$

Small values of Λ correspond to large values of SSA/SSE; and under H_0,

$$F_A = \frac{\text{SSA}/(r-1)}{\text{SSE}/(r-1)(c-1)}$$

is distributed as $F[r-1, (r-1)(c-1)]$.

Similar considerations produce an analogous test statistic for the hypothesis that there is no factor B effect. The usual ANOVA obtained with statistical software gives the sums of squares for both tests:

$$\text{SSTotal} = \sum\sum (Y_{ij} - \bar{Y})^2 = \text{SSA} + \text{SSB} + \text{SSE}.$$

Notice, however, that the statistics F_A and F_B cannot be claimed to be independent, since they involve the same denominator.

EXAMPLE 14.9c Consider these artificial data, chosen so that the arithmetic is simple enough for you to reproduce easily:

	B_1	B_2	B_3	$\bar{Y}_{i\cdot}$	$\hat{\theta}_i$
A_1	3	5	4	4	-7
A_2	11	10	12	11	0
A_3	16	21	17	18	7
$\bar{Y}_{\cdot j}$	10	12	11	11	
$\hat{\phi}_j$	-1	1	0		

The sum of squares about the grand mean $\bar{Y} = 11$ is 312, and the ANOVA table is as follows:

Source	d.f.	SS	MS	F
Factor A	2	294	147	49
Factor B	2	6	3	1
Error	4	12	3	--
Total	8	312	--	

The large value of F_A reflects the rather obvious differences among the responses at the levels of factor A. ∎

When there is more than one observation in at least one cell, it is possible to sort out an interaction component from what would otherwise be part of "experimental error." The reason we can do this is that the variability within cells provides a way of estimating the error variance that is valid no matter what the factor and interaction effects may be. This estimate serves as a basis for judging whether other components of variability are real effects or simply random variation.

Data consist of independent responses Y_{ijk}, the kth replicate in the cell defined by the ith level of factor A and the jth level of factor B, and we assume the model that includes an interaction term [see (1)-(3)]:

$$Y_{ijk} = \mu + \theta_i + \phi_j + \xi_{ij} + \epsilon_{ijk},$$

where $\epsilon_{ijk} \sim \mathcal{N}(0, \sigma^2)$. We assume also that in each cell there are the same number of observations, say s. Cell means are $\bar{Y}_{ij\,.}$, and the grand mean is

$$\bar{Y} = \tfrac{1}{rcs} \sum \sum \sum Y_{ijk} = \tfrac{1}{s} \sum \sum \bar{Y}_{ij\,.}.$$

The row and column means are

$$\bar{Y}_{i\,.\,.} = \tfrac{1}{c} \sum \bar{Y}_{ij\,.}, \quad \bar{Y}_{.\,j\,.} = \tfrac{1}{r} \sum \bar{Y}_{ij\,.}.$$

We define the *error sum of squares* in terms of within cells deviations—deviations from the cell means:

$$\mathrm{SSE} = \sum_i \sum_j \sum_k (Y_{ijk} - \bar{Y}_{ij\,.})^2.$$

Since the inner sum in SSE is the numerator of a sample variance from a normal population, it follows that

$$\sum_{k=1}^{s} \frac{(Y_{ijk} - \bar{Y}_{ij\,.})^2}{\sigma^2} \sim \mathrm{chi}^2(s-1),$$

independent of any restrictions on θ, ϕ, or ξ. And since these variables are independent, their sum over i and j is again a chi-square variable:

$$\frac{\mathrm{SSE}}{\sigma^2} \sim \mathrm{chi}^2[rc(s-1)].$$

Moreover, MSE = SSE$/(rcs - rc)$ is an unbiased estimator for σ^2, whatever the factor or interaction effects there may be.

The joint maximum likelihood estimators for the various parameters are found much as before, with these results:

$$\hat{\mu} = \bar{Y}, \quad \hat{\theta}_i = \bar{Y}_{i..} - \bar{Y}, \quad \hat{\phi}_{.j} = \bar{Y}_{.j.} - \bar{Y},$$

$$\hat{\xi}_{ij} = \bar{Y}_{ij.} - \bar{Y}_{i..} - \bar{Y}_{.j.} + \bar{Y}, \quad \text{and} \quad \hat{\sigma}^2 = \frac{SSE}{rcs}.$$

Notice that $\hat{\xi}_{ij}$ is the same function of the cell means as what was used in defining what we termed SSE when there was only one observation per cell. The sum of squares of these estimates of the interaction components ξ_{ij} is termed the **interaction sum of squares**:

$$\text{SSInter} = \sum\sum (\bar{Y}_{ij.} - \bar{Y}_{i..} - \bar{Y}_{.j.} + \bar{Y})^2.$$

Mean squares are obtained by dividing sums of squares by corresponding degrees of freedom.

To test the null hypothesis $\theta_i = 0$, we now have SSE as a basis of comparison for SSA, and use the test ratio

$$F_A = \frac{\text{MSA}}{\text{MSE}} = \frac{\text{SSA}/(r-1)}{\text{SSE}/(rcs - rc)},$$

with numerator and denominator degrees of freedom as here displayed. The ANOVA table is as follows:

Source	Sum of squares	Degrees of freedom
Factor A	$\sum cs(\bar{Y}_{i..} - \bar{Y})^2$	$r - 1$
Factor B	$\sum rs(\bar{Y}_{.j.} - \bar{Y})^2$	$c - 1$
Interaction	$\sum\sum s(\bar{Y}_{ij.} - \bar{Y}_{i..} - \bar{Y}_{.j.} + \bar{Y})^2$	$(r-1)(c-1)$
Error	$\sum\sum\sum (Y_{ijk} - \bar{Y}_{ij.})^2$	$rc(s - 1)$
Total	$\sum\sum\sum (Y_{ijk} - \bar{Y})^2$	$rcs - 1$

In practice, a statistician will usually first test $\xi_{ij} = 0$ using the statistic

$$F_{\text{inter}} = \frac{\text{MSInter}}{\text{MSE}} = \frac{\text{SSInter}/[(r-1)(c-1)]}{\text{SSE}/(rcs - rc)},$$

and then only test $\theta_i = 0$ if it is concluded that there is insufficient evidence to reject the hypothesis of no interaction. This is not that the test for factor A effect is invalid, but that if there *is* interaction, it is hard to interpret what $\theta_i = 0$ means. (On the other hand, in successive applications of two F-tests with test ratios that are not independent, it is

534   Chapter 14   Linear Models

hard to know just what the level of the test for the factor effect is.)

While it is true that one can learn more and with more precision with the larger amount of data, still it is useful and perhaps remarkable that with only the single observation per cell it is possible to get at a treatment effect in the presence of another factor. And perhaps this design should be contrasted with another technique—that of holding all but the one variable fixed to determine a treatment effect. In that approach, any conclusion about a treatment effect is only applicable when the other variables are fixed at the levels under which the responses are observed.

PROBLEMS

14-43. Calculate the F-statistic for testing the hypothesis of no mean difference among seed types with the data in the table at the beginning of §14.9.

14-44. Find the error sum of squares for the randomized block data shown below. Explain what could account for this and what conclusion about the factor effect would follow from the F-test:

16	20	15
14	18	13
12	16	11

14-45. Fill in some of the details of the development on page 530:
(a) Obtain the joint maximum likelihood estimates of the parameters.
(b) Obtain the partitioning of $\sum (Y_{ij} - \mu - \theta_i - \phi_j)^2$ described there.
(c) Find the likelihood ratio for testing H_0: $\theta_i = 0$ for all i.
(d) Justify the statement about the distribution of F_A.
(e) Obtain a formula for F_B, for testing H_0: $\phi_j = 0$ for all j.

14-46. Show that the paired sample t-test for comparing means (§9.7) is a special case of the F-test in a randomized block design in which the block is a matched pair.

14-47. Develop the m.l.e.'s, the decomposition of a sum of squares, and (with Cochran's theorem) the null distributions and d.f.'s of the sums of squares shown in the ANOVA table for the case of several observations per cell.

14-48. Obtain these calculation formulas for a two-way classification with one observation per cell:

$$\text{SSA} = \frac{1}{r}\sum_i\left(\sum_j Y_{ij}\right)^2 - C, \quad \text{SSB} = \frac{1}{c}\sum_j\left(\sum_i Y_{ij}\right)^2 - C,$$

where

$$C = \frac{1}{rc}\left(\sum\sum Y_{ij}\right)^2.$$

14-49. Consider the data shown below. [Data of this sort are best handled by computer, but these were doctored so as to be fairly manageable by hand (and hand calculator) calculation. Responses are a day's output.] Test the hypothesis of no interaction, and calculate the F-statistics for testing factor effects.

		\multicolumn{4}{c}{Machine}
		1	2	3	4
		59	61	48	47
	1	43	49	47	40
		63	52	58	51
		60	49	40	38
Operator	2	51	52	48	50
		48	55	56	41
		63	58	46	48
	3	60	48	53	50
		57	56	51	50

14.10 Other designs

Suppose three factors have been identified that should be taken into account, at r, c, and s levels respectively. A treatment is then defined by a combination of factor levels, and there are rcs of these. If t observations are taken with each treatment, the total number of observations is $rcst$. But this number can easily involve too much time, money, or experimental material. For instance, if $r = c = s = t = 3$, the product is 81. Rather than write out the analysis of variance for this case we'll mention that there are ways of reducing the amount of data needed—although any reduction in the amount of data will necessarily be at the sacrifice of precision. One way is to use what is called a *fractional factorial design*. In these designs, certain systematically chosen cells can be left with no observations, but in such a way that it is still possible to estimate the main effects of interest. Another possibility is that of an *incomplete block design*, in which the subjects in the blocks are fewer than the number of treatments to be included.

The only technique we'll present here is that of a *Latin square* design for the case of three treatments. Suppose, to take a concrete situation, there are three factors: machines, operators, and time periods, with four "levels" for each. A complete factorial design would require at least 64 observations,— many more if we had several observations per cell. A Latin square design uses only 16 factor-level combinations, that is, 16 of the 64 possible treatments. These combinations are chosen so that each operator works once in each time period on each machine.

A *Latin square* is a square array of (Latin) letters, say A, B, C, D, with the property that no letter appears more than once in any column or in any row. Such arrays have been extensively derived, studied, and tabulated. One such example is the following, in which each row after the first is obtained by successive cyclic permutations :

		\multicolumn{4}{c}{Time period}
		1	2	3	4
	1	A	B	C	D
Operator	2	B	C	D	A
	3	C	D	A	B
	4	D	A	B	C

Other such squares are obtainable from this one by permuting the four columns and rows 2, 3, and 4. But another square, one which cannot be obtained from the above as a permutation, is this :

		\multicolumn{4}{c}{Time period}
		1	2	3	4
	1	A	B	C	D
Operator	2	B	D	A	C
	3	C	A	D	B
	4	D	C	B	A

To carry out the experiment, one Latin square would be selected at random from all possible squares. (There are 576, when $s = 4$.) For a given combination of operator and time period, the machine to be used is determined by the letter in the square in that cell. The observations Y

(day's output, say) are 16 in number: $Y_{ij(k)}$, the third subscript being determined from the first two by the particular Latin square being used.

We assume a model of this form:

$$Y_{ij(k)} = \mu + \alpha_i + \beta_j + \gamma_k + \epsilon_{ij(k)},$$

where $i = 1, \ldots, s$, $j = 1, \ldots, s$, and k is determined by i and j. (There is no way to get at any interaction, so if it is present, it has to be in the error term.) As a slight variation in notation, let $\bar{Y}_{(k)}$ denote the average of the s observations taken at level k of the third factor. The ANOVA table is shown on below. The test statistic for a row effect is

$$F_R = \frac{\text{SSRow}/(s-1)}{\text{SSE}/(s-1)(s-2)}.$$

Source	Sum of squares	Degrees of freedom
Row	$s \sum (\bar{Y}_{i\cdot} - \bar{Y})^2$	$s - 1$
Column	$s \sum (\bar{Y}_{\cdot j} - \bar{Y})^2$	$s - 1$
Letters	$s \sum (\bar{Y}_{(k)} - \bar{Y})^2$	$s - 1$
Error	$\sum \sum (Y_{ij(k)} - \bar{Y}_{i\cdot} - \bar{Y}_{\cdot j} - \bar{Y}_{(k)} + 2\bar{Y})^2$	$(s-1)(s-2)$
Total	$\sum \sum (Y_{ij(k)} - \bar{Y})^2$	$s^2 - 1$

The value of F_R is to be assessed with reference to its null distribution, namely, $F[(s-1), (s-1)(s-2)]$. Details of the ANOVA, degrees of freedom, and independence are left as exercises for the student.

An experiment with four factors at s levels each can be designed similarly using a *Latin-Greco square* and requiring only s^2 observations. In such a square, each cell defined by levels of the first two factors is given a Latin letter and a Greek letter so that each level of each factor appears exactly once with each level of each other factor. One 3×3 Latin-Greco square is the following:

Factor 2:

		1	2	3
Factor 1:	1	$A\alpha$	$B\beta$	$C\gamma$
	2	$B\gamma$	$C\alpha$	$A\beta$
	3	$C\beta$	$A\gamma$	$B\alpha$

The area of experimental design is far too extensive to be treated here, and this is as far as we go.

PROBLEMS

14-50. To compare a treatment (insulin) in four doses: A, standard .6 units; B, standard 1.2 units; C, unknown .6 units; D, unknown 1.2 units, four rabbits are used on four days. Suppose we take rows as blocks, each being a day (Apr. 23, 25, 26, 27), and columns as blocks, each being a rabbit. The treatment is assigned using a Latin square, as follows, in which along with the letter of a treatment is given the measured response (blood sugar after 50 min.):

B 24	C 46	D 34	A 48
D 33	A 58	B 57	C 60
A 57	D 26	C 60	B 45
C 46	B 34	A 61	D 47

Construct the ANOVA table and test the hypothesis of no treatment effect.

14-51. Obtain the decomposition of SSTotal as given in the ANOVA table in §14.10 and check the degrees of freedom.

14-52. There are two "standard" squares (i.e., with first row A B C D) that are not obtainable from those shown in §14.10 by permutations. Find them.

14-53. Show: $E(\text{SS "Letters"}) = (s-1)\sigma^2 + s\sum \gamma_k^2$.

15
Decision Theory

Many problems of statistical inference, notably those in business, industry, and government, are essentially decision problems. In these, the state of nature is unknown, but a decision must be made—one whose consequences depend not only on the decision but on the unknown state of nature. Such problems are statistical when there are data that give partial information about the unknown state.

Following the development of the Neyman-Pearson theory of tests and the von Neumann theory of games in the 1930's, A. Wald was led in the 1940's to formulate the statistical decision problem in the setting of game theory, providing a unifying structure in which many statistical problems can be profitably studied.

In statistical decision theory, a problem of inference is viewed as a two-person game; one player is the statistician and the other, "Nature." The statistician must choose a strategy not knowing what strategy Nature has chosen, and the game's payoff is the consequence of these choices. We begin by studying the decision problem when no data are available, and then show how to incorporate observed data into the decision process.

15.1 Convex sets

Points $P = (x_1, ..., x_k)$ can be treated as vectors, and combined by linear operations as vectors. Recall that vectors are added by adding components, and multiplied by a scalar quantity by multiplying each component by that scalar. Two vectors are equal if and only if their respective components are equal, and the zero vector is a vector of zeros: $\mathbf{0} = (0, 0, ..., 0)$.

A **linear combination** of vectors $P_1, ..., P_n$ is a combination of the form

$$a_1 P_1 + \cdots + a_n P_n = \sum a_i P_i.$$

A linear combination of two points in which the coefficients sum to 1 is a point P on the *line* through the two points:

$$P = \gamma P_1 + (1 - \gamma) P_2. \tag{1}$$

The vector equation (1) summarizes the k equations

$$\begin{cases} x_1 = \gamma x_{11} + (1 - \gamma) x_{21}, \\ \quad \vdots \\ x_k = \gamma x_{1k} + (1 - \gamma) x_{2k}. \end{cases}$$

These are parametric equations of the line through the points P_1 and P_2, and those points are given by $\gamma = 1$ and $\gamma = 0$, respectively. If the real number γ is unrestricted, then (1) defines the infinite line. If we impose the restriction $0 \leq \gamma \leq 1$, then (1) includes only the points on the line segment joining P_1 and P_2. Each component of P is a weighted average of the corresponding components of P_1 and P_2.

Consider now linear combinations $\sum a_i P_i$ in which the coefficients are probabilities: $a_i \geq 0$, $\sum a_i = 1$. These are *convex* combinations. They arise in mechanics in determining the center of gravity of a set of mass points. Thus, given masses m_i at P_i, the center of gravity is at

$$\text{c.g.} = \sum \frac{m_i}{M} P_i, \quad M = \sum m_i.$$

Since the relative masses m_i/M are nonnegative and sum to 1, the c.g. is a convex combination of the P_i. Any convex combination can be interpreted as the c.g. of a mass distribution over those points, and this interpretation is useful because of our intuitive appreciation of the notion of "center of gravity."

The definition of a convex set in the plane was introduced in Chapter 9. That definition is applicable to sets in k dimensions:

DEFINITION A set in $\Re^k$ is *convex* if and only if the entire line segment joining any pair of its points lies within the set.

THEOREM 1 *The set C of convex combinations of points $P_1, \ldots, P_n$ is a convex set.*

To prove this, consider two convex combinations of the points P_i, $Q = \sum \alpha_i P_i$ and $R = \sum \beta_i P_i$, where $\alpha_i \geq 0$, $\beta_i \geq 0$, $\sum \alpha_i = 1$, and $\sum \beta_i = 1$. We need to show that the entire line segment between Q and R is also in C. Any point S on that line segment is a convex combination of Q and R:

$$S = \gamma Q + (1-\gamma)R = \gamma \sum \alpha_i P_i + (1-\gamma) \sum \beta_i P_i$$
$$= \sum [\gamma \alpha_i + (1-\gamma)\beta_i] P_i,$$

where $0 \leq \gamma \leq 1$. This linear combination of the points P_i is in fact a convex combination of them, since then

$$\gamma \alpha_i + (1-\gamma)\beta_i \geq 0, \text{ and } \sum [\gamma \alpha_i + (1-\gamma)\beta_i] = \gamma + (1-\gamma) = 1.$$

Thus, S is indeed a point of C, as was to be shown.

DEFINITION The convex set of all convex combinations of P_1, ..., P_n is the convex set *generated* by those points. The smallest convex set containing any given collection of points is called their *convex hull*.

The convex set generated by n points in the plane is a polygon, which can be thought of as formed by a rubber band stretched to include all of the n points and then released to rest on pegs inserted at those points. The convex set generated by n points in 3-space is a polyhedron.

EXAMPLE 15.1a The five points (1, 1), (1, 2), (2, 2), (2, 4), (4, 0) are plotted in Figure 15-1 as P_1, ..., P_5. The convex set generated by these five points is the shaded quadrilateral—the set of centers of gravity.

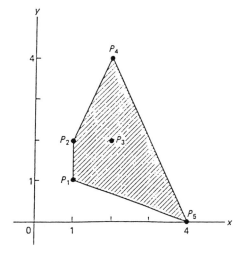

Figure 15-1

Since P_3 is inside the set generated by the other four points, it can be represented as a convex combination of those four points. No point *outside* the quadrilateral can be a convex combination or balance point. ■

15.2 Utility and subjective probability

Decision-making in the face of uncertainty can only be rational if one takes into account the consequences of decisions and the probabilities of the various possible states of nature. To have a quantitative basis for a choice of action, the decision-maker needs to measure consequences on a numerical scale and to assume some specific distribution for the state of nature. These are necessarily subjective determinations. In each case, we would want to ask two questions: Is there such a thing? And how do we find it out? We'll only give brief consideration to the first question—and even there we'll beg the question, referring the student to more detailed sources.

Subjective probabilities for θ, as viewed by the decision-maker, can be shown to exist provided one accepts certain axioms about relative likelihoods. To give these we adopt some temporary notation: $E \prec F$ means that E is less likely than F; $E \succ F$, that E is more likely than F; $E \sim F$, that E and F are equally likely; and $E \precsim F$, that F is at least as likely as E.

AXIOMS FOR RELATIVE LIKELIHOOD

1. Either $E \prec F, E \succ F,$ or $E \sim F$.

2. Suppose $E_1 E_2 = \emptyset$ and $F_1 F_2 = \emptyset$. If $E_i \precsim F_i$ for $i = 1, 2$, then $E_1 \cup E_2 \precsim F_1 \cup F_2$. Moreover, if either $E_1 \prec F_1$ or $E_2 \prec F_2$, then $E_1 \cup E_2 \prec F_1 \cup F_2$.

3. For every event E, $\emptyset \precsim E$.

4. $\emptyset \prec \Omega$.

5. If $E_1 \supset E_2 \supset \cdots$, and $F \precsim E_n$ for all n, then $\bigcap E_n \succsim F$.

A sixth axiom involves the notion of a "uniform distribution" on $[0, 1]$. (Since we are still trying to define a probability, this has to be defined in terms of relative likelihood.) A random variable X on Ω with values on $[0, 1]$ is said to have a uniform distribution provided that the relation $[X \in I_1] \precsim [X \in I_2]$ holds if and only if the length of I_1 does not exceed the length of I_2, for every pair of subintervals I_1 and I_2 of $[0, 1]$.

6. There exists a uniform distribution on $[0, 1]$.

It can be shown[1] that Axioms 1-6 imply a unique distribution on Ω that satisfies the axioms of a probability space (§2.4).

Turning now to consequences of decisions, we must point out that loss or gain is not necessarily monetary. It can be as vague as loss of prestige, comfort, or goodwill. The term **utility** has come to be used to denote a certain measure of gain (or negative loss) that seems appropriate for decision problems. More specifically, utility is a real-valued function $u(\mathcal{P})$ on prospects $\mathcal{P}$ which satisfies the following properties. In stating them we'll use the symbol $\succsim$ to mean "at least as desirable," the symbol $\sim$ to mean "equally desirable," and $\succ$ to mean "more desirable."

PROPERTIES OF UTILITY

1. Utility is bounded: $u(\mathcal{P}) \leq K < \infty$.

2. If $\mathcal{P}_1 \succ \mathcal{P}_2$, then $u(\mathcal{P}_1) \geq u(\mathcal{P}_2)$.

3. If $\mathcal{P}$ is the random prospect of facing $\mathcal{P}_1$ with probability α and $\mathcal{P}_2$ with probability $1 - \alpha$, then $u(\mathcal{P}) = \alpha u(\mathcal{P}_1) + (1 - \alpha) u(\mathcal{P}_2)$.

We denote the random prospect in (3), called a *mixture* of $\mathcal{P}_1$ and $\mathcal{P}_2$, by $[\mathcal{P}_1, \mathcal{P}_2]_\alpha$.

It can be shown that one who accepts certain axioms for preferences has a uniquely defined utility function $u(\mathcal{P})$ satisfying properties (1)-(3).[2]

PREFERENCE AXIOMS

1. For any $\mathcal{P}$ and $\mathcal{Q}$, either $\mathcal{P} \sim \mathcal{Q}$, $\mathcal{P} \succ \mathcal{Q}$, or $\mathcal{P} \prec \mathcal{Q}$.

2. If both $\mathcal{P} \succsim \mathcal{Q}$ and $\mathcal{Q} \succsim \mathcal{R}$, then $\mathcal{P} \succsim \mathcal{R}$.

3. If $\mathcal{P} \succ \mathcal{Q}$, then for any probability p and any prospect $\mathcal{R}$,
$$[\mathcal{P}_1, \mathcal{R}]_p \succ [\mathcal{Q}, \mathcal{R}]_p.$$

4. Given prospects $\mathcal{P} \succ \mathcal{Q} \succ \mathcal{R}$, there are mixtures defined by probabilities α and β such that
$$\mathcal{P} \succ \{\mathcal{P}, \mathcal{R}\}_\alpha \succ \mathcal{Q} \succ [\mathcal{P}, \mathcal{R}]_\beta \succ \mathcal{R}.$$

---

[1] M. DeGroot, *Optimal Statistical Decisions*, New York: McGraw-Hill (1970), p.77ff.

[2] See B. W. Lindgren, *Elements of Decision Theory*, New York; Macmillan Publ. Co. (1971) p. 233ff, or M. DeGroot, op. cit. In fact, Axiom 3 is needed in a stronger form

Axiom 1 seems quite reasonable as a requirement. If one cannot order prospects, there is little hope for a rational decision. Nor is Axiom 2, transitivity of preferences, unreasonable. Axiom 3 says that improving one of the prospects in a finite mixture improves the mixture. Axiom 4 says that there is no prospect $\mathcal{P}$ so wonderful that the slightest chance of encountering it instead of $\mathcal{R}$ is better than any ordinary $\mathcal{Q}$; and that there is no $\mathcal{R}$ so terrible that the slightest chance of encountering it instead of $\mathcal{P}$ is worse than any ordinary $\mathcal{Q}$.

EXAMPLE 15-2a Consider a game with payoff $\$2^X$, where X is the number of trials it takes to get heads with a fair coin. The expected payoff is

$$E(2^X) = \sum_{1}^{\infty} 2^k \cdot \frac{1}{2^k} = \infty.$$

Yet people seem unwilling to pay more than a few dollars to play this game. (This is the "St. Petersburg paradox.") One explanation lies in the fact that a person's utility for money is generally not proportional to the amount. For instance, suppose the utility function is $u(x) = x/2^b$ if $x \leq 2^b$, and $u(x) = 1$ for $x > 2^b$. The expected utility is

$$E[u(\text{payoff})] = \sum_{1}^{b} \frac{2^x}{2^b} \cdot \frac{1}{2^x} + \sum_{b+1}^{\infty} \frac{1}{2^x} = \frac{1}{2^b}(b+1).$$

The entry fee, in dollars, should equal this expected utility, or $b+1$ dollars. ∎

In the rest of this chapter, consequences are to be considered as measured on a utility scale, and probabilities for the state of nature are the subjective probabilities of the decision-maker. But before proceeding with the study of decision problems we'll point out how the notion of utility can help explain why people will enter into unfair bets.

A bet resulting in a new total capital M_W is called a **fair bet** if the expected total capital after the bet is equal to the initial capital M_0:

$$M_0 = pM_W + (1-p)M_L.$$

The utility U_B of the random prospect consisting of taking this fair bet is the expected value of the utility of the capital after the bet:

$$U_B = pu(M_W) + (1-p)u(M_L).$$

That is,

15.2 Utility and subjective probability

$$\begin{pmatrix} M_0 \\ U_B \end{pmatrix} = p \begin{pmatrix} M_W \\ u(M_W) \end{pmatrix} + (1-p) \begin{pmatrix} M_L \\ u(M_L) \end{pmatrix}.$$

This says that the point (M_0, U_B) is a convex combination of $(M_W, u(M_W))$ and $(M_L, u(M_L))$ and therefore lies on the line segment joining those two points. The fair bet will be worthwhile (on the utility scale) if the curve representing utility as a function of capital is below that line segment, as in Figure 15-2. That is, the bet is worth taking if $U_B > u(M_0)$.

On the other hand, this same bet (new capital M_W with probability p and new capital M_L with probability $1 - p$) is disadvantageous in terms of money (and advantageous to the one offering the bet) when the initial capital is M', where

$$M' > pM_W + (1-p)M_L,$$

which implies that the expected net gain will be negative:

$$p(M_W - M') + (1-p)(M_L - M') < 0.$$

In Figure 15-2 we show the initial capital M' such that $U_B = u(M')$.

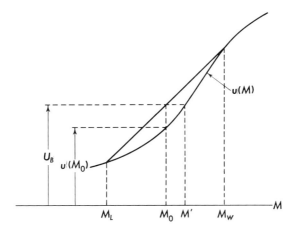

Figure 15-2

If the initial capital were any greater than M', the bet would not be worth taking, because then the utility of the bet would be smaller than the utility of the initial capital.

The utility function for money $u(M)$ shown in Figure 15-2 is *convex*, or concave up, over the interval between M_L and M_W. Where it is concave down, even a fair bet would be a loser on average—in terms of utility. A person's utility for money is typically concave up for very small M, approximately linear for intermediate M, and concave down for large M. That is, a capital consisting few pennies is no better than a capital of 0; but four billion dollars is perhaps more than half as valuable as eight—in terms of utility.

PROBLEMS

15-1. Determine which of the following restrictions on (x, y) define convex sets in the plane:

(a) $x^2 + y^2 \leq 1$
(b) $2x^2 + y^2 \geq 1$
(c) $y > x^2$
(d) $xy > 0$
(e) $xy = 0$
(f) $xy > x > 0$
(g) $x + 2y \geq 3$
(h) $x + y \leq 1$
(i) $x + 2y = 3$, $0 < x < 1$
(j) $x^2 + y^2 \leq 4$, $y > 1$
(k) $xy > 0$, $x + y > 4$

15-2. Show algebraically (rather than geometrically) that the plane set defined by $x + y \leq 1$, $y \geq 0$, and $x \geq 0$ is convex.

15-3. Find the convex combination of the four points $(0, 4, 0)$, $(3, 1, 0)$, $(0, 0, 4)$, and $(4, 3, 5)$ in which the line $x = y = z$ issuing from the origin first strikes the convex set generated by these four points.

15-4. Show that the set of convex combinations of points $P_1, \ldots, P_m$ in n-space is the smallest convex set containing these points. (That is, show that each convex combination is included in any convex set containing them all.)

15-5. A man needs \$30 to buy a ticket to a football game he very much wants to see but can scrape up only \$15 around the house. Devise a graph of utility vs. money that would be appropriate in his situation. In terms of this utility scale, find his utility in tossing a coin with a friend for \$15, double or nothing. (Is his utility improved with this bet?)

15-6. Suppose, in the St. Petersburg game of Example 15.1b, the utility for money is $u(x) = \log_2 x$. [This is unbounded, but some are willing to let utility be unbounded.] Would this resolve the paradox?

15-7. Consider a bet with final capital \$10 (win) or \$0 (lose), and the utility function $u(x) = \frac{1}{9}(x - 1)$ for $x \geq 1$, and $u(x) = 0$ for $x < 1$ (x in dollars).
(a) Find the utility of taking the bet.
(b) Is there any probability of success so unfavorable that the bet should not be taken, given that the initial capital is (a) \$1? (b) \$2?
(c) What probability of winning makes the bet "fair?"
(d) Given the probability in (c), find the largest initial capital for which

the bet would not lose utility.

15-8. From Axioms 1 and 2 for relative likelihoods, show that $E \precsim F$ if and only if $E^c \succsim F^c$.

15.3 Loss, regret, and mixed actions

We now define a **loss function** $\ell(a, \theta)$ as the negative utility in taking action a when Nature is in state θ. If the state space Ω is finite, consisting of states $\theta_1, \theta_2, \ldots, \theta_k$, then for each action a the losses $\ell(\theta_i, a)$ can be taken as coordinates of a point $(L_1, \ldots, L_k)$ in k-space, where $L_i = \ell(\theta_i, a)$. So each action a in the action space $\mathcal{A}$ is represented by one such point.

From the decision-maker's point of view, it might reasonably be argued that decisions should be based not on losses, but on **regrets**:

$$r(\theta, a) = \ell(\theta, a) - \min_a \ell(\theta, a).$$

This is the amount in excess of the smallest possible loss, for a particular θ, that one incurs by taking action a. Thus, one "regrets" having taken a when the amount $r(\theta, a)$ could have been avoided if the state θ had been known. It is also known as "opportunity loss."

Because the complications that usually attend important, real problems tend to obscure the exposition, our first examples are without a specific setting. They involve only two states of nature, so as to permit representation in a plane figure.

EXAMPLE 15.3a Consider a decision problem in which the action space is $\mathcal{A} = \{a_1, a_2, a_3\}$ and the parameter space, $\Theta = \{\theta_1, \theta_2\}$. Suppose losses are as follows:

	a_1	a_2	a_3
θ_1	6	8	12
θ_2	7	4	1

To calculate corresponding regrets we subtract 6 (minimum loss for θ_1) from each loss in the first row and 1 from each loss in the second row.

	a_1	a_2	a_3
θ_1	0	2	6
θ_2	6	3	0

Figure 15-3 shows the three actions in terms of losses as the points $(6,7)$, $(8,4)$ $(12,1)$. It also shows them in terms of regrets as the points $(0, 6)$, $(2, 3)$, $(6, 0)$.

Observe that none of the three available actions is better (in terms of either loss or regret) than the other two for *both* states of nature: The actions are not simply ordered. Surely, it would be desirable to have some scheme for representing each action with a *single* number; this would provide an ordering of the actions so that the action which is "best" (in that sense) could be chosen. ∎

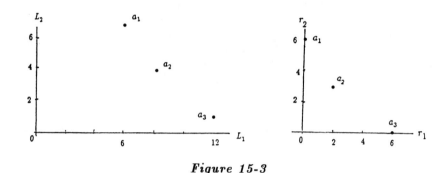

Figure 15-3

In treating hypothesis-testing as decision-making, we introduced the notion of a *randomized* test. In decision theory it is convenient to use this device of randomization—to define a **mixed action** as a probability distribution on the action space. If the action space $\mathcal{A}$ has m (*pure*) actions, a mixed action is a vector of probabilities, $\mathbf{p} = (p_1, ..., p_m)$. To make a decision we choose a probability distribution for $\mathcal{A}$ and take the specific action that results upon doing the experiment of chance defined by that distribution.

We then judge the consequences of a mixed action according to the *expected* loss, $E[\ell(\theta, a)]$. When there are k states in Θ, the vector of expected losses can be written as follows:

$$\begin{pmatrix} L_1 \\ \vdots \\ L_m \end{pmatrix} = p_1 \begin{pmatrix} \ell(\theta_1, a_1) \\ \vdots \\ \ell(\theta_k, a_1) \end{pmatrix} + \cdots + p_m \begin{pmatrix} \ell(\theta_1, a_m) \\ \vdots \\ \ell(\theta_k, a_m) \end{pmatrix}.$$

Since $p_i \geq 0$ and $\sum p_i = 1$, the point representing the mixed action $\mathbf{p}$ is a convex combination of the points that represent the pure actions. The set of all points that represent mixed actions is the convex set generated by the pure actions.

When we deal with regrets in place of losses, the points that represent regrets are simply translations of the points that represent losses—translations by the amount $\min_a \ell(\theta, a)$, which does not depend on a. When the actions are assigned probabilities, and we average with respect to the distribution defined by these probabilities, we obtain

$$E[r(\theta, a)] = E[\ell(\theta, a)] - \min_a \ell(\theta, a).$$

Thus, the mixed action points are translated along with the pure action points.

EXAMPLE 15.3b For the three-action decision problem of Example 15.3a, a mixed action is a 3-vector (p_1, p_2, p_3), and the pure actions are the particular special cases $(1, 0, 0)$, $(0, 1, 0)$, and $(0, 0, 1)$. The set of mixed actions is shown in Figure 15-4 as the shaded set of points whose coordinates are expected regrets, the convex hull of the points representing pure actions with regret coordinates. ∎

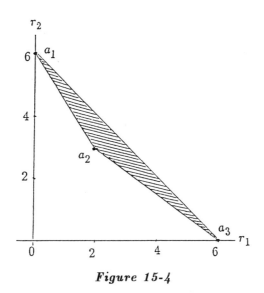

Figure 15-4

15.4 Minimax actions

We have seen that actions are not simply ordered according either to loss or to regret, but we need such an ordering in order to choose among them—some way of assigning a single number to each action. One way is to follow the *minimax principle*: Expect the worst and act accordingly: For each action, find the maximum possible loss under the various states of nature and assign this maximum to the action. The action to take is the one with the smallest maximum loss.

In choosing from a set of *mixed* actions, we apply the minimax procedure to the *expected* losses. This is in accord with the principle that the utility of a random prospect is calculated as an expected value.

EXAMPLE 15.4a Consider again the loss table of Example 15.3a and the corresponding regret table of Example 15.4a, each with an added row of maxima over θ:

	a_1	a_2	a_3
θ_1	6	8	12
θ_2	7	4	1
max	7	8	12

	a_1	a_2	a_3
θ_1	0	2	6
θ_2	6	3	0
max	6	3	6

The minimax *loss* is 7, with action a_1; the minimax *regret* is 3, with action a_2. This example illustrates the fact that in applying the minimax principle, the action we take depends on whether we apply it to losses or to regrets.

A graphical determination of the minimax action is enlightening. In Figure 15-5 we again represent each action as a point whose coordinates are corresponding regrets. A point (r_1, r_2) is above the line $r_1 = r_2$ when the larger coordinate is r_2 and below that line when it is r_1.

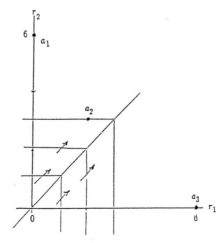

Figure 15-5

To choose between two actions represented by points both above the line $r_1 = r_2$, we take the lower one—the one first encountered by a horizontal line moving upward from below. To choose between two actions represented by points both below the line $r_1 = r_2$, we take the one farther to the left—the one first encountered by a vertical line moving to the right from the left. To choose among points, some below and some above the line $r_1 = r_2$, we take the one first met by a wedge moving up and to the right with its vertex on the 45° line, shown (for the regret case) in Figure 15-5.

The same graphical procedure works for finding the mixed action with the minimum maximum loss or minimum maximum regret. The set of mixed actions is shown again in Figure 15-6, with actions plotted according to regrets. From this figure it is evident that the wedge first strikes the set of mixed actions at the point where the line $x = y$ meets the set, in this case at a mixture of a_2 and a_3—at the point

$$\alpha \begin{pmatrix} 2 \\ 3 \end{pmatrix} + (1 - \alpha) \begin{pmatrix} 6 \\ 0 \end{pmatrix},$$

where α is found by equating coordinates: $2\alpha + 6 - 6\alpha = 3\alpha$, or $\alpha = 6/7$. With this minimax mixed action, the expected regret is $3\alpha = 18/7$. ∎

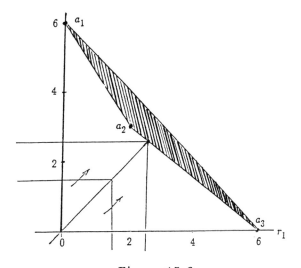

Figure 15-6

The graphical procedure illustrated in Example 15-4a can be applied in any situations in which there are just two states of nature, but finding minimax procedures is not that easy when there are more than two states. A more general approach is to set up the problem as one of linear programming,[3] but we'll not present this here. A different graphical approach can be used when there are just two possible actions and a finite number of states, as we illustrate in the next example.

EXAMPLE 15-4b Consider a problem with two actions a and four states θ. A *mixed* action $(p, 1-p)$ is defined by the probability p, and expected losses are

$$E[\ell(\theta, a)] = p\ell(\theta, a_1) + (1-p)\ell(\theta, a_2).$$

We then plot these expected losses, one for each state θ, as a function of p. The maximum loss for each p occurs for the θ represented by the line that is highest at that point, and we can easily calculate the p for which that maximum is a minimum.

Suppose losses are those in the following table, shown with expected losses for each θ:

	θ_1	θ_2	θ_3	θ_4
a_1	4	2	1	-1
a_2	0	-1	5	2
E(loss)	$4p$	$3p-1$	$5-4p$	$2-3p$

The expected losses are plotted against p in Figure 15-7, and the maximum over θ (as a function of p) is indicated by the heavy line. The minimum of the maximum expected loss occurs at $p = 5/8$, the value for which $5 - 4p = 4p$. ∎

Minimax actions are optimal in the theory of two-person, zero-sum games, in that by using such a strategy one can be assured of no more loss than the minimum maximum—no matter what the opponent (Nature, in our decision problems) does. Similarly, the opponent has a strategy that is optimal; and with it the opponent can be assured that the first player (the decision-maker) can do no better than what can be achieved using the minimax strategy. In statistical problems, where the "opponent" is Nature, it may be unduly pessimistic to think that Nature is pursuing an optimal strategy against us. Nevertheless, as we'll soon see, a minimax procedure is a Bayes procedure for some prior. (We've seen Bayes procedures in Chapters 8 and 9.)

---

[3] See B. W. Lindgren, *Elements of Decision Theory*, New York: Macmillan Publishing Co., 1971, p. 233ff.

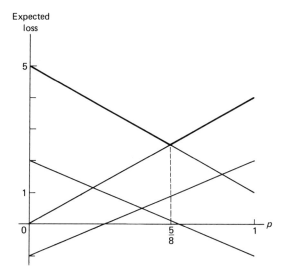

Figure 15-7

PROBLEMS

15-9. Find the minimax loss and minimax regret (pure) actions, given the following loss table:

	a_1	a_2	a_3	a_4
θ_1	2	4	1	0
θ_2	-3	0	1	2
θ_3	-1	5	-1	-2

15-10. Investigate minimax strategies, pure and mixed, using (a) loss, and (b) regret, for these losses:

	a_1	a_2
θ_1	10	15
θ_2	10	0

15-11. Consider the losses shown in the following table:

	a_1	a_2	a_3	a_4	a_5
θ_1	-1	1	0	2	5
θ_2	3	2	5	3	2

(a) Plot the action points and find the minimax mixed action and the

minimum maximum expected loss associated with this action.
(b) Obtain the table of regrets and find the minimax regret action.

15-12. Consider a problem with state space $\Omega = [0, 1]$, action space $A = [0, 1]$, and loss function $\ell(\theta, a) = (\theta - a)^2$. Find the minimax action.

15-13. A lot of five articles is to be accepted or rejected. The state θ is the number of defective articles in the lot: $\theta = 0, 1, ..., 5$. Assume losses equal to the number of good articles in a lot, if we reject it, and twice the number of defectives in a lot, if we accept it. Find the minimax mixed action.

15.5 Bayes actions

A minimax approach may not be the wisest, especially when one is convinced that the "worst" for which it prepares us has only a remote chance of being the actual state. Assigning "chances" to states of nature is what Bayesians do.

We now adopt this approach, describing our beliefs about θ in a probability distribution, so that the state of nature is a random variable, Θ. If the distribution is discrete, it is defined by a probability function $g(\theta)$. The loss incurred in taking action a is a random variable, $\ell(\Theta, a)$, with expected value

$$B(a) = E_g[\ell(\Theta, a)] = \sum \ell(\theta, a) g(\theta), \tag{1}$$

called the *Bayes loss* associated with action a. If Θ has a continuous distribution, with p.d.f. $g(\theta)$, the Bayes loss is an integral:

$$B(a) = \int \ell(\theta, a) g(\theta) d\theta.$$

The Bayes loss $B(a)$ provides us with a numerical ordering of the available actions. A *Bayes action* is one that minimizes the Bayes loss.

Suppose the decision maker is willing to consider mixed actions. If A is the random action defined by a mixing distribution $p(a)$ on the action space, the Bayes loss is

$$\mathcal{B}(p) = E_p[E_g\{\ell(\Theta, A)\}]. \tag{2}$$

In the case of a discrete action space, a *mixed* action is a probability vector $\mathbf{p}$: Take action a_j with probability p_j. The Bayes loss is

$$B(\mathbf{p}) = E_{\mathbf{p}}B(a) = \sum_i \sum_j \ell(\theta_i, a_j) p_j g(\theta_i)$$

$$= \sum_j \left\{ \sum_i \ell(\theta_i, a_j) g(\theta_i) \right\} p_j.$$

This is a convex combination of the pure-action Bayes losses.

When there are just two states of nature, a graphical procedure can be used to determine the Bayes action. Although this procedure is not general, it lends valuable insight and suggests properties that hold more generally.

EXAMPLE 15.5a For purposes of illustration, we return to the loss table of Example 15.4a and assume prior probabilities $g_1 = .7$ and $g_2 = .3$ for θ_1 and θ_2, respectively. The loss table is repeated here with a last row giving the Bayes losses $B(a)$. A sample calculation of a Bayes loss: $B(a_1) = .7 \times 6 + .3 \times 7 = 6.3$. This happens to be the smallest of the three $B(a)$'s, so a_1 is the Bayes action for the given prior. Similar calculations show that a_2 is the Bayes action for the prior $(.5, .5)$, and a_3 is the Bayes action for the prior $(.2, .8)$.

	a_1	a_2	a_3
θ_1	6	8	12
θ_2	7	4	1
$B(a)$	6.3	6.8	8.7

It helps to see what is going on geometrically. Consider first the prior $(.7, .3)$. For an action with losses (L_1, L_2),

$$B(a) = .7L_1 + .3L_2.$$

Thus, all actions with a particular Bayes loss K satisfy the equation

$$.7L_1 + .3L_2 = K,$$

represented by a family of lines with slope $-.7/.3$, one line for each K. To find the Bayes action, think of a line with this slope being moved up from below the set of points representing the actions until it first encounters an action point. That point represents the Bayes action, since other actions lie on lines of the family with larger values of K—larger Bayes losses. Figure 15-8 shows lines with slopes $-7/3$, -1, and $-2/8$, as they first encounter a_1, a_2, and a_3, respectively. Observe, incidentally but importantly, that when we filled in the convex hull of the three pure actions to include mixed actions, the procedure we follow to find Bayes actions always produces a pure action as a Bayes action.

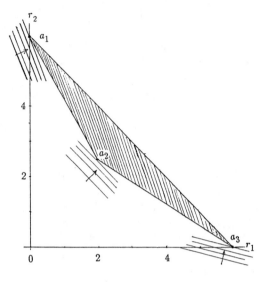

Figure 15-8

Notice, however, that there is a prior (for instance) that defines a line with slope $-3/2$, and this is the slope of the line segment joining $(6,7)$ and $(8,4)$. Any action point on this line segment has the same Bayes loss as the endpoints, so they are all Bayes actions. Still, this same Bayes loss is achieved at the endpoints, and we don't require a mixed action to achieve it. ■

It is true in general, as in the above example, that as concerns Bayes solutions, there is no gain in using mixed actions in preference to pure actions:

THEOREM 2 *The minimum Bayes loss using only pure actions is the same as the minimum Bayes loss using mixed actions:*

$$\min_a B(a) = \min_{\mathbf{p}} \mathcal{B}(\mathbf{p}).$$

To show this we observe first that among the distributions defined by $\mathbf{p}$ are the singular distributions with all the probability on the individual actions a; and this means that the left side is no smaller than the right side. To get the inequality in the other direction, we use the fact that average of any random quantity cannot be smaller than the minimum:

$$\min_a B(a) \le E_\mathbf{p} B(A) = \mathcal{B}(\mathbf{p}).$$

Since this holds for all **p**, the inequality holds for the **p** that minimizes the right hand side, as we wished to show.

The result of Theorem 2 is evident in the case of Example 15.5a from the graphical procedure we used there. That procedure also suggests that the Bayes action would be the same using *regret* as using loss, since the method does not depend on the location of the origin. In fact, this is always the case: The expected regret, given any p.d.f. $g(\theta)$ for Θ, is

$$\int r(\theta, a) g(\theta)\, d\theta = \int \ell(\theta, a) g(\theta)\, d\theta - \int \min_a \ell(\theta, a) g(\theta)\, d\theta,$$

and the second term on the right is independent of a. Thus, minimizing expected loss minimizes expected regret.

In the next example we illustrate another graphical method of finding a Bayes solution for the case of two states of nature, a method that points out a relationship between the minimax and Bayes principles.

EXAMPLE 15.5b We return to the problem of the preceding example, and an arbitrary distribution for nature: $P(\theta_1) = \gamma$, $P(\theta_2) = 1 - \gamma$. The Bayes regrets, plotted in Figure 5-9, are as follows:

$$B(a_1) = 0\gamma + 6(1-\gamma) = 6 - 6\gamma$$
$$B(a_2) = 2\gamma + 3(1-\gamma) = 3 - \gamma$$
$$B(a_3) = 6\gamma + 0(1-\gamma) = 6\gamma.$$

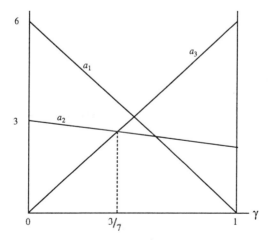

Figure 15-9

We see from the figure that if $0 \leq \gamma < 3/7$, the Bayes action is a_3; if $3/7 \leq \gamma \leq 3/5$, the Bayes action is a_2; and if $3/5 \leq \gamma \leq 1$, the Bayes action is a_1. When $\gamma = 3/7$, the Bayes regret of $18/7$ is the largest that would be incurred using Bayes actions. The distribution defined by this γ is said to be the *least favorable* distribution for Θ. We notice, too, that this largest Bayes regret of $18/7$ (using a Bayes action) is equal to the minimax regret! (See Example 15.4a.) ∎

In treating the decision problem as a "zero-sum" game, a game in which one player's loss is the other player's gain, the losses for the decision-maker are the negatives of the losses for nature. Thus, by inverting the graph in Figure 15-8 we get the graph of losses to nature for mixed actions. Using the graphical technique to determine a distribution for Nature that maximizes the minimum Bayes loss (the least favorable to the decision maker) is equivalent to finding a mixed action for nature that minimizes her maximum loss. A theorem in the theory of games states that by taking minimax mixed actions, each player is acting optimally, and the minimax loss for one is just the negative of the minimax loss for the other. That is why, in the example we have studied, the maximum of the minimum Bayes losses is equal to the expected loss incurred with the minimax mixed action.

15.6 Admissibility

In a general decision problem, actions are not simply ordered. However, if action a_1 results in losses that are less than those incurred with action a_2, for *all* states of nature, then there is clearly no point in considering a_2 at all.

DEFINITION Action a_1 is said to *dominate* action a_2 if and only if $\ell(a_1, \theta) \leq \ell(a_2, \theta)$ for all θ. The domination is *strict* if the inequality is strict for some θ.

DEFINITION An action is said to be *admissible* if and only if it is not strictly dominated by any other action.

(These definitions apply to mixed as well as to pure actions.) In the case of two states of nature, an action represented in terms of losses by the point P dominates all actions in the upper right quadrant with P at the vertex.

EXAMPLE 15-6a Again we take a decision problem with just two states of nature so that we can see the graphical representation. Consider this table of losses:

	a_1	a_2	a_3	a_4	a_5
θ_1	1	2	3	4	5
θ_2	5	1	5	3	0

In Figure 15-10 we see that there is no dominance relation among a_1, a_2, and a_4, but that a_1 strictly dominates a_3. Action a_4 is dominated by each action on the heavy portion of the lower left boundary of the convex set of all mixed actions, and action a_2 dominates all actions in the portion of that set which lie in the upper quadrant to its upper right. The only admissible actions are convex combinations of a_1 and a_2 and convex combinations of a_2 and a_5, but of course there are convex combinations of a_1, a_2, and a_5 that are not admissible. ∎

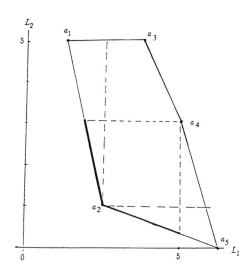

Figure 15-10

PROBLEMS

15-14. Referring to Problem 15-13, find the Bayes action,
(a) given that the distribution of M is uniform.
(b) given $M \sim \text{Bin}(5, 1/2)$.
(c) given $M \sim \text{Bin}(5, p)$, with p unknown.

15-15. Find the Bayes action for the loss table in Problem 15-9, repeated here for convenience, given probabilities $(\frac{1}{6}, \frac{1}{3}, \frac{1}{2})$ for $(\theta_1, \theta_2, \theta_3)$.

	a_1	a_2	a_3	a_4
θ_1	2	4	1	0
θ_2	-3	0	1	2
θ_3	-1	5	-1	-2

15-16. Given the following loss table (from Problem 15-11) and probabilities w and $1-w$ for θ_1 and θ_2, respectively, express the Bayes action as a function of w and find the least favorable distribution for Θ.

	a_1	a_2	a_3	a_4	a_5
θ_1	-1	1	0	2	5
θ_2	3	2	5	3	2

15-17. Given that $\Omega = \mathcal{A} = [0, 1]$, as in Problem 15-12, find the Bayes action if $\ell(\theta, a) = (\theta - a)^2$ and Θ has the p.d.f.
   (a) $g(\theta) = 1$.   (b) $g(\theta) = 6\theta(1-\theta)$.

15-18. Show that if an action a is inadmissible, there is a Bayes action that does not involve a.

15-19. Given the loss table of Problem 15-15, look for dominance relations, and show that a_2 and a_3 are not admissible in the class of mixed actions.

15.7 The risk function

When there is uncertainty about the state of nature, and there is an experiment whose distribution is defined by that state, the obvious thing to do in an attempt to make better decisions is to perform that experiment, resulting in data. Experimentation is usually not free, and it is necessary to balance the cost of experimentation and the potential gain of information.

When data are available that provide information about the unknown state of nature, a decision procedure should be based on those data. For the moment, let Z represent either the available data or a statistic calculated from the data. A **statistical decision function** is a function d from the data space $\mathcal{Z}$ to the action space $\mathcal{A}$: $a = d(z)$. The task of the statistician as decision-maker is to choose a suitable decision function d.

Inasmuch as Z is random, the action to be taken is random, so the loss is a random variable: $\ell(\theta, d(Z))$. This depends on θ both explicitly, in the first argument, and implicitly, in the distribution of Z. (If the distribution of Z did not depend on θ, then Z would not provide information about θ.) Since we are measuring losses as negative utility, we base our analyses on the expected value of the loss; this is the **risk function** (defined in §9.10 for tests):

$$R(\theta, d) = E_\theta[\ell(\theta, d(Z))].$$

This depends on θ because θ is an argument of the loss function, and also because the average E_θ is with respect to the distribution of Z as defined by θ.

In defining the risk function, we could have used (as some do) the regret function in place of the loss function. It is noteworthy that the expected regret turns out to be the same thing as what would be obtained by "regretizing" the risk defined in terms of loss:

$$E_\theta[r(\theta, d(Z))] = R(\theta, d) - \min_d R(\theta, d). \qquad (1)$$

Showing (1) in generality is left as an exercise for the reader, an exercise in inequalities and minima.

With the availability of data, we have converted the problem from one of selecting an action, given a loss function ℓ, to that of selecting a decision function d, given a risk function R. Mechanically, this is the same kind of problem, and we can apply either the minimax principle or the Bayesian approach to finding a decision rule. Thus, the decision function d^* that minimizes the maximum value of $R(\theta, d)$ over θ is a minimax rule. When we compare the minimax expected loss achieved using d^*, with the minimax loss using the minimax action, we find (as the difference) the worth of the data. Similarly, given a p.f. or p.d.f. $g(\theta)$ for Θ, the *Bayes risk* for the rule d is

$$B(d) = E_g[R(\Theta, d)].$$

The function $\tilde{d}$ that minimizes $B(d)$ is a Bayes decision rule.

EXAMPLE 15.7a We continue the decision problem and loss function of earlier examples (15.3a, 15.4a), but now suppose that we can observe a random variable Z with two possible values, a variable whose distribution depends on θ in known way. For convenient reference, we repeat the regret table:

	a_1	a_2	a_3
θ_1	0	2	6
θ_2	6	3	0
max	6	3	6

Suppose the distribution of Z is given by the following probabilities:

	θ_1	θ_2
z_1	.8	.1
z_2	.2	.9

Since Z has two possible values and $\mathcal{A}$ consists of three possible actions, there are 3^2 or nine functions from Z to $\mathcal{A}$. These are as follows:

	d_1	d_2	d_3	d_4	d_5	d_6	d_7	d_8	d_9
If $Z = z_1$, take action	a_1	a_2	a_3	a_1	a_2	a_1	a_3	a_2	a_3
If $Z = z_2$, take action	a_1	a_2	a_3	a_2	a_1	a_3	a_1	a_3	a_2

Observe that rules d_1, d_2, d_3 are just the pure actions. To calculate the risk $R(\theta, d)$ based on regret, we proceed as follows: For $\theta = \theta_1$ and $d = d_5$,

$$R(\theta_1, d_5) = E_{\theta_1}[r(\theta_1, d_5(Z))]$$

$$= r(\theta_1, d_5(z_1))f(z_1 \mid \theta_1) + r(\theta_1, d_5(z_2))f(z_2 \mid \theta_1)$$

$$= r(\theta_1, a_2) \times .8 + r(\theta_1, a_1) \times .2 = 2 \times .8 + 0 \times .2 = 1.6.$$

There remain 17 more such calculations of values of $R(\theta, d)$, resulting in the following table of risks:

	d_1	d_2	d_3	d_4	d_5	d_6	d_7	d_8	d_9
θ_1	0	2	6	.4	1.6	1.2	4.8	2.8	5.2
θ_2	6	3	0	3.3	5.7	.6	5.4	.3	2.7

The minimax pure decision rule is d_6. Figure 15-11 shows the convex set of all decision rules, represented by points whose coordinates are (R_1, R_2) the risks under θ_1 and θ_2. The original no-data rules (d_1, d_2, d_3) and their mixtures are shown as the shaded convex set. Note that with good use of the data one can generally do better. Some rules are worse, making poor use of the data.

From Figure 15-11 we can see that the minimax mixed rule is a mixture of d_4 and d_6. We choose the mixture

$$p\begin{pmatrix}.4\\3.3\end{pmatrix}+(1-p)\begin{pmatrix}1.2\\.6\end{pmatrix}$$

with the property that the coordinates are equal:

$$.4p+1.2(1-p)=3.3p+.6(1-p),$$

or $p = 6/35$. Using the data has allowed us to reduce the minimax risk from 3 to 19.8/19, or just over 1. The difference represents the worth of the data.

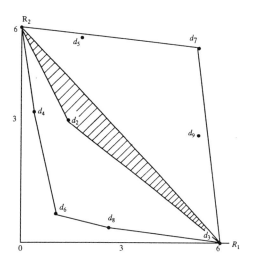

Figure 15-11

The Bayes decision rule, as usual, depends on the initial distribution we assume for Θ. If $\gamma = P(\theta_1)$, the Bayes risk for a decision rule d with risk coordinates (R_1, R_2) is

$$B(d) = \gamma R_1 + (1-\gamma) R_2.$$

The slope of the line segment joining d_1 and d_4 is $-27/4$; and if $\gamma = 27/31$, all d's on that line segment are Bayes decision rules. If $\gamma > 27/31$, the Bayes rule is d_1, and if γ is not too much smaller than $27/31$, the Bayes rule is d_4. With similar calculations for the other line segments forming the lower left boundary we obtain the following results:

If $\gamma > 27/31$, the Bayes rule is d_1.
If $27/35 < \gamma < 27/31$, the Bayes rule is d_4.
If $3/19 < \gamma < 27/35$, the Bayes rule is d_6.
If $3/35 < \gamma < 3/19$, the Bayes rule is d_8.
If $\gamma < 3/35$, the Bayes rule is d_3.

At the endpoints of these various γ-intervals, there are many Bayes decision rules—mixtures of the pure rules that define the endpoints.

If $\gamma = \frac{1}{2}$, the Bayes rule is d_6, and the Bayes risk is

$$B(d_6) = \tfrac{1}{2}(1.2 + .6) = .9.$$

Compare this with the Bayes loss using the Bayes pure action (from Example 15.5a), which is d_3: $B(d_3) = \tfrac{1}{2}(6 + 0) = 3$. The difference represents the value of the data. ∎

EXAMPLE 15.7b Consider testing H_0: $X \sim \mathcal{N}(0, 4)$ against H_1: $X \sim \mathcal{N}(1, 4)$, using a sample of size $n = 4$, and a critical region C. Assume this loss function:

$$\ell(0, a) = \begin{cases} A \text{ if } a = \text{rej. } H_0, \\ 0 \text{ if } a = \text{acc. } H_0. \end{cases} \quad \ell(1, a) = \begin{cases} 0 \text{ if } a = \text{rej. } H_0, \\ B \text{ if } a = \text{acc. } H_0. \end{cases}$$

The risk function is then

$$R(\theta, C) = \begin{cases} A \cdot P(C \mid H_0) = A\alpha, & \theta = 0, \\ B \cdot P(C^c \mid H_1) = B\beta, & \theta = 1. \end{cases}$$

The most powerful critical regions are of the form $\bar{X} > K$, with $\alpha = 1 - \Phi(K)$ and $\beta = \Phi(K - 1)$. The risks (R_0, R_1) are plotted in Figure 15-12 for the case $A = 2B = 2$. Because these critical regions are most powerful, the corresponding points constitute the lower left boundary of the set of all tests and are admissible. A minimax test corresponds to the point where $A\alpha = B\beta$, or $2[1 - \Phi(K)] = \Phi(K - 1)$. A trial and error solution yields $K \doteq .804$, and the corresponding point (R_0, R_1) is shown in Figure 15-12 as the point where the line $R_0 = R_1$ meets the curve representing the tests $\bar{X} > K$.

If we assume prior probabilities $g_0 = .3$ and $g_1 = .7$ (and again with $A = 2B = 2$), the Bayes risk for a critical region C is $B(C) = .6\alpha + .7\beta$. The most powerful critical regions are $C_K = [\bar{X} > K]$, and for these,

$$B(C_K) = .6[1 - \Phi(K)] + .7\,\Phi(K - 1).$$

This is minimized at $K = .5 - \log(7/6)$ where $R_0 \doteq .73$ and $R_1 \doteq .26$. ∎

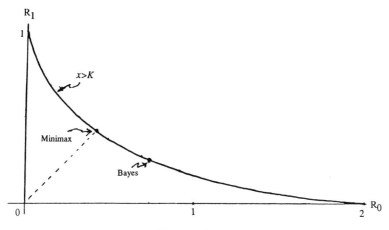

Figure 15-12

15.8 Using the posterior

We have seen how to take data into account by calculating the risk function. For a Bayesian solution, we applied a given a prior distribution for Θ to the risk function in order to find the Bayes risk for a particular decision rule d:

$$B(d) = E_g[R(\Theta, d)]$$
$$= \int R(\theta, d) g(\theta) \, d\theta = \int \int \ell(\theta, d(z)) f(z \mid \theta) g(\theta) \, dz \, d\theta.$$

According to Bayes' theorem, $f(z \mid \theta) g(\theta) = h(\theta \mid z) f(z)$, so we can rewrite the Bayes risk (with a change of order of integration) as a weighted integral of the expected posterior loss:

$$B(d) = \int f(z) \left\{ \int \ell(\theta, d(z)) h(\theta \mid z) d\theta \right\} dz = \int f(z) \, E_h[\ell(\Theta, d(z))] \, dz.$$

A Bayes procedure is one that minimizes $B(d)$, and from the last integral it is clear [since $f(z)$ is nonnegative] that if we minimize the expected posterior loss $E_h[\ell(\Theta, d(z))]$ for each z, then $B(d)$ will be as small as possible.

The impact of what we have just seen is that we can find the Bayes action corresponding to an observed value $Z = z$ by applying the *posterior* distribution to the original loss function. If we do this for each z, we'll have found a decision rule $a = d(z)$ that minimizes $B(d)$—the same rule that is obtained by applying the *prior* distribution to the risk function. But the beauty of this procedure is that we need only find the action that is appropriate for the z that we actually observe! We don't need to calculate an optimal action for a value of Z has a value observe. [As H. Chernoff says, we cross only the bridge we come to.]

Applying the posterior to the loss function is precisely what we did in our encounter with Bayesian methods in Chapters 8 and 9.

EXAMPLE 15.8a In Example 15.7a we found the Bayes decision rule for prior probabilities $g_1 = g_2 = \frac{1}{2}$: If $Z = z_1$, take action a_1; if $Z = z_2$, take action a_3. We now use the approach just described, first calculating the posterior probabilities when we observe $Z = z_1$:

$$h_1 = h(\theta_1 \mid z_1) = \frac{f(z_1 \mid \theta_1)g_1}{f(z_1 \mid \theta_1)g_1 + f(z_1 \mid \theta_2)g_2} = \frac{.8 \times .5}{.8 \times .5 + .1 \times .5} = \frac{8}{9}.$$

And of course, $h_2 = 1/9$. With these we find the expected losses for the three possible actions: 6/9, 19/9, and 48/9 for a_1, a_2, and a_3, respectively. The smallest is 6/9, so we take action a_1. If z_1 is what we observe, we need go no farther! But if for some reason we want the whole decision rule, we also apply the posterior probabilities corresponding to $Z = z_2$: (2/11, 9/11), to the losses for the three actions: 54/11, 31/11, 12/11. The smallest is for a_3, and we see that this approach again gives us d_6 as the Bayes rule. ∎

In this last example, with very small state space, data space, and action space, the advantage of applying the posterior to the original loss function is not dramatic. However, what we have done is to replace a problem in the calculus of variations (minimizing a functional over a set of functions) by a problem in ordinary calculus (minimizing a function over a set of numbers). This is usually a considerable simplification.

PROBLEMS

15-20. In estimating the parameter θ in $\text{Exp}(1/\theta)$, suppose we use the quadratic loss function $\ell(\theta, a) = (a - \theta)^2$ and a random sample of size n.
(a) Find the risk function for each of the following rules: $d_1(\mathbf{X}) = \bar{X}$, $d_2(\mathbf{X}) = X_n$, $d_3(\mathbf{X}) = \bar{X} + 1$, and $d_4(\mathbf{X}) = n\bar{X}/(n+1)$.
(b) Is there any dominance relation among the d_i?

15-21. A loss table and distributions for an observation Z are as follows:

	a_1	a_2	a_3
θ_1	0	3	4
θ_2	4	2	0
θ_3	5	6	3

	z_1	z_2
θ_1	.2	.8
θ_2	.4	.6
θ_3	.7	.3

Find all decision rules and corresponding risk functions, and then:
 (a) Find the Bayes rule and the value of the data, assuming a uniform distribution for the three states.
 (b) Find the minimax pure decision rule.
 (c) Find the maximum likelihood rule.

15-22. Consider the states θ_1: "rain" and θ_2: "no rain"; and the actions a_1: "stay at home," a_2: "go out, without an umbrella," and a_3: "go out with an umbrella." Assume these losses:

	a_1	a_2	a_3
θ_1	4	5	2
θ_2	4	0	5

Suppose a rain indicator or weather report has the reliability indicated in the following distributions for Z, which is a report of z_1: "rain" or z_2: "no rain."

	z_1	z_2
θ_1	.8	.2
θ_2	.1	.9

 (a) Find the nine decision rules and calculate the table of risks.
 (b) Plot them according to their risks under θ_1 and θ_2.
 (c) Find the minimax pure and minimax mixed decision rules.
 (d) Find the Bayes rule if a prior assessment (from looking out the window, say) is that there is a 50-50 chance of rain.

15-23. Suppose the state Θ and action A are numbers on the interval $[0, 1]$. Given the loss function $\ell(\theta, a) = (\theta - a)^2$, find the risk function for each of the following decision rules using data $Z \sim \text{Bin}(2, \theta)$:
 (a) $d_1 = Z/2$. (b) $d_2 = (Z+1)/4$.
Does one rule dominate the other? Which is better if, before observing Z, we have a uniform distribution for θ?

15-24. In the setting of Problem 15-22, suppose the observation Z has no uncertainty: The indicator says "rain" when it is going to rain and "no rain" when it is not going to rain. Calculate the risks; conclusion?

15-25. Again referring to Problem 15-22, suppose the weather report is unrelated to θ: $f(z_1 \mid \theta_1) = f(z_1 \mid \theta_2) = p$. Calculate some risks and make a conjecture as to the value of the data.

15-26. Referring to Problem 15-21, again assuming a uniform prior distribution over the three states, obtain the posterior distribution for θ when $Z = z_1$, and find the corresponding Bayes action. Verify that this is the action you would take had you been using the Bayes decision rule found in Problem 15-21.

15-27. Referring to Problem 15-22, suppose the rain indicator says "rain." Find the posterior for Θ when the prior odds on rain are 7:3, and verify that the Bayes action determined the posterior odds agrees with the action you would take if you followed the Bayes decision rule with these prior odds.

15.9 Randomized decision rules

In §15.3, we defined a mixed decision rule and in §9.4 we discussed randomized tests, which are actually decision rules. In both cases we introduced an extraneous experiment of chance, and we now take up the connection between these two types of randomization.

DEFINITION A *behavioral decision rule* assigns a probability distribution on the action space $\mathcal{A}$ to each outcome z of the data Z. It is carried out by sampling to obtain $Z = z$, and then performing the experiment assigned to that Z-value to see what action to take.

DEFINITION A *mixed decision rule* is a distribution on the space of decision functions $d(\cdot)$. It is carried out by performing the experiment defined by that distribution to obtain a particular rule d, obtaining data $Z = z$, and then taking the action assigned to z by d.

Applying each rule, given data $Z = z$, leads to an action. The question is whether there is a correspondence between mixed and behavioral rules. First consider an example.

EXAMPLE 15.9a Suppose there are three available actions; $\mathcal{A} = \{a_1, a_2, a_3\}$ and an observation Z which can take on one of two possible values. We repeat here the table that defines the nine (pure) decision rules from §15.7:

	d_1	d_2	d_3	d_4	d_5	d_6	d_7	d_8	d_9
If $Z = z_1$, take action	a_1	a_2	a_3	a_1	a_2	a_1	a_3	a_2	a_3
If $Z = z_2$, take action	a_1	a_2	a_3	a_2	a_1	a_3	a_1	a_3	a_2

A mixed decision rule is a probability vector $(q_1, \ldots, q_9)$, assigning probability q_i to rule d_i. We then calculate as follows:

$$P(a_1 \mid z_1) = q_1 + q_4 + q_6 = \phi_1(z_1)$$
$$P(a_2 \mid z_1) = q_2 + q_5 + q_8 = \phi_2(z_1)$$
$$P(a_3 \mid z_1) = q_3 + q_7 + q_9 = \phi_3(z_1)$$
$$P(a_1 \mid z_2) = q_1 + q_5 + q_7 = \phi_1(z_2)$$
$$P(a_2 \mid z_2) = q_2 + q_4 + q_9 = \phi_2(z_2)$$
$$P(a_3 \mid z_2) = q_3 + q_6 + q_8 = \phi_3(z_2).$$

Clearly, $\phi_1 + \phi_2 + \phi_3 = 1$ for each z, so for each z we have defined a probability vector (ϕ_1, ϕ_2, ϕ_3) on the action space—a behavioral decision rule.

Conversely, suppose a behavioral rule is given—a probability vector for $\mathcal{A}$ assigned to each z: $[\phi_1(z), \phi_2(z), \phi_3(z)]$. To be specific, let it be defined as follows:

$$\phi_1(z_1) = .2, \quad \phi_2(z_1) = .5, \quad \phi_3(z_1) = .3,$$
$$\phi_1(z_2) = .6, \quad \phi_2(z_2) = .3, \quad \phi_3(z_2) = .1.$$

Setting these probabilities equal to the above sums of q's we have six equations in nine unknowns, a system with many solutions. One solution is the vector $(.12, .15, .03, .06, .30, .02, .18, .05, .09)$, obtained as joint probabilities when the components ϕ_i define independent marginal distributions. ∎

The procedure in this example is easily generalized to cases in which the actions and the possible values of Z are finite in number: $\mathcal{A} = \{a_1, \ldots, a_J\}$ and $\mathcal{Z} = \{z_1, \ldots, z_K\}$. There are $J^K = R$ decision functions $d_r(\cdot)$—functions from $\mathcal{Z}$ to $\mathcal{A}$. A *mixed* decision rule is a probability vector $\mathbf{q} = (q_1, \ldots, q_R)$. For such a rule $\mathbf{q}$ we can calculate the probabilities of the various actions, given an observation $Z = z_k$:

$$\phi_k(a_j) = P(A = a_j \mid z_k) = \sum_r d_{rj}(z_k) q_r,$$

where

$$d_{rj}(z) = \begin{cases} 0 & \text{if } d_r(z) \neq a_j, \\ 1 & \text{if } d_r(z) = a_j. \end{cases}$$

These RK ϕ's define a behavioral rule: $\phi(a) = [\phi_1(a), ..., \phi_K(a)]$, where for each a, $0 \leq \phi_k(a) \leq 1$ and $\sum \phi_k(a) = 1$. The risk function for the rule $\mathbf{q}$ is

$$R(\theta, \mathbf{q}) = \sum_r \sum_z \ell(\theta, d_r(z)) f(z \mid \theta) q_r$$

$$= \sum_r \sum_j \ell(\theta, a_j) \Big\{ \sum f(z \mid \theta) \Big\} q_r,$$

where the inner sum is taken over the set S_{rj} of z's such that $d_r(z) = a_j$:

$$\sum_{S_{rj}} f(z \mid \theta) = \sum_z d_{rj}(z) f(z \mid \theta).$$

Interchanging order of summations, we see that the risk function for $\mathbf{z}$ is the same as the risk function corresponding to the behavioral rule ϕ:

$$R(\theta, \mathbf{z}) = \sum_j \ell(\theta, a_j) \Big\{ \sum_z \Big(\sum_r d_{rj}(z) q_r \Big) f(z \mid \theta) \Big\}$$

$$= \sum_j \ell(\theta, a_j) \sum_z \phi_j(z) f(z \mid \theta).$$

Conversely, suppose we begin with a behavioral rule ϕ, a probability K-vector, where each component is a function of a. The components of a mixed decision rule $\mathbf{q}$ are probabilities q_r assigned to the pure decision rules d_r, where $d_r(z_k) = a_{j_k}$, where each of the R r's corresponds to one of the R distinct vectors $\mathbf{j} = (j_1, ..., j_K)$, each j_k being one of the integers 1 to J. Define the probabilities q_r as products:

$$q_r = \phi_{j_1}(z_1) \cdots \phi_{j_K}(z_K).$$

It can be verified that these define a probability vector, and that starting with this mixture one is led (by the development of the preceding paragraph) back to the given behavioral rule.

15.10 Monotone problems and procedures

We consider problems in which the relevant family of distributions is indexed by a single real parameter θ and has a monotone likelihood ratio. (See §9.12.) We call the problem monotone if, in addition, the loss function has a monotone character, defined as follows.

A regret function $r(\theta, a)$ defines a function of θ for each action—a family of regret curves, each identified by a particular action. These curves will lie above the θ-axis, or on it for some states θ. When a regret is 0 at some θ, the corresponding action is optimal for that state. The set of θ-values for which the regret is zero for a particular action is the *optimality set* for that action. A regret function r is a **monotone regret function** if the regret curves it defines satisfy these conditions:

1. The optimality sets for the various actions are intervals (finite, infinite, single point, or empty) which constitute a partition of the parameter space.
2. For each action, the regret curve does not decreas as θ moves away from the optimality set in either direction.
3. Any two regret curves intersect at most once: The difference $r(\theta, a_1) - r(\theta, a_2)$ changes sign at most once for given a_1 and a_2.

Figure 15-13 shows examples of monotone regret functions. Figure 15-13(a) shows the regret $(\theta - a)^2$, used in estimation problems, for just three values of the action a. For each a there is only one parameter value $(\theta = a)$ for which a is optimal, and these points of zero regret partition the parameter space. It is also clear that conditions (2) and (3) above are satisfied. Figures 15-13(b) and (c) show monotone regret functions appropriate for the two- and three-action cases, respectively.

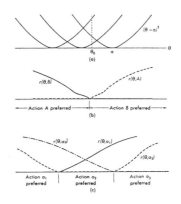

Figure 15-13

Figure 15-14 shows regret functions that are *not* monotone, for the three- and two-action problems. In 15-14(a), the first two conditions are fulfilled but (3) is not, since $r(\theta, a_1) - r(\theta, a_2)$ changes sign twice. In 15-14(b), all three conditions are violated.

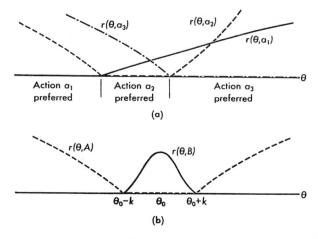

Figure 15-14

EXAMPLE 15.10a Lots of a certain product (say, eggs) are to be classified into three grades according to the average lot quality θ: $\theta > \theta_2$ for grade A, $\theta_1 < \theta < \theta_2$ for grade B, and $\theta < \theta_1$ for grade C. The classification is to be done on the basis of a sample from each lot. Suppose the penalties for misclassification are 1 unit if the quality assigned is off by one grade, 2 if it is off by two grades, and 0 if it is correct. This is a regret function and is satisfied by the three conditions for a monotone regret. ∎

To describe the known results concerning monotone problems, we need to define the notion of a **monotone procedure**. Suppose a decision is to be based on data for which the likelihood ratio is monotone in terms of the statistic $T = t(\mathbf{X})$. The decision function $d(\mathbf{X})$ is said to be *monotone* provided that for any two given samples $\mathbf{X}$ and $\mathbf{Y}$, the inequality $t(\mathbf{X}) \leq t(\mathbf{Y})$ is equivalent to the statement that the optimality set for the action $d(\mathbf{X})$ is not to the right of the optimality set for $d(\mathbf{Y})$. This means simply that the axis of values of $t(\mathbf{X})$ is partitioned by the decision rule d in a manner that corresponds to the partition of the parameter axis by the optimality sets.

Thus, when there are just two actions, the parameter axis is partitioned by the regret function into two sets, $\theta < \theta^*$ and $\theta > \theta^*$ (with $\theta = \theta^*$ in either); and a monotone procedure is defined by a value t^* that divides the t-axis into $T < t^*$ and $T > t^*$, the former inequality calling for the action corresponding to $\theta < \theta^*$, and the latter, for the action corresponding to $\theta > \theta^*$. If there are three available actions, a monotone rule is defined by points t_1 and t_2 such that $T < t_1$ calls for the action that is best for small θ, $T > t_2$ calls for the action that is best for large θ, and $t_1 < T < t_2$ calls for the action that is best for intermediate θ. And so on, for any finite number of actions. In an estimation problem, a monotone procedure takes, as an estimator for θ, a monotone increasing function of $t(\mathbf{X})$.

The action corresponding to one of the $n-1$ boundary points (in the case of n actions) can be assigned by putting the boundary point into one or the other of the contiguous sets, either by specifying this at the outset or by the toss of a (suitably biased) coin. In the latter case, the rule is a behavioral decision rule, defined by a set a set of boundary points and a set of probabilities, one for each boundary point.

EXAMPLE 15-10b Returning to the (egg) classification problem of Example 15-10a, suppose that the classification is to be made on the basis of the weight of a single egg taken from the lot. If this weight X is normally distributed with given variance and mean μ, the family of distributions of X has a monotone likelihood ratio in terms of X, since these distributions are in the exponential family with $R(x) = x$. A monotone procedure is then defined by the constants x_1, x_2, where $x_1 < x_2$, and carried out as follows: If $X < x_1$, classify the lot as C; if $x_1 < X < x_2$, classify it as B, and if $X > x_2$, classify it as A. [If X is indeed normal, the events $X = x_i$ have probability 0. In practice, because of round-off, one would need to define probabilities p_1 and p_2 for randomizing at the boundary points.] ∎

The class of all monotone procedures has been found to be *essentially complete* for monotone problems with finitely many actions, and for the monotone estimation problem.[4] That is, for any procedure *not* in the class there is one *in* the class that is at least as good (as measured by risk) for all θ. It can also be shown, subject to increasingly less general conditions as n increases, that monotone procedures for a monotone problem with n actions are admissible.[5] (That is, not only do the monotone procedures include essentially all of the admissible ones, they include nothing but admissible ones.)

PROBLEMS

15-28. Refer to the decision problem given in Problem 15-22.
(a) Express the minimax mixed decision rule obtained in that problem

---

[4] These results are obtained by S. Karlin and H. Rubin in "The theory of decision procedures for distributions with monotone likelihood ratios," *Ann. Math. Stat.* 27, 272-299 (1956).

[5] Ibid.

as a behavioral decision rule.

(b) Express the following behavioral rule as a mixture of nonrandomized decision function: If $Z = z_1$, toss two coins and take action a_3 unless two heads turn up, in which case take action a_3. If $Z = z_2$, toss two coins and take action a_3 unless two heads turn up, in which case take action a_3.

15-29. An observation X is normal with unit variance and mean $\mu = -2$ when $\theta = \theta_1$, $\mu = 0$ when $\theta = \theta_2$, and $\mu = 2$ when $\theta = \theta_3$. For possible actions A, B, C, losses are as follows:

	θ_1	θ_2	θ_3
A	0	1	2
B	1	0	1
C	2	1	0

Consider these four decision rules:
$d_1(X) = A$ if $X < -1$, B if $-1 < X < 1$, and C if $X > 1$,
$d_2(X) = A$ if $X < -1.5$, B if $-1.5 < X < 1.5$, and C if $X > 1.5$,
$d_3(X) = A$ if $X > 0$, and C if $X < 0$,
$d_4(X) = B$ if $X < -1$, C if $-1 < X < 1$, and A if $X > 1$.
Obtain the risk function for each rule. Which ones are monotone?

15-30. Verify what is claimed to be verifiable at the end of §15.9.

TABLES

   I. Values of the Standard Normal Distribution Function   576
  Ia. Percentiles of the Standard Normal Distribution   578
  Ib. Two-tail Probabilities for the Standard Normal Distribution   578
  II. Poisson Distribution Function   579
 IIIa. Tail-probabilities of the t-distribution   582
 IIIb. Percentiles of the t-distribution   585
 IVa. 95th Percentiles of the F-distribution   586
 IVb. 99th Percentiles of the F-distribution   587
  Va. Tail Areas of the Chi-square Distribution   588
  Vb. Percentiles of the Chi-square Distribution   590
  VI. One-sample Signed-rank Tail-probabilities   591
 VII. Two-sample Wilcoxon Rank-sum Tail-probabilities   592
VIII. One-sample Kolmogorov-Smirnov Critical Values   594
  IX. Two-sample Kolmogorov-Smirnov Acceptance Limits   595
   X. Tail-probabilities of the Lilliefors Distribution   596
  XI. Expected Values of Order Statistics from $\mathcal{N}(0, 1)$ 597
 XII. Selected Percentage Points of the Shapiro-Wilk W 598
XIII. Distribution of the Standardized Range, $W = R/\sigma$ 599

TABLE I Values of the Standard Normal Distribution Function

$$\Phi(z) = \int_{-\infty}^{z} \frac{1}{\sqrt{2\pi}} \exp\left(-\tfrac{1}{2}u^2\right) du = P(Z \leq z)$$

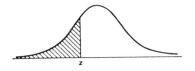

z	0	1	2	3	4	5	6	7	8	9
−3.	.0013	.0010	.0007	.0005	.0003	.0002	.0002	.0001	.0001	.0000
−2.9	.0019	.0018	.0017	.0017	.0016	.0016	.0015	.0015	.0014	.0014
−2.8	.0026	.0025	.0024	.0023	.0023	.0022	.0021	.0021	.0020	.0019
−2.7	.0035	.0034	.0033	.0032	.0031	.0030	.0029	.0028	.0027	.0026
−2.6	.0047	.0045	.0044	.0043	.0041	.0040	.0039	.0038	.0037	.0036
−2.5	.0062	.0060	.0059	.0057	.0055	.0054	.0052	.0051	.0049	.0048
−2.4	.0082	.0080	.0078	.0075	.0073	.0071	.0069	.0068	.0066	.0064
−2.3	.0107	.0104	.0102	.0099	.0096	.0094	.0091	.0089	.0087	.0084
−2.2	.0139	.0136	.0132	.0129	.0126	.0122	.0119	.0116	.0113	.0110
−2.1	.0179	.0174	.0170	.0166	.0162	.0158	.0154	.0150	.0146	.0143
−2.0	.0228	.0222	.0217	.0212	.0207	.0202	.0197	.0192	.0188	.0183
−1.9	.0287	.0281	.0274	.0268	.0262	.0256	.0250	.0244	.0238	.0233
−1.8	.0359	.0352	.0344	.0336	.0329	.0322	.0314	.0307	.0300	.0294
−1.7	.0446	.0436	.0427	.0418	.0409	.0401	.0392	.0384	.0375	.0367
−1.6	.0548	.0537	.0526	.0516	.0505	.0495	.0485	.0475	.0465	.0455
−1.5	.0668	.0655	.0643	.0630	.0618	.0606	.0594	.0582	.0570	.0559
−1.4	.0808	.0793	.0778	.0764	.0749	.0735	.0722	.0708	.0694	.0681
−1.3	.0968	.0951	.0934	.0918	.0901	.0885	.0869	.0853	.0838	.0823
−1.2	.1151	.1131	.1112	.1093	.1075	.1056	.1038	.1020	.1003	.0985
−1.1	.1357	.1335	.1314	.1292	.1271	.1251	.1230	.1210	.1190	.1170
−1.0	.1587	.1562	.1539	.1515	.1492	.1469	.1446	.1423	.1401	.1379
−.9	.1841	.1814	.1788	.1762	.1736	.1711	.1685	.1660	.1635	.1611
−.8	.2119	.2090	.2061	.2033	.2005	.1977	.1949	.1922	.1894	.1867
−.7	.2420	.2389	.2358	.2327	.2297	.2266	.2236	.2206	.2177	.2148
−.6	.2743	.2709	.2676	.2643	.2611	.2578	.2546	.2514	.2483	.2451
−.5	.3085	.3050	.3015	.2981	.2946	.2912	.2877	.2843	.2810	.2776
−.4	.3446	.3409	.3372	.3336	.3300	.3264	.3228	.3192	.3156	.3121
−.3	.3821	.3783	.3745	.3707	.3669	.3632	.3594	.3557	.3520	.3483
−.2	.4207	.4168	.4129	.4090	.4052	.4013	.3974	.3936	.3897	.3859
−.1	.4602	.4562	.4522	.4483	.4443	.4404	.4364	.4325	.4286	.4247
−0	.5000	.4960	.4920	.4880	.4840	.4801	.4761	.4721	.4681	.4641

TABLE I (cont.)

z	0	1	2	3	4	5	6	7	8	9
.0	.5000	.5040	.5080	.5120	.5160	.5199	.5239	.5279	.5319	.5359
.1	.5398	.5438	.5478	.5517	.5557	.5596	.5636	.5675	.5714	.5753
.2	.5793	.5832	.5871	.5910	.5948	.5987	.6026	.6064	.6103	.6141
.3	.6179	.6217	.6255	.6293	.6331	.6368	.6406	.6443	.6480	.6517
.4	.6554	.6591	.6628	.6664	.6700	.6736	.6772	.6808	.6844	.6879
.5	.6915	.6950	.6985	.7019	.7054	.7088	.7123	.7157	.7190	.7224
.6	.7257	.7291	.7324	.7357	.7389	.7422	.7454	.7486	.7517	.7549
.7	.7580	.7611	.7642	.7673	.7703	.7734	.7764	.7794	.7823	.7852
.8	.7881	.7910	.7939	.7967	.7995	.8023	.8051	.8078	.8106	.8133
.9	.8159	.8186	.8212	.8238	.8264	.8289	.8315	.8340	.8365	.8389
1.0	.8413	.8438	.8461	.8485	.8508	.8531	.8554	.8577	.8599	.8621
1.1	.8643	.8665	.8686	.8708	.8729	.8749	.8770	.8790	.8810	.8830
1.2	.8849	.8869	.8888	.8907	.8925	.8944	.8962	.8980	.8997	.9015
1.3	.9032	.9049	.9066	.9082	.9099	.9115	.9131	.9147	.9162	.9177
1.4	.9192	.9207	.9222	.9236	.9251	.9265	.9278	.9292	.9306	.9319
1.5	.9332	.9345	.9357	.9370	.9382	.9394	.9406	.9418	.9430	.9441
1.6	.9452	.9463	.9474	.9484	.9495	.9505	.9515	.9525	.9535	.9545
1.7	.9554	.9564	.9573	.9582	.9591	.9599	.9608	.9616	.9625	.9633
1.8	.9641	.9648	.9656	.9664	.9671	.9678	.9686	.9693	.9700	.9706
1.9	.9713	.9719	.9726	.9732	.9738	.9744	.9750	.9756	.9762	.9767
2.0	.9772	.9778	.9783	.9788	.9793	.9798	.9803	.9808	.9812	.9817
2.1	.9821	.9826	.9830	.9834	.9838	.9842	.9846	.9850	.9854	.9857
2.2	.9861	.9864	.9868	.9871	.9874	.9878	.9881	.9884	.9887	.9890
2.3	.9893	.9896	.9898	.9901	.9904	.9906	.9909	.9911	.9913	.9916
2.4	.9918	.9920	.9922	.9925	.9927	.9929	.9931	.9932	.9934	.9936
2.5	.9938	.9940	.9941	.9943	.9945	.9946	.9948	.9949	.9951	.9952
2.6	.9953	.9955	.9956	.9957	.9959	.9960	.9961	.9962	.9963	.9964
2.7	.9965	.9966	.9967	.9968	.9969	.9970	.9971	.9972	.9973	.9974
2.8	.9974	.9975	.9976	.9977	.9977	.9978	.9979	.9979	.9980	.9981
2.9	.9981	.9982	.9982	.9983	.9984	.9984	.9985	.9985	.9986	.9986
3.	.9987	.9990	.9993	.9995	.9997	.9998	.9998	.9999	.9999	1.0000

Note 1: If a normal variable X is not "standard," its values must be "standardized": $Z = (X - \mu)/\sigma$, i.e., $P(X \leq x) = \Phi[(x - \mu)/\sigma]$.

Note 2: For "two-tail" probabilities, see Table Ib.

Note 3: For $z \geq 4$, $P(Z > z) \doteq \dfrac{1}{z\sqrt{2\pi}} e^{-z^2/2}$

Note 4: Entries opposite 3 and −3 are for 3.0, 3.1, 3.2, etc., and −3.0, −3.1, etc., respectively.

TABLE 1a Percentiles of the Standard Normal Distribution		TABLE 1b Two-Tail Probabilities for the Standard Normal Distribution	
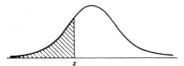		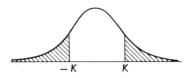	
$P(Z \leq z)$	z	$P(|Z| > K)$	K
.001	−3.0902	.001	3.2905
.005	−2.5758	.002	3.0902
.01	−2.3263	.005	2.80703
.02	−2.0537	.01	2.5758
.03	−1.8808	.02	2.3263
.04	−1.7507	.03	2.1701
.05	−1.6449	.04	2.0537
.10	−1.2816	.05	1.9600
.15	−1.0364	.06	1.8808
.20	−.8416	.08	1.7507
.30	−.5244	.10	1.6449
.40	−.2533	.15	1.4395
.50	0	.20	1.2816
.60	.2533	.30	1.0364
.70	.5244		
.80	.8416		
.85	1.0364		
.90	1.2816		
.95	1.6449		
.96	1.7507		
.97	1.8808		
.98	2.0537		
.99	2.3263		
.995	2.5758		
.999	3.0902		

TABLE II Poisson Distribution Function:

$$F(c) = \sum_{k=0}^{c} \frac{m^k e^{-m}}{k!} \quad (m = \text{expected value})$$

c	.02	.04	.06	.08	.10	.15	.20	.25	.30	.35	.40
0	.980	.961	.942	.923	.905	.861	.819	.779	.741	.705	.670
1	1.000	.999	.998	.997	.995	.990	.982	.974	.963	.951	.938
2		1.000	1.000	1.000	1.000	1.000	.999	.998	.996	.994	.992
3							1.000	1.000	1.000	1.000	.999
4											1.000

c	.45	.50	.55	.60	.65	.70	.75	.80	.85	.90	.95
0	.638	.607	.577	.549	.522	.497	.472	.449	.427	.407	.387
1	.925	.910	.894	.878	.861	.844	.827	.809	.791	.772	.754
2	.989	.986	.982	.977	.972	.966	.959	.953	.945	.937	.929
3	.999	.998	.998	.997	.996	.994	.993	.991	.989	.987	.984
4	1.000	1.000	1.000	1.000	.999	.999	.999	.999	.998	.998	.997
5					1.000	1.000	1.000	1.000	1.000	1.000	1.000

c	1.0	1.1	1.2	1.3	1.4	1.5	1.6	1.7	1.8	1.9	2.0
0	.368	.333	.301	.273	.247	.223	.202	.183	.165	.150	.135
1	.736	.699	.663	.627	.592	.558	.525	.493	.463	.434	.406
2	.920	.900	.879	.857	.833	.809	.783	.757	.731	.704	.677
3	.981	.974	.966	.957	.946	.934	.921	.907	.891	.875	.857
4	.996	.995	.992	.989	.986	.981	.976	.970	.964	.956	.947
5	.999	.999	.998	.998	.997	.996	.994	.992	.990	.987	.983
6	1.000	1.000	1.000	1.000	.999	.999	.999	.998	.997	.997	.995
7					1.000	1.000	1.000	1.000	.999	.999	.999
8									1.000	1.000	1.000

c	2.2	2.4	2.6	2.8	3.0	3.2	3.4	3.6	3.8	4.0	4.2
0	.111	.091	.074	.061	.050	.041	.033	.027	.022	.018	.015
1	.355	.308	.267	.231	.199	.171	.147	.126	.107	.092	.078
2	.623	.570	.518	.469	.423	.380	.340	.303	.269	.238	.210
3	.819	.779	.736	.692	.647	.603	.558	.515	.473	.433	.395
4	.928	.904	.877	.848	.815	.781	.744	.706	.668	.629	.590
5	.975	.964	.951	.935	.916	.895	.871	.844	.816	.785	.753
6	.993	.988	.983	.976	.966	.955	.942	.927	.909	.889	.867
7	.998	.997	.995	.992	.988	.983	.977	.969	.960	.949	.936
8	1.000	.999	.999	.998	.996	.994	.992	.988	.984	.979	.972
9		1.000	1.000	.999	.999	.998	.997	.996	.994	.992	.989
10				1.000	1.000	1.000	.999	.999	.998	.997	.996
11							1.000	1.000	.999	.999	.999
12									1.000	1.000	1.000

TABLE II Poisson Distribution Function (cont.)

c	4.4	4.6	4.8	5.0	5.2	5.4	5.6	5.8	6.0	6.2	6.4
0	.012	.010	.008	.007	.006	.005	.004	.003	.002	.002	.002
1	.066	.056	.048	.040	.034	.029	.024	.021	.017	.015	.012
2	.185	.163	.143	.125	.109	.095	.082	.072	.062	.054	.046
3	.359	.326	.294	.265	.238	.213	.191	.170	.151	.134	.119
4	.551	.513	.476	.440	.406	.373	.342	.313	.285	.259	.235
5	.720	.686	.651	.616	.581	.546	.512	.478	.446	.414	.384
6	.844	.818	.791	.762	.732	.702	.670	.638	.606	.574	.542
7	.921	.905	.887	.867	.845	.822	.797	.771	.744	.716	.687
8	.964	.955	.944	.932	.918	.903	.886	.867	.847	.826	.803
9	.985	.980	.975	.968	.960	.951	.941	.929	.916	.902	.886
10	.994	.992	.990	.986	.982	.977	.972	.965	.957	.949	.939
11	.998	.997	.996	.995	.993	.990	.988	.984	.980	.975	.969
12	.999	.999	.999	.998	.997	.996	.995	.993	.991	.989	.986
13	1.000	1.000	1.000	.999	.999	.999	.998	.997	.996	.995	.994
14				1.000	1.000	1.000	.999	.999	.999	.998	.997
15							1.000	1.000	.999	.999	.999
16									1.000	1.000	1.000

c	6.6	6.8	7.0	7.2	7.4	7.6	7.8	8.0	8.5	9.0	9.5
0	.001	.001	.001	.001	.001	.001	.000	.000	.000	.000	.000
1	.010	.009	.007	.006	.005	.004	.004	.003	.002	.001	.001
2	.040	.034	.030	.025	.022	.019	.016	.014	.009	.006	.004
3	.105	.093	.082	.072	.063	.055	.048	.042	.030	.021	.015
4	.213	.192	.173	.156	.140	.125	.112	.100	.074	.055	.040
5	.355	.327	.301	.276	.253	.231	.210	.191	.150	.116	.089
6	.511	.480	.450	.420	.392	.365	.338	.313	.256	.207	.165
7	.658	.628	.599	.569	.539	.510	.481	.453	.386	.324	.269
8	.780	.755	.729	.703	.676	.648	.620	.593	.523	.456	.392
9	.869	.850	.830	.810	.788	.765	.741	.717	.653	.587	.522
10	.927	.915	.901	.887	.871	.854	.835	.816	.763	.706	.645
11	.963	.955	.947	.937	.926	.915	.902	.888	.849	.803	.752
12	.982	.978	.973	.967	.961	.954	.945	.936	.909	.876	.836
13	.992	.990	.987	.984	.980	.976	.971	.966	.949	.926	.898
14	.997	.996	.994	.993	.991	.989	.986	.983	.973	.959	.940
15	.999	.998	.998	.997	.996	.995	.993	.992	.986	.978	.967
16	.999	.999	.999	.999	.998	.998	.997	.996	.993	.989	.982
17	1.000	1.000	1.000	.999	.999	.999	.999	.998	.997	.995	.991
18				1.000	1.000	1.000	1.000	.999	.999	.998	.996
19								1.000	.999	.999	.998
20									1.000	1.000	.999
21											1.000

TABLE II Poisson Distribution Function (cont.)

c	10.0	10.5	11.0	11.5	12.0	12.5	13.0	13.5	14.0	14.5	15.0
2	.003	.002	.001	.001	.001	.000					
3	.010	.007	.005	.003	.002	.002	.001	.001	.000		
4	.029	.021	.015	.011	.008	.005	.004	.003	.002	.001	.001
5	.067	.050	.038	.028	.020	.015	.011	.008	.006	.004	.003
6	.130	.102	.079	.060	.046	.035	.026	.019	.014	.010	.008
7	.220	.179	.143	.114	.090	.070	.054	.041	.032	.024	.018
8	.333	.279	.232	.191	.155	.125	.100	.079	.062	.048	.037
9	.458	.397	.341	.289	.242	.201	.166	.135	.109	.088	.070
10	.583	.521	.460	.402	.347	.297	.252	.211	.176	.145	.118
11	.697	.639	.579	.520	.462	.406	.353	.304	.260	.220	.185
12	.792	.742	.689	.633	.576	.519	.463	.409	.358	.311	.268
13	.864	.825	.781	.733	.682	.628	.573	.518	.464	.413	.363
14	.917	.888	.854	.815	.772	.725	.675	.623	.570	.518	.466
15	.951	.932	.907	.878	.844	.806	.764	.718	.669	.619	.568
16	.973	.960	.944	.924	.899	.869	.835	.798	.756	.711	.664
17	.986	.978	.968	.954	.937	.916	.890	.861	.827	.790	.749
18	.993	.988	.982	.974	.963	.948	.930	.908	.883	.853	.819
19	.997	.994	.991	.986	.979	.969	.957	.942	.923	.901	.875
20	.998	.997	.995	.992	.988	.983	.975	.965	.952	.936	.917
21	.999	.999	.998	.996	.994	.991	.986	.980	.971	.960	.947
22	1.000	.999	.999	.998	.997	.995	.992	.989	.983	.976	.967
23		1.000	1.000	.999	.999	.998	.996	.994	.991	.986	.981
24				1.000	.999	.999	.998	.997	.995	.992	.989
25					1.000	.999	.999	.998	.997	.996	.994
26						1.000	1.000	.999	.999	.998	.997
27								1.000	.999	.999	.998
28									1.000	.999	.999
29										1.000	1.000

TABLE IIIa Tail-probabilities of the t-distribution

	Degrees of freedom									
t	4	5	6	7	8	9	10	11	12	13
1.0	.186	.181	.178	.175	.173	.172	.170	.169	.169	.168
1.1	.166	.160	.157	.154	.152	.150	.149	.147	.146	.146
1.2	.145	.142	.138	.135	.132	.130	.129	.128	.127	.126
1.3	.131	.125	.121	.117	.115	.113	.111	.110	.109	.108
1.4	.117	.110	.105	.102	.100	.098	.096	.095	.093	.092
1.5	.104	.097	.092	.089	.086	.084	.082	.081	.080	.079
1.6	.092	.085	.080	.077	.074	.072	.070	.069	.068	.067
1.7	.082	.075	.070	.066	.064	.062	.060	.059	.057	.056
1.8	.073	.066	.061	.057	.055	.053	.051	.050	.049	.048
1.9	.065	.058	.053	.050	.047	.045	.043	.042	.041	.040
2.0	.058	.051	.046	.043	.040	.038	.037	.035	.034	.033
2.1	.052	.045	.040	.037	.034	.033	.031	.030	.029	.028
2.2	.046	.040	.035	.032	.029	.028	.026	.025	.024	.023
2.3	.042	.035	.031	.027	.025	.023	.022	.021	.020	.019
2.4	.037	.031	.027	.024	.022	.020	.019	.018	.017	.016
2.5	.033	.027	.023	.020	.018	.017	.016	.015	.014	.013
2.6	.030	.024	.020	.018	.016	.014	.013	.012	.012	.011
2.7	.027	.021	.018	.015	.014	.012	.011	.010	.010	.009
2.8	.025	.019	.016	.013	.012	.010	.009	.009	.008	.008
2.9	.022	.017	.014	.011	.010	.009	.008	.007	.007	.006
3.0	.020	.015	.012	.010	.009	.007	.007	.006	.006	.005
3.1	.018	.013	.011	.009	.007	.006	.006	.005	.005	.004
3.2	.017	.012	.009	.008	.006	.005	.005	.004	.004	.003
3.3	.015	.011	.008	.007	.005	.005	.004	.004	.003	.003
3.4	.014	.040	.007	.006	.004	.003	.003	.002	.002	.002
3.5	.013	.009	.006	.005	.004	.003	.003	.002	.002	.002
3.6	.012	.008	.006	.004	.003	.003	.002	.002	.002	.002
3.7	.011	.007	.005	.004	.003	.002	.002	.002	.002	.001
3.8	.010	.006	.004	.003	.003	.002	.002	.001	.001	.001
3.9	.009	.006	.004	.003	.002	.002	.001	.001	.001	.001
4.0	.008	.005	.004	.003	.002	.002	.001	.001	.001	.001
4.5	.006	.003	.002	.001	.001	.001	.001	.000	.000	.000
5.0	.004	.002	.001	.001	.000	.000	.000	.000	.000	.000

TABLE IIIa Tail-probabilities of the t-distribution (cont.)

	Degrees of freedom								
t	14	15	16	17	18	19	20	21	22	23
1.0	.167	.167	.166	.166	.165	.165	.165	.164	.164	.164
1.1	.145	.144	.144	.143	.143	.143	.142	.142	.142	.141
1.2	.125	.124	.124	.123	.123	.122	.122	.122	.121	.121
1.3	.107	.107	.106	.105	.105	.105	.104	.104	.104	.103
1.4	.092	.091	.090	.090	.089	.089	.088	.088	.088	.087
1.5	.078	.077	.077	.076	.075	.075	.075	.074	.074	.074
1.6	.066	.065	.065	.064	.064	.063	.063	.062	.062	.062
1.7	.056	.055	.054	.054	.053	.053	.052	.052	.052	.051
1.8	.047	.046	.045	.045	.044	.044	.043	.043	.043	.042
1.9	.039	.038	.038	.037	.037	.036	.036	.036	.035	.035
2.0	.033	.032	.031	.031	.030	.030	.030	.029	.029	.029
2.1	.027	.027	.026	.025	.025	.025	.024	.024	.024	.023
2.2	.023	.022	.021	.021	.021	.020	.020	.020	.019	.019
2.3	.019	.018	.018	.017	.017	.016	.016	.016	.016	.015
2.4	.015	.015	.014	.014	.014	.013	.013	.013	.013	.012
2.5	.013	.012	.012	.011	.011	.011	.011	.010	.010	.010
2.6	.010	.010	.010	.009	.009	.009	.009	.008	.008	.008
2.7	.009	.008	.008	.008	.007	.007	.007	.007	.007	.006
2.8	.007	.007	.006	.006	.006	.006	.006	.005	.005	.005
2.9	.006	.005	.005	.005	.005	.005	.004	.004	.004	.004
3.0	.005	.004	.004	.004	.004	.004	.004	.003	.003	.003
3.1	.004	.004	.003	.003	.003	.003	.003	.003	.003	.003
3.2	.003	.003	.003	.003	.002	.002	.002	.002	.002	.002
3.3	.003	.002	.002	.002	.002	.002	.001	.001	.001	.001
3.4	.002	.002	.002	.002	.001	.001	.001	.001	.001	.001
3.5	.002	.002	.001	.001	.001	.001	.001	.001	.001	.001
3.6	.001	.001	.001	.001	.001	.001	.001	.001	.001	.001
3.7	.001	.001	.001	.001	.001	.001	.001	.001	.001	.000
3.8	.001	.001	.001	.001	.001	.000	.000	.000	.000	.000
3.9	.001	.001	.001	.001	.000	.000	.000	.000	.000	.000
4.0	.001	.001	.001	.000	.000	.000	.000	.000	.000	.000
4.5	.000	.000	.000	.000	.000	.000	.000	.000	.000	.000
5.0	.000	.000	.000	.000	.000	.000	.000	.000	.000	.000

TABLE IIIa
Tail-probabilities of the t-distribution (cont.)

t	\multicolumn{10}{c}{Degrees of freedom}									
	24	25	26	27	28	29	30	35	40	∞
1.0	.164	.163	.163	.163	.163	.163	.163	.162	.162	.159
1.1	.141	.141	.141	.141	.140	.140	.140	.139	.139	.136
1.2	.121	.121	.120	.120	.120	.120	.120	.119	.119	.115
1.3	.103	.103	.103	.102	.102	.102	.102	.101	.101	.097
1.4	.087	.087	.087	.086	.086	.086	.086	.085	.085	.081
1.5	.073	.073	.073	.073	.072	.072	.072	.071	.071	.067
1.6	.061	.061	.061	.061	.060	.060	.060	.059	.059	.055
1.7	.051	.051	.051	.050	.050	.050	.050	.049	.048	.047
1.8	.042	.042	.042	.042	.041	.041	.041	.040	.040	.036
1.9	.035	.035	.034	.034	.034	.034	.034	.033	.032	.029
2.0	.023	.023	.023	.023	.022	.022	.022	.022	.021	.018
2.1	.019	.019	.018	1018	.018	.018	.018	.017	.017	.014
2.3	.015	.015	.015	.015	.015	.014	.014	.014	.013	.011
2.4	.012	.012	.012	.012	.012	.012	.011	.011	.011	.008
2.5	.010	.010	.010	.009	.009	.009	.009	.009	.008	.006
2.6	.008	.008	.008	.007	.007	.007	.007	.007	.006	.005
2.7	.006	.006	.006	.006	.006	.006	.006	.005	.005	.004
2.8	.005	.005	.005	.005	.005	.004	.004	.004	.004	.003
2.9	.004	.004	.004	.004	.004	.004	.003	.003	.003	.002
3.0	.003	.003	.003	.003	.003	.003	.003	.002	.002	.001
3.1	.002	.002	.002	.002	.002	.002	.002	.002	.002	.001
3.2	.002	.002	.002	.002	.002	.002	.002	.001	.001	.001
3.3	.001	.001	.001	.001	.001	.001	.001	.001	.001	.001
3.4	.001	.001	.001	.001	.001	.001	.001	.001	.001	.000
3.5	.001	.001	.001	.001	.001	.001	.001	.000	.000	.000
3.6	.001	.001	.001	.000	.000	.000	.000	.000	.0001	.000
3.7	.000	.000	.000	.000	.000	.000	.000	.000	.000	.000
3.8	.000	.000	.000	.000	.000	.000	.000	.000	.000	.000
3.9	.000	.000	.000	.000	.000	.000	.000	.000	.000	.000
4.0	.000	.000	.000	.000	.000	.000	.000	.000	.000	.000
4.5	.000	.000	.000	.000	.000	.000	.000	.000	.000	.000
5.0	.000	.000	.000	.000	.000	.000	.000	.000	.000	.000

TABLE IIIb Percentiles of the t-distribution

df	\multicolumn{8}{c}{Percent}
	60	70	80	86	90	95	97.5	99	99.5
1	.325	.727	1.38	1.96	3.08	6.31	12.7	31.8	63.7
2	.289	.617	1.06	1.39	1.89	2.92	4.30	6.96	9.92
3	.277	.584	.978	1.25	1.64	2.35	3.18	4.54	5.84
4	.271	.569	.941	1.19	1.53	2.13	2.78	3.75	4.60
5	.267	.559	.920	1.16	1.48	2.01	2.57	3.36	4.03
6	.265	.583	.906	1.13	1.44	1.94	2.45	3.14	3.71
7	.263	.549	.896	1.12	1.42	1.90	2.36	3.00	3.50
8	.262	.546	.889	1.11	1.40	1.86	2.31	2.90	3.36
9	.261	.543	.883	1.10	1.38	1.83	2.26	2.82	3.25
10	.260	.542	.879	1.09	1.37	1.81	2.23	2.76	3.17
11	.260	.540	.876	1.09	1.36	1.80	2.20	2.72	3.11
12	.259	.539	.873	1.08	1.36	1.78	2.18	2.68	3.06
13	.259	.538	.870	1.08	1.35	1.77	2.16	2.65	3.06
14	.258	.537	.868	1.08	1.34	1.76	2.14	2.62	2.98
15	.258	.536	.866	1.07	1.34	1.75	2.12	2.60	2.95
16	.258	.535	.865	1.07	1.34	1.75	2.12	2.58	2.92
17	.257	.534	.863	1.07	1.33	1.74	2.11	2.57	2.90
18	.257	.534	.862	1.07	1.33	1.73	2.10	2.55	2.88
19	.257	.533	.861	1.07	1.33	1.73	2.09	2.54	2.86
20	.257	.533	.860	1.07	1.32	1.72	2.09	2.53	2.84
21	.257	.532	.859	1.06	1.32	1.72	2.08	2.52	2.83
22	.256	.532	.858	1.06	1.32	1.72	2.07	2.51	2.82
23	.256	.532	.858	1.06	1.32	1.71	2.07	2.50	2.81
24	.256	.531	.857	1.06	1.32	1.71	2.06	2.49	2.80
25	.256	.531	.856	1.06	1.32	1.71	2.06	2.48	2.79
26	.256	.531	.856	1.06	1.32	1.71	2.06	2.48	2.78
27	.256	.531	.855	1.06	1.31	1.70	2.05	2.47	2.77
28	.256	.530	.855	1.06	1.31	1.70	2.05	2.47	2.76
29	.256	.530	.854	1.06	1.31	1.70	2.04	2.46	2.76
30	.256	.530	.854	1.05	1.31	1.70	2.04	2.46	2.75
35	.255	.529	.852	1.05	1.31	1.69	2.03	2.44	2.72
40	.255	.529	.851	1.05	1.30	1.68	2.02	2.42	2.70
∞	.253	.524	.842	1.04	1.28	1.64	1.96	2.33	2.58

TABLE IVa 95th Percentiles of the F-distribution

	\multicolumn{13}{c}{Numerator Degrees of Freedom}
	1	2	3	4	5	6	8	10	12	15	20	24	30
1	161	200	216	225	230	234	239	242	244	246	248	249	250
2	18.5	19.0	19.2	19.2	19.3	19.3	19.4	19.4	19.4	19.4	19.4	19.5	19.5
3	10.1	9.55	9.28	9.12	9.01	8.94	8.85	8.79	8.74	8.70	8.66	8.64	8.62
4	7.71	6.94	6.59	6.39	6.26	6.16	6.04	5.96	5.91	5.86	5.80	5.77	5.75
5	6.61	5.79	5.41	5.19	5.05	4.95	4.82	4.74	4.68	4.62	4.56	4.53	4.50
6	5.99	5.14	4.76	4.53	4.39	4.28	4.15	4.06	4.00	3.94	3.87	3.84	3.81
7	5.59	4.74	4.35	4.12	3.97	3.87	3.73	3.64	3.57	3.51	3.44	3.41	3.38
8	5.32	4.46	4.07	3.84	3.69	3.58	3.44	3.35	3.28	3.22	3.15	3.12	3.08
9	5.12	4.26	3.86	3.63	3.48	3.37	3.23	3.14	3.07	3.01	2.94	2.90	2.86
10	4.96	4.10	3.71	3.48	3.33	3.22	3.07	2.98	2.91	2.85	2.77	2.74	2.70
11	4.84	3.98	3.59	3.36	3.20	3.09	2.95	2.85	2.79	2.72	2.65	2.61	2.57
12	4.75	3.89	3.49	3.26	3.11	3.00	2.85	2.75	2.69	2.62	2.54	2.51	2.47
13	4.67	3.81	3.41	3.18	3.03	2.92	2.77	2.67	2.60	2.53	2.46	2.42	2.38
14	4.60	3.74	3.34	3.11	2.96	2.85	2.70	2.60	2.53	2.46	2.39	2.35	2.31
15	4.54	3.68	3.29	3.06	2.90	2.79	2.64	2.54	2.48	2.40	2.33	2.29	2.25
16	4.49	3.63	3.24	3.01	2.85	2.74	2.59	2.49	2.42	2.35	2.28	2.24	2.19
17	4.45	3.59	3.20	2.96	2.81	2.70	2.55	2.45	2.38	2.31	2.23	2.19	2.15
18	4.41	3.55	3.16	2.93	2.77	2.66	2.51	2.41	2.34	2.27	2.19	2.15	2.11
19	4.38	3.52	3.13	2.90	2.74	2.63	2.48	2.38	2.31	2.23	2.16	2.11	2.07
20	4.35	3.49	3.10	2.87	2.71	2.60	2.45	2.35	2.28	2.20	2.12	2.08	2.04
21	4.32	3.47	3.07	2.84	2.68	2.57	2.42	2.32	2.25	2.18	2.10	2.05	2.01
22	4.30	3.44	3.05	2.82	2.66	2.55	2.40	2.30	2.23	2.15	2.07	2.03	1.98
23	4.28	3.42	3.03	2.80	2.64	2.53	2.37	2.27	2.20	2.13	2.05	2.01	1.96
24	4.26	3.40	3.01	2.78	2.62	2.51	2.36	2.25	2.18	2.11	2.03	1.98	1.94
25	4.24	3.39	2.99	2.76	2.60	2.49	2.34	2.24	2.16	2.09	2.01	1.96	1.92
30	4.17	3.32	2.92	2.69	2.53	2.42	2.27	2.16	2.09	2.01	1.93	1.89	1.84
40	4.08	3.23	2.84	2.61	2.45	2.34	2.18	2.08	2.00	1.92	1.84	1.79	1.74
60	4.00	3.15	2.76	2.53	2.37	2.25	2.10	1.99	1.92	1.84	1.75	1.70	1.65

(Denominator Degrees of Freedom — row labels at left)

This table is adapted from Table XVIII in *Biometrika Tables for Statisticians*, Vol. I, 1954, by E. S. Pearson and H. O. Hartley, originally prepared by M. Merrington and C. M. Thompson, with the kind permission of the editor of *Biometrika*.

TABLE IVb 99th Percentiles of the F-distribution

	\multicolumn{13}{c}{Numerator Degrees of Freedom}
	1	2	3	4	5	6	8	10	12	15	20	24	30
1	4050	5000	5400	5620	5760	5860	5980	6060	6110	6160	6210	6235	6260
2	98.5	99.0	99.2	99.2	99.3	99.3	99.4	99.4	99.4	99.4	99.4	99.5	99.5
3	34.1	30.8	29.5	28.7	28.2	27.9	27.5	27.3	27.1	26.9	26.7	26.6	26.5
4	21.2	18.0	16.7	16.0	15.5	15.2	14.8	14.5	14.4	14.2	14.0	13.9	13.8
5	16.3	13.3	12.1	11.4	11.0	10.7	10.3	10.1	9.89	9.72	9.55	9.47	9.38
6	13.7	10.9	9.78	9.15	8.75	8.47	8.10	7.87	7.72	7.56	7.40	7.31	7.23
7	12.2	9.55	8.45	7.85	7.46	7.19	6.84	6.62	6.47	6.31	6.16	6.07	5.99
8	11.3	8.65	7.59	7.01	6.63	6.37	6.03	5.81	5.67	5.52	5.36	5.28	5.20
9	10.6	8.02	6.99	6.42	6.06	5.80	5.47	5.26	5.11	4.96	4.81	4.73	4.65
10	10.0	7.56	6.55	5.99	5.64	5.39	5.06	4.85	4.71	4.56	4.41	4.33	4.25
11	9.65	7.21	6.22	5.67	5.32	5.07	4.74	4.54	4.40	4.25	4.10	4.02	3.94
12	9.33	6.93	5.95	5.41	5.06	4.82	4.50	4.30	4.16	4.01	3.86	3.78	3.70
13	9.07	6.70	5.74	5.21	4.86	4.62	4.30	4.10	3.96	3.82	3.66	3.59	3.51
14	8.86	6.51	5.56	5.04	4.69	4.46	4.14	3.94	3.80	3.66	3.51	3.43	3.35
15	8.68	6.36	5.42	4.89	4.56	4.32	4.00	3.80	3.67	3.52	3.37	3.29	3.21
16	8.53	6.23	5.29	4.77	4.44	4.20	3.89	3.69	3.55	3.41	3.26	3.18	3.10
17	8.40	6.11	5.18	4.67	4.34	4.10	3.79	3.59	3.46	3.31	3.16	3.08	3.00
18	8.29	6.01	5.09	4.58	4.25	4.01	3.71	3.51	3.37	3.23	3.08	3.00	2.92
19	8.18	5.93	5.01	4.50	4.17	3.94	3.63	3.43	3.30	3.15	3.00	2.92	2.84
20	8.10	5.85	4.94	4.43	4.10	3.87	3.56	3.37	3.23	3.09	2.94	2.86	2.78
21	8.02	5.78	4.87	4.37	4.04	3.81	3.51	3.31	3.17	3.03	2.88	2.80	2.72
22	7.95	5.72	4.82	4.31	3.99	3.76	3.45	3.26	3.12	2.98	2.83	2.75	2.67
23	7.88	5.66	4.76	4.26	3.94	3.71	3.41	3.21	3.07	2.93	2.78	2.70	2.62
24	7.82	5.61	4.72	4.22	3.90	3.67	3.36	3.17	3.03	2.89	2.74	2.66	2.58
25	7.77	5.57	4.68	4.18	3.86	3.63	3.32	3.13	2.99	2.85	2.70	2.62	2.54
30	7.56	5.39	4.51	4.02	3.70	3.47	3.17	2.98	2.84	2.70	2.55	2.47	2.39
40	7.31	5.18	4.31	3.83	3.51	3.29	2.99	2.80	2.66	2.52	2.37	2.29	2.20
60	7.08	4.98	4.13	3.65	3.34	3.12	2.82	2.63	2.50	2.35	2.20	2.12	2.03

Denominator Degrees of Freedom

This table is adapted from Table XVIII in *Biometrika Tables for Statisticians*, Vol. I, 1954, by E. S. Pearson and H. O. Hartley, originally prepared by M. Merrington and C. M. Thompson, with the kind permission of the editor of *Biometrika*.

TABLE Va Tail Areas of the Chi-square Distribution

χ^2	d.f. 1	d.f. 2	d.f. 3	χ^2	d.f. 4	d.f. 5	d.f. 6	χ^2	d.f. 7	d.f. 8	d.f. 9
3.0	.083	.223	.392	8.0	.092	.156	.238	13.0	.072	.112	.163
3.2	.074	.202	.362	8.2	.085	.146	.224	13.2	.067	.105	.154
3.4	.065	.183	.334	8.4	.078	.136	.210	13.4	.063	.099	.145
3.6	.038	.165	.308	8.6	.072	.126	.197	13.6	.059	.093	.137
3.8	.051	.150	.284	8.8	.066	.117	.185	13.8	.055	.087	.130
4.0	.045	.135	.261	9.0	.061	.109	.174	14.0	.051	.082	.122
4.2	.040	.122	.241	9.2	.056	.101	.163	14.2	.048	.077	.115
4.4	.036	.111	.221	9.4	.052	.094	.152	14.4	.045	.072	.109
4.6	.032	.105	.204	9.6	.048	.087	.143	14.6	.041	.067	.103
4.8	.028	.091	.187	9.8	.044	.081	.133	14.8	.039	.063	.097
5.0	.025	.082	.172	10.0	.040	.075	.125	15.0	.036	.059	.091
5.2	.023	.074	.158	10.2	.037	.070	.116	15.2	.034	.055	.086
5.4	.020	.067	.145	10.4	.034	.065	.109	15.4	.031	.052	.081
5.6	.018	.061	.133	10.6	.031	.060	.102	15.6	.029	.048	.076
5.8	.016	.055	.122	10.8	.029	.055	.095	15.8	.027	.045	.071
6.0	.014	.050	.111	11.0	.027	.051	.088	16.0	.025	.042	.067
6.2	.013	.045	.102	11.2	.024	.046	.082	16.2	.023	.040	.063
6.4	.011	.041	.094	11.4	.032	.044	.077	16.4	.022	.037	.059
6.6	.010	.037	.086	11.6	.021	.041	.072	16.6	.020	.035	.055
6.8	.009	.033	.079	11.8	.019	.038	.067	16.8	.019	.032	.052
7.0	.008	.030	.072	12.0	.017	.035	.062	17.0	.017	.030	.049
7.2	.007	.027	.066	12.2	.016	.032	.058	17.2	.016	.028	.046
7.4	.007	.025	.060	12.4	.015	.030	.054	17.4	.015	.026	.043
7.6	.006	.022	.050	12.6	.013	.027	.050	17.6	.014	.024	.040
7.8	.005	.020	.050	12.8	.012	.025	.046	17.8	.013	.023	.038
8.0	.005	.018	.046	13.0	.011	.023	.043	18.0	.012	.021	.035
8.2	.004	.017	.042	13.2	.010	.022	.040	18.2	.011	.020	.033
8.4	.004	.015	.038	13.4	.009	.020	.037	18.4	.010	.018	.031
8.6	.003	.014	.035	13.6	.009	.018	.034	18.6	.010	.017	.029
8.8	.003	.012	.032	13.8	.008	.017	.032	18.8	.009	.016	.027
9.0	.003	.011	.029	14.0	.007	.016	.030	19.0	.008	.014	.024
9.2	.002	.010	.027	14.2	.007	.014	.027	19.2	.008	.014	.024
9.4	.002	.009	.024	14.4	.006	.013	.025	19.4	.007	.013	.022
9.6	.002	.007	.022	14.6	.006	.012	.024	19.6	.006	.012	.021
9.8	.002	.007	.020	14.8	.005	.011	.022	19.8	.006	.011	.019
10.0	.001	.007	.019	15.0	.005	.010	.020	20.0	.006	.010	.018
10.2	.001	.006	.017	15.2	.004	.010	.019	20.2	.005	.010	.017
10.4	.001	.006	.015	15.4	.004	.009	.017	20.4	.005	.009	.016
10.6	.001	.005	.014	15.6	.004	.009	.017	20.6	.004	.008	.015
10.8	.001	.005	.013	15.8	.003	.007	.015	20.8	.004	.008	.014
11.0	.001	.004	.012	16.0	.003	.007	.014	21.0	.004	.007	.013

TABLE Va (cont.)

χ^2	d.f. 10	11	12		13	d.f. 14	15		16	d.f. 17	18
17.0	.074	.108	.150	21.0	.073	.102	.137	25.0	.070	.095	.125
17.2	.070	.102	.142	21.2	.071	.097	.131	25.2	.066	.090	.120
17.4	.066	.097	.135	21.4	.065	.092	.125	25.4	.063	.086	.114
17.6	.062	.091	.128	21.6	.062	.087	.119	25.6	.060	.082	.109
17.8	.058	.086	.122	21.8	.059	.083	.113	25.9	.057	.078	.104
18.0	.055	.082	.116	22.0	.055	.079	.108	26.0	.054	.074	.100
18.2	.052	.077	.110	22.2	.052	.075	.103	26.2	.051	.071	.095
18.4	.049	.073	.104	22.4	.049	.071	.098	26.4	.049	.067	.091
18.6	.046	.069	.099	22.6	.047	.067	.093	26.6	.046	.064	.087
18.8	.043	.065	.093	22.8	.044	.064	.088	26.8	.044	.061	.083
19.0	.040	.061	.089	23.0	.042	.060	.084	27.0	.041	.058	.079
19.2	.038	.058	.084	23.2	.039	.057	.080	27.2	.039	.055	.075
19.4	.035	.054	.079	23.4	.037	.054	.076	27.4	.037	.052	.072
19.6	.033	.051	.075	23.6	.035	.051	.072	27.6	.035	.050	.068
19.8	.031	.048	.071	23.8	.033	.048	.069	27.8	.033	.047	.065
20.0	.029	.045	.067	24.0	.031	.046	.065	28.0	.020	.045	.062
20.2	.027	.043	.063	24.2	.029	.043	.062	28.2	.030	.043	.059
20.4	.026	.040	.060	24.4	.028	.041	.059	28.4	.028	.040	.056
20.6	.024	.038	.057	24.6	.026	.039	.056	28.6	.027	.038	.053
20.8	.023	.036	.053	24.8	.025	.037	.053	28.8	.025	.036	.051
21.0	.021	.033	.050	25.0	.023	.035	.050	29.0	.024	.035	.048
21.2	.020	.031	.048	25.2	.022	.033	.047	29.2	.023	.033	.046
21.4	.018	.029	.045	25.4	.020	.031	.045	29.4	.021	.031	.044
21.6	.017	.028	.042	25.6	.019	.029	.042	29.6	.020	.029	.042
21.8	.016	.026	.040	25.8	.018	.026	.038	29.8	.019	.028	.039
22.0	.014	.024	.038	26.0	.017	.026	.038	30.0	.018	.026	.037
22.2	.013	.023	.035	26.2	.016	.024	.036	30.2	.017	.025	.036
22.4	.012	.021	.033	26.4	.015	.023	.034	30.4	.016	.024	.034
22.6	.012	.020	.031	26.6	.014	.022	.032	30.6	.015	.022	.032
22.8	.011	.019	.029	26.8	.013	.020	.030	30.8	.014	.021	.030
23.0	.010	.018	.028	27.0	.012	.019	.029	31.0	.013	.020	.029
23.2	.009	.017	.026	27.2	.012	.018	.027	31.2	.013	.019	.027
23.4	.009	.016	.025	27.4	.011	.017	.026	31.4	.012	.018	.026
23.6	.008	.015	.023	27.6	.010	.016	.024	31.6	.011	.017	.025
23.8	.008	.014	.022	27.8	.010	.015	.023	31.8	.011	.016	.023
24.0	.004	.013	.020	28.0	.009	.014	.022	32.0	.010	.015	.022
24.2	.004	.012	.019	28.2	.008	.013	.020	32.2	.009	.014	.021
24.4	.004	.011	.018	28.4	.008	.013	.019	32.4	.009	.013	.020
24.6	.003	.010	.017	28.6	.007	.012	.018	32.6	.008	.013	.019
24.8	.003	.010	.016	28.8	.007	.011	.017	32.8	.008	.012	.018
25.0	.003	.009	.015	29.0	.007	.010	.016	33.0	.007	.011	.017

TABLE Vb Percentiles of the Chi-square Distribution

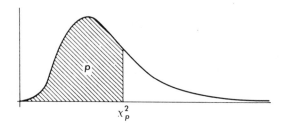

Degrees of freedom	$\chi^2_{.005}$	$\chi^2_{.01}$	$\chi^2_{.025}$	$\chi^2_{.05}$	$\chi^2_{.10}$	$\chi^2_{.20}$	$\chi^2_{.30}$	$\chi^2_{.50}$	$\chi^2_{.70}$	$\chi^2_{.80}$	$\chi^2_{.90}$	$\chi^2_{.95}$	$\chi^2_{.975}$	$\chi^2_{.99}$	$\chi^2_{.995}$
1	.000	.000	.001	.004	.016	.064	.148	.455	1.07	1.64	2.71	3.84	5.02	6.63	7.88
2	.010	.020	.051	.103	.211	.446	.713	1.39	2.41	3.22	4.61	5.99	7.38	9.21	10.6
3	.072	.115	.216	.352	.584	1.00	1.42	2.37	3.66	4.64	6.25	7.81	9.35	11.3	12.8
4	.207	.297	.484	.711	1.06	1.65	2.20	3.36	4.88	5.99	7.78	9.49	11.1	13.3	14.9
5	.412	.554	.831	1.15	1.61	2.34	3.00	4.35	6.06	7.29	9.24	11.1	12.8	15.1	16.7
6	.676	.872	1.24	1.64	2.20	3.07	3.83	5.35	7.23	8.56	10.6	12.6	14.4	16.8	18.5
7	.989	1.24	1.69	2.17	2.83	3.82	4.67	6.35	8.38	9.80	12.0	14.1	16.0	18.5	20.3
8	1.34	1.65	2.18	2.73	3.49	4.59	5.53	7.34	9.52	11.0	13.4	15.5	17.5	20.1	22.0
9	1.73	2.09	2.70	3.33	4.17	5.38	6.39	8.34	10.7	12.2	14.7	16.9	19.0	21.7	23.6
10	2.16	2.56	3.25	3.94	4.87	6.18	7.27	9.34	11.8	13.4	16.0	18.3	20.5	23.2	25.2
11	2.60	3.05	3.82	4.57	5.58	6.99	8.15	10.3	12.9	14.6	17.3	19.7	21.9	24.7	26.8
12	3.07	3.57	4.40	5.23	6.30	7.81	9.03	11.3	14.0	15.8	18.5	21.0	23.3	26.2	28.3
13	3.57	4.11	5.01	5.89	7.04	8.63	9.93	12.3	15.1	17.0	19.8	22.4	24.7	27.7	29.8
14	4.07	4.66	5.63	6.57	7.79	9.47	10.8	13.3	16.2	18.2	21.1	23.7	26.1	29.1	31.3
15	4.60	5.23	6.26	7.26	8.55	10.3	11.7	14.3	17.3	19.3	22.3	25.0	27.5	30.6	32.8
16	5.14	5.81	6.91	7.96	9.31	11.2	12.6	15.3	18.4	20.5	23.5	26.3	28.8	32.0	34.3
17	5.70	6.41	7.56	8.67	10.1	12.0	13.5	16.3	19.5	21.6	24.8	27.6	30.2	33.4	35.7
18	6.26	7.01	8.23	9.39	10.9	12.9	14.4	17.3	20.6	22.8	26.0	28.9	31.5	34.8	37.2
19	6.83	7.63	8.91	10.1	11.7	13.7	15.4	18.3	21.7	23.9	27.2	30.1	32.9	36.2	38.6
20	7.43	8.26	9.59	10.9	12.4	14.6	16.3	19.3	22.8	25.0	28.4	31.4	34.2	37.6	40.0
21	8.03	8.90	10.3	11.6	13.2	15.4	17.2	20.3	23.9	26.2	29.6	32.7	35.5	38.9	41.4
22	8.64	9.54	11.0	12.3	14.0	16.3	18.1	21.3	24.9	27.3	30.8	33.9	36.8	40.3	42.8
23	9.26	10.2	11.7	13.1	14.8	17.2	19.0	22.3	26.0	28.4	32.0	35.2	38.1	41.6	44.2
24	9.89	10.9	12.4	13.8	15.7	18.1	19.9	23.3	27.1	29.6	33.2	36.4	39.4	43.0	45.6
25	10.5	11.5	13.1	14.6	16.5	18.9	20.9	24.3	28.2	30.7	34.4	37.7	40.6	44.3	46.9
26	11.2	12.2	13.8	15.4	17.3	19.8	21.8	25.3	29.2	31.8	35.6	38.9	41.9	45.6	48.3
27	11.8	12.9	14.6	16.2	18.1	20.7	22.7	26.3	30.3	32.9	36.7	40.1	43.2	47.0	49.6
28	12.5	13.6	15.3	16.9	18.9	21.6	23.6	27.3	31.4	34.0	37.9	41.3	44.5	48.3	51.0
29	13.1	14.3	16.0	17.7	19.8	22.5	24.6	28.3	32.5	35.1	39.1	42.6	45.7	49.6	52.3
30	13.8	15.0	16.8	18.5	20.6	23.4	25.5	29.3	33.5	36.2	40.3	43.8	47.0	50.9	53.7
40	20.7	22.1	24.4	26.5	29.0	32.3	34.9	39.3	44.2	47.3	51.8	55.8	59.3	63.7	66.8
50	28.0	29.7	32.3	34.8	37.7	41.3	44.3	49.3	54.7	58.2	63.2	67.5	71.4	76.2	79.5
60	35.5	37.5	40.5	43.2	46.5	50.6	53.8	59.3	65.2	69.0	74.4	79.1	83.3	88.4	92.0

Note: For degrees of freedom $k > 30$, use $\chi_p^2 = \frac{1}{2}(z_p + \sqrt{2k-1})^2$, where z_p is the corresponding percentile of the standard normal distribution.

This table is adapted from Table VIII of *Biometrika Tables for Statisticians*, Vol. 1, 1954, by E. S. Pearson and H. O. Hartley, originally prepared by Catherine M. Thompson, with the kind permission of the editor of *Biometrika*.

TABLE VI One-sample Signed-rank Tail-probabilities

Sample size

c	5	6	7	8	9	10	11	12	13	14	15	c
0	.031	.016	.008	.004	.002	.001	.000	.000	.000	.000	.000	0
1	.062	.031	.016	.008	.004	.002	.001	.000	.000	.000	.000	1
2	.094	.047	.023	.012	.006	.003	.001	.001	.000	.000	.000	2
3	.156	.078	.039	.020	.010	.005	.002	.001	.001	.000	.000	3
4	.219	.109	.055	.027	.014	.007	.003	.002	.001	.000	.000	4
5		.156	.078	.039	.020	.010	.005	.002	.001	.001	.000	5
6		.219	.109	.055	.027	.014	.007	.003	.002	.001	.000	6
7			.148	.074	.037	.019	.009	.005	.002	.001	.001	7
8			.188	.098	.049	.024	.012	.006	.003	.002	.001	8
9			.234	.125	.064	.032	.016	.008	.004	.002	.001	9
10				.156	.082	.042	.021	.010	.005	.003	.001	10
11				.191	.102	.053	.027	.013	.007	.003	.002	11
12				.230	.125	.065	.034	.017	.009	.004	.002	12
13					.150	.080	.042	.021	.011	.005	.003	13
14					.180	.097	.051	.026	.013	.007	.003	14
15					.213	.116	.062	.032	.016	.008	.004	15
16						.138	.074	.039	.020	.010	.005	15
17						.161	.087	.046	.024	.012	.006	17
18						.188	.103	.055	.029	.015	.008	18
19						.216	.120	.065	.034	.018	.009	19
20							.139	.076	.040	.021	.011	20
21							.160	.088	.047	.025	.013	21
22							.183	.102	.055	.029	.015	22
23							.207	.117	.064	.034	.018	23
24								.113	.073	.039	.021	24
25								.151	.084	.045	.024	25
26								.170	.095	.052	.028	26
27								.190	.108	.059	.032	27
28								.212	.122	.068	.036	28
29									.137	.077	.042	29
30									.153	.086	.047	30
31									.170	.097	.053	31
32									.188	.108	.060	32
33									.207	.121	.068	33
34										.134	.076	34
	15	25	28	36	45	55	66	78	91	105	120	

TABLE VII Two-sample Wilcoxon Rank-sum Tail-Probabilities

m \ n \ c	4 \ 4	4 \ 5	4 \ 6	4 \ 7	4 \ 8	4 \ 9	4 \ 10
10	.014	.008	.005	.003	.002	.001	.001
11	.029	.016	.010	.006	.004	.003	.002
12	.057	.032	.019	.012	.008	.006	.004
13	.100	.056	.033	.021	.014	.010	.007
14	.171	.096	.057	.036	.024	.017	.012
15	.243	.143	.086	.055	.036	.025	.018
16		.206	.129	.082	.055	.038	.027
17			.176	.115	.077	.053	.038
18				.158	.107	.074	.053
19				.206	.141	.099	.071
20					.184	.130	.094
21						.165	.120
M	36	40	44	48	52	56	60

m \ n \ c	5 \ 5	5 \ 6	5 \ 7	5 \ 8	5 \ 9	5 \ 10
15	.004	.002	.001	.001	.000	.000
16	.008	.004	.003	.002	.001	.001
17	.016	.009	.005	.003	.002	.001
18	.028	.015	.009	.005	.003	.002
19	.048	.026	.015	.009	.006	.004
20	.075	.041	.024	.015	.009	.006
21	.111	.063	.037	.023	.014	.010
22	.155	.089	.053	.033	.021	.014
23	.210	.123	.074	.047	.030	.020
24		.165	.101	.064	.041	.028
25		.214	.134	.085	.056	.038
26			.172	.111	.073	.050
27			.216	.142	.095	.065
28				.177	.120	.082
28					.149	.103
M	55	60	65	70	75	80

Notes: 1. m is the size of the smaller sample.
2. The entry opposite c is the cumulative tail probability.

$$P(R \leq c) = P(R \geq M - c),$$

where R is the rank sum for the smaller sample, and M is the sum of the minimum and maximum values of R. [Note that $\frac{M}{2}$ = m.v.(R).]

3. For m or $n > 10$, use a normal approximation (see Section 12.5).

Table VII (continued)

m / n / c	6 / 6	6 / 7	6 / 8	6 / 9	6 / 10
24	.008	.004	.002	.001	.001
25	.013	.007	.004	.002	.001
26	.021	.011	.006	.004	.002
27	.032	.017	.010	.006	.004
28	.047	.026	.015	.009	.005
29	.066	.037	.021	.013	.008
30	.090	.051	.030	.018	.011
31	.120	.069	.041	.025	.016
32	.155	.090	.054	.033	.021
33	.197	.117	.071	.044	.028
34		.147	.091	.057	.036
35		.183	.114	.072	.047
36			.141	.091	.059
37			.172	.112	.074
38				136	.090
39				.164	.110
M	78	84	90	96	102

m / n / c	7 / 7	7 / 8	7 / 9	7 / 10
34	.009	.005	.003	.002
35	.013	.007	.004	.002
36	.019	.010	.006	.003
37	.027	.014	.008	.005
38	.036	.020	.011	.007
39	.049	.027	.016	.009
40	.064	.036	.001	.012
41	.082	.047	.027	.017
42	.104	.060	.036	.022
43	.130	.076	.045	.028
44	.159	.095	.057	.035
45	.191	.116	.071	.044
46		.140	.087	.054
47		.168	.105	.067
48		.198	.126	.081
49			.150	.097
50			.176	.115
M	105	112	119	126

m / n / c	8 / 8	8 / 9	8 / 10
46	.010	.006	.003
47	.014	.008	.004
48	.019	.010	.006
49	.025	.014	.008
50	.032	.018	.010
51	.041	.023	.013
52	.052	.030	.017
53	.065	.037	.022
54	.080	.046	.027
55	.097	.057	.034
56	.117	.069	.042
57	.139	.084	.051
58	.164	.100	.061
59	.191	.118	.073
60		.138	.086
61		.161	.102
M	136	144	152

m / n / c	9 / 9	9 / 10
59	.009	.005
60	.012	.007
61	.016	.009
62	.020	.011
63	.025	.014
64	.031	.017
65	.039	.022
66	.047	.027
67	.057	.033
68	.068	.039
69	.081	.047
70	.095	.056
71	.111	.067
72	.129	.078
73	.149	.091
74	.170	.106
M	171	180

m / n / c	10 / 10
74	.009
75	.012
76	.014
77	.018
78	.022
79	.026
80	.032
81	.038
82	.045
83	.053
84	.062
85	.072
86	.083
87	.095
88	.109
89	.124
M	210

TABLE VIII One-sample Kolmogorov-Smirnov Critical Values

	\multicolumn{5}{c}{*Significance level*}
Sample size	.20	.15	.10	.05	.01
1	.900	.925	.950	.975	.995
2	.684	.726	.776	.842	.929
3	.565	.597	.642	.708	.829
4	.494	.525	.564	.624	.734
5	.446	.474	.510	.563	.669
6	.410	.436	.470	.521	.618
7	.381	.405	.438	.486	.577
8	.358	.381	.411	.457	.543
9	.339	.342	.368	.409	.486
10	.322	.342	.368	.409	.486
11	.307	.326	.352	.391	.468
12	.295	.313	.338	.375	.450
13	.284	.302	.325	.361	.433
14	.274	.292	.314	.349	.418
15	.266	.283	.304	.338	.404
16	.258	.274	.295	.328	.391
17	.250	.266	.286	.318	.380
18	.244	.259	.278	.309	.370
19	.237	.252	.272	.301	.361
20	.231	.246	.264	.294	.352
25	.21	.22	.24	.264	.32
30	.19	.20	.22	.242	.29
35	.18	.19	.21	.23	.27
Large	$\dfrac{1.07}{\sqrt{n}}$	$\dfrac{1.14}{\sqrt{n}}$	$\dfrac{1.22}{\sqrt{n}}$	$\dfrac{1.36}{\sqrt{n}}$	$\dfrac{1.63}{\sqrt{n}}$

TABLE IX Two-sample Kolmogorov-Smirnov Acceptance Limits

		\multicolumn{11}{c}{Sample Size n_1}
		1	2	3	4	5	6	7	8	9	10	12	15
Sample Size n_2	1	* *	* *	* *	* *	* *	* *	* *	* *	* *	* *		
	2		* *	* *	* *	* *	* *	* *	7/8 *	16/18 *	9/10 *		
	3			* *	* *	12/15 *	5/6 *	18/21 *	18/24 *	7/9 8/9		9/12 11/12	
	4				3/4 *	16/20 *	9/12 10/12	21/28 24/28	6/8 7/8	27/36 32/36	14/20 16/20	8/12 10/12	
	5					4/5 4/5	20/30 25/30	25/35 30/35	27/40 32/40	31/45 36/45	7/10 8/10		10/15 11/15
	6						4/6 5/6	29/42 35/42	16/24 18/24	12/18 14/18	19/30 22/30	7/12 9/12	
	7							5/7 5/7	35/56 42/56	40/63 47/63	43/70 53/70		
	8								5/8 6/8	45/72 54/72	23/40 28/40	14/24 16/24	
	9									5/9 6/9	52/90 62/90	20/36 24/36	
	10										6/10 7/10		15/30 19/30
	12											6/12 7/12	30/60 35/60
	15												7/15 8/15

Note 1: Reject H_0 if $D = \max |F_{n_2}(x) - F_{n_1}(x)|$ exceeds the tabulated value. The upper value gives a level at most .05 and the lower at most .01.

Note 2: Where * appears, do not reject H_0 at the given level.

Note 3: For large values of n_1 and n_2, the following approximate formulas may be used:

$$\alpha = .05: \quad 1.36\sqrt{\frac{n_1 + n_2}{n_1 n_2}}.$$

$$\alpha = .01: \quad 1.63\sqrt{\frac{n_1 + n_2}{n_1 n_2}}.$$

This table is derived from F. J. Massey, "Distribution Table for the Deviation Between Two Sample Cumulatives," *Ann. Math. Stat.* **23**, 435–41 (1952). Adapted with the kind permission of the author and the *Ann. Math. Stat.* Formulas for large-sample sizes were given by N. Smirnov, "Tables for Estimating the Goodness of Fit of Empirical Distributions," *Ann. Math. Stat.* **19**, 280–81 (1948).

TABLE X Tail-probabilities of the Lilliefors Distribution (for testing normality)

n	Right-tail probability*					
	.20	.15	.10	.05	.01	.001
5	.289	.303	.319	.343	.397	.439
6	.269	.281	.297	.323	.371	.424
7	.252	.264	.280	.304	.351	.402
8	.239	.250	.265	.288	.333	.384
9	.227	.238	.252	.274	.317	.365
10	.217	.228	.241	.262	.304	.352
11	.208	.218	.231	.251	.291	.338
12	.200	.210	.222	.242	.281	.325
13	.193	.202	.215	.234	.271	.314
14	.187	.196	.208	.226	.262	.305
15	.181	.190	.201	.219	.254	.296
16	.176	.184	.195	.213	.247	.287
17	.171	.179	.190	.207	.240	.279
18	.167	.175	.185	.202	.234	.273
19	.163	.170	.181	.197	.228	.266
20	.159	.166	.176	.192	.223	.260
25	.143	.150	.159	.173	.201	.236
30	.131	.138	.146	.159	.185	.217
40	.115	.120	.128	.139	.162	.189
100	.074	.077	.082	.089	.104	.122
400	.037	.039	.041	.045	.052	.061

*Assumes that in calculating the K-S statistic, the population mean and variance are replaced by the sample mean and variance, respectively.

TABLE XI Expected Values of Order Statistics from $\mathcal{N}(0, 1)$

n	$n-9$	$n-8$	$n-7$	$n-6$	$n-5$	$n-4$	$n-3$	$n-2$	$n-1$	n	n
2										.564	2
3									0	.846	3
4									.297	1.029	4
5								0	.495	1.163	5
6								.202	.642	1.163	6
7							0	.353	.757	1.352	7
8							.153	.473	.852	1.424	8
9						0	.275	.572	.932	1.485	9
10						.123	.376	.656	1.001	1.539	10
11					0	.225	.462	.729	1.062	1.586	11
12					.103	.312	.537	.793	1.116	1.629	12
13				0	.190	.388	.603	.850	1.164	1.668	13
14				.088	.267	.456	.662	.901	1.208	1.703	14
15			0	.165	.335	.516	.715	.848	1.248	1.736	15
16			.077	.234	.396	.570	.763	.990	1.285	1.766	16
17		0	.146	.295	.451	.619	.807	1.030	1.319	1.794	17
18		.069	.208	.351	.502	.665	.848	1.066	1.350	1.820	18
19	0	.131	.264	.402	.548	.707	.886	1.099	1.380	1.844	19
20	.062	.187	.315	.448	.590	.745	.921	1.131	1.408	1.867	20

Note: Entry under "$n - i$" is the expected value of the ith smallest in a random sample of size n.

TABLE XII Selected Percentage Points of the Shapiro-Wilk W

	Significance level			
n	.01	.05	.10	.50
5	.675	.777	.817	.922
10	.776	.842	.869	.940
15	.815	.878	.903	.954
20	.858	.902	.921	.962
35	.919	.943	.952	.976
50	.935	.953	.963	.981
75	.956	.969	.973	.986
99	.967	.976	.980	.989

TABLE XIII Distribution of the Standardized Range, $W = R/\sigma$

	Sample Size										
	2	3	4	5	6	7	8	9	10	12	15
$a_n = E(W)$	1.128	1.693	2.059	2.326	2.534	2.704	2.847	2.970	3.078	3.258	3.472
$b_n = \sigma_w$	.853	.888	.880	.864	.848	.833	.820	.808	.797	.778	.755
$W_{.005}$	.01	.13	.34	.55	.75	.92	1.08	1.21	1.33	1.55	1.80
$W_{.01}$	.02	.19	.43	.66	.87	1.05	1.20	1.34	1.47	1.68	1.93
$W_{.025}$	.04	.30	.59	.85	1.06	1.25	1.41	1.55	1.67	1.88	2.14
$W_{.05}$	.09	.43	.76	1.03	1.25	1.44	1.60	1.74	1.86	2.07	2.32
$W_{.1}$	.18	.62	.98	1.26	1.49	1.68	1.83	1.97	2.09	2.30	2.54
$W_{.2}$	.36	.90	1.29	1.57	1.80	1.99	2.14	2.28	2.39	2.59	2.83
$W_{.3}$	.55	1.14	1.53	1.82	2.04	2.22	2.38	2.51	2.62	2.82	3.04
$W_{.4}$	.74	1.36	1.76	2.04	2.26	2.44	2.59	2.71	2.83	3.01	3.23
$W_{.5}$	.95	1.59	1.98	2.26	2.47	2.65	2.79	2.92	3.02	3.21	3.42
$W_{.6}$	1.20	1.83	2.21	2.48	2.69	2.86	3.00	3.12	3.23	3.41	3.62
$W_{.7}$	1.47	2.09	2.47	2.73	2.94	3.10	3.24	3.35	3.46	3.63	3.83
$W_{.8}$	1.81	2.42	2.78	3.04	3.23	3.39	3.52	3.63	3.73	3.90	4.09
$W_{.9}$	2.33	2.90	3.24	3.48	3.66	3.81	3.93	4.04	4.13	4.29	4.47
$W_{.95}$	2.77	3.31	3.63	3.86	4.03	4.17	4.29	4.39	4.47	4.62	4.80
$W_{.975}$	3.17	3.68	3.98	4.20	4.36	4.49	4.61	4.70	4.79	4.92	5.09
$W_{.99}$	3.64	4.12	4.40	4.60	4.76	4.88	4.99	5.08	5.16	5.29	5.45
$W_{.995}$	3.97	4.42	4.69	4.89	5.03	5.15	5.26	5.34	5.42	5.54	5.70

This table is adapted from Tables XX and XXII in *Biometrika Tables for Statisticians*, Vol. I, 1954, by E. S. Pearson and H. O. Hartley, with the kind permission of the editor of *Biometrika*.

References and Further Reading

Anderson, T. W. *An Introduction to Multivariate Statistical Analysis.* New York: John Wiley & Sons, Inc., 1958.

Berger, J. *Statistical Decision Theory and Bayesian Analysis,* 2nd Ed., New York: Springer Verlag, 1985.

Berger, R., and Casella, G. *Statistical Inference.* Pacific Grove, Cal.: Wadsworth & Brooks/Cole, 1990.

Bickel, P. and K. Doksum, *Mathematical Statistics.* Oakland, Cal,: Holden-Day, Inc., 1977.

Bishop, Y. A., S. E. Fienberg, and P. W. Holland. *Discrete Multivariate Analysis: Theory and Practice.* Cambridge, Mass: M. I. T. Press, 1975.

Cox, D. R., and D. V. Hnkley. *Theoretical Statistics.* London: Chapman & Hall, Ltd., 1974.

Cramér, H. Mathematical Methods of Statistics. Princeton, N. J.: Princeton University Press, 1946

DeGroot, M. H. *Optimal Statistical Decisions.* New York: McGraw-Hill Book Company, 1970.

Feller, W. *An Introduction to Probability Theory and its Applicaitons,* Vol. 1, 3rd Ed., 1968; Vol 2, 1966. New York: John Wiley & Sons, Inc.

Ferguson, T. S. *Mathematical Statistics.* New York: Academic Press, Inc., 1967.

Fisher, R. A. *Statistical Methods and Scientific Inference,* 3rd Ed. New York: Hafner Press, 1973.

Fisher, R. A. *Statistical Methods for Research Workers.* Edinburgh: Oliver & Boyd, Ltd., 1936.

Fisher, R. A. *The Design of Experiments.* Edinburgh: Oliver and Boyd, Ltd., 1951.

Gibbons, J. *Nonparametric Statistical Inference.* New York: McGraw-Hill Book Co., 1971.

Kendall, M. and Stuart, A. *The Advanced Theory of Statistics*

Lehman, E. *Nonparametrics: Statistical Methods Based on Ranks.* San Francisco: Holden-Day Inc., 1975.

Lehman, E. *Testing Statistical Hypotheses*, 2nd Ed. Pacific Grove, Cal.: Wadsworth & Brooks/Cole, 1986.

Lehman, E. *Theory of Point Estimation*. New York: John Wiley & Sons, 1983.

Lindgren, B. *Elements of Decision Theory*. New York, Macmillan Publishing Co., 1971.

Rao, C. R. *Linear Statistical Inference and Its Applications*, 2nd Ed. New York: John Wiley & Sons, Inc., 1973.

Searle, S. R. *Linear Models*. New York: John Wiley & Sons, Inc., 1971.

Scheffé, H. *The Analysis of Variance*. New York: John Wiley & Sons, Inc., 1959.

Wald, A. *Sequential Analysis*. New York: John Wiley & Sons, Inc., 1950.

Weisberg, S. *Applied Regression Analysis*, 2nd Ed. New York: John Wiley & Sons, Inc., 1985.

Wilks, S. S. *Mathematical Statistics*. New York: John Wiley & Sons, Inc., 1962

Zacks, S. *The Theory of Statistical Inference*. New York: John Wiley & Sons, Inc., 1973.

Answers to Problems

Chapter 1

1-1. (a) 11,880 (b) 495 (c) 161,700 (d) 27,720
  (e) $n^3 - 3n^2 + 2n$ (f) $\frac{1}{6}(n^3 - 3n^2 + 2n)$
1-2. k^m
1-3. 30
1-4. (a) 216 (b) 2^n (c) 5040
1-5. (a) 1024 (b) 56 (c) 252 (d) 36 (e) 100
1-6. (a) 720 (b) 72 (c) 72 (d) 240
1-7. (a) 120 (b) 60
1-8. (a) 2,598,960 (b) 1287 (c) 9216 (d) 36
1-9. 7,059,052
1-10. (a) 120 (b) 112 (c) 64
1-11. (a) 560 (b) 105
1-18. (a) FG (b) $\emptyset$ (c) E
1-23. Ω, $\emptyset$, $\{a\}$, $\{b\}$, $\{c\}$, $\{d\}$, $\{a,b\}$, $\{a,c\}$, $\{a,d\}$, $\{b,c\}$,
  $\{b,d\}$, $\{c,d\}$, $\{a,b,c\}$, $\{a,c,d\}$, $\{a,b,d\}$, $\{b,c,d\}$
1-25. (a) 6 (b) 11 (c) 15 (d) 5
1-26. (a) 20 (b) 4 (c) 4 (d) 8 (e) 12
1-27. (a) 3 (b) 4 (c) 39 (d) 25 (d) 52 (e) 10 (f) 6 (g) 18
1-28. {HHHH}, {HHHT, HHTH, HTHH, THHH},
  {HHTT, HTHT, HTTH, TTHH, THTH, THHT},
  {TTTH, TTHT, THTT, HTTT}, {TTTT}
1-29. (a) {12}, {13}, {23, 14}, {15, 24}, {25, 34}, {35}, {45}
  (b) {12, 13, 23, 14}
1-30. (a) [0, 1) (b) (−1, 1)
1-31. (a) Unit circles (b) Straight lines with slope 2
  (c) Hyperbolas (c) Pairs of parallel lines with slope 1
1-32. (a) $\{(x, y, z) \mid x < y < z\}$. [Hard to draw, it's 1/6 of $\Re^3$.]
  (b) $\{(1, 2, 3), (1, 3, 2), (2, 1, 3), (2, 3, 1), (3, 1, 2), (3, 2, 1)\}$

(c) Sextuples each consisting of six permutations of 3 numbers.
1-33. (a) $\emptyset$, Ω, $\{a,e\}$, $\{c,d\}$, $\{b,e\}$, $\{b\}$, $\{e\}$, $\{a\}$, $\{b,c,d,e\}$,
$\{a,c,d,e\}$, $\{a,b,c,d\}$, $\{a,b,e\}$, $\{c,d,e\}$, $\{a,c,d\}$
(b) $X(\omega) = 0$ on $\{a\}$, 1 on $\{b\}$, 2 on $\{c,d\}$, 3 on $\{e\}$,—or any other four distince values on the four partition sets.
1-34. Circles on the hemisphere, parallel to the base plane, at heights
$h = \sqrt{1 - r^2}$ above the plane.

Chapter 2

2-1. (a) 1/6   (b) 5/6   (c) 11/12   (d) 14/15
2-2. (a) 1/4   (b) 1/4   (c) 1/6   (d) 1/6   (e) 11/12
2-3. 1/20
2-4. (a) 1/4   (b) 1/12   (c) 1/24
2-5. 2/5
2-6. 1/3
2-7. (a) 3/95   (b) 1
2-8. .11 (approx.)
2-9. (a) .423 (b) .0475 (c) .00144 (d) .00198 (e) .003532 (f) .000014
2-10. No—one in 25,827,165; approx 1:1525 for 4, and 1:89677 for 5.
2-12. .91
2-13. If your prob. is 3/8, fair is $3 to $5—but you only put up $2.
If his prob. is 6/8, fair is $6 to $2—but he puts up only $5.
2-15. 3/8
2-16. 11/30
2-25. (a) 15/16   (b) 39/64   (c) 5/16
2-26. $(\frac{9}{16}, \frac{3}{16}, \frac{3}{16}, \frac{1}{16})$
2-27. (b) 11/16
2-28. (a) 1/4   (b) 1/2   (c) 1/2
2-29. (a) $1 - e^{-4}$ (b) e^{-2} (c) .2325
2-30. $1 - \cos\theta_0$
2-31. (a) 1/2   (b) 1/8   (c) 3/4   (d) $1 - \pi/16$ (e) 1/8
(f) 3/4   (g) 1/4   (h) 7/8   (i) 0
2-32. (a) 1/3   (b) 1/2
2-33. .6
2-34. (a) $P(\{b,c\}) = \frac{1}{4}$, $P(\{b,c,d,e\}) = P(\{a,d,e,\}) = \frac{3}{4}$, $P(\{d,e\}) = \frac{1}{2}$
2-35. (a) $\dfrac{P(E)}{P(E) + P(F)}$
2-36. (a) 1/26   (b) 11/221   (c) 6/25   (d) 1/13
2-37. (a) 1/33   (b) 1/33   (c) 1/11   (d) 1/3   (e) 1/10305
2-38. (a) 1/2   (b) 1/2
2-39. 1/4
2-40. 1/8

2-41. (a) 1/2 (b) 1/2 (c) 3/7

2-42. (a) $\dfrac{(M)_3(N-M)_2}{(N)_5} = \dfrac{\binom{M}{3}\binom{N-M}{2}}{\binom{5}{3}\binom{N}{5}}$

2-43. (a) 1/2 (b) 1/3
2-45. .0194
2-46. 1:15:30
2-47. 3/10
2-48. Let $p = P(\text{names Mark})$: $P(\text{Matt escapes} \mid \text{names Mark}) = \dfrac{p}{1+p}$.
If $p = 1/2$, new probability is the same as the old.

Chapter 3

3-1.
x	0	1	2	3
$f(x)$	1/8	3/8	3/8	1/8

3-2. (b) 1/4 (c) 2/3

3-3. (a) 0's on main diagonal, 1/12 for each off diagonal entry.

(b)
x	1	2	3	4
$f(x)$	1/4	1/4	1/4	1/4

(and same for Y)

(c)
z	3	4	5	6	7
$f(z)$	1/6	1/6	2/6	1/6	1/6

(d), (e)

	L: 2	3	4	
1	1/6	1/6	1/6	1/2
S: 2	0	1/6	1/6	1/3
3	0	0	1/6	1/6
	1/6	1/3	1/2	1

(f) 1/2

3-4.
x	0	1	3
$f(x)$	2/6	3/6	1/6

3-5.
x	2	3	4	5	6	7	8	9	10	11	12
$36f(\cdot)$	1	2	3	4	5	6	5	4	3	2	1
y	−5	−4	−3	−2	−1	0	1	2	3	4	5

3-6. $f(x) = \dfrac{\binom{2}{x}\binom{10}{3-x}}{\binom{12}{3}}$, $x = 0, 1, 2,$

3-7. (b) 5/6 (c) 1/12 (d) 1/2 (e) 1/2
3-8. $f(0, 0) = 16/36$, $f(0, 1) = f(1, 0) = 8/36$, $f(2, 2) = 2/36$,
$f(0, 2) = 1/36$, $f(2, 0) = 1/36$
(a) 20/36 (b) $f(0) = 25/36$, $f(1) = 10/36$, $f(2) = 1/36$ (X & Y)
3-11. (a) 0 (b) 1/2 (c) 1/12 (d) 0

Answers to Problems

3-12. (a) 0   (b) 1/4   (c) 1/2
 (d) $Q_1 = \frac{1}{2}$, $Q_3 =$ any number on $[1, 2]$, median $= 1$
 (e) 3/4   (f) 1/4   (g) 1/2
3-13. (a) 0   (b) 1/4   (c) .96   (d) .6   (e) $1/\sqrt{2}$
3-14. (a) $\pi/3$ (b) $1 - 1/\sqrt{2}$
3-16. (a) $\dfrac{\theta r^2}{2\pi}$, for $0 < \theta < 2\pi$, $0 < r < 1$
3-18. (a) $\frac{1}{2}(1+x)$, $-1 < x < 1$ (b) 1/4
3-19. (a) 1/4   (b) $F(x) = \begin{cases} \frac{1}{2}(1+x)^2, & -1 < x < 0 \\ 1 - \frac{1}{2}(1-x)^2, & 0 < x < 1 \end{cases}$
3-20. (a) $f(x) = \dfrac{1/\pi}{1+x^2}$ (b) Med. $= 0$, $Q_1 = -1$, $Q_3 = 1$ (c) .1476
3-21. (a) $1 - e^{2y}$, $y > 0$ (b) e^{-2a}
3-22. .113
3-24. (a) $k = \frac{1}{2}$ (b) $k = 6$ (c) $\frac{1}{2}$
3-25. (a) $F(x) = x^2/4$, $0 < x < 2$; 1/16
 (b) $F(x) = 3x^2 - 2x^3$, $0 < x < 1$; 1/2
 (c) $F(x) = \begin{cases} \frac{1}{2} e^x, & x < 0, \\ 1 - \frac{1}{2} e^{-x}, & x > 0, \end{cases}$ $P(0 < X < 1/2) \doteq .197$

3-26. (a) 1/2   (b) $f(\lambda) = \frac{2}{\pi}\sqrt{4 - \lambda^2}$, $-2 < \lambda < 2$ (for both X and Y)
 (c) 3/4   (d) $\lambda/4$, $0 < \lambda < 4$
3-27. (a) $f(\lambda) = e^{-\lambda}$, $\lambda > 0$ (for both X and Y)   (b) $1 - 5e^{-4}$
 (c) $(1 - e^{-x})(1 - e^{-y})$, $x > 0$, $y > 0$ (d) 0
3-28. (a) 0   (b) $x + y$ (on the square)   (c) $\frac{1}{2}(x^2 + x)$ (e) 1/3
 (d) $f(\lambda) = \lambda + \frac{1}{2}$, $0 < \lambda < 1$, for both X and Y (exchangeable).
3-30. $F_{\Theta, \Phi}(\theta, \phi) = \dfrac{\phi(1 - \cos\theta)}{2\pi}$, $0 < \theta < \pi/2$, $0 < \phi < 2\pi$.
3-31. $\alpha = 1/4$, $F_1(x) =$ unit step at $x = 1$, $F_2(x) = \begin{cases} \frac{2}{3}x, & 0 < x < 1 \\ \frac{1}{3}x, & 2 < x < 3 \end{cases}$
3-32.

x	-2	0	4
$f(x)$	1/3	1/3	1/3

3-33. (a) $\mathcal{U}(0, 2)$ (b) $f(y) = (2\sqrt{y})^{-1}$, $0 < y < 1$.   (c) $\mathcal{U}(0, 1)$
3-34. $F(y) = y$, $0 < y < 1$.
3-36. $f(u, v) = 1/12$ at $(3,-1)$, $(3,1)$, $(4,-2)$, $(4,2)$, $(5,-3)$, $(5,-1)$,
       $(5,1)$, $(5,3)$, $(6,-2)$, $(6,2)$, $(7,-1)$, $(7,1)$.
 $f_U(u) = 1/6$ for $u = 3, 4, 6, 7$, and $1/3$ for $u = 5$.
 $f_V(v) = \frac{1}{12}(4 - |v|)$, $v = -3, -2, -1, 1, 2, 3$.

3-37. $\theta e^{-\theta v}$, $v > 0$
3-38. $1 - e^{-z} - ze^{-z}$, $z > 0$
3-39. w^2, $0 < w < 1$
3-40. $f(u) = \begin{cases} u, & 0 < u < 1 \\ 2 - u, & 1 < u < 2 \end{cases}$

3-41. (a)

(X or Y)	0	1	2	3
	8/27	12/27	6/27	1/27

(b)

x	0	1
$f(x \mid 2)$	1/2	1/2

(c)

u	0	1	2	3
$f(u)$	1/27	6/27	12/27	8/27

(d)

x	0	1	2	3
$f(x)$	1/8	3/8	3/8	1/8

3-42. (a) They are the same.

(b)

v	-2	-1	0	1	2
$f(v)$	1/36	1/9	5/18	1/3	1/4

c)

u	3	5
$f(u \mid 1)$	1/2	1/2

3-43. (a) $f(\lambda) = 2(1 - \lambda)$, $0 < \lambda < 1$ (for both)

(b) $f(x \mid y) = \dfrac{1}{1 - y}$, $0 < x < 1 - y$

3-44. (a) Uniform on $\left(-\sqrt{3}, \sqrt{3}\right)$ (b) $1/\sqrt{3}$

3-45. $f_{Y \mid x}(y) = \dfrac{2(1 - x - y)}{(1 - x)^2}$, $0 < y < 1 - x$ [for $X \mid y$ exchange x & y]

3-46. $f_{X \mid y}(x) = \dfrac{x + y}{\frac{1}{2} + y}$, $0 < x < 1$

3-49. (a) .25389 (b) $\doteq$ .73
3-50. (a) $f(x, y) = 1/y$, $0 < x < y < 1$

(b) $-\log x$, $0 < x < 1$ (c) $\dfrac{-1}{y \log x}$, $0 < x < y < 1$

3-51. $72x^7(1 - x)$
3-52. (b) independent; others, dependent.
3-53. Indepependent in each case.
3-54. (a) $p^{\Sigma x_i}(1 - p)^{3 - \Sigma x_i}$, $x_i = 0, 1$ (b) $e^{-\Sigma x_i}$, $x_i > 0$
3-55. $f(x_1, x_2) = 1/36$ for x_1 and $x_2 = 1, 2, 3, 4, 5, 6$.
3-56. Uniform on the product set determined by A and B
3-57. $f(x, y) = 1/12$ for $x = 1, 3$ and $f(x, y) = 1/6$ for $x = 2$

3-59. (a) $f(\theta, \phi) = \dfrac{\sin \theta}{2\pi}$, $0 < \phi < 2\pi$, $0 < \theta < \pi/2$

3-61. (a) 1/2 (b) 1/3
3-67. $f(k, m) = 1/2^m$, $m \geq k$, $k = 1, 2, \ldots$
3-70. Exchangeable in each case
3-71. Exchangeable in each case except (a)

Chapter 4

4-1. (a) 3/2 (b) 1 (c) 1/3
4-2. 5/2
4-3. 21/2
4-4. (a) 4 (b) 5/3
4-5. (a) 2 (b) 1
4-6. $3 - k/2$, $k = 1, 2, 3$
4-8. (a) 0 (b) 0 (c) 1/2
4-9. 1
4-10. 1
4-11. 5/18
4-12. 3/4
4-14. (a) 1/3 (b) 1/2
4-15. 1/2
4-16. The sum is divergent.
4-17. (a) $\dfrac{e^{tb} - e^{ta}}{t(b-a)}$ (b) $\dfrac{1}{1-t}$, $t < 1$
4-18. (a) 1/2 (b) 1/3
4-19. 1/20
4-20. (a) $y/2$ (b) 1/3 (c) 1/6
4-21. 0
4-22. Converse is false—Problem 4-21 provides a counter-example.
4-23. (a) 2 (b) 1 (c) 12 (d) 1/3
4-24. $(1-t)^{-2}$
4-25. $E(Y \mid x) = x/3$ $(0 < x < 1)$, $EY = 1/6$
4-27. (a) 3/4 (b) 1 (c) 25/99
4-28. $1/\sqrt{18}$
4-29. (a) 2 (b) 1/6 (c) 1/20
4-30. $\mu = 7$, $\sigma^2 = 35/6$
4-31. (a) $\mu_X = \mu_Y = 2$, $\sigma_X^2 = 2/3$, $\sigma_Y^2 = 1/2$
 (b) $\text{var}(X \mid 1) = \text{var}(X \mid 3) = 2/9$, $\text{var}(X \mid 2) = 8/9$ (c) $\frac{2}{3} = \frac{5}{9} + \frac{1}{9}$
4-32. 0, 1
4-33. (a) $\mu_4' - 4\mu\mu_3' + 6\mu^2\mu_2' - 3\mu^4$
4-34. .96
4-38. 1/3 vs. 2/9 when $b = 3$, $1/\sqrt{2}$ vs. 1 when $b = \sqrt{2}$
4-39. $\text{var } X = 7/180$
4-40. (a) $\mu_X = 2$, $\mu_Y = 1$, $\sigma_X^2 = 1$, $\sigma_Y^2 = 2/3$, $\sigma_{X,Y} = 0$
 (b) 0 (c) 3/8 (d) no (e) yes
4-41. 1/2
4-42. .488

Answers to Problems 609

4-43. $\dfrac{-M}{N(N-1)}\left(1-\dfrac{M}{N}\right)$

4-44. (a) 14 (b) 1

4-45. Means (1, 3), variances (2, 18), covariance $= -8$

4-46. 1

4-51. .00458

4-52. 1.12 mi.

4-53. $B_0^2 \operatorname{var} \alpha + A_0^2 \operatorname{var} \beta$

4-54. $\psi(t) = e^{bt}, \ \mu'_k = b^k, \ \mu_k = 0$

4-55. 2

4-56. $\dfrac{e^t - 1}{t}, \ \mu = 1/2, \ \sigma^2 = 1/12$

4-57. [Let $X = a + (b-a)Y$.] $\psi_X(t) = \dfrac{e^{bt} - e^{at}}{t(b-a)}$.

4-58. $e^{t^2/2}$

4-59. $\exp[at + b^2 t^2 / 2]$

4-60. (a) $(1-t)^{-n}$ (b) $\mu = \sigma^2 = n$

4-61. $\psi(t) = 1 + t + \dfrac{7}{12}t^2 + \cdots; \ \mu = 1, \ \sigma^2 = 1/6$

4-62. $\psi(t) = e^{t^2}; \ \mu = 0, \ \sigma^2 = 2$

4-63. (a) No: $\dfrac{1}{1+t^2} = 1 - t^2 + \cdots$ would imply a negative variance.

(b) No: $1 + t^3 + t^6 + \cdots$ would imply $\mu = \sigma^2 = 0$, and $\psi(t) = 1$.

4-64. $\mu = 1, \ \sigma^2 = 2$

4-65. $(1-t^2)^{-1}$ [see Problem 4-55]

4-66. $P(X=1) = P(X=-1) = p/2, \ P(X=0) = 1 - p$

4-67. $(2-t)^{-1}$; 2, 1

4-68. $K = 1/e, \ \mu = \sigma^2 = 1$

4-69. $(2-t)^{-n}$

4-70. $\psi(t) = (1-2t)^{-n/2}, \ \operatorname{var} Y = 2n$

4-71. 25/216

4-72. .063

4-73. $e^{iat} \phi_X(bt)$

4-74. $\dfrac{e^{it} - 1}{it}$

4-75. $e^{-t^2/2}$

4-76. $\exp(i t m - |at|)$

4-77. $[\phi_X(t/n)]^n$

4-78. (a) $(1 - it/n)^{-n}$ (b) $e^{-t^2/(2n)}$ (c) $e^{-|t|}$

4-79. $f(x, y) = e^{-x-y}, \ x > 0, \ y > 0$

4-80. $\psi_X(s) = e^{-s^2}, \ \psi_Y(t) = e^{-t^2}$ (not independent)

4-81. $\psi_{X,Y}(s,t) = \dfrac{1}{(1-t)(1-t-s)}, \ \psi_X(s) = \dfrac{1}{1-s}, \ \psi_X(t) = \dfrac{1}{(1-t)^2}$

4-82. $\psi_{U,V}(s, t) = \psi_{X,Y}(as + ct, bs + dt)$

Chapter 5

5-6. .0062
5-7. .087
5-8. 142.6
5-9. .92

Chapter 6

6-1. Ber(p), where $p = P(X > 1)$
6-2. (a) 1/16   (b) .989   (c) .981   (d) 1/8
6-3. $1 - 1/12^4$
6-4. $16/3^7$
6-5. (a) Bin(50, $\frac{1}{2}$)
6-6. (a) 5, $\sqrt{4.75}$ (b) .76
6-7. $(n)_k p^k$
6-9. (a) 5   (b) 40.2
6-11. (a) .46   (b) .08   (c) .97
6-12.

x	0	1	2	3	4	5
Exact	.3487	.3874	.1937	.0574	.0112	.0015
Approx.	.29	.41	.24	.052	.005	.0001

6-13. .0242

6-14. (a) $\dfrac{\binom{4}{x}\binom{6}{5-x}}{\binom{10}{5}}$, $x = 0, 1, 2, 3, 4$ (b) $\mu = 2$, $\sigma^2 = 2/3$

6-15. .07022 vs .0729
6-16. (a) 3   (b) 1
6-17. .26156 (normal approx., .26)

6-18. $\dfrac{\binom{5}{k}\binom{15}{5-x}}{\binom{20}{5}}$

6-19. 1848
6-20. (a) 1/4   (b) $X \sim$ Hyper(10, 5, 3), $Y \sim$ Hyper(10, 3, 3)   (c) 2.4
6-21. .183
6-23. Hyper $(n_1 + n_2, n_1, m)$
6-24. (a) $(5/6)^4$ (b) 6   (c) 9/2
6-25. (a) 00   (b) .9556
6-26. 3 or 4

6-27. (a) 32   (b) 14
6-28. $k!q^{k-1}/p^k$
6-30. (a) .889   (b) .080   (c) .135
6-31. .184
6-32. (a) .879   (b) .607   (c) .52
6-33. (a) .1465   (b) $\left(\frac{2T}{3}\right)^k \frac{e^{-2T/3}}{k!}$ (c) .0357
6-34. (a) .449   (b) .677
6-35. (a) .974   (b) .074 (Poi. approx.) or .072 (binom. approx.)
6-36. $9e^3$
6-37. (a) .066   (b) .0628
6-38. Poisson: .908, binomial: .9106
6-39. (a) .265 (Poisson)   (b) .046 (normal)   (c) .964 (Poisson)
6-44. .713
6-45. $\text{Bin}\left(m, \frac{t_1}{t_1 + t_2}\right)$
6-47. $\frac{1}{\lambda}\log 2$
6-48. (a) e^{-3} (b) .423   (c) .632   (d) 6 min   (e) 2 min
6-49. (a) 7.5 days   (b) 4   (c) .67   (d) .692
6-50. Exp(.8)
6-54. 3/2
6-57. .712
6-58. (a) 120   (b) $\frac{945}{32}\sqrt{\pi}$ (c) 4/15   (d) $\frac{3}{32}\sqrt{2\pi}$ (e) $\frac{8}{81}$
       (f) $\frac{1}{30030}$ (g) $\frac{3\pi}{512}$ (h) 1/12
6-62. $\frac{15}{64}\sqrt{\frac{\pi}{2}}$
6-64. $\frac{\lambda}{\alpha - 1}$
6-65. $\frac{s}{t - 1}$
6-69. (a) .7734   (b) .1336   (c) 104   (d) 48   (e) 1120   (f) 8.65, 11.35
6-72. 20.5, 5.93
6-73. $\sigma\sqrt{2/\pi}$
6-74. $\frac{1}{\sqrt{2\pi v}}e^{-v/2}$, $v > 0$
6-75. (a) $\frac{1}{\pi}\exp\left(-\frac{x^2 + y^2}{2}\right)$ (b) $\frac{1/\pi}{1 + u^2}$
6-76. $\frac{2^k}{\sqrt{\pi}}\Gamma(k + \frac{1}{2})$
6-80. (a) 69   (b) .159

6-83. (a) $\dfrac{1}{\sqrt{2\pi}\sigma}\exp\left\{\dfrac{-(\log x - \mu)^2}{2\sigma^2}\right\}$, $x > 0$ (b) $e^{\mu + \sigma^2}$, $(e^{\sigma^2} - 1)e^{2\mu + \sigma^2}$

6-84. (a) $\dfrac{r}{\sigma^2}\exp\left\{-\dfrac{r^2}{2\sigma^2}\right\}$, $r > 0$; $\sqrt{\dfrac{\pi\sigma^2}{2}}$, $2\sigma^2(1 - \pi/4)$

(b) $\sqrt{\dfrac{2}{\pi}}\left\{\dfrac{u^2}{\sigma^3}\right\}\exp\left(-\dfrac{u^2}{2\sigma^2}\right)$, $u > 0$

6-85. $\binom{n}{x_Y,\ x_N,\ x_U} p_Y^{x_Y} p_N^{x_N} p_U^{x_U}$

6-86. .0556

6-87. Multinomial, with these cell probabilities:
 .0013, .0215, .1359, .3413, .3413, .1359, .0215, .0113

6-88. .0794

6-89. .03845

Chapter 7

7-1. (a) $(b - a)^{-n}$, $a < x_i < b$ (b) $(2\pi\sigma^2)^{-n/2}\exp\left\{\dfrac{-1}{2\sigma^2}\sum(x_i - \mu)^2\right\}$

(c) $p^{\Sigma x_i}(1 - p)^{n - \Sigma x_i}$, $x_i = 0, 1$. (d) $\dfrac{m^{\Sigma x_i}}{\Pi x_i!}e^{-nm}$, $x_i = 0, 1, 2, \ldots$

(e) $(1 - p)^{\Sigma x}p^n$, $x_i = 0, 1, 2, \ldots$ (f) $\prod\{\pi[1 + (x_i - \theta)^2]\}^{-1}$

7-2. $\dfrac{\binom{M}{\Sigma x_i}\binom{N - M}{n - \Sigma x_i}}{\binom{n}{\Sigma x_i}\binom{N}{n}}$

7-3. $\binom{n}{y_1,\ \ldots,\ y_k} p_1^{y_1} \cdots p_k^{y_n}$, $y_i = 0, 1,\ldots, n$, $\sum y_i = n$

7-4. (b) $\bar{X} = 75$, $S = 12.98$ (c) $\tilde{X} = 80$, midrange $= 72$
 (d) $Q_1 = 62$, $Q_3 = 86$ (e) $R = 42$, $IQR = 24$
 (g) $\bar{X} = 75$, $S = 13.39$

7-5. $\bar{X} = \pm 3$

7.6 (a) 50, 286   (b) 23, 8

7-7. (b) 11.2

7-8. (a) 13.406   (b) 1.344   (c) 1.11   (d) 13.5

7-12. (a) $f(x, y) = 1/6$, $(x, y) = (1, 2), (1, 3), (1, 4), (2, 3), (2, 4), (3, 4)$
 (b) $f_R(1) = 1/2$, $f_R(2) = 1/3$. $f_R(3) = 1/6$
 (c) $f_{\bar{X}}(y) = 1/6$ for $y = 3/2, 2, 3, 7/2$, $f_{\bar{X}}(5/2) = 2/6$.

7-13. .0456

7-14. $E(S^2) = 1/12$, var $S^2 = \dfrac{1}{360}\left(\dfrac{2}{n} + \dfrac{1}{n^2} - \dfrac{3}{n^3}\right)$

7-15 (a) .242   (b) .023

7-16. (a) 0   (b) .6

7-17. (a) $Y \approx N\left(\mu'_k, \dfrac{\mu'_{2k} - \mu'^2_k}{n}\right)$

7-21. (a) .103   (b) .109

7-22. .414

7-26. (a) $\dfrac{1}{n+1}$, $\dfrac{n}{(n+2)(n+1)^2}$ (b) $\dfrac{n}{n+1}$, $\dfrac{n}{(n+2)(n+1)^2}$

7-27. (a) $ET = \dfrac{1}{n\lambda}$ (b) $\dfrac{33}{18\lambda}$

(c) $\dfrac{1}{\lambda}\left(\dfrac{1}{n} + \dfrac{1}{n-1} + \cdots + \dfrac{1}{1}\right)$

7-28. (a) $\dfrac{n!}{(k-1)!(k-1)!} y^{k-1}(1-y)^{k-1}$, $0 < y < 1$, where $n = 2k - 1$.

(b) $EY = \dfrac{1}{2}$, $\text{var } Y = \dfrac{1}{4(n+2)}$.

7-29. (a) $\dfrac{a+bn}{n+1}$, $\dfrac{n(b-a)^2}{(n+2)(n+1)^2}$

(b) $f_R(u) = \dfrac{n(n-1)}{(b-a)^n}(b-a-u)u^{n-2}$, $0 < u < b - a$

7-32. (b) $20r^3(1-r)$, $0 < r < 1$

7-33. (a) $2n(n-1)\displaystyle\int_{-\infty}^{x}[F(2x-s) - F(s)]^{n-2}f(s)f(2x-s)\,ds$

(b) $f(x) = \begin{cases} n2^{n-1}x^{n-1}, & 0 < x < .5, \\ n2^{n-1}(1-x)^{n-1}, & .5 < x < 1. \end{cases}$

7-34. (b) $\dfrac{1}{n+1}$

7-35. $K[F(u)]^{j-1}[F(v) - F(u)]^{k-j-1}[1 - F(v)]^{n-k}f(u)f(v)$, $u < v$,

where $K = \dfrac{n!}{(j-1)!(k-j-1)!(n-k)!}$.

7-36. $\tilde{X} \approx N\left(0, \dfrac{\pi^2}{4n}\right)$, $Q_1 \approx N\left(\theta-1, \dfrac{3\pi^2}{4n}\right)$, $Q_3 \approx N\left(\theta+1, \dfrac{3\pi^2}{4n}\right)$

7-37. $\tilde{X} \approx N\left(\dfrac{\log 2}{\lambda}, \dfrac{1}{n\lambda^2}\right)$, $Q_1 \approx N\left(\dfrac{\log(4/3)}{\lambda}, \dfrac{1}{3n\lambda^2}\right)$, $Q_3 \approx N\left(\dfrac{\log 4}{\lambda}, \dfrac{3}{n\lambda^2}\right)$

7-38. (a) $L(\mu, \sigma^2) = \sigma^{-n}\exp\left\{-\dfrac{1}{2\sigma^2}\sum(X_i - \mu)^2\right\}$, $\sigma > 0$

(b) $L(\lambda) = \lambda^n e^{-\lambda\sum X_i}$, $\lambda > 0$ (c) $L(\lambda) = e^{-n\lambda}\lambda^{\sum X_i}$, $\lambda > 0$

(d) $L(p) = p^n(1-p)^{\sum X_i - n}$, $0 < p < 1$

7-39. $L(\theta) = \displaystyle\prod_{i=1}^{n}[1 + (X_i - \theta)^2]^{-1}$

7-40. (a) Row opposite z_i gives corresponding likelihood function.
7-41. $M = 9$
7-42. (a) $(\bar{X}, V)$ (b) $1/\bar{X}$ (c) $\bar{X}$ (d) $1/\bar{X}$
7-43. $B^n(\theta)\exp[Q(\theta)\sum R(x_i)]$
7-44. $\hat{\theta} = \bar{X}$, defining the same model as $\hat{\lambda}$ in 7-38(b)
7-45. $L(\theta) = \begin{cases} 1, & X_{(n)} - \frac{1}{2} < \theta < X_{(1)} + \frac{1}{2} \\ 0, & \text{elsewhere} \end{cases}$
7-47. $L(p) = p^5(1-p)^7$, $0 < p < 1$
7-48. Partition sets of T are of the form $\{(a, b), (b, a)\}$. Partition sets of Y consist of lines with slope -1; this is a reduction of the partition defined by T, so Y is a function of T.
7-49. (a) Given A: $(0, 0, 1)$; given B: $(2/3, 1/3, 0)$, for all θ
(b) The indicator function of $A = \{a, b\}$; it is sufficient.
(c) No—e.g., $P(c \mid T = 1)$ is $2/3$ under θ_1, but $1/4$ under θ_2.
7-50. For $\bar{X}$, planes: $\sum X_i = b$; for $\bar{X}^2$, pairs of planes: $\sum X_i = \pm b$.
7-51. (a) Each is a set of the six permutations of a particular order statistic, e.g., $\{124, 142, 421, 412, 241, 214\}$
(b) 123, 124, 125, 134, 135, 145, 234, 235, 245, 345 (equally likely)
(c) (i) $\{123\}, \{124, 134\}, \{125, 135, 145\}, \{234\}, \{235, 245\}, \{345\}$
(ii) $\{123, 124, 134, 125, 135, 145\}, \{234, 235, 245\}, \{345\}$
(iii) $\{123\}, \{124, 134, 234\}, \{125, 135, 145, 235, 245, 345\}$
(iv) $\{123, 234, 345\}, \{124, 134, 235, 245\}, \{125, 135, 145\}$
7-52. $\sum X_i$ is sufficient.
7-53. $f(\mathbf{x} \mid \lambda, \sum X_i = k) = \binom{k}{x_1, \ldots, x_n}\frac{1}{n^k}$ for $\sum x_i = k$
[That is, the distribution is multinomial $(k; \frac{1}{n}, \frac{1}{n}, \ldots, \frac{1}{n})$.]
7-54. $f(a, b \mid N \geq 3, X_1 + X_2 = 4) = \begin{cases} 1/3, & \text{if } a + b = 4, \\ 0, & \text{if } a + b \neq 4 \end{cases}$
$f(a, b) \mid N = 2, X_x + X_2 = 4) = \begin{cases} 1 & \text{if } a = b = 2, \\ 0 & \text{otherwise} \end{cases}$
7-55. $\sum X_i$ in each case
7-56. $\sum X_i$
7-57. $\sum Z_i^2$
7-59. $(X_{(1)}, \sum X_i)$
7-60. (e) $\sum(X_i - \mu)^2$; for the others, $\sum X_i$ is minimal sufficient.
7-61. (a) $\sum Z_i^2$ (b) $\sum U_i^2$
7-62. $\sum X_i$ = number of "successes"
7-63. $\{a, b\}, \{c\}$
7-64. The sample frequencies of the a_i are minimal sufficient.
7-66. $(\sum X_i, \sum X_i^2)$
7-67. $\prod X_i$

7-68. (a) n (b) $\frac{n}{pq}$ (c) $\frac{n}{\lambda}$ (d) $\frac{n}{p^2 q}$

7-70. $I(\theta) = [h'(\theta)]^2 \tilde{I}(\xi)$

7-71. $2n/\sigma^2$

7-73. $I(\theta) = \left(\frac{B'}{B}\right)^2 - \frac{B''}{B} + \frac{B'}{B}\frac{Q''}{Q'}$

7-74. Given a: $(\frac{20}{38}, \frac{18}{38})$; given b: $(\frac{20}{62}, \frac{42}{62})$

7-75. $N(\nu, \tau^2)$

7-76. $\text{Gam}(\alpha, \beta)$

7-77. $\text{Gam}(\alpha, \beta)$

7-78. $\text{Beta}(r, s)$

7-79. Either way, posterior for (f_0, f_1) is $(\frac{50}{77}, \frac{27}{77})$ given (a, a), $(\frac{50}{113}, \frac{63}{113})$ given either (a, b) or (b, a), and $(\frac{50}{197}, \frac{147}{197})$ given (b, b).

7-81. $M - 1 \sim \text{Bin}(N - 1, p)$

7-82. $N(\bar{X}, 1/n)$

7-83. Same as 7-82.

Chapter 8

8-1. $\sum a_i = 1$

8-4. m.s.e.$(T_1) = \frac{pq}{n}$, m.s.e.$(T_2) = \frac{npq}{(n+2)^2} + \left(\frac{1-2p}{n+2}\right)^2$

8-5. (a) $2\bar{X}$; m.s.e. $= \frac{\theta^2}{3n}$ (b) $\frac{n+1}{n} X_{(n)}$; m.s.e. $= \frac{\theta^2}{n(n+2)}$

8-6. $T = 1$ if $X = 0$; $T = 0$, otherwise

8-8. (a) $\sqrt{\frac{\hat{p}(1-\hat{p})}{n}}$ (b) $\frac{\bar{X}}{\sqrt{n}}$

8-9. $\sqrt{\frac{S_T^2}{n_T} + \frac{S_C^2}{n_C}}$

8-10. (a) $\frac{n}{n-1}\lambda$ (b) $\frac{n-1}{n\bar{X}}$; m.s.e. $= \frac{\lambda^2}{n-2}$

8-11. $T = 1$ if $X = 0$; $T = 0$, otherwise

8-13. 9/10

8-14. (a) $\frac{3}{n+2}$ (b) $\frac{n+2}{3n}$, 1/3

8-15.(ii) (a) $I_n(\mu) = n$ (b) $I_n(p) = \frac{n}{pq}$ (c) $I_n(\theta) = n/\theta^2$
(d) $I_n(\lambda) = n/\lambda$ (e) $I_n(\theta) = \frac{n}{\theta(\theta-1)}$

8-19. $\frac{n-2}{n}$

8-20. (b) $\sum X_i$ (c) $\frac{n-1}{\sum X - 1}$

8-28. $1/\bar{X}$

8-29. $\pi/(2\bar{X}^2)$
8-30. $\bar{X}/(1-\bar{X})$
8-31. (a) $2\bar{X}$
8-35. $\bar{X}/t_0$
8-36. $1/\bar{T}$
8-37. (a) $(\bar{X}, \bar{Y})$ (b) $\sqrt{\sum Y_i / \sum X_i}$ (not sufficient)
8-38. $\sum X_i^2/n$
8-39. (b) $\dfrac{n}{2(n+1)}$, $1/2$
8-40. $\hat{\alpha} = \bar{Y} - \beta\bar{X}$, $\hat{\beta} = \dfrac{\sum x_i Y_i - n\bar{x}\bar{Y}}{\sum x_i^2 - n\bar{X}^2}$
8-41. $\dfrac{n+1}{\beta + \sum X_i}$
8-42. $8/21$
8-43. $25/6$
8-44. 4
8-45. (a) $2.30 \pm .028$ (b) $2.30 \pm .036$ (c) $2.30 \pm .010$ (d) $2.30 \pm .013$
8-46. 664
8-47. $.204 < p < .256$
8-48. $.407 < p < .766$
8-49. $\bar{X} \pm \dfrac{k^2}{2n} \pm \sqrt{\dfrac{k^2 \bar{X}}{n} + \dfrac{k^4}{4n^2}}$, where $k = z_{1-\frac{\eta}{2}}$.
8-50. $X \pm 6.314$
8-51. (a) $.198 < \lambda < .573$ (b) $.257$
8-52. (a) $4.505 < \mu < 19.5$ (b) 18.07 (c) $.5476$ (d) $.9994$
8-53. $1.454 < \sigma < 3.495$
8-54. (a) $9.29 < \mu < 9.35$ (b) $.0517 < \sigma < .0976$
     (c) $(\mu - 9.318)^2 < .181\sigma^2$, $.00267 < \sigma^2 < .00953$
8-55. (a) $10.1, 8$ (b) $10.192, 104$
8-56. (a) $10.2 \pm .98$ $(.99)$ (b) $10.2 \pm .196$ $(.954)$
8-57. $a_n S$
8-58. $\dfrac{a_n}{a_n^2 + b_n^2} R$
8-59. $\dfrac{1}{2^{\gamma/2-1}\Gamma(\nu/2)} y^{\gamma-1} e^{-y^2/2}$, $y > 0$

Chapter 9

9-1. S, C, C, S, C
9-2. $(\tfrac{5}{3})^n (\tfrac{3}{7})^Y$
9-3. (a) $\Lambda = 2^{n/2} \exp(-\tfrac{1}{4}\sum X_i^2)$ (b) $\sum X_i^2 \sim \chi^2(n)$

9-4. Λ^*: (2/3, 3, 1/3, 3/2, 1); z_3, z_1, z_5, z_4, z_2

9-5. $\Lambda^{2/n} = xe^{-x+1}$, where $x = \overline{X^2}/\theta_0$

9-6. $\Lambda^{1/n} = ue^{-u+1}$, where $u = \lambda_0 \overline{X}$

9-7. (a) $\Lambda = \begin{cases} 1/8, & Y = 0, 3, \\ 27/32, & Y = 1, 2 \end{cases}$ (b) $f_\Lambda(1/8) = 1/4$, $f_\Lambda(27/32) = 3/4$

9-8. $Z = \dfrac{\overline{X}_1 - \overline{X}_2}{\sqrt{S_1^2/n_1 + S_2^2/n_2}}$

9-9. 1.96

9-10. .44

9-11. .045 (one-sided) or .09 (two-sided)

9-12. .0544

9-13. (a) .33   (b) No (.33 > .05)   (c) .33 (one sided)

9-14. $Z \doteq 9$, $P \doteq 4 \times 10^{-19}$

9-15. $P = .0006$

9-16. (a) $P = .151$ (b) $P = .062$

9-17. .016

9-18. (a) $p_0^{25} p_1^{50} (1 - p_0 - p_1)^{25}$ (b) $\Lambda = .00277$, $P = .003$

9-19. .0456

9-20. (b)

	H_0		H_A			
M	0	1	2	3	4	5
P(rej)	0	.4				
P(acc)			.3	.1	0	0

(a),(c)

Y	C_1	C_2	C_3	C_4	C_5	C_6	C_7	C_8
0	R	R	R	A	R	A	A	A
1	R	R	A	R	A	R	A	A
2	R	A	R	R	A	A	R	A
α	1	1	1	.1	1	.4	0	0
β	0	.4	.6	.3	1	1	.9	1

9-21. C_1

9-22. (a) Reject if $Y > 2$, accept if $Y = < 2$,   (b) .40
reject with probability .107 if $Y = 2$

9-23. (a)

z	z_1	z_2	z_3	z_4
Λ	2/5	3	1/4	4/3

(d) .9 vs .3 ($\{z_3, z_1\}$ is better)

(b), (c)

C	$\emptyset$	$\{z_3\}$	$\{z_3, z_1\}$	$\{z_3, z_1, z_4\}$	Ω
α	0	.1	.3	.7	1

9-24. (a) $\phi(t) = 1$ if $t < 2$, 0 if $t > 2$, and .5 if $t = 2$.

(b) $\sum_{t < 2} f(t) + \frac{1}{2}f(2)$

9-25. $-2 \log \Lambda = 5.99$; yes

9-26. (a) Rej. $\theta = 1$ if $\sum X_i < 5.45$ or > 15.7; $\alpha \doteq .10$
(b) Rej. $\theta = 2$ if $\sum X_i < 9.59$ or > 34.2

9-27. $\mu > 12.15$

9-28. (a) $K = \lambda_0 + 1.645\sqrt{\lambda_0/25}$ (b) $\lambda > \bar{X} + \frac{1.645^2}{50} - .329\sqrt{\bar{X}}$

9-29. $\bar{X} > K_1$ or $\bar{X} < K_2$ $(K_1 < K_2)$

9-30. (a) $X > 1$ (b) $X > 0$ or $X < -1$ (c) $X > -1$ (d) $1.4 < X < 6$

9-31. $|X| > \sqrt{\frac{1}{K} - 1}$

9-32. (a) $1/2$ (b) $p^8(1-p)^2$ (c) .115 (d) $7/2^7$ (e) .5

9-33. (a) .04 (b) .03754

9-34. (a) .16 (b) .068 (c) .125 (d) .68

9-35. .125

9-36. .92

9-37. .377

9-38. (a) .044 (b) .11

9-39. (a) $K_1 = 3.83$, $K_2 = 12.4$ (b) .19, .26

9-40. (b) $S^2 > \sigma_0^2 + z_{1-\alpha}\sqrt{2\sigma_0^4/n}$ (c) $P = .0228$

9-41. $Z \doteq 5.1$, $P < .0001$

9-42. $F = 1.13$, (28, 19) d.f.; $P > .05$ $[P \doteq .4]$

9-43. $\dfrac{2n^2(m + n - 2)}{m(n - 2)^2(n - 4)}$, for $F(m, n)$.

9-47. $t = 2.45$ (8 d.f.), $P = .02$

9-50. $.56 < \mu_2 - \mu_1 < 9.44$

9-51. 76.3 ± 45.1

9-52. $t \doteq 4.0$ (5 d.f.), $P = .005$

9-53. $t \doteq 3.9$ (39 d.f.), $P < .001$

9-54. $t \doteq -.29$ (not significant), and $t \doteq 2.9$ (significant): Omitting the observation farthest from 114 rejects $\mu = 114$; the t-test is is not appropriate for such badly skewed data.

9-55.
α	1	.75	.5	.75	.25	.5	.25	0
β	0	.64	.32	.04	.96	.68	.36	1

(same order of C's as in 9-20)

9-56. $\alpha = .208$, $\beta = .268$

9-57. (a) $\alpha = 1 - k$, $\beta = k^2$ (c) $k = .39$

9-58. (a) $\alpha = \Phi(-\sqrt{n}K/2)$, $\beta = \Phi(\sqrt{n}(K-1)/2)$
(b) $n = 86$, $K \doteq .5$ (c) $n = 52$, $K = .645$

9-59. $\alpha = .506$, $\beta = .494$

9-60. $\bar{X} > 1/2$

9-61. $\alpha = .035$, $\beta = .30$

9-62. Reject H_0 if $Y \geq 6$, accept H_0 if $Y \leq 4$, and reject H_0 with probability $\gamma = .16$ if $Y = 5$.

9-63. (a) C: $\emptyset$, $\{z_3\}$, $\{z_3, z_1\}$, $\{z_3, z_1, z_5\}$, $\{z_3, z_1, z_5, z_4\}$, Ω.
(b) $\{z_3, z_1\}$ is better (power .6 as compared with .2 for $\{z_4\}$).

9-65. (a) $\sum_1^n (X_i - \mu_0)^2 > K$ when $\sigma_0 < \sigma_1$ (b) $n = 68$, $K = 390.9$

9-67. $\sum_1^n X_i < K$

9-68. Reject H_0 if $\prod X_i > K$

9-69. $\alpha = \sqrt{1 - 2K/3}$

9-71. $\Phi[\sqrt{n}(\lambda - 1)]$; H_0: $\lambda \leq \lambda_0$ vs. H_A: $\lambda > \lambda_0$

9-72. (a) $\Phi\left\{\dfrac{20(p - .52)}{\sqrt{p(1 - p)}}\right\}$; H_0: $p \leq p_0$ vs. H_A: $p > p_0$

(b) $\Phi\left\{\dfrac{20(p - .52)}{\sqrt{p(1 - p)}}\right\} + \Phi\left\{\dfrac{20(.48 - p)}{\sqrt{p(1 - p)}}\right\}$; H_0: $p = p_0$ vs. H_A: $p \neq p_0$

9-73. (a) $\pi(\theta) = 1 - 2^{-\theta}$, $0 < \theta < 1$ (b) $K = .9$

9-74. (a) $\Phi(\mu\sqrt{n} - K)$ (b) $n = 271$

9-75. (a) $\pi = \frac{1}{2}(\pi_1 + \pi_2)$

(b) $\phi(\mathbf{x}) = 1$ for $\mathbf{x} \in C_1 C_2$; 0 for $\mathbf{x} \in C_1' C_2'$; $\frac{1}{2}$, elsewhere.

9-76. $\pi(\sigma^2) = 1 - F_{\chi^2}\left\{\dfrac{(n - 1)K}{\sigma^2}\right\}$

9-77. (a) $\alpha = .1$
(b) $\pi(\mu) = \Phi[5(\mu_0 - \mu) - 1.645] + \Phi[5(\mu - \mu_0) - 1.645]$

9-78. (a) $1/4$, $1/4$, $1/3$, 1 (b) $\{z_1, z_2\}$ (c) $2/3$ vs. $1/2$ for all $\theta \in H_A$

9-79. $C = .0307$, $D = .00975$

9-80. $\pi(\sigma_2^2/\sigma_1^2) = 1 - F(3.44\sigma_2^2/\sigma_1^2) + F(.29\sigma_2^2/\sigma_1^2)$, [where F is the c.d.f. of $F(8, 8)$]

9-81. $EY = 2\lambda + k$, $\operatorname{var} Y = 2k + 8\lambda$

9-82. $3/4$

9-84. $\bar{X} > K$

9-85. $\bar{X} > K$

9-86. (a) $\pi(\theta) = \frac{1}{2} - \frac{1}{\pi}\tan^{-1}(1 - \theta)$

9-90. $\Phi\left(\dfrac{100p - 61}{10\sqrt{p(1 - p)}}\right) + \Phi\left(\dfrac{39 - 100p}{10\sqrt{p(1 - p)}}\right)$

9-91. Reject H_0 if $\sum X_i \geq 59$

Chapter 10

10-1. $\chi^2 = 5.6$ (3 df), $P = .133$
10-2. $\chi^2 = 6.55$ (3 df), $P = .088$
10-3. (a) $\chi^2 = 12.5$ (2 df), $P = .002$ (b) $\chi^2 = 0$, $\Lambda = 1$
10-4. $Z = 1.41$ ($P \doteq .08$—weak evidence against H_0)
10-5. (a)-(c)

Combination	#(patterns)	Prob.	χ^2	Λ
0 0 6	3	1/243	12	.00137
0 1 5	6	12/243	7	.02048
0 2 4	6	30/243	4	.0625
1 2 3	6	120/243	1	.5926
0 3 3	3	20/243	3	.0878
1 1 4	3	30/243	3	.25
2 2 2	1	30/243	0	1

(e) 11/16

10-6. $\chi^2 \doteq 7$ (2 df); combining last 2 cells gives $\chi^2 = 2.2$ (1 df), $P \doteq .14$
10-7. $\chi^2 = 2.66$ (3 df); or better, combine cells 3 & 4: $\chi^2 = .196$ (2 df)
10-10. $\chi^2 = 6.2$ (1 df), $P = .013$
10-11. $\chi^2 = 12.86$ (6 df), $P = .045$ ($-2 \log \Lambda = 12.6$)

10-12. (a) $\begin{array}{|ccc|} \hline .5 & 1 & .5 \\ .5 & 1 & .5 \\ \hline \end{array}$ (b) $\begin{array}{|ccc|} \hline 1 & 1 & 0 \\ 0 & 1 & 1 \\ \hline \end{array}$ $\begin{array}{|ccc|} \hline 0 & 1 & 1 \\ 1 & 1 & 0 \\ \hline \end{array}$, Prob. = 1/3, $\Lambda = 1/4$

(c) $\begin{array}{|ccc|} \hline 0 & 2 & 0 \\ 1 & 0 & 1 \\ \hline \end{array}$ $\begin{array}{|ccc|} \hline 1 & 0 & 1 \\ 0 & 2 & 0 \\ \hline \end{array}$, Prob. = 1/6, $\Lambda = 1/16$

10-15. A: $\begin{array}{|cc|} \hline 4 & 0 \\ 5 & 3 \\ \hline \end{array}$ B: $\begin{array}{|cc|} \hline 3 & 1 \\ 6 & 2 \\ \hline \end{array}$ C: $\begin{array}{|cc|} \hline 2 & 2 \\ 7 & 1 \\ \hline \end{array}$ D: $\begin{array}{|cc|} \hline 1 & 3 \\ 8 & 0 \\ \hline \end{array}$

χ^2:   2   0   2   8
$-2 \log \Lambda$: 2.9   0   1.9   9.0
Prob.:   .255   .509   .218   .018

Critical regions: $\{D, C\}$, $\alpha = .236$; $\{D, A\}$, $\alpha = .273$

10-16.

Pattern for 0	χ^2	$-2 \log \Lambda$	Prob.
0 0 4	12	15.3	1/165
1 3 0	5.7	6.3	32/165 ($P = .2$)
2 0 2	4	4.2	36/165
1 1 2	.75	.73	96/165

10-18. (a) .1155   (b) 1
10-19. (a) 4.06   (b) 900
10-20. .267, .0745
10-21. $\chi^2 = 32/11$, $P \doteq .09$
10-25. $\chi^2 = 19.6$, $P < .001$
10-26. (a) $\begin{array}{|cc|} \hline p^2 & p(1-p) \\ \hline \end{array}$ (b) .4056   (c) .00593

$p(1-p)$ $(1-p)^2$ (d) $3.96 + 1.81 = 5.77$

10-27. H_0: Independence, $\chi^2 = 2.97$ (1 df)

10-33. (a) $-2\log \Lambda_{40} = 30.7$ (4 df) $P < .001$

(b) $-2\log \Lambda_{30} = 13.42$ (3 df) $P < .005$

(c) $-2\log \Lambda_{20} = .77$ (3 df)

Chapter 11

11-1. (a) $\frac{1}{8} < \Lambda_n < \frac{9}{2}$ (b) $.339n - 1.50 < \sum_1^n X_i < 1.085 + .339n$

(c) Reject H_0 at the 8th stage.

11-2. [Continue sampling if $\frac{1}{18} < \frac{3^n}{2^n + Y_n} < \frac{19}{2}$, where $Y_n = \#(H)$.]

11-3. (a) $5n - 13.25 < \sum_1^n X_i < 5n + 13.25$

(b) $5n - 6.77 < \sum_1^n X_i < 5n + 9.36$

(c) $7n - 4.417 < \sum_1^n X_i < 7n + 4.417$

11-4. Continue sampling if $\frac{n - \log B}{\log 2} < \sum_1^n X_i < \frac{n - \log A}{\log 2}$

11-7. Reject H_0 at the 10th stage ($\Lambda_{10} = 69.4$).

11-8. $\alpha = \frac{A(B-1)}{B-A}$, $\beta = \frac{1-A}{B-A}$

11-9. (a) $p = \dfrac{1 - \left(\dfrac{1-p_0}{1-p_1}\right)^h}{\left(\dfrac{p_0}{p_1}\right)^n - \left(\dfrac{1-p_0}{1-p_1}\right)^h}$ (b) $E(Z) = p\log\dfrac{p_0}{p_1} + (1-p)\log\dfrac{1-p_0}{1-p_1}$

(c)

h	$-\infty$	-1	0	1	∞
p	0	.5	.603	.7	1
$\pi(p)$	0	.05	.5	.95	1

11-10. 30.4, 32.2

11-11. $\dfrac{\log 19[1 - 2\pi(p)]}{\log(5/3) + p\log(3/7)}$

11-13.

θ	0	1	1.44	2	∞
$\pi(\theta)$	1	.9	.44	.05	0
$E(N \mid \theta)$	0	7.74	13.5	10.3	3.25

Chapter 12

12-1. $f_{X', Y'}(u, v) = f_{X, Y}(u - h, v - k)$

12-2. $f_{U,V}(u,v) = \frac{1}{3} f\left(\frac{u+v}{3}, \frac{-u+2v}{3}\right)$

12-3. $f_Z(z) = -\log z, \quad 0 < z < 1$

12-4. $f_W(w) \propto \dfrac{w^{\alpha-1}(1-w)^{\beta-1}}{B(\alpha, \beta)}$

12-5. $f_{X/Y}(u) \propto \dfrac{u^{k/2-1}}{(1+u)^{\frac{k+m}{2}}}, \quad u > 0$

12-6. $U = \dfrac{X+Y-2}{\sqrt{6}}, \quad V = \dfrac{X-Y}{\sqrt{2}}$ (not a unique answer)

12-7. $f_{U,V}(u,v) = \frac{1}{2}, \quad 0 < u+v < 2, \quad 0 < u-v < 2$

12-9. $\dfrac{1}{2\pi\sqrt{5}} \exp\left\{-\frac{9}{10}[x^2 - \frac{4}{9}x(y-4) + (y-4)^2]\right\}$

12-10. (a) $\sigma_Y^2 = \frac{2}{7}, \mu_Y = 0, \sigma_X^2 = \frac{4}{7}, \mu_X = 0, \sigma_{X,Y} = \frac{2}{7}$ (b) $\dfrac{\sqrt{7}}{2\pi}$
  (c) $x^2 - xy + 2y^2 = k$ (ellipses) (d) $X|y \sim N(\frac{y}{2}, \frac{1}{2}), \quad Y|x \sim N(\frac{x}{4}, \frac{1}{4})$

12-11. $E(Y \mid x) = \frac{x}{2} + 1, \quad E(X \mid y) = y - 1$

12-12. Bivariate normal: $\mu_X = \mu_Y = 0, \sigma_X^2 = 5, \sigma_Y^2 = 10, \sigma_{X,Y} = 1$

12-13. $y = 2 - x/2$

12-19. (a) 47   (b) 92   (c) 65, r.m.s.p.e. $= \sigma_Y = 12$

12-20. $P = .03$

12-21. (a) $E(Y \mid x) = \frac{1}{2}(1 - |x|)$ (b) 1/3

12-22. (a),(b) $E(Y \mid x) = 1 + x$

12-23. $.985 < p < .991$

12-28. $\boldsymbol{\mu}_Y = \mathbf{A}\boldsymbol{\mu}_X + \mathbf{b}, \quad \mathbf{M}_Y = \mathbf{A}\mathbf{M}_X\mathbf{A}'$

12-29. $f_X(\mathbf{x}) = n!, \quad x_i > 0, \quad \sum x_i < 1$

12-30. $f_Z(\mathbf{z}) = (z_2 z_3^2 \cdots z_n^{n-1}) e^{-z_n}, \quad z_n > 0, \quad 0 < z_i < 1, \quad i = 1, \ldots, n-1$

12-31. (b) $\sqrt{2/5}$

(c) $\boldsymbol{\mu}_Y = \begin{pmatrix} 8 \\ 0 \\ 2 \end{pmatrix}, \quad \mathbf{M}_Y = \begin{pmatrix} 13 & 9 & 11 \\ 9 & 13 & 11 \\ 11 & 11 & 11 \end{pmatrix}$

(d) $\psi(s_2, s_3) = \exp[2s_3 + \frac{1}{2}\{13s_1^2 + 22s_1 s_2 + 11s_2^2\}]$

$f(y_2, y_3) = \dfrac{1}{2\pi\sqrt{22}} \exp\left\{-\frac{1}{2}[\frac{1}{2}y_2^2 - y_2(y_3 - 2) + \frac{13}{22}(y_3 - 2)^2]\right\}$

Chapter 13

13-1. $1/6$

13-2. $12(1-y_2)^2, \quad 0 < y_1 < y_2 < 1$

13-3. $n = 38$

13-4. (a) .393   (b) .057

13-5. $f(\mathbf{y}) = n!$, $y_i > 0$, $\sum y_i < 1$

13-6. p.d.f. $= (n-1)!/y_n^{n-1}$, $0 < y_1 < \cdots < y_n$

13-7. (a),(b)

r	2	3	4	5	6
$f(r)$	.1	.2	.4	.2	.1

(c) $Er = 4$, $\sigma_r^2 = 1.2$

13-8. (a) $Z = .93$ (b) $Z = .15$

13-9. (a) $r = 11$, $Er = 13$, $\sigma_r = 2.4$ (no evidence against H_0)
 (b) $s = 13$, $Es = 47/3$, $\sigma_s = 1.99$ (little evidence against H_0)
 (c) $r = 10$; $13 \le s \le 15$ (still little evidence)

13-10. 2 1 4 3 5, 2 1 3 5 4, 1 3 2 5 4

13-12. $\dfrac{2(n-1)}{3(n-4)}$

13-14. (a) 2 (b) $(1-p)^7(28p^2 + 7p + 1)$ (c) $\phi(y) = \begin{cases} 1, & y = 0, 1, 2, \\ .062, & y = 3, \\ 0, & \text{otherwise.} \end{cases}$

13-15. $\alpha = 1/2^6$ (both tests)

13-16. (a) .09 (b) .020

13-17. (a) $W = 2W_+ - n(n+1)/2$ (b) 0, $\tfrac{1}{6}n(n+1)(2n+1)$

13-18. About 94%

13-21. (a) $249 < m < 287$ (b) $255 < m < 278$

13-23. (a) $R_Y = 20$, $P = .13$ (b) $3c_1 = 1.504$, $P = .13$ (c) $P = .107$

13-24. (a) $P = .00138$ (b) $R_N = 75$, $P = .0019$

13-25. (a) $P = .0146$ (b) $P = .014$

13-28. (a) 12349, 12358, 12367, 12457, 13456

13-30. $.050 \doteq .051$, $.0093 \doteq .008$

13-31. (a) 1 (b) .864

13-32. $c_1 = 1.04$, $P = .0017$

13-33. $D_n = .277$, $P > .20$

13-34. $D_n = 8/160$, $P > .20$

13-35. $D_n = .78$, $P < .01$

13-36. .85

13-38. $D_n = .196$, $P \doteq .15$

13-39. $D_n = .169$, $P \doteq .20$

13-41. (a) $D = .8 > 19/30$ (rej. H_0 at $\alpha = .01$) (b) .001

Chapter 14

14-1. $y = \tfrac{1}{34}(53 + 6x)$

14-2. (a) $\tilde{\gamma} = \dfrac{\Sigma xY}{\Sigma x^2}$

(b) $\tilde{\alpha} = \dfrac{\Sigma z^2 \Sigma xY - \Sigma xz \Sigma zY}{\Sigma x^2 \Sigma z^2 - (\Sigma xz)^2}$, $\tilde{\beta} = \dfrac{\Sigma x^2 \Sigma zY - \Sigma xz \Sigma xY}{\Sigma x^2 \Sigma z^2 - (\Sigma xz)^2}$

Answers to Problems

14-3 (a) $\mathbf{X} = \begin{pmatrix} x_1^2 \\ \vdots \\ x_n^2 \end{pmatrix}$, $\beta = \gamma$ (b) $\beta = \begin{pmatrix} \alpha \\ \beta \end{pmatrix}$, $\mathbf{X} = \begin{pmatrix} x_1 & z_1 \\ \vdots & \vdots \\ x_n & z_n \end{pmatrix}$

(c) $\beta = \begin{pmatrix} \beta_0 \\ \beta_1 \\ \beta_2 \end{pmatrix}$, $\mathbf{X} = \begin{pmatrix} 1 & x_1 & x_1^2 \\ \vdots & \vdots & \vdots \\ 1 & x_n & x_n^2 \end{pmatrix}$

14-4. $\beta_0 = 2$, $\beta_1 = 0$, $\beta_2 = -1$

14-5. $(\sum Y_i, \sum x_i Y_i, \sum Y_i^2)$

14-6. $N\!\left(\alpha + \beta(x - \bar{x}), \sigma^2 \left\{\frac{1}{n} + \frac{(x-\bar{x})^2}{SS_{xx}}\right\}\right)$

14-7. $y = \frac{1}{3}(32 + 12.2x)$; $(2.91, 5.23)$

14-8. $y = 5.27 + .8658x$

14-14. (a) $\hat{\beta} = \sum x_i Y_i / \sum x_i^2$

(b) $\hat{\beta} \sim N\!\left(\beta, \frac{\sigma^2}{\Sigma x^2}\right)$ (c) $n\hat{\sigma}^2/\sigma^2 \sim \text{chi}^2(n-1)$

14-15. No ($SS_{xx} \to 1$, so $\text{var}\hat{\beta} \to \sigma^2$), unless $\sigma^2 = 0$.

14-17. $.181 \pm .129$

14-19. (a) $\hat{Y}_0 \pm t_{1-\frac{\gamma}{2}}(n-2) \times \text{s.e.}(\hat{Y}_0)$,

(b) $\hat{Y}_0 \pm t_{1-\frac{\gamma}{2}}(n-2) \times \text{s.e.p.}(\hat{Y}_0)$

14-20. (a) 608.4 ± 18.85 (b) 608.4 ± 47.05

14-21. $T = \dfrac{\bar{Y} - \alpha_0}{S/\sqrt{n}}$

14-22. Reject β_0 when $\dfrac{(\hat{\beta} - \beta_0)^2}{\hat{\sigma}^2/\Sigma x_i^2} > K$.

14-24. $F = 5$ $(1, 5)$ d.f.; $P \doteq .066$

14-26. Works only if n_i is independent of i.

14-27. (a) 43.6

(b) $n = 28$, Reg.d.f. $= 3$, SSReg $= 818.28$

(c)

Source	d.f.	SS	MS
Regression	3	818.28	272.76
Error	24	24.026	1.0011
Total	27	842.31	

Source	Individual SS	Cumulative SS	Cum d.f.	Cum MS	Adj. R^2
x_2	754.620	754.62	1	754.62	.8959
x_1	57.760	812.38	2	406.18	.9645
$x_1 x_2$	5.901	818.28	3	272.76	.9715
Residual	24.026	824.38	27		

14-28. Determinant of the normal equations is 0 [(x, z)'s are collinear].
14-32. With x = freq.: $Y = .381 - .0359x + .00987\,x^2$; $F = 13.14$, (2, 10) d.f.; Adj.$R^2 = .669$ (Adj.$R^2 = .62$ with Y on $\log x$)
14-33. $F = \infty$: reject H_0 at any level of significance.
14-34. (a) $F = 8.42$ (2, 12) d.f.; $P \doteq .0052$
    (b) $.96 \pm .51$ (c) $\Delta \bar{X} \pm .7865$, for each case
14-39. Reject H_0 when $S_p^{2n}/\prod S_i^{2n_i} > K$.
14-42. $\Delta \bar{X} \pm .7842$ in each case.
14-43. $F_{\text{Tr}} = 4.11$ ($P \doteq .067$) [$F_B = 9.8$ ($P \doteq .013$)]
14-44. All $\hat{\epsilon}_i = 0$, so SSE $= 0$ (Reject H_0 at any level.)

14-45. (e) $F_B = \dfrac{\text{SSB}/(c-1)}{\text{SSE}/(r-1)(c-1)}$

14-49. $F_{\text{int}} = .22$ (not signif), $F_{\text{op}} = 1.62$ (not signif),
$$F_{\text{mach}} = 4.77\ (P \doteq .01)$$

14-50.

Source	SS	df	MS	F	
Rows	408	3	136	3.81	$P = .077$
Columns	504	3	168	4.71	$P = .051$
Letters	1224	3	408	11.44	$P = .007$
Error	214	6	107/3	---	
Total	2350	15			

14-52.   A B C D   A B C D
       B A D C   B A D C
       C D A B   C D B A
       D C B A   D C A B

Chapter 15

15-1. All but b, d, e
15-3. $(\frac{2}{9}, \frac{4}{9}, \frac{3}{9}, 0)$, yielding the point $(\frac{4}{3}, \frac{4}{3}, \frac{4}{3})$
15-5. $u(\text{bet}) = 1/2$ [with $u(M) = 1$ for $M > 16$, and 0 otherwise]
15-6. Perhaps: expected utility $= 2 = u(\$4)$
15-7. (a) $u(\text{bet}) = P(\text{win})$ (c) $1/10$ (d) $\$1.9$
    (b) (i) Take all bets. (ii) Take bet if $p \geq 1/9$.
15-9. a_3 (loss), a_1 (regret)
15-10. (a) a_1
    (b) pure, a_2; mixed: a_2 with prob. 2/3, a_1 with prob. 1/3.
15-11. (a) a_2 with prob. γ, a_5 with prob. $1 - \gamma$, $\frac{3}{4} \leq \gamma \leq 1$;
    minmax loss $= 2$
    (b) a_1 with prob. 2/3, a_2 with prob. 1/3
15-12. $a = 1/2$

15-13. Pass with prob. 1/3, reject with prob. 2/3
15-14. a_1
15-15. a_2 if $w \leq 1/3$, a_1 if $w \geq 1/3$; $w = 1/3$ is least favorable.
15-16. (a) Reject (b) Reject (c) Pass if $p \leq 1/3$, reject if $p \geq 1/3$
15-17. (a) 1/2 (b) 1/2
15-19. a_1 dominates a_2; a_3 is dominated by mixtures of (a_1, a_4), e.g. (.5, .5), so a_2 and a_3 are not admissible.
15-20. (a) θ^2/n, θ^2, $\theta^2/n+1$, $\theta^2/(n+1)$
  (b) $d_4 \gg d_1 \gg d_2$; d_3 beats d_2 for some θ but not all.
15-21. (a) If z_1, take a_3; if z_2, take a_1. Worth of data = .2/3
  (b), (c) Same as (a)
15-22. (c) Pure: a_3 if z_1, a_2 if z_2 Mixed: If z_1, take a_3; if z_2, take a_3 with prob. 7/17, and a_2 with prob. 10/17
  (d) If z_1, take a_3; if z_2, take a_2.
15-23. (a) $\frac{1}{2}\theta(1-\theta)$ (b) $\frac{1}{16}[(1-2\theta+\theta^2)$
  Neither dominates the other. If $\theta \sim \mathcal{U}(0,1)$, d_2 is better.
15-24. All others dominated by: If indicator says rain, take umbrella, if not, go without umbrella. (Believe the data.)
15-25. All rules are mixtures of no-data rules, so data have no value.
15-26. $(\frac{2}{13}, \frac{4}{13}, \frac{7}{13})$, take a_1
15-27. Posterior odds are 56:3 if z_1 (take a_3), and 14:27 if z_2 (take a_2).
15-28. (a) If $Z = z_1$, take a_2; if $Z = z_2$, take a_3 with prob. 7/17 and a_2 with prob. 10/17.
  (b) (d_8, d_9) with probabilities $(\frac{1}{4}, \frac{3}{4})$.
15-29. Partial answer:
$$R(\theta_1, d_1) = R(\theta_3, d_1) = .1600, R(\theta_2, d_1) = .3174$$

Index

Absolute moments 105
Addition law 28, 29
Admissibility 558
Algebra of sets 8
Alternative hypothesis 292
Analysis of variance 502
ANOVA table 502
Arrangements 3
Association, measures of 380
Asymptotic distribution
   of sample moments 211
   of sample percentiles 218
   of a sum 140
Asymptotic efficiency 264
Asymptotic relative efficiency 465
   of the Fisher-Yates test 477
   of sign tests 465 465
   of the signed-rank test 467
   of the rank sum test 476
Average: *see* Mean
Average sample number 403
Axioms for preferences 543
Axioms for probability 27

Bayes actions 554
Bayes estimates 279
Bayes loss 314, 395, 554
Bayes risk 561
Bayes test 313, 339
Bayes' theorem 37
Behavioral decision rule 568

Behrens-Fisher problem 327
Bernoulli populations 148
Bernoulli process 144
Beta distribution 175
Beta function 175
Bets 544
Bias 254
Binomial coefficient 7
Binomial distribution 145
   approximation by normal 147
   approximation by Poisson 163
   relation to hypergeometric 153
Binomial theorem 7
Bivariate distribution function 51
Bivariate normal density 418
Bivariate normal distribution 417
Bivariate transformation 410
Blocking 527
Bonferroni inequality 29
Borel sets 13
Box plot 203
Box-Muller transformation 417
Bridge deck 6

Categorical data 357ff
Cauchy distribution 191
Central limit theorem 140
Central moments 104
Characteristic function 128
Chebyshev inequality 110
Chi-square distribution 182

Chi-square test 361
Circular normal distribution 417
Class interval 200
Classification models 489, 517
Cochran's theorem 441
Coefficient of determination 502
Combinations 4
Comparisons
   of discrete populations 371
   of distribution functions 385
   of locations 325
   of means 300
   of proportions 359
   of variances 318
Complement of a set 8
Complete class 573
Completely randomized design 524
Completeness 267
Composite hypothesis 292
Conditional density 72
Conditional distribution 69, 422, 439
Conditional independence 389
Conditional mean 95
Conditional probability 33
Conditional test 378
Conditional variance 108
Conditionality principle 251
Confidence bound 283
Confidence level 282
Confidence interval 282
   for the median 467
   for μ 282, 288
   for p 284, 358
   for regression coefficients 499
   for σ 286, 290
Confidence regions 286
Conjugate families 247
Consistent estimator 270
Contingency tables 369ff
Continuity correction 147
Continuity theorem 129

Continuous distributions 31, 54
Continuous random variables 54
Continuous random vector 58
Contour curves 16
Contrasts 522
Control chart 317
Controlled variable 489
Correlation coefficient
   population 116
   sample 427
Convergence, types of 135
Convex combinations 540
Convex hull 541
Convex set 331, 539
Countable aditivity 28
Covariance 113
Covariance matrix 433
Cramér-Rao inequality 262
Cramér-von Mises statistic 483
Critical region 307
Cross-product ratio 390
Cumulative distribution function
   see Distribution function

Decision function 560
Decision rules 307
Degrees of freedom 441
DeMorgan's laws 14
Density function 31, 55
Density of a distribution 31
Difference, second 51
Difference of sets 11
Discrete distributions 30
Discrete probability space 30
Discrete random variable 43
Disjoint sets 10
Distribution-free statistic 363, 445, 450
Distribution function
   bivariate 51
   multivariate 52
   univariate 47
Dominance 558

Double sampling 395

Efficiency 260, 262
Elementary outcome
Empty set 8
Error sizes 329
Errors, types I and II 318
Errors, random 116
Errors in testing 318
Essential completeness 573
Estimation 253ff
Estimator 253
Event 14
Exchangeable variables 85
Expansion theorem 131
Expectation 90
Expected value
  of a function 91, 99
  of a matrix 432
  of a random variable 90, 94
  of a sum 92, 102
Exponential distribution 167
Exponential family 188, 238, 263, 336

F-distribution 319
Factor level 528
Factorial experiment 528
Factorial moment generating function 124
Factorial moments 124
Factorization criterion 231
Fair bet 544
Field of sets 12
Finite additivity 27
Fisher, R. A. 303
Fisher information 240
Fisher's exact test 376
Fisher-Yates test 477
Fixed effects model 524
Frechét inequality 262
Frequency distribution 200
Functions on a sample space 15
Functions of random variables 63

Gamma distribution 173
Gamma function 173
Gauss-Markov theorem 510
Gaussian distribution: see Normal distribution
Generating functions 120
Geometric distribution 154
Gompertz distribution 171
Goodness of fit
  continuous 479
  discrete 361

Hazard 171
Hierarchy
  of loglinear models 387
  of regression models 512
Histogram 200
Homogeneity
  of means 325, 519
  of probabilities 371
  of variances 319
Hypergeometric distribution 150
Hypothesis 291

Improper prior 249
Independence,
  conditional 389
  joint 389
  of $\bar{X}$ and S^2 212
  tests for 371
Independent events 36 83
Independent experiments 37
Independent random variables 77, 80, 81
Independent random vectors
Indicator function 16
Information 240
Information inequality 262
Insufficient reason
Interaction 528
Interquartile range 202
Intersection of sets 8
Interval estimate 281
Inversion formula 129

Iterated expectation 101

Jacobian 430
Joint density function 58
Joint distribution function 45, 51
Jump function 48

Khintchine's theorem 138
Kolmogorov-Smirnov
   one-sample test 479
   two-sample test 485

Laplace distribution
Latin squares 536
Latin-greco squares 537
Law of large numbers 21, 137
Law of total probability 28
Least favorable distribution 558
Least squares, method of 491
Level curves 16
Level of confidence 282
Level of significance 311
Likelihood 220
Likelihood equation 225
Likelihood function 221, 222
Likelihood principle, 251
Likelihood ratio 296, 297
Likelihood ratio test 297, 366
Lilliefors test 483
Limit theorems 135
Linear combination 539
Linear models 490
Linear regression 493
Linear transformation 64
Linearity, test for 505
Linearity of expectations 102
Loglinear models 385
Lognormal distribution 186
Loss function 547
Lottery 19

Mann-Whitney statistic 474
Marginal distribution 45, 52, 417
Mathematical expectation 89

Maximum likelihood
   estimator 275
   principle 225, 252
   properties 277
Maxwell distribution 186
McNemar's test 380
Mean (value)
   population
   sample 203
   tests for 301, 322
Mean deviation 106
Mean squared error 254
Mean squared prediciton error 425
Mean squared successive
   difference 455
Measurable function 15
Median
   asymptotic distribution 218
   population 51
   sample 202
   tests for 460, 461
Median test 471
Method of moments 272
Midrange 202
Minimal sufficiency 235
Minimax principle 549
Minimum variance estimate
Mixed action 548
Mixed decision rule 568
Mixed moments
Mixture distribution 96
Moment generating function 120, 131
Moments
   of a population 104
   of a sample 202
Moments, distribution of 208
Moments, method of 272
Monotone likelihood ratio 352
Monotone regret function 571
Monotone power function 352
Monotone procedure 572

Monte Carlo method
Most powerful test 332
Multinomial distribution 187
Multiple regression 508
Multiplication principle 2
Multiplication rule 35
Multivariate
   density 61
   distribution function 53
   distributions 409ff
   generating functions 131
   normal distributions 435

Negative binomial distribution 156
Nested models 387,512
Neyman-Pearson lemma 335
Noncentral distributions 347
Nonmeasurable function 15
Nonparametric inference 445ff
Normal distribution 178, 179
Normal equations 491
Normal scores 457
Normal scores test 477
Null hypothesis 242

Observed significance level 311
Operating characteristic 341
Opportunity loss 547
Order statistic 402
   distribution of 215, 446
   sufficiency of 232

P-value 302
Pairing 327, 348
Parallel axis theorem 107, 204
Partition 11, 15
Partitions, counting 5
Percentile
   population 50
   sample 202
Permutations 3
Permutation test 456, 470
Personal probability 24, 542

Pivotal quantity 282
Pointwise convergence 135
Poisson distribution 162
Poisson postulates 161
Poker hands 6, 26
Pooled variance 326
Population
Positive definite matrix 433
Posterior distribution 246 565
Power 332
Power funciton 341
Prediction 504
Principle
   conditionality 252
   likelihood 251
   sufficiency 250
Prior distribution 220, 246
Probability
   axioms 27
   density function 55
   element 57, 59
   function 44
   generating function 125
   interval 284
   space 19
Product set 81

Quadratic form 433
Quadratic loss 279
Quantile 50
Quartile 50

Random effects model 524
Random errors 116
Random failure model 168
Random numbers 208
Random orientations 32
Random sample 197
Random sampling
Random selection 21
Random variable 43ff
Random vector 58
Randomization test 455
Randomized block design 527

Randomized decision rule 568
Randomized test 309 337
Randomness, tests for 452ff
Range, sample 202
Rank of a quadratic form 441
Rank tests 461, 474
Rankits 457
Rankit plot 484
Rao-Blackwell theorem 266
Rayleigh distribution 186
Reduction of a partition 228
Regression effect 428
Regression function 95
Regression models 489
Regret 547
Relative efficiency 259
Relative likelihood 542
Reliability 171
Residuals 493
Risk function 345, 561
Rotations 415
Run tests 452

St. Petersburg paradox 544
Sample,
   distribution 198
   distribution function 199
   mean 203
   median 202
   moments 204, 209
   standard deviation 204
   variance 204, 213
Sample, random 197
Sample space 1, 8
Sampling 195ff
Sampling distribution 206
Sampling without replacement
   150, 157
Schwarz inequality 262
Score function 262
Sequential likelihood ratio test
   395
Sequential procedures 395ff

Serial correlation 456
Set algebra 8
Set operations 8, 9
Shapiro-Wilk statistic 484
Shearing transformation
Sherman's test 483
Sigma algebra: see Borel field
Sign test 459
Signed-rank test 461
Significance testing 300
Significance level 309
Simple hypothesis 292
Simple random sampling 197
Simulation 207
Sizes of errors in testing 329
Slutzky's theorem 137
Smirnov test 482
Standard deviation
   population 106
   sample 204, 285
Standard error
   difference in means 259, 326
   of the mean 257
   of the fitted value
   of regression coefficients 499
   of a predicted value 505
Standard normal distribution 178
Standard score 112, 296
Statistic 199
Statistical decision function 560
Statistical hypothesis 292
Stieltjes integral 94
Strong likelihood principle 251
Statistical significance 303
Subjective probability 24
Sufficiency 228
   in estimation 265
   in testing 343
Sufficiency principle 250
Sufficient statistic 230
Support of a distribution 58

t-distribution 288, 323
Test of a hypothesis 291
Test statistic 294
Three-way classifications 386
Ties, in rank tests 464
Tolerance interval 449
Total probability, law of 28
Transformations 409, 430
Trend alternative 452
Trinomial theorem 7
Two-way classificaitons 369, 385
Types I and II errors

UMP test 350
Unbiased estimator 254
Unbiased test 354
Uniform distribution 30, 57, 59
Uniformly most powerful test 350
Union of sets 9
Uniqueness theorem 129
Utility 541

Variance
  of the mean 209
  of a population 107
  of a sample 204
  of a sum 115
Vector representation 118
Venn diagram 10
Wald, A. 395
Weak law of large numbers 137
Weibull distribution 171
Wilcoxon test
  one-sample 461
  two-sample 473

Z-score 112, 296